Springer Proceedings in Complexity

Springer Complexity

Springer Complexity is an interdisciplinary program publishing the best research and academic-level teaching on both fundamental and applied aspects of complex systems—cutting across all traditional disciplines of the natural and life sciences, engineering, economics, medicine, neuroscience, social, and computer science.

Complex Systems are systems that comprise many interacting parts with the ability to generate a new quality of macroscopic collective behavior the manifestations of which are the spontaneous formation of distinctive temporal, spatial, or functional structures. Models of such systems can be successfully mapped onto quite diverse "real-life" situations like the climate, the coherent emission of light from lasers, chemical reaction–diffusion systems, biological cellular networks, the dynamics of stock markets and of the Internet, earthquake statistics and prediction, freeway traffic, the human brain, or the formation of opinions in social systems, to name just some of the popular applications.

Although their scope and methodologies overlap somewhat, one can distinguish the following main concepts and tools: self-organization, nonlinear dynamics, synergetics, turbulence, dynamical systems, catastrophes, instabilities, stochastic processes, chaos, graphs and networks, cellular automata, adaptive systems, genetic algorithms, and computational intelligence.

The three major book publication platforms of the Springer Complexity program are the monograph series "Understanding Complex Systems" focusing on the various applications of complexity, the "Springer Series in Synergetics", which is devoted to the quantitative theoretical and methodological foundations, and the "SpringerBriefs in Complexity" which are concise and topical working reports, case-studies, surveys, essays, and lecture notes of relevance to the field. In addition to the books in these two core series, the program also incorporates individual titles ranging from textbooks to major reference works.

More information about this series at http://www.springer.com/series/11637

Şefika Şule Erçetin
Editor

Chaos, Complexity and Leadership 2014

Springer

Editor
Şefika Şule Erçetin
International Science Association (ISCASS)
Ankara, Turkey

Faculty of Education
Hacettepe University
Ankara, Turkey

ISSN 2213-8684 ISSN 2213-8692 (electronic)
Springer Proceedings in Complexity
ISBN 978-3-319-18692-4 ISBN 978-3-319-18693-1 (eBook)
DOI 10.1007/978-3-319-18693-1

Library of Congress Control Number: 2015951455

Springer Cham Heidelberg New York Dordrecht London

Printed on acid-free paper

Springer International Publishing AG Switzerland is part of Springer Science+Business Media (www.springer.com)

Şefika Şule Erçetin: *To my lovely country and my family . . .*

Preface

Leadership is one of the most important aspects in human development and perhaps the most influential social institution in any society, since in general, it prevails over a set of beliefs, values, norms and understanding from the adult generation to its youth. Frankly, it stands out as one of the institutions humanity has created to serve certain needs, and, like all human institutions, it responds to certain changes in the environment. Accordingly, the institution of leadership is influenced by chaotic and complex aspects of social systems. Indeed leadership, chaos and complexity are inextricably interwoven with each other. Leadership is thus chaotic and complex notion; every leadership style exists within chaotic and complex social systems.

Given this ontological and epistemological conundrum, thus, it became imperative for us to explore further the nature of chaos; complexity is applied to social sciences and especially leadership studies. *Chaos, Complexity and Leadership 2014* is a result of efforts, which try to give new insights for a multidisciplinary perspective to understand conventional notions with chaos and complexity theory, undertaken by a wide range of scholars and professionals from all over the world.

This book captures a wide range of researches and professionals' writings in regard to chaos and complexity and how these ascribe to leadership in the contemporary society. The concept of leadership, both in the conventional and education systems, has been expansively explored. Meanwhile some papers and articles in the book paid particular attention and tribute to issues of chaos and complexity from a theoretical viewpoint, whilst others engaged an exploration of leadership in view of chaos and complexity that abound.

The book is intended to impress upon the reader that linearity, as was known to influence systems, no longer applies to leadership the same way and thus there is a paradigm shift to a completely new perspective about the main elements of leadership. This book was equally motivated by the academic desire that we and others had to further the view that chaos and complexity are not necessarily negative as wrongly perceived; since within that chaos and complexity is an organised structure that transforms systems. Thus, it becomes imperative for the reader and researcher coupled with leaders of different walks to understand that leadership ought to be conceived from the perspective of non-linearity, if a comprehensive concept is to be realised.

Developing a clear understanding of the concept of chaos and complexity was never going to be easy, whilst exploration of the same from the perspective of leadership was even more daunting. Along the same line dissecting the concepts of linearity and non-linearity challenges many researchers. Nevertheless, authors were selected from an ideal setting based on knowledge, experience and prior involvement in exploration of chaos and complexity as well as leadership.

We would like to thank the International Science Association for all the effort that led to the realisation of this book. All the authors who spared their time to develop chapters or papers that made this book a possibility are also appreciated.

Having spent a great deal of time developing the concept and structure of the book, identifying authors, orientating them, scanning through their manuscripts, and finally evaluating them, we feel this work will be a groundbreaking one in the realm of abrogating the curtain on the unexplored concept of chaos and complexity in leadership. Yet again, we feel that readers would enjoy studies related to chaos, complexity, leadership and non-linearity at all.

Ankara, Turkey Şefika Şule Erçetin

Contents

Chapter 1
Understanding Recep Tayyip Erdoğan and Turkey with Plasma as a Metaphor of the Fourth State of Matter

Şefika Şule Erçetin

Abstract This study is only concerned with and limited to "Recep Tayyip Erdoğan's leadership and Turkey" as one of the unlimited topics discussed in an international context and Turkey regarding Recep Tayyip Erdoğan and Turkey itself. There is need to emphasize the fact that it is not easy at all to understand and explain Recep Tayyip Erdoğan's leadership and Turkey. Why would you say so, because trying to understand and explain "Recep Tayyip Erdoğan's leadership and Turkey" with ordinary concepts, approaches, and preconceptions, is not only difficult to correct but may be impossible and indeed a complex and an unusual case of perception altogether. This makes us seek a different, peculiar, original conception, and approach. Thus, because of the close relationship between leadership and some properties of the fourth state of matter, I have adopted plasma as a metaphor to explore Recep Tayyip Erdoğan's leadership; and I have endeavored to negotiate such a jig-saw within a limited context.

1.1 Introduction

Recep Tayyip Erdoğan, Turkey and other political issues have been discussed intensively in local and international contexts for the last 12 years. When this article has been written, there are 4 months left to general in Turkey (June 2015), which will be the first time the AK Party goes into an election without Erdoğan during this period. Whether the expression will not be like other presidents or not remains to be seen since, as known, Erdoğan has kept the presidential system so active

Ş.Ş. Erçetin (✉)
International Science Association (ISCASS), Ankara, Turkey

Faculty of Education, Hacettepe University, Ankara, Turkey
e-mail: sefikasule@gmail.com

Ş.Ş. Erçetin (ed.), *Chaos, Complexity and Leadership 2014*,
Springer Proceedings in Complexity, DOI 10.1007/978-3-319-18693-1_1

and put his on agenda throughout this period. This indeed makes his leadership, approach and views even the more prominent. This discussion includes a scale of instability point, positive and negative extremes, superficial or deep contexts, love and hate contexts, anger and gratitude context or different constructive and destructive emotions towards him. At this point in this discussion, it is important to emphasize the fact that I have offered a part of this study to the debate but our analysis has only captured some of the key idiocies intended. This study is a product of efforts to explain preconceptions and understanding without judgment of the 9 elections within about 12 years in the history of the Republic. At the same time, the study explores Turkey's 12th President, Recep Tayyip Erdoğan, a leader of Turkey first directly elected by 51.8 % of the vote in terms of his leadership style and performance. This effort is different, original, with an original conception as well as some properties of matter and thus, the result of our approach is a bold experiment in the academic sense for adoption of the fourth state of matter – plasma as a metaphor for examination of leadership.

At the same time, this work is a brave initiative in the context of academic stance against those who have claimed that Recep Tayyip Erdoğan, is a dictator and even those who attended his invitation from some artists has been lynched through social media in the world and in Turkey.

Come on now, let's talk about plasma as a fourth state of matter to understand Erdoğan's leadership.

1.2 Plasma as a Fourth State of Matter

Plasma refers to an ionized gas consisting of positive ions and free electrons in proportions resulting in more or less no overall electric charge, typically at low pressures (as in the upper atmosphere and in fluorescent lamps) or at very high temperatures (as in stars and nuclear fusion reactors). Plasma is the largest component of blood, making up about 55 % of its overall content. It's mainly made of water and surrounds the blood cells, carrying them around the body.

Plasma helps maintain blood pressure and regulates body temperature. It contains a complex mix of substances used by the body to perform important functions. These substances include minerals, salts, hormones and proteins. Three important proteins found in plasma are:

- albumin
- clotting (coagulation) factors
- immunoglobulins

The concept of Plasma began appearing in physics literature since the early 1920s. Czech scientist Johannes Purkinje used the concept of Plasma for the first time, in our language meaning 'blood fluid, serum' ([4], p. 1). In 1927, Nobel

Prize-winning chemist Irving Langmuir defined "plasma" as neutral, ionized gas; then in subsequent studies the concept of Plasma has been proposed as the "fourth state of matter" ([15], p. 12). Research on plasma has produced results such as, the ionosphere which reflects radio waves and the discovery of the radiation belts around the world, the understanding that the universe is largely composed of plasma. These results suggest that future possible energy sources evaluated as "controlled thermonuclear fusion" have been of great interest since their birth since the 1960s. These have been used especially in areas such as defense "of high power laser" which has led to the beginning of a massive production in the field ([10], p. 7):

Plasma is analyzed again as having a different structure from the principles of solids, liquids and gases. Solid, liquid and gaseous states I often experience in everyday life are well known structures. However there are key questions to pose here; what plasma? How close to us could this fourth state of matter be in everyday life? How much are you aware of the matter in this case? Introduction to the literature of the plasma concept seems a century ago. However, explaining the formation of the universe using the "big bang model" shows that the presence of the scope of plasma is about 15 billion years ago, based on the big bang. The debate on the structure of matter is as old as human history. Together with the big bang explosion, the first state of the universe in which everything is so hot is that of plasma. Later, when the universe cools, it turns into gas, liquid, and finally solid state respectively. If we are to look at this as the first state of matter, then Plasma should be known with a very high temperature. As this terrific energy cools, it gives life to the other states of matter ([6] pp. 1–3).

Matter's structures are classified according to the arrangement of atoms and molecules; the temperature classification is associated with different states. At room temperature the solid state atoms are less active than the same in liquid state atoms. Mobility in the given internal structure of the heat substance or matter can be improved. Heating a solid material increases mobility in the microstructure and this means allow the passage of the liquid. If the temperature continues to deliver substances in the liquid, the distance between atoms or molecules and the mobility agent gradually increases and is converted into gas. Electrons in the gas move very quickly but around the atoms. If ionization starts, it continues to be necessary to heat the gas; stop the rotation of the electrons around the nucleus which is severed because it moves freely in the gas so that the free electrons of the inner structure of the material are observed that consist of positive and neutral core. This state is defined as the state of the plasma. In the plasma state, the atoms of the material are fragmented and become a system consisting of positively charged ions and electrons that move constantly. Free floating electrons in the plasma, take heat from the material and converts it back to sit substance trajectory in order to maintain the original state and accordingly the restructuring process (recombination) is realized. Karadeniz [14] has formulated this case as follows:

Melting

Solid + E Melting

Liquid

Solidification

Evaporation

Liquid+ E Evaporation

Gas

Condensation

Ionization

Gas+ E Ionization

Plasma

Recombination

As the temperature increases, solid melts to liquid, liquid evaporates to gas, plasma makes a transition into the ionizing gas; when temperature decreases, the above substances are said to return to their original states respectively.

In general, the characteristics of Plasma are as follows ([11]: 1, [17]: 7–12, Anatolia Plasma Technology Energy Center, [1, 8, 13]: 8–15, [3]: 7, [2, 5, 12]):

1. Whereas liquid, solid and gas phase molecules are formed by one of the atoms and ion species; nuclear and electrons in the plasma are composed of all of these types of particles. In such a case, the solid does not behave as liquids and gases. In ordinary matter, electrons show a tendency to connect to an atom. One remains free from time to time but it is patched to another core immediately. However, any electrons in the plasma are not bound to one core.
2. Gases at high temperatures become positively charged ions and free electrons. If you have a gas collection composed of free electrons, charged ions, neutral atoms and molecules, it is in a state of plasma. Plasma is a community of ionized gas totally or partially.
3. The decomposition, ionization of plasma and continuous restructuring which is the inverse of these events may occur.

   (a) And the sum of the ionization plasma decomposition which occurs in the combustion plasma for constant power must be constantly equal to reconstruction power.

(b) If decomposition and ionization restructuring is greater than the plasma then power is increased and explosion is a possibility.
(c) Plasma ends with more restructuring from disassociation and ionization.

4. The plasma is electrically neutral. However, unlike gas,

(a) Because of free electrons and ions of electricity that have stayed, transmission of heat energy is more clear since it is very well known that silver conductive is highly conductive than copper
(b) Electrical and magnetic fields interact and this creates an electrical field.

5. The gases tend to fill the gaps in a plasma collection.
6. The chemical reaction takes place faster than the gas in the plasma.
7. Plasma contains substantial freedom of movement and interaction when compared with solids.

Particles in the plasma are always in constant interaction. Coulomb forces between charged particles in the plasma are effective even at long distances. Thus, all particles are in constant interaction with other nearby particles in the plasma. Due to the above, these interactions show a collective harmony among themselves. Coulomb interaction between the particles occurs in the form of gas, but away from the radius of the sum of the two particles. The interaction of particles outside this distance is ignored.

1.3 Leadership and Plasma as a Metaphor

Leadership, just as in the entire human history, is one of the most debated issues in everyday life and scientific fields today [7]. Each new work still attempts to explain, understand and describe leadership. Leadership makes us feel the presence of other people in our lives every day and our presence is also felt strongly in everyday lives of others. This will basically continue to be so for as long as I have leaders and the led. Leadership as a phenomenon and concept is changing, evolving getting differentiated with humanity.

In a certain period of time, the presence of concepts agreed upon to explain the nature of life in a particular culture and perceptions of leadership in a particular area causes the emergence of theories based on the concept of leadership. Indeed, at any given time, the results of changes occurring in concepts and theories are reshaped with an effect on each other.

In this context there is need to accept that leadership is a dynamic and multivariate phenomenon and concept such as culture or time. For example, whatever the time period, political, military, cultural, scientific and other fields where leadership is at the forefront, the approaches related to leadership when examined are simply at the level of perceptions, facts and concepts. Sometimes extraordinary leaders such as Recep Tayyip Erdoğan, have characteristics that are known for their demonstrated leadership models and trying to explain and understand the concept in the ordinary manner may prove to be shallow and indeed limited. This situation may not work in other fields of science where concepts and conditions can facilitate detection and understanding of leaders with different insights. It also led to the brand new and original literature that seeks to provide concepts and solutions. In this study the "plasma" concept is borrowed from physics because it possesses some features that cab used to evaluate and understand Recep Tayyip Erdoğan, and Turkey's leadership. The concept of Plasma is considered as multi-dimensional in leadership and management literature. However, until this day, the first and only national and international study in leadership and management literature that places the concept of plasma into context is that of Erçetin et al. [8] entitled "multi-dimensional approach to leadership in a chaotic environment" which is a chapter in an edited book.

In the study, "plasma" is defined as a new leadership profile which is formed in the context of circumstances to cope with the chaotic situation. Basic assumptions adopted in the study are as follows:

1. It is not easy to examine leadership, and how leaders should behave in the context of chaos.
2. Reducing or increasing the effects to the system makes it go out of chaos to order and thus more passable; these two situations are based on the two concepts being substantially circular.
3. Chaos to order depends on the order in chaos. If you increase the degree of one, you can trigger the formation of the other.
4. The discussion of leadership from the perspective of chaos requires a multidimensional approach.
5. Each individual leadership type and model is compatible with the specific situation in certain measure and value.
6. How different parts that make up each whole also put forward the different types of leadership does not apply to all situations.
7. It is not true, as is said in this regard, to suggest that it is possible to set a leadership type compatible with all situations
8. Leadership in chaos there is need to create different compositions of leadership required for each case.
9. Leadership in chaos is based on developing the actual knowledge of when to use which type of leadership.
10. It is possible to decide on whether to apply a leadership model with a proper analysis of the situation we find ourselves in.

In the context of these basic assumptions, the following are expected from the plasma leader ([8]: 100):

1. Be aware of the system.
2. To explore the basic elements that affect the system.
3. To explore the relationship of these factors affecting the system periodically.
4. Take into account the power of network in decision making.

The most important thing is to draw a multidimensional leadership profile beyond usual leadership traits which includes all types and models with the ability to be used in situations beyond all models and types of leadership.

In the same study another step was added to the earlier "multi-dimensional perceptual model of leadership" developed by Erçetin and Düzer [9]. The final version of the model discusses leadership in terms of changes occurring whilst following a process of changing perceptions; and how the leadership situation at a certain time can be analyzed as well as what should be given special consideration. Six leadership statuses emerged based on perceptions of the leader and followers: "implicit", "potential", "reluctant", "uninspiring", "dull" and "ideal" ([8]: 94). Now at this point, the six leadership status in the multidimensional perceptual model of leadership are blended with the four states and properties of matter (Table 1.1).

Table 1.1 Forms of leadership and states of matter

Leadership forms	States of matter
Implicit leadership:	Solid state
Perceives other stakeholders as potential leaders to be passive, timid and with a low level of self-esteem	Regular, fixed, immobile molecular structure.
It is possible to move to different leadership situations in an appropriate environment.	Strong ties to feel it.
Embers leadership:	
The ideal leadership perception here is passive with shyness and low self-esteem, but after a certain period of time and due to various reasons, there is decrease in such leadership perceptions.	State change (liquid or gas)
It is possible to move to different leadership situations in an appropriate environment.	
Reluctant leadership:	Liquid state
There are some interaction problems with the followers	Surface tension: different interaction in liquid surface and the interior.
There is also gradual decline in the perception of leadership potential.	
Uninspiring leadership:	
The ideal leadership perception here is passive with shyness and low self-esteem, but after a certain period of time and due to various reasons, there is decrease in such leadership perceptions.	
Potential leadership:	The Gaseous state
Perceives other stakeholders as potential leaders who are active with a high level of self-esteem	Effective molecular structure.
	It makes you feel weak ties.
Communication and interaction with each other is a reality	There is energy and interaction potential, but the current lack of energy can convert it
Ideal leadership:	Plasma state
The awareness and differences of potential leaders contrasts with those of the neutrals.	Ionized molecular structure (which takes energy).
Leadership perceptions and self-esteem of themselves and others are at a high level.	Balanced and neutral structure with energy at a high level.
Powerful, fast, intense communication and interaction.	
The product of this interaction expressing total energy as well as values that make up the gravity lead to a common vision; the vision to realize the power of faith and energy	

Ideally, plasma leadership states can be examined by the properties of plasma. For example, Ideal states of plasma leadership should be defined as ideal at all different levels and in all different forms of leadership.

At this stage, the properties of plasma are not only the leader and leadership, but also a sense of who adopts the idea, the country of one revealing the context, and how society should highlight and use the idea to describe the regional location.

In essence, I have evaluated the concept of plasma up to this point; I have indeed tried to create a theoretical and conceptual framework.

To understand the "leadership of Recep Tayyip Erdoğan and Turkey", it is important to explain the concept of Plasma in its original setting which offers a genuine insight and approach. I discuss this new approach to this in the next section.

1.4 Understanding Turkey and Recep Tayyip Erdoğan

Turkey is positioned between eastern and western zones consisting of three sides surrounded by the sea, in other words, the basic dynamics of the history of the country's geopolitical position as a geographic source is that of high temperature. Throughout history, regions and countries need their own internal dynamics such as different levels of temperature as well as the influence of other factors. The critical point here which is perhaps often ignored or not taken into account is the heat which is now joined by the complex subatomic structural energy which reveals simple scattered structures. When forces in high temperature are simplified and combined, the super force is revealed.

Different types of particles in the plasma such as nuclear and electrons have been used to refer to the different ethnic and religious groups in the Turkish society. Turkey is a well known state of ethnic and religious diversity in the context of natural heritage and leadership. Turkey's most important and most critical feature is the Ottoman Empire. Turkey is currently in more than 10 political and geographical boundaries of Turks, Kurds, Circassians, Laz, Arab, Georgian, Bosnian and Albanian ethnic groups; with the exception of Islam being the major religious denomination, there are a number of different religions and sects within Islam's intellectual, institutional and corporate spectrum. As a leader, Recep Tayyip Erdoğan of Turkey has exhibited a high level of awareness about the dynamics based on these characteristics in the region. On 16 June 2009, during a Parliamentary group discussion, Erdoğan revealed this clearly and unequivocally when he said: "Turkey is a country that has for centuries lived together and intertwined with different cultures, different ethnic groups, different denominations, and different beliefs. These differences are richness for us, and all these different elements are an essential aspect of this country. Accordingly, another advantage of these elements is that one cannot have priority over another".

One of the most important aspects of politics for change during the last 12 years has been the ethnic and religious groups. Development of rights of ethnic and religious groups have been controversial issue in Turkey. There have been ups and downs in politics about ethnic and religious groups. Although there have been criticism towards Ak Party, there have positive steps in development of rights of ethnic and religious group in Turkey. Improving the rights of the different ethnic and religious groups has led to debate as a democratic necessity. In this sense, courage and policies put forth are very important. A Circular issued by the Prime Minister Recep Tayyip Erdoğan's office on 13 May 2010 and covered by the Official Gazette No. 27580 is also indicative of the commitment to the introduction of plasma leadership potential which ensures that all the different parts of society live in peace within the structure. "According to our Constitution the principle of equality; Turkish citizens belonging to non-Muslim minorities living in our country, like all citizens of the Republic of Turkey, they are an integral part of the next national culture and identity, they possess the ability to live and maintain their own identity and culture. Failure to remove their difficulties in this citizens States in front of the business and operations, lack of prejudice to the rights of the relevant legislation as it is necessary to our state and they being a part of the Turkish nation is of great importance and critical to our progress." As expected, it is not only legal arrangemrnt within 12 years. Ak Party and Recep Tayyip Erdoğan initiated various social and cultural developments in rights of ethnic and religious.

Another very important example is the solution process. This process refers to a period in which memorization of Turkish political history was disrupted and the adopted approaches and understanding was fragmented. The process was started by the Democratic Initiative. Erdoğan, speaking at the address to nation in October of 2009, said the following about it. "We started this process with a democratic initiative name which is actually a multi-faceted process of National Unity. Because this issue is a matter of a nation, a state issue. The focus of this initiative is not just terrorism; but the well-being of our country; all obstacles to the brotherhood of our people are the subject of this initiative. This initiative is an expression of love, a peace initiative and a brotherhood opening. This is a social refreshment initiative; an awareness initiative."

Later this movement defined as the solution process is not only relevant to Turkey, but also a crucial step for regional and global politics. It can be said that with this process Erdoğan put his entire political career and everything at risk. In this context, the long-term resolution process refers to a situation where no one dared to address the complex challenges of society requiring intensive labor and effort from their onset to this day. Erdoğan's speech in Midyat, Mardin in 2013, coined the following words of the solution process openly and reveals the basic philosophy of the subject. "Our ancient civilization has never made such a difference, separation has never happened. Precision is always protected especially in our culture and our traditions. Whoever makes ethnic nationalism can do so, but he is in mischief is after tribulation. All that however is here say. Somehow we have stood as one in this land for a thousand years and with Allah's will, we shall still be one and alive, we shall always remain big and strong on this land." The process and the efforts put forth are another indication of Erdoğan's extraordinary vision as a leader. If the process is successful, it will be very critical to Turkey since all major problems that have occurred over the years in every sense of heavy and intense distress will be solved once and for all. This will be an important step in the region and an example for world peace at the same time. With time, the resolution process continues with the occasional flash point of near-explosion and go-around.

As I mentioned earlier, plasma implies freedom of movement and interaction. In such a case, the solid does not behave as liquids and gases. Electrons in ordinary matter always show a tendency to connect to an atom. With time one remains free. But almost immediately one gets patched up to another core. However, electrons in the plasma are not bound to one core. As already discussed here the first feature of plasma is consistent with Turkey's changing and evolving regional and global role. At the same time the leadership of Recep Tayyip Erdoğan has seen a change in the vision and mission that embody the state of Turkey's foreign policy. In this regard, Erdoğan perceives the need to undertake leadership monitoring in Turkey's foreign policy. In fact, on 7th December, 2009 in the United States meeting on Turkey in World Politics, Erdoğan argued that "We can talk in the dock, you can talk with the West, at the same time we are a country that can talk to the north and south. We cannot assume destruction of one another while we are in contact with all of them; we also meet with all of them." As in many areas, Turkey's foreign policy is influenced by Erdoğan's winds of change and vision. There is no doubt, changing in Turkish foreign policy is theoretically built by Prime Minister and former Minister of Foreign Affairs Ahmet Davutoğlu but fostered and encouraged by Recep Tayyip Erdoğan.

At this point, Erdoğan emerged as the most important leader in every discourse by determining foreign policy as one of the most important values. The potential power behind the Turkish people was assured with the slogan of "self-confidence". Turkey's new foreign policy, which is the balance of power in this axis as well as the Middle East, and Africa, has come to the point of being a source of inspiration in South America and Central Asia. "Turkey's Prime Minister Erdoğan, who inspired the sound of the silent, the oppressed and the patron of those left homeless in their own land is a friendly regime spokesman who is suppressed by political leaders" [16]. Similar to this case also, the interaction of electric and magnetic fields in the plasma can be mimicked by creating an electrical field. Likewise, Turkey has established institutions and organizations to support development in health, economic and educational fields regardless of their settings and backgrounds all over the world.

Disassociation, ionization in the plasma and the restructuring events that are the opposite of these events occur tend to occur or take place. Here restructuring of plasma is a necessity for constantly being equal to the sum of the actual plasma since disassociation and ionization in the plasma is critical for combustion in stationary power. This feature can be used by many people to analyze and understand the structure of the strong bond between Erdoğan and his followers which is constantly renewed and hard to detect. Recep Tayyip Erdoğan's mayoral era of leadership has been perhaps the link between the most remarkable features of today and throughout the process of being Turkey's first elected president. Erdoğan constantly carries a plane which is equal to total disassociation and ionization restructuring. At this level he renews his colleagues and team, political life itself, and projects without compromising any stance. For example, in this issue is related to the Turkish flag. The respect that has been extended to it during functions and places where people gather as well as the protection it is accorded have surprised many a leader the world over. This behavior has aroused both great love and admiration for Turkey; it also aroused curiosity of many people all over the world who watched in awe. Just like in plasma, Erdoğan's deep relationship with his followers, increases and is like an explosion which results into undeniable influence of personal characteristics that shape leadership.

Relationship between Recep Tayyip Erdoğan and his followers is interesting and mystical. It is very often to be observed thousands of crying people when Recep Tayyip Erdoğan recite a poem. So, any sole dynamic is not enough to understand relationship between Recep Tayyip Erdoğan and his followers. A research done between people of different ages, gender and educational qualification, some of the concepts most commonly used to define President Recep Tayyip Erdoğan as a leader are as follows: "patriotic, The World Leader, brave, honest, believing, Internally motivated, Compassionate, the Merciful Realistic Solution Focused, Action Man, Heart of Man, voice of the orphans." It can be said that both emotional bond and pragmatic relationship Erdoğan enjoys with his followers exceeds and makes him different from other leaders. Erdoğan looks at his followers as his own, sees them as a part of the family; and does not separate himself from them. In the words of the song adapted and used as a selection of music for a long time, it can be emphasized that Erdoğan has made it clear that leadership qualities can be realized. Like lyrics of political campaign "Together we walked in this way, we walked under the rain with ..." Erdoğan does not keep his leadership independent from his followers and indeed quite the opposite; he perceives his success in his toughest political life via the support of followers. Therefore, his relationship with followers in terms of sharing feelings, thoughts and vision is intense, fast and at a very high level. On that point on 12 December 1997, Erdoğan was sent to prison where he stayed for 4 months; however the intensity of emotions following the reading of a poem at the poetry can be emphasized as the most interesting thing in those most critical times. The most recent example is of 27 August 2014 when Erdoğan was invited as the founding President of the AKP's first extraordinary congress of the party to give a farewell speech. During the party, a section of the poem, "*Dear, oh dear, dearest*" was read in his famous speech on the request of a young Turkish poet Sezai Karakoç, the brainchild of the said poem.

Beloved
The Most Beloved
Oh, Beloved.
Do not prolong my banishment on earth,
Are there any news from the birds of your country?
There is a spring rising up even from the graves
What comes out of the dead man when there is the Beloved?
There is an existence apart from existence and inexistence
I am not the only guilty one
There is an evil eye that ravages me
There are lines to sing adoring that song

No matter what they do, there is a will that comes from the sky
What happens if the day falls there is an architect who fixes the night
What If I am burnt, there is a fortress made of my ashes
There is a victory that grows in each failure
You have the key to reach the secret of all secrets
There is a vein in your heart that calls your banishment back
I never lose hope of you,
There is a plane in your heart named compassion.
Beloved
The Most Beloved
Oh, Beloved

Erdoğan is a powerful leader who is simple and clear enough despite the complexity at play, flexible and ready enough to make a timely transformation of the system, creative and sharing enough to transform the amount of wealth available, and a determinant of uncertainties abound. In this sense, he is a decisive and visionary leader who cannot be rejected by the audience. In this sense, Erdoğan exhibits a leadership profile which is above and beyond the leadership styles in other states of matter I know and may have considered in the context of this article. Thus, with such qualities, they leave behind their opponents in almost every election. This makes his supporters in The Republic of Turkey's 12 year continuous management to enjoy all contrasting structures, old stereotype problems, fast interaction, and instant variable agenda reminiscent of plasma dynamic structure, from parts of the energy to "collective harmony" as well as strong "plasma energy", which reflect "national will-stability".

As I emphasized at the beginning, this study was a bold experiment in academic sense; yet in terms of academic output it is equally a unique. The study should be regarded as a preamble and groundbreaking one in adopting a technical approach in trying to explain "Recep Tayyip Erdoğan's leadership and Turkey" in a different, unique, and original insight, again it must be remembered, that this study is of great importance not only to Turkey but the world in general.

References

1. Akar T (nd) Basics and characteristics of plasma as the fourth state of matter http://www.fizikportali.com/forum/maddenin-4-hali-plazma-ve-temel-ozellikleri-t943.0.html. Accessed 18 Aug 2013
2. Anadolu Plasma Technology Energy Center; taken from the address http://www.anadoluplazma.com/plazmagazfark.aspx. On 09/11/2014. Ankara
3. Aygun Y (2010) Low energy plasma focus fusion devices for construction and neutronic measurements. Unpublished PhD thesis, Ankara University, Graduate School of Science, Engineering Physics Department, Ankara
4. Bellan MP (2006) Fundamentals of plasma physics. Cambridge University Press, Cambridge
5. Bittencourt JA (2004) Fundamentals of plasma physics. Springer, New York. http://www.bilgiustam.com/maddenin-dorduncu-hali-plazma/. Accessed 10 Sept 2013. Dordrecht/Heidelberg/New York
6. Eliezer S, Eliezer Y (2001) The fourth state of matter. The Institute of Physics, London

7. Erçetin ŞŞ (2000) Spiral leader in vision. Nobel Publishing, Ankara
8. Erçetin ŞŞ, Açıkalın SN, Bülbül MŞ (2013) Multi-dimensional approach to leadership in chaotic environments. In: Banerjee S (ed) Chaos and complexity theory for management: nonlinear dynamics. IGI Global, Hershey, pp 89–104
9. Erçetin ŞŞ, Düzer A (2008) Multidimensional perceptual leadership model. World Appl Sci J 3(1):25–33
10. Fitzpatrick R (2014) address the http://www.ucc.ie/academic/physics/tony/py4107lec01-12.pdf was taken on 10 Oct 2014
11. Fridman A (2008) Plasma chemistry. Cambridge University Press, Cambridge
12. Gurnett D, Bhattacharjee A (2005) Introduction to plasma physics: with space and laboratory applications. Cambridge University Press, Cambridge
13. Inan US, Gołkowski M (2011) Principles of plasma physics for scientists and engineers. Cambridge University Press, New York
14. Karadeniz S (1990) Plasma technique. TMMOB Mechanical Engineers forum, Publication No: 137, Ankara
15. Kikuchi M (2011) Frontiers physics in fusion and fusion research. Springer, London
16. Küçükcan T (2011) New Turkey's foreign policy dynamism. Sabah Newspaper on 6 August
17. Smirnov BM (2012) Fundamentals of ionized gases: basic topics in plasma physics. Wiley-VCH Verlag GmbH & Co. KGaA, Weinheim, Germany.

Chapter 2
Language and Discourse as a Leadership Tool in Chaos Environment

Mehmet Şişman

Abstract Language is a tool of communication and agreement between people and also a management and leadership tool. Organizational and managerial actions are based largely on language and discourse. Phenomenon known as organization has been formed the basis of communication and discourse since the formation stage and organization has continued its existence with discourse. People who are coming together around common goals for organization and the social reality of the organization as a result of dialogue based on the language. Leadership process like other managerial actions largely consist of actions based on language. In this regard, language of the leaders, the world which is created with the concepts that make up that language are extremely important for followers. The language keeps an important place in the contemporary debate on leadership. Accordingly, leadership is defined as a language game that consists of largely language-based action. Although there are hundreds of definitions on leadership in the literature, one of the outstanding issues according to these definitions is that leadership is seen as a process of influence. Language is one of the affect tools. The basic way of understanding and expressing organizational life and managerial actions are analyze language structure, language products and discourse format of people who are in organizational life and actions. Aim of this study is that draw attention to the role of language as a tool of management and leadership for understanding organizational life and subjective aspects of organizational life by trying to show the importance of language and discourse in analyzing organizations. In crisis situations, discourse and the language that managers' used are also extremely effective for the overcome the crisis and for the end of the chaos. Managers can used language and discourse as an affect tool in positively for exit from the chaos and for the formation of new order and also managers can also cause in an increase in chaos and the organization to become dysfunctional.

M. Şişman (✉)
Higher Education Council, Ankara, Turkey
e-mail: msisman51@gmail.com

Ş.Ş. Erçetin (ed.), *Chaos, Complexity and Leadership 2014*,
Springer Proceedings in Complexity, DOI 10.1007/978-3-319-18693-1_2

2.1 Introduction

Organizations can be conceptualize in different ways, accordingly different organizational images may occur. Organization can be defined with different metaphors and concepts such as machines, organisms, social systems, family, teams, theater etc. These mentioned metaphors are commonly used to identify organization. These metaphors that having their own concepts are useful to understand the nature of organization and management. Furthermore, in recent years organizations can be conceptualize as narratives, discourse and speech areas. The discourse analysis which has long been used in linguistics, has been used extensively in the managerial and organizational research since the recent times [1]. The objectives, structures and functions of the organization has different meaning depending on the different conceptualizations and metaphors used. In the past, more hard concepts were used for analysis of organizations, soft concepts such as culture, language and discourse has been neglected.

Although in past such metaphors as machines and organisms were benefited for analysis of organizations, subjective aspect and meaning world of organization are more emphasized in new developed metaphors and paradigms related to the organization. Another approach for analysis of organization is symbolic approach. In this regard, organizations are seen as social systems, symbolic systems consisting of a shared sense, cultures and symbols [2]. According to this, people within an organization form the environment in which they are in based on social interaction through symbols and symbolic processes [3]. In this way, symbols in an organization refer to inherent value structure under the observed features and elements of organization, important beliefs and feelings [2]. One of these symbols is language. Language is an advanced symbols system. Organizations in terms of interpretive and symbolic perspective consist of symbols such as myths, sagas, ceremonies, stories etc. that people share [4]. They also are basic elements to understand the organization and organizational culture. Symbols in terms of schools give tips on how administrators and teaching staff see purpose and function of education and the nature of teaching [5].

Each metaphor for the organization creates a different organizational image. In this regard, Morgan [6] analyzed metaphors in detail. Discussions regarding on organizations and management in last decades are more culture and language centered. Accordingly, organizations has been started to conceptualize in different ways, such as symbol systems, discursive spaces and linguistic systems. In this chapter, language and discourse as a leadership and affect tool are discussed and also impact of discourse and language used by leaders in the chaos on followers are emphasized.

2.2 Discourse and Language as a Leadership Tool in Organizations

People interact with each other through the exchange of mutually meanings and messages. This interaction is performed with sounds, words, non-verbal behavior (verbal communication) in such way. Organizational communication is the process of creating shared meanings according to the expectations of sent and received messages, continuing this meanings and changing when it is needed. Organization and organizational continuity are not mentioned without communication [7]. Individuals in the process of interaction and communication are developed around shared values and common sense, "we" sense and common "identity". Language forms the basis for the management as well as the organization. The importance of language in understanding and explaining the nature of the organization and management can be considered from various perspectives. These can be listed as follows [8]:

1. Leaders of words and adjectives that have been used to identify organizational problems and in making decisions determines the fundamental approach to the nature of organization and management.
2. Some leadership researches conducted in recent years has been devoted to how leaders use the language. For example, visionary leaders use language in determining organizational values, in explaining them to members and in adoption of envisaged vision of followers.
3. The language is an affect and motivating tool of human. Therefore, success of a leader is depends on the success of using language in influencing and motivating others. Accordingly, some researchers see leadership as a language game.
4. The language is an influence and persuasion tool both in realization of change and in reducing resistance to change during organizational change process. The phenomenon of change is actually a discourse pattern. Therefore a discourse is substituted with another discourse in the change process.
5. The boundaries between units within the organization are also a language product. Different groups and units within an organization can speak different languages. Every profession has its own set of jargon. Learn a profession is to learn the language of the profession in a way.
6. Solving conflicts in organizations, meetings, negotiations and so on actions consist of entirely language based activities. In these processes unsaids are also importans as saids. This issue is worth serious attention in research in terms of organizational policies, organizational culture and group dynamics.
7. The language is also considered issue in terms of organizations, in motivation, integration and joining theory.
8. Sound as a language element is also an important language element. Therefore, organizations can be seen in different concepts, as places that have unique speech and hearing ability.

In the light of contemporary debates, in new leadership conceptualizations, leaders are expected to be an inspiration, excitement and motivation source for group members, be sensitive to the beliefs and values of the group, following leaders by believing. At the same time, leaders are expected to be a human being that can create a new discourse to establish a different and better world, can create a new ethical and moral values and enables integration of followers around these. Leadership such as administrative actions are consist of the largely language-based action. In this regard, the language of the leader, illustrated world with the concept behind the language are extremely important in terms of the followers.

Language is not only an element of culture but also a tool for people that perception, identification, interpretation and construction of physical and social world in which people live in. People try to understand, interpret, recognize and explain the world that live in by language and concepts and metaphors located in the language. Although there are hundreds of definitions on leadership, one of the prominent definition is that leadership is seen as an influence process. Language is the most important tool in these influencing.

Affect power of the language on people has a large literature. The language of leadership can be considered from different perspectives. The first is that identifying in any meaningful way organization and the goal of the organization. Here, message to give of the leader to the members is extremely important, it is also the process of creating a conceptual framework for goals. The second one is that leaders added an emotional power and content to messages which created in the form of language based. This depends on the ability to use the symbolic language of leaders.

2.3 Language and Discourse in Terms of Educational and School Management

There are some metaphors that reinforce the sense of reality in every language. Metaphors help individual to understand and explain that the historical and cultural environment of individuals located in. Metaphors in the field of education and school management are seen as image plays to develop research hypothesis, to search and explain management practise, to represent alternative perspectives, to organize and schematized views, to named and interpret observations and data, to develop theory [9].

The speaking language in a school is useful in goals and plans of the school, to identify success and failures, to explain what can confirmed and can not be [5]. Metaphors and similes in a language are linguistic structures that help theorists, as well as to practitioners in understand, explain and identify phenomenons and develop theories and models. Furthermore, metaphors explain how a school administrators interpret roles in the school, how conceptualize the school, how reflect their own values and beliefs to the application [9]. Clues on core values of the school can provide with discourse analysis to be made in schools.

Management language used in verbal communication as well as written communication, words in this language, text format, appearance, color of ink, the celebration of any success in school, written rules and regulations that set about school give powerful messages to other people. Written materials also give clues about the values that the administrator has [10]. Some words, metaphors, slogans, legends, stories, anecdotes, mottos etc. that can be used to influence the members of the school community by school leaders as well as their clothes, gestures, attitudes and behaviors reinforce certain images, values and emotions about the school. Thus, these symbols school provide integration between the members constituting the school community as well as to show joint efforts of all members for the realization of shared school goals [11].

Like every organization each school also has a unique archive of myth. This archive contains school-related narratives, mythes and legends. They can be real or not but they should not be kept away for analysing and understanding school life. These define boundaries between individuals and units in schools, the role and status, indicate the solution format of the problems, reflect the conflicts and can be used as a tool of sovereignty and legitimacy. Schools as well as other organizations describe through language and discourse analysis [12].

Most of the educational and administrative actions in schools are based on the action form of the language. More than one discourse can be in schools. Different sub-cultures of the school have different languages and discourses. Students and teachers carry the language codes that they have acquired from their living areas and sub-cultures to the school. Therefore, teachers and students from different social and economic environment can develop a different language and discourse in themselves. Besides these, different administrative language and discourse can be mentioned in a school [12].

People that participating discourse generate and transmit their messages through discourse. Discourse in mentioned process not only reflect existing situation of the group that join the process, but also social identity of the group is attempted to maintain or reproduce by discourse. But if the dominant discourse patterns in the school and classroom and the discourse patterns in the family or in other areas of social life are different, conflict may be between discourses. Before everything else democratization of discourse in social living spaces are needed the entrenchment of democracy in schools and in society. This requires the distribution of power between the social groups and redefining and regulated relationships based on power. Because the dominant language and discourse within different social groups are shaped by dominant groups and owned power.

In schools, different patterns of discourse can take place in different areas, between different groups. For example, the traditional discourse between teachers and students in schools are discourse patterns that are based on subject-object distinction and students are object and teachers are subject. In this case, roles, tasks and responsibilities of teachers and students are shaped by this discourse pattern. So, discourse determine that, which people what to say and what to do. Dominant norms in social culture in societies organized in hierarchical form provide shaping discourse in the school and classroom environment according to this.

Administrators emphasizes the importance of values in school with the ceremonies and rituals in some important days in the school. These activities that provide a common experience for people at school can become a sacred ritual when shared values are highlighted. The ceremonies strengthen common beliefs and values that are shared at school. Parties, meetings, school opening and closing ceremonies, retirement ceremonies are important opportunity to emphasize some values. These are actions for socialize, meet, integration and provide motivation. In addition, successes are also celebrated with these types of activities, non-approved behavior and situations are described, school-related heroism and success stories are expressed. The stories that are told about manager or by managers express the expectations from followers and members [10].

2.4 Using Language and Discourse of Leaders in Chaos Environment

Suggested solutions of traditional management literature are insufficient in understanding organizations and solving organizational and managerial problems in the changing times. The main purpose of traditional organizational-managerial sciences is that making predictions about organizational actions and behaviors and to develop recommendations for leaders to control those actions and behaviors. This situation is a result of linear view organizations and management. Chaos theory which has been tried to adapt from nature of science to the social sciences and in this context to organizational and managerial sciences has been challenged to some assumptions of traditional management science [13].

According to the chaos theory, chaos has sensitive to initial point. The sensitive dependence on initial conditions shows how a small change at one place or moment in a nonlinear system can result in large differences to a later state in the system [14]. Leadership and chaos can be considered fundamentally interconnected concepts [15]. Leaders in the chaos environment can cause to the disappearance of chaos with language and discourse to be used and also they can cause to increase chaos. Leadership is process that build a consensus and coalitions between different forms of discourse and to strengthen the common meanings. One word to be used by school and class leader in a school or classroom environment can upset everything in a moment. Poet Yunus Emre's following couplet summarize very nice the function of language and discourse in a conflict and chaos environment:

Word is, war end/word is, to lose the head
Word is, poisons meal/word is, honey with butter

When the organization envisions the future and possible probabilities, it will challenge change and uncertainty, and will make the chaos an opportunity to success [16]. Even in the extreme cases of organizations that sought to bring chaos and confusion, their purpose was generally to clear the present situation and bring a new order [17]. So, if principals cannot manage crisis situations and chaos in work environment, that organization should not function [16]. As a result, language and discourse that used by managers in crisis situations are extremely effective to overcome the crisis, to create a new vision, to end of the chaos. Administrators can use language and discourse to create a new order after the chaos as an influence tool in the positive sense, as well as using language and discourse can be caused to increase chaos and to lose the function of the organization.

Although managers have not been directly involved in organizational processes affect directly or indirectly to people involved in these processes. Managers in one school have direct and indirect effects on school results and outcomes. Language and discourse used by managers affect behaviors, attitudes, motivations of members of the school community and the realization level of school purposes. Leaders can increase individual and organizational success with their using language and discourse, as well as they can decrease motivation of members and individual and organizational performance.

2.5 Conclusion

Language is a fundamental tool that affecting and motivating people. Motivation crisis is one of the most important problems that observed in organizations and schools. Rewarding and encouraging language and discourse are required instead of external control-related language and discourse in school management. Building their own social reality of the school depends on the participation of all school community members on equal terms to discourse process. Managers in school should lead to develop a new corporate language for the democratization of discourse in school.

Discourse is not only build social realities of the social group, but also is an important determinant in the process of social change. When performed on behalf of a number of schools to change and restructuring is viewed, desired project is seen that some of the new concepts frequently used. When we look at some projects for the realization of change and restructuring in schools, it is seen that a number of new concepts commonly used. However, some of these concepts that borrowed from other cultures and areas can not find a place in the concept world of the people. There will be no more benefit than performs for change and restructuring unless attend on account of the subjective aspects of school life, unless school to form the desired new language and discourse, unless shared the desired pattern of new coporate language and discourse in school by the school community in school development process.

Schools aren't static and linear systems but nonlinear and dynamic systems. First of all, a new language and discourse and new mental models are required in the restructuring process in schools. Openness, participation and dialogue based discourse pattern should be preferred instead of official rhetoric based, authoritative and sanctioned discourse pattern in schools. School leaders are also required to assume a leadership role in this process.

References

1. Alvesson M, Karreman D (2000) Varieties of discourse: on the study of organizations through discourse analysis. Hum Relat 53(9):1125–1149
2. Dandridge TC (1983) Symbols' function and use. In: Pondy LR et al (eds) Organizational symbolism. Jai Press Inc, London, pp 63–79
3. Gahmberg H (1990) Metaphor management: on the semiotics of strategic management. In: Turner BA (ed) Organizational symbolism. De Gruyter, Berlin, pp 151–157
4. Şişman M, Turan S (2004) Örgütsel semboller ve eğitimde sembolik liderlik. Eğitim Yönetimi 37:96–117
5. Gordon D (1992) The symbolic dimension of administration for effective schools. Paper presented at the annual meeting of the American Educational Research Association, San Francisco, p 25, 20–24 April
6. Morgan G (1986) Images of organization. Sage, Beverly Hills
7. Schall MS (1983) A communication-rules. Approach to organizational culture. Adm Sci Q 28:557–581
8. Boje D, Oswick C, Ford J (2004) Language and organization: the doing of discourse. Acad Manage Rev 29:571–577
9. Bredeson PW (1985) An analysis of metaphorical perspectives of school principals. Educ Adm Q 21(1):29–50
10. Deal TE, Peterson KD (2009) The principal's role in shaping school culture, pitfalls, paradoxes & promises, 2nd edn. Wiley, San Francisco
11. Kelly BE, Bredeson PV (1991) Measures of meaning in a public and in a parochial school: principals as symbol managers. J Educ Adm 29(3):6–22
12. Şişman M (2004) Örgütler Açısından Dil ve Söylem: Okul Yaşamının Yeniden Yorumlanmasına Doğru. Akdeniz Üniversitesi Eğitim Fakültesi Dergisi 1(1):122–132
13. Lissack MR (1999) Complexity: the science, its vocabulary, and its relation to organization's emergence. J Complex Issues Organ 1(1):110–126
14. Açıkalın SN, Erçetin SS (2014) Great illusion in twenty-first century-chaos knocking door. In: Banerjee S, Erçetin SS (eds) Chaos, complexity and leadership 2012, Springer proceedings in complexity. Springer, Cham, New York doi:10.1007/978-94-007-7362-2_1
15. Erçetin ŞŞ, Açıkalın ŞN, Bülbül MŞ (2013) A multi-dimensional approach to leadership in chaotic environments. In: Banerjee S (ed) Chaos and complexity theory for management: nonlinear dynamics, vol 5. IGI Global, Hershey, pp 89–104
16. Erçetin ŞŞ, Kayman EA (2014) How to be a quantum leader in an intelligent organization? In: Erçetin ŞŞ, Banerjee S (eds) Chaos, complexity and leadership 2012. Springer, Cham, New York
17. Çakar U, Alakavuklar ON (2013) Chaotic essence inside the organizational reality. In: Banerjee S (ed) Chaos and complexity theory for management: nonlinear dynamics, vol 7. IGI Global, Hershey, pp 147–163

Chapter 3
Chaotic Awareness and Simulacra in the Recent Emergence of the Self-Organized Multitude

K. Gediz Akdeniz and Nicholas Anastasopoulos

Abstract Order and the power of cosmos came to an end with the advent of chaos theory. Thus, the deconstruction of the dualities of modernity, as reflected in the cosmos-chaos and order-disorder binary dualities, are in effect. According to the Chaotic Awareness Simulation theory, the simulation of such deconstructions could lead to the emergence of non-predictable changes and transformations in responsive societies. Complex System principles provide us with a methodological tool in an effort to understand the multiple phenomena taking place recently in regard to certain movements. In this study, following a new anarchy definition and the Chaotic Awareness Simulation theory, recent mass protests are being examined in the simulation world and semantic differences are being investigated, with the Istanbul Gezipark protest being used as a case study.

3.1 Introduction

Our times have been experiencing massive citizen's mobilizations signalling a systemic change, as evidenced by a record of 843 protests globally between 2006 and 2013, a number that far exceeds any other period in modern history [12]. The Occupy movement and particularly the Occupy Wall Street protests, the street movements in the various western European capitals, the 14 M and the Plaza movement in Spain, the protests in the Athenian streets and squares in Greece, the World Cup protests in Brazil, the Arab Spring protests in various Arab countries, and recently the Istanbul Gezipark protests in Turkey, all have provided fertile ground in social sciences, political studies and philosophy for speculative analyses and theorization. The concept of *zuhur*, a Turco-Arabic term (deriving from the word Zahara) is being used here as synonymous to *emergence*, [15] in order to refer to an

K.G. Akdeniz (✉)
Department of Physics, Istanbul University, Istanbul, Turkey
e-mail: gakdeniz@istanbul.edu.tr

N. Anastasopoulos
School of Architecture, National Technical University of Athens, Athens, Greece
e-mail: nanastasopoulos@arch.ntua.gr

Ş.Ş. Erçetin (ed.), *Chaos, Complexity and Leadership 2014*,
Springer Proceedings in Complexity, DOI 10.1007/978-3-319-18693-1_3

uncertain state, and an identity condition applicable to the Arab world, but not valid in modern societies of the west [2, 3]. The particular type of emergence embodied in zuhur is being seen in the cultural and historic context of the Gezipark protests, in order to express a transition of the Turkish society that this phenomenon indicates, in this moment in history [5, 8].

3.2 Societies as Complex Systems

Given the sweeping magnitude and scope of recent movements and protests the world over, it is legitimate to compare them to previous moments in history, which have been discussed in revolutionary terms. Even though it may be an easy trap to view recent global movements as a unique case, Wallerstein and Graeber remind us that the "world revolution of 1789," followed by the "world revolution of 1848," saw revolutions break out almost simultaneously in 50 countries, from Wallachia to Brazil.[1] In addition, according to David Graeber such revolutions, which are often fueled by *utopian* visions for better societies, are seizures of power by popular forces aiming at transformation of a country's very nature of the political, social, and economic system [11]. History tells us that they hardly ever succeed in taking power immediately, but their effects may be deeply felt in societal structures and far-reaching in time. Such readings of revolutions help us understand their significance and the complex mechanisms at work, affecting existing structures at different places and moments in time, and at a global scale. They also inform our understanding of societies as *complex systems*, constantly affecting each other, altogether altering their environments and the framework of their functions, organizing and operating mechanisms, simultaneously adapting to the new conditions that emerge. Complex systems demonstrate 'path dependence', possess a structure and self-organize when far from equilibrium, or at the edge of chaos. As a result of self-organization in systems emergent properties may appear. In this theoretical framework, the terms of the multitude and of *emergence*, describing complex behaviors of individuals capable of behaving as single entities and simultaneously as temporary members of a larger whole in contemporary societies, as well as of emergence, a concept which attempts to understand the spontaneous birth of a new behavior, contribute to this analysis [14].

To describe a community that dynamically reinvents itself by continuously learning and adapting in response to environmental changes harks to theoretical

[1]In a similar pattern, according to Graeber, the Russian Revolution of 1917 was a world revolution which was responsible for what later came to be known as the New Deal, the birth of the European welfare states, as well as, of course, for communism itself in USSR, Eastern Europe, China and Cuba. The most recent world revolution of global magnitude is credited to be what is known as the '60s with the May'68 events, etc.—which, broke out almost everywhere, seized power nowhere, but nonetheless gradually changed everything in the course of history, politics and culture with far-reaching effects which we still experience today.

developments in the philosophy of science that matured in the 1990s known as *Complexity Theory*. An appropriate theoretical tool appears to be the *Complex Adaptive Systems* domain. A *Complex Adaptive System* is an open system whose elements interact dynamically and non-linearly, exhibiting unpredictable behaviors while it co-evolves with the environment and is being affected by positive and negative feedback loops [9]. In this theory, complex systems possess monitoring mechanisms and adaptive capabilities in response to internal dysfunctions (governance, imbalances, etc.) or external crises (environmental, geopolitical, etc.) to initiate change either through gradual adaptation or if necessary, though abrupt self-organized revolt.

3.3 Global Protests and Movements as a Parallel Universe

In recent decades, unjust financial and political governance at a global level and policies in favor of the 1 % supported by the global economic elite, have been ruining the economy to the detriment of the poor and the middle class. We have come to understand that the economic crisis is a *hyper reality*, an artificial phenomenon of the banking system and the global market economy, completely detached from the reality of the physical world. Because of this discrepancy, Occupy Wall Street, the European Occupy Movements and the World Cup protests in Brazil lend themselves for a reading through simulation with *BS theory* (simulation theory after Baudrillard). Significant changes have been recorded in the rise of communitarianism, alternative economies and experimental forms of autonomy in Greece, Spain and Italy [6]. Such changes may be responsible for radically altering in time current modern life concepts of governance in a "simulation world", as in the *new anarchy* example [4, 5] and for alternative forms of self-organization [16]. Nevertheless, in all these movements their reality principle is that of modern society. Since Brazilian people are strongly engaged with football, the World Cup has been the right target for protests and for an activist mobilization in Brazil. Still, the football world itself proved to be not sensitive enough, while its organizational structure and its relations within society are formed by rules of the market economy. Therefore, its simulation will be hard to cause any significant change in the Brazilian society, while the governments in Brazil may be more careful in the future regarding matters of economy and finance for the Brazilian people.

On the other hand, the identity, behavior and dynamics found in the Arab Spring insurrections have different properties than the aforementioned movement protests. Primarily mobilized by a common goal, which was the desire for the dissolution of an authoritarian regime, they have been described as chaotic [1, 10]. They might also be seen as an example of complex utopian events in the *simulation world*. Such *simulacra* though may not easily cause any hyper reality in abruptly transforming non-modern societies towards western human behaviors [4]. Furthermore, according to BS theory [7], since the Arab countries are a Muslim majority and non-modern societies, the *simulation reality principle* of the movements in Arab countries is not currently their own reality. Thus, one can predict by BS theory that simulacra would not cause the change of traditions of authoritarian regimes in Arab countries any time soon. Also what is being experienced in today's Arab world (particularly in Syria, Egypt and Tunisia) is not a surprising development for CAS (Complex Awareness Simulation) theory [5]. In Tahrir Square for example, the Islamic movement of the "Muslim Brothers" strongly dominated the other protest groups with a fundamentalist political agenda. This means that, while most of the other activist groups were globally simulated in a modern context by non-Arab social media and by digital civil society social media and other processes, the complex correlations of the activist groups in the Arab Spring event proved to not be adequate in raising a chaotic awareness reality principle. These observations lead us to conclude that the chaotic awareness as a reality principle in the Arab Spring, has remained subdued and in all likelihood, it is not going to be causing substantial change towards new revolutionary forms of governance in the Arab World any time soon.

3.4 The Istanbul-Gezipark Protests (May–June 2013)

In some non-governmental forms of organization and protest behaviors, Gezipark looks similar to the Occupy street movements [10]. But in both BS and CAS theories the reality principle is an essential component in social movements and if we follow these simulation theories, the reality principle of Gezipark is not dominated by a modern reality, like in Occupy street movements' cases. On the contrary, by focusing on its nature of spontaneous behaviors, according to CAS theory, its reality principle is that of a *chaotic awareness*. Chaotic awareness may be interpreted as a capacity for self-organization and for maintaining an awareness of the whole, the purpose and constrains in a situation, entailing complex circumstances. This was the case of a public square demonstration comprising diverse micro-activist groups and individual minority voices, which represented different identities, were not affiliated with any political party or ideology and possessed no centralized macro-group leaderships, thus forming the condition of a multitude [13]. Through the use of innovative critique of capitalism including humor, and a variety of performative and creative practices, activist groups and civil society organizations displayed a high level of complexity within their self-organized and constrained conditions in protesting against the authoritarian government practices. In defending initially the environment, the multiple voices united and were transformed into a quest for participatory democracy. Utilizing some of the best innovative practices from other world protests, Gezipark was characterized by a large number of alternative strategies, including performative practices, signing and practice of yoga by massive numbers as pacifist techniques [8]. All these happened while the public space was being redefined, along with the members of society themselves. *Zuhur* thus emerged, when the Gezipark as chaotic awareness reality became simulated as disorder, by both international and state-operated media channels and TV. Similar to the butterfly effect, it (as Zuhur in CAS theory) caused transformations in all elements of Turkish Society to non-measurable and perhaps non-reversible states, with yet undefined and non-scalable invariant representations. It can be claimed as the first New Anarchy experiment in Simulation world, for a re-evaluation of Turkish Society and Turkish Politics [5].

There is a discrepancy being revealed here, between emerging maturity and desire for emancipation in Turkish society, against consistently authoritarian governance, lack of transparency, censoring social media and the internet threats and outbursts of financial and corruption scandals. The sophistication of the Turkish multitude as the plethora of creative and innovative, radical, self-organized events testify to the emergence of higher and maturing societal characteristics.

3.5 Conclusions

As this becomes increasingly evident in most societies, respective governments appear to be incapable or unwilling to acknowledge the need for a transition to a more inclusive and participatory way of governance. First of all, by the critiques of the BS theory we understand that the Occupy Protests, the Athens Demonstrations and the World Cup Protests have similarities in having globally economic character in their principles of reality. Therefore, the emergence of these protest movements in the simulation world will be the same simulacra in the context of modern world. Hyper realities would be new life styles and new life goals to cover for the economic crisis. Such common simulacra in modern societies could cause solidarity to lead the way to a new "western-type" humanist homogenization and direct democracy in simulation world. But they will never cause any provocation impact in the non-modern societies.

On the other hand, according to both BS and CAS theories, the initial structure of the Arab Spring was not an appropriate reality principle for the emergence of simulacra or zuhur in the Arab World because; it lost its ability to overcome the operations of non-Arab international organizations and the strong Islamic dynamics in simulation world. Thus, it is still remaining in the difficulty of finding new roles to improve modernizations using their own realities of the Arab World. But interpreted through the zuhur/emergence definition in CAS theory, Gezipark distinguishes itself as a case of *zuhur*, contrary to the Occupy protest movement, including the Arab Spring. Both BS and CAS theories remind us that protest movements cannot be considered as the suitable strategy for solidarity in the world against global economic powers. The simulation of these protests do not have additive properties in the human homogenization and global aspirations for democracy. To force such solidarity through protest movement could obstruct unpredictable worldwide revolutions in non-modern societies. Gezipark confirmed itself as *zuhur* and already caused non-measurable change in all levels of Turkish society [5]. It has also caused unpredictable social evolutions in world politics and at all structures within Turkish organizations as well as the Turkish activist movements [8]. World politics already started to revise their reality principle of understanding of the Islamic world, and Turkey has become a key player for the European Union.

Last but not least, according to CAS theory the "orientalist other" as perceived by the Europeans and Western Civilization and attributed to the Turks may have ended with the impact of Gezipark, which appeared as a black-hole moment in social media. This search leads us to interpret the status of Gezipark event, in simulation world as a new stage for open democracy. This discourse could also lead us to realize the possibility of a rising global solidarity of these protests against the simulation world of social media, the internet, TV, etc., as has been demonstrated with alternative forms of organization.

References

1. Açıkalın ŞN, Bölücek C (2014) Understanding Arab Spring with chaos theory—uprising or revolution. In: Banerjee S, Erçetin ŞŞ, Tekin A (eds) Chaos theory in politics. Springer, Netherlands, pp 29–47
2. Akdeniz KG (2010) Disorder in complex human systems. Proceedings of the conference. In: Fritzsch H, Phua K (eds) Honor of Murray Gell-Mann's 80th birthday quantum mechanics, elementary particles, quantum cosmology and complexity. World Scientific Publishing, Singapore/Hackensack, pp 630–637
3. Akdeniz KG (2012) Is Arab Spring a complex utopia? In: Banerjee S, Erçetin ŞŞ, Tekin A (eds) Chaos theory in politics. Springer, Netherlands
4. Akdeniz KG (2013) New communitarianism movements and complex utopia? In: Banerjee S, Erçetin ŞŞ, Tekin A (eds) Chaos theory in politics. Springer, Netherlands
5. Akdeniz KG (2014) Chaotic awareness in Gezipark, chaos? In: Banerjee S, Erçetin ŞŞ, Tekin A (eds) Chaos theory in politics. Springer, Netherlands
6. Anastasopoulos N (2014) The rise of communitarianism and other alternative movements from the Athenian crises. In: Russell E, Gallardo P (eds) Yesterday's tomorrows on utopia and dystopia. Cambridge Scholars Publishing, Newcastle upon Tyne
7. Baudrillard J (1995) Simulacra and simulation, 1st edn. University of Michigan Press, New York
8. Eligüzel AH, Tangün YA (2014) An essay about the association of anarchist islands and complexity science. Paper presented in I. Philosophy students conference, 3–4 May 2014, METU, Ankara, Turkey
9. Espinosa A, Jon W (2011) A complexity approach to sustainability: theory and application. Imperial College Press, London/Hackensack
10. Göle N (2013) Gezi – anatomy of a public square movement. Insight Turkey 15(3):7–14
11. Graeber D (2012) The democracy project: a history, a crisis, a movement. Allen Lane/Penguin, London/New York
12. Gutierrez B (2014) Around the world in 843 protests: living the most revolutionary times in history. Nation of change. http://www.nationofchange.org/around-world-843-protests-living-most-revolutionary-times-history-1397917299. Accessed 20 Apr 2014
13. Hardt M, Antonio N (2005) Multitude: war and democracy in the age of empire. Penguin, New York
14. Miller P (2010) The smart swarm: how understanding flocks, schools, and colonies can make us better at communicating, decision making, and getting things done. Avery, New York
15. Sawyer RK (2005) Social emergence: societies as complex systems. Cambridge University Press, New York
16. Žižek S (2012) Remarks on occupy Wall Street. In: Francois Roche (ed) Any corp, Log. No. 25, pp 118–120 (Summer 2012)

Chapter 4
A Holistic Approach to Comprehending the Complexity of the Post-growth Era: The Emerging Profile

Gandolfo Dominici, Vasja Roblek, and Rosa Lombardi

Abstract The goal of this study is to investigate the problems caused by the rapid growth of the global economy, coupled with high population growth and the excessive exploitation of natural resources. The global economy will not be able to maintain its current speed of growth in the long term, so a paradigm shift in production and consumption is necessary, if the collapse of ecosystems and the concurrent depletion of natural resources are to be avoided. This is the reason why capitalism needs to take a turn towards a sustainable and naturally harmonized development model.

4.1 Introduction

Growth may be one of the most frequently used words in the speeches and statements of politicians, economists, and managers, where it is employed with reference to social and economic (sustainable) development [26, 48, 50, 83].

The degree of social development [64, 78, 79] is based on different forms of organization, and not on periods or areas of specific events. Developmental stages are not necessarily tied to specific technologies. For example, pastoral nomads may be members of small tribes or organized under a centralized leadership. Social development does not take place along a fixed line that passes through the incidental historical pattern of development stages [59].

G. Dominici (✉)
Business Systems Laboratory, Roma, Italy

Department S.E.A.S., Polytechnic School, University of Palermo, Palermo, Italy
e-mail: gandolfo.dominici@libero.it

V. Roblek
Independent Researcher and Management Consultant, Trzic, Slovenia
e-mail: vasja.roblek@gmail.com

R. Lombardi
Business Administration, Link Campus University, Rome, Italy
e-mail: r.lombardi@unilink.it

Ş.Ş. Erçetin (ed.), *Chaos, Complexity and Leadership 2014*,
Springer Proceedings in Complexity, DOI 10.1007/978-3-319-18693-1_4

In the development process, the challenges of the environment present themselves both within ourselves and in our relations with neighboring organizations. The evolution of a diversification process is based on changes in the new organization. In this way, companies adapt to different environments and varying organizational issues.

The unsustainable and turbulent economic and social conditions are key issues for the debate about growth in the twenty-first century economy.

Humanity needs to focus on the problems of sustainable growth [19] above the capacity of natural ecosystems, caused by the rapid expansion of global production, coupled with high population growth and the excessive exploitation of natural resources.

The increased use of natural resources is strongly influenced by human activity and by economic policies aimed at increasing current economic indicators (such as GDP, level of employment, and consumption per capita); this approach is resulting in a higher demand for natural resources [17, 56]. Due to the rapid growth of production on a global scale in the twenty-first century, we have arrived at a turning point where the Earth's natural resources are no longer able to continually absorb the levels of pollution produced by the economy and cannot regenerate the natural resources used by humans [43]. World economic growth is based on the accelerated consumption of resources that, given the present level of technology, are unsustainable by the global ecosystem in terms of material and energy. This leads the global economy into an unbalanced state [33]. In other words, the global indicators of consumption of key natural resources (e.g., land footprint, water footprint, carbon footprint, and overall material use) are increasing, and have significantly exceeded the biological capacity of the planet [39] since the 1980s, when the human pressures on the utilization of natural resources exceeded the level of global recovery for the first time [60].

The wider adoption of the sustainable operations techniques by businesses would significantly advance the holistic approach with connected knowledge activities [68, 71]. Such holistic models include the promotion of sustainable development; they applying the sustainable paradigm beyond its environmental dimensions [80]. Consequently, a valid sustainable strategy should include both an economic and environmental perspective and the development of the sociopolitical and cultural aspects of development [42].

In this direction, the aim of this study is to investigate sustainable development through a holistic approach, in order to understand the global processes of growth. Through a review of the literature on sustainable development, we create a holistic framework by choosing a set of indicators based on the social, economic, environmental, and health perspectives.

The study has the following structure: After the introduction, Sect. 4.2 describes the (sustainable) growth concept for the capitalist economy through the postgrowth era. Section 4.3 describes the postgrowth era and the relationships between the environment, economics, and society. Section 4.4 proposes a discussion on a sustainable and naturally harmonized development model. Section 4.5 illustrates the conclusions of the study, describes its limitations, and suggests future research.

4.2 From Capitalist Economy to the (Sustainable) Growth Concept in the Postgrowth Era

Economic historiography describes the nineteenth and twentieth centuries as a period of economic growth of the markets [62]. The first industrial revolution enabled the development of industries such as textiles, pharmaceuticals, chemicals, steel, and transport. The second industrial revolution brought about the introduction of electrification, which had in the twentieth century a significant impact on the development of information and communication technologies. The economic growth caused by industrialization was felt among the population in the second half of the nineteenth century with the first division of profits throughout the economy; wage growth began to match productivity growth [1].

With the industrialization of Western Europe, the average income per person that was estimated as being about 90 % of its value in Africa today. Life expectancy in Western Europe and Japan in 1800 was about 40 years, and in a period of economic growth in the nineteenth and twentieth century there was a more than sixfold increase in the population [63].

The introduction of social reforms in the first half of the twentieth century affected industrialized countries and led to a general increase in wages. There was an increase in investment in developing school systems [7], which resulted in the growth of the educational infrastructure. The period after the Second World War was characterized by the creation of jobs, which demanded higher and more specialized education [1]. This period was considered the golden age of growth and prosperity for the middle class.

The rise of computer technologies at the beginning of the 1980s led to the transformation of traditional heavy industry into the technological development-oriented economy [75].

At the beginning of the twenty-first century, the theory of the innovative economy asserted that knowledge was main vehicle for economic growth. The rate of economic growth in the innovation economy therefore depends on the products or services that arise from knowledge [3].

In the second decade of the twenty-first century, the advent of the third industrial revolution has brought new technological development, including advanced robotics, which has increasingly impacted both on the job migration caused by globalization and on the increasing loss of jobs. Frey and Osborne [35] argued that jobs are at a high risk of being automated in 47 % of the occupational categories into which work is customarily sorted. The substitution of labor with automated production presents benefits to the owners of capital. This has been proceeding apace since the 1980s in a higher concentration of resources to capital, while the share of the world's income for retribution to employees has fallen [84].

In 2008, the financial crisis showed how short-term profitability mindsets and related strategies, policies, and actions of individuals and individual organizations are the cause of ecological and ethical disaster [27, 61, 81]. These events have contributed to the conclusions that most organizations operate on business models identified as unsustainable [11].

Two major economic players, the US and the EU, have begun to lose their advantages in the field of innovation [5, 18, 24, 87] and in certain high-tech sectors. Brazil, Russia, India, China, South Korea, Turkey, Indonesia, and so on are newcomers on the global markets [67]. In recent years, the disposable income in China and India has soared to 8 % a year, as opposed to a mere 2 % in the US and 1 % in the Japan. In line with high economic growth and digital adoption in these emerging markets, their middle class are gain higher levels of income.

This downturn has also caused crises in the oil market, beginning in Fall 2014, on account of the depressed demand induced by the continuing global recession, together with the increased oil extraction from nontraditional sources.

In the world economy, oil is a crucial factor. Iran, Venezuela, and Russia—all members of OPEC—have economies based on the export of commodities. They consequently ensure social peace using high prices of oil, and are presently faced with recession and currency devaluation. The Eurozone is currently showing lower rates of economic growth and deflation. It is expected that the world economy will grow by 2.9 % in 2015. The USA will be the best-performing rich economy in 2015, with growth forecast at 3.3 %. The world economy will be held back by weakness in the euro area and Japan, and by slower growth in emerging markets. China will see its expansion slow to 7 % in 2015 [28].

This turbulent scenario calls for solutions and structural reforms. The future scarcity of fossil fuels means that the exponential growth in consumption and GDP that is based on them will no longer be possible. Structural reforms include promotion of long-term planning to ensure continuity in policy through political changes. Humanity also needs to realize that the Earth no longer provides the stock of natural resources and fossil fuels which, even today, form the basis of economic growth [31]. New energy planning must be based on renewable energy sources. The creation of new jobs in clean and innovative technologies (the low-carbon economy) will both affect sustainability and reduce unemployment.

With a zero or even negative growth rate, decreasing profits in competitive markets are likely to lead to the transformation of the final phase of capitalism in a postcapitalist society [25] based on sustainable consumption governance and nonprofit businesses [10, 14, 69].

4.3 The Relation Between Environment, Economics, and Society in the Postgrowth Era

4.3.1 Welfare and Sustainable Development Without Growth

Beginning from the concept of social wellbeing founded on economic, social, and environmental factors [8, 23, 37], we focus here on the social factors of consumption and economic wellbeing (on the aggregate level). We particularly consider the factors of livability, environmental quality, quality of life, and sustainability.

In this regard, we aim to answer to the following questions:

How do livability, environmental quality, quality of life, and sustainability interact with each other?
What areas of operation, and their aspects, are essential to these factors?

4.3.2 Sustainable Development: Definition

The joint analysis of consumption and sustainable development presented in the Brundtland Commission report [90] allows for the following definition:

> development that meets the needs of the present without compromising the ability of future generations to meet their own needs.

In this way, the connection between sustainable development, the growth of a country's well-being, and human consumption is mainly founded on the needs of the world's poor and on the limitations imposed by current technology and social organization on the environment's ability to meet present and future needs [90].

4.3.3 Country Welfare Indicators

At the end of the nineteenth century, minimum wages were implemented in a number of advanced economies in order to prevent exploitation and social conflict. The International Labour Organization (ILO) considers that the minimum wage is a fundamental right of workers and an international standard followed by a 90 % of its member states. Today in Europe, 15 of the 19 Eurozone countries (all but Austria, Cyprus, Finland, and Italy) and 22 of the 28 EU Member States (excluding also Denmark and Sweden) have a legally prescribed minimum wage. Minimum wages also exist in three non-European G7 states: the US, Japan, and Canada. A minimum wage should not be too high relative to productivity in the state, since this would have a negative impact on competitiveness, leading to job losses and an increased burden of social support on the state budget [89]. The political viability of the minimum wage results from the imbalance of power between labor and capital. Without minimum wage policies, there would be the possibility of a permanent downward pressure on wages, which would undermine the effective functioning of the labor market. Minimum wages, therefore, are not in conflict with the labor market: they are a prerequisite for the effective functioning of the labor market.

The establishment of the minimum wage can increase productivity on three levels:

- by increasing employee motivation to work;
- by reducing staff turnover;
- by encouraging employers to introduce modern technology, investing in knowledge and better qualified workers.

Table 4.1 Country welfare indicators

Group of indicators	Indicators
Social (S)	S1: School enrolment rate
	S2: Internet connectedness rate
	S3: IT literacy rate
Economic (Ec)	Ec1: Per capita GDP
	Ec2: Per capita energy consumption
	Ec3: GDP of the population employed in agriculture
	Ec4: Exports of manufactured goods
	Ec5: Share of manufacturing industries in the GDP
Environmental (En)	En1: Ecological carrying capacity
	En2: OECD green growth indicator
	En3: Ecological footprint
	En4: Carbon footprint
Health (H)	H1: Life expectancy at birth
	H2: Number of nurses per 1000 people
	H3: Number of pharmacists per 1000 people
	H4: Number of dentists per 1000 people
	H5: Number of physicians per 1000 people

The importance of the minimum wage is evident in the case of rich, developed Germany. In Germany prior to the enforcement of the minimum wage (€8.5 per hour), 17 % of employees were receiving lower wages [16].

In terms of the ratio between the minimum wage and the nominal monthly per capita GDP, Germany (at 49.88 %) is close the United Kingdom (49.22 %), the Netherlands (52.03 %), Ireland (51.50 %), France (56.21 %) and Belgium (55.78 %). Lower ratios are shown by the most significant non-European G7 states: the USA (30.11 %), Japan (28.79 %), and Canada (37.26 %) [58].

A more holistic welfare approach should be based both on a wider set of indicators of social and economic development (Ec indicators, S indicators) as well as environmental and health development (En indicators, H indicators) as shown in Table 4.1 [32, 34, 91].

The progress of welfare through the generations can thus be investigated in the same manner as a country's social, economic, and environmental development [65, 77].

The traditional development model is characterized by material and energetic elements and focuses on fossil fuels, mass consumption, and the disposal of waste [40, 75]. In economically developed countries, indicators of income [55] and population growth in GDP are no longer sufficient as measures of overall progress, quality of life, or individual satisfaction [2]. Other indicators show that growth in GDP is also correlated with increasing environmental pressures [57, 66].

New indicators, such as indicators of happiness (e.g., the human development index) can measure the well-being of an economy or a nation. Frey and Stutzer [36, 38] show that in 1957, 35 % of the US population was very happy, while by the mid-1990s, only 33 % was very happy, even though the income of the population had doubled. In this sense, culture [30] is a very relevant factor in increasing wealth and in reducing poverty—especially in Asian countries where there is still cultural discrimination against women.

Toffler [85] predicted the emergence of the internet new economy [74], influencing the rise of the third wave of capitalism [76] that began in the early 1990s. Rapid technological developments, cheaper computer equipment, and the increased availability of broadband internet connection have been the factors influencing the change of life and work style [53]. Technology and economic growth have become inseparable during the period of transition to the new innovation economy [54]. With the advent of the information society [86], digital literacy has become a relevant key indicator of development.

In this context, the "Malthusian" idea of ecological carrying capacity has been recognized in the management of organizations. Ecological carrying capacity can be defined as the maximum density of individuals of a species that an ecosystem can support [51]. The ecological carrying capacity is identified through measures to protect certain habitats, defining the upper and lower limits of the population [70]. Such indicators have sometimes assisted in making decisions about the protection of endangered species [22, 46, 72].

The Malthusian approach can be used in a number of different ways. The first purpose is the evaluation of the carrying capacity of the human population, justifying policies that suspend the human right to reproduce. The second purpose is directed to the oversight of spending and demand for resources and the total human impact of consumption and production on the environment.

Other indicators for evaluating production policies and consumption uses are as follows:

- per capita ecological footprint (EF);
- ecological footprint analysis (EFA).

In this field, ecological footprint is recognized as a measure of human demand on the Earth's ecosystems. So, the surface of land and water (including natural resources) must be considered in measuring production outcomes [44].

Comparing information on the ecological footprint of global production and consumption, Wackernagel et al. [88] highlighted that, while in 1961, we needed an ecological footprint of 0.7 ha per person for our production and consumption needs, in 2007, this figure had already reached 2.7. Hence, the organic load-carrying capacity of our planet and our lifestyle are no longer naturally harmonized [45].

Ecological footprints are also related to carbon footprints [47]. In particular, for each product it can be determined how much greenhouse gas was emitted over the course of its life, including production, consumption, and disposal [92]. The Kyoto Protocol defined carbon footprint as a measure of the impact of human activities on the environment. The carbon footprint approach is gaining ground, and with increasing environmental awareness is becoming a benchmark for consumers.

Although the carbon footprint can also be used as basis for sustainable policies [15], it is difficult to collect all the information needed, since it appears that only 128 of the 4,609 largest companies listed on the world's stock exchanges disclose the most basic information about how they meet their social and ecological responsibilities [4, 6, 41]. More than 60 % of the world's largest listed companies currently fail to disclose their emissions; three quarters are not transparent about their water consumption. The number of large listed companies that disclose their energy use increased by 88 % from 2008 to 2012; however, there was only a 5 % rise from 2011 to 2012. A similar reporting slowdown is occurring for other first-generation indicators [20].

4.4 The Sustainable and Naturally Harmonized Development Model

According to the industrial model of capitalism that has led to the economic and environmental crisis [49], the modeling of sustainable and naturally harmonized development needs to be focused on the global economy, including both developed and developing countries.

On one hand, the world economy is now much greater than its natural base; on the other hand, the rapid development of developing countries represents a threat to the ecosystem. In terms of quantities used of basic commodities, such as wheat, fertilizers, steel, and coal, China has overtaken the United States to become the world's largest consumer [12]. Hence, developing countries, just like as economically developed countries, need a different development model.

We argue that contemporary capitalism should take a new direction towards a sustainable and naturally harmonized development model (Fig. 4.1) or it will soon

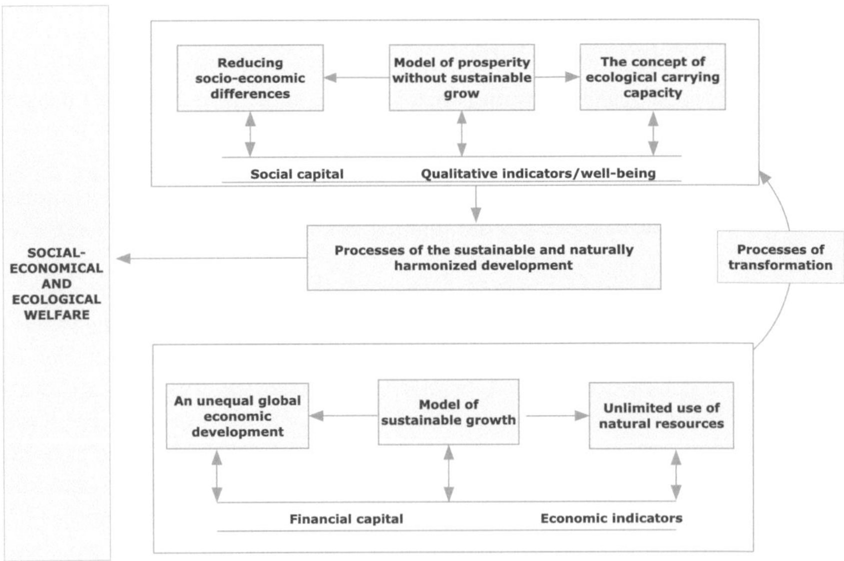

Fig. 4.1 A sustainable and naturally harmonized development model

face collapse. The flaws of the present-day capitalist model are investigated by Piketty [73] in the light of the proposition that, without a radical transformation of the financial system, working more and working harder will not ensure prosperity. The neoliberal ideology punishes everyone who is not an owner of capital. Policy-makers and NGOs, at the macrolevel, as well as companies and investors on the microlevel, need to conceive of value creation in a holistic sense when formulating strategy and allocating dwindling resources—particularly as they seek to build long-term value [20].

The world economy remains dependent on natural resources, even after its transition to an information society. In this scenario, politicians (mis)understand economic development to be constituted of nothing more than statistical increases in economic indicators; they are not effectively focusing on the deteriorating health of the global ecosystem. Alternative measurements of well-being may help replace the present indicators of growth to which policymakers are addicted [82].

Innovative companies are invited to develop services based on the sharing economy, creating a shift in the culture of possession of goods. The most relevant example of this is the web portal Airbnb, designed to promote the trust-economy. Airbnb targets seekers and owners of accommodation who wish to rent their free space (an entire flat or just a room). Thus, individuals share their excess space with strangers and earn money.

Another field is car and apartment rental. For example, the company Uber, established in 2009 and now operating in 45 countries, is based on a web application that allows users to find a private car for transportation needs. In 2014, this service was the target of protests by taxi services in France, Germany, and the United Kingdom. The taxi drivers were protesting against what they saw as an unfair business practice, and one which might also pose risks to passengers [29].

Brown [12, 13] argued that markets must recognize environmental issues, and that the current prices of goods do not include environmental costs. A new process of market creation is thus necessary to include ecological costs.

In this regard, one of the key challenges to the concept of prosperity without growth is the development of a new macroeconomics of sustainable development [9].

Daly [21] emphasized the key role of environmental conditions in terms of the unexploited stock of natural capital and the maintenance of low levels of material and energy flows within the regeneration and the assimilation capacity of the ecosystem. The macroeconomics for sustainability should not be considered a model for maintaining economic stability by increasing environmental pressure. The required paradigm shift implies no longer assuming the material consumption will grow. The shift must thus be from quantitative to qualitative economic development [21].

A sustainable economy [19] is based on the integration of social, human, physical, and environmental capital. The engine of growth is thus not only the greening of the economy through new technologies that produce fewer emissions and consume fewer resources, but also focusing on investment in knowledge in order to better use human capital. A sustainable society is related to a complete transformation of the global energy system, from centralized to decentralized, and from nonrenewable to renewable energy sources [52]. Policymakers should be aware that, for advanced economies, prosperity without growth is not a utopian dream but a financial and environmental necessity [52].

4.5 Conclusions, Limitations, and Future Research

The capitalist system cannot continue to be based on mere profit and to be measured by quantitative statistical data alone; a radical shift is needed to avoid its collapse. It is necessary to be aware that every natural process of growth is limited by the ecological carrying capacity. Recognizing the environmental and social costs of the current paradigm of consumption is a way towards sustainably transitioning to a global culture of sustainability. The paradigm shift from maximizing profits to maximizing well-being is still in progress. The development of the market economy towards maximizing sustainable prosperity is proceeding in both developed and developing societies.

The main limit of this study derives from the missing support for empirical evidence of sustainable development through the application of the indicators in the four groups. From this perspective, future research should be aimed at verifying the changes in sustainable development, both investigating new data and information and comparing social, environmental, economic, and health indicators on the international level.

Acknowledgement To cope with regulation on Academic publishing we declare that although this study is based on a collaborative effort by all authors, Sects. 4.1, 4.3.1 and 4.3.2 can be attributed to Gandolfo Dominici; Sects. 4.2, 4.3.3 and 4.4 can be attributed to Vasja Roblek and Sect. 4.5 to Rosa Lombardi.

References

1. Allen RC (2009) The British industrial revolution in global perspective. Cambridge University Press, Cambridge
2. Anderson V (2014) Alternative economic indicators. Routledge, London
3. Antonelli C (2003) The economics of innovation, new technologies, and structural change. Routledge, London
4. Arvidsson S (2011) Disclosure of non financial information in the annual report: a management team perspective. J Intellect Cap 12(2):277–300
5. Arora A, Merges RP (2004) Specialized supply firms, property rights and firm boundaries. Ind Corp Chang 13(3):451–475
6. Banghoj J, Plenborg T (2008) Value relevance of voluntary disclosure in the annual report. Account Finance 48(2):159–180
7. Becker G (1993) Human capital. A theoretical and empirical analysis, with special reference to education. University of Chicago Press, Chicago
8. Berkman LF, Kawachi I, Glymour M (eds) (2014) Social epidemiology. Oxford University Press, Oxford

9. Bermejo R (2014) Sustainable development in the Brundtland report and its distortion. In: Handbook for a sustainable economy. Springer, Berlin
10. Blauwhof FB (2012) Overcoming accumulation: is a capitalist steady-state economy possible? Ecol Econ 84:254–261
11. Boons F, Montalvo C, Quist J, Wagner M (2013) Sustainable innovation, business models and economic performance: an overview. J Clean Prod 45:1–8
12. Brown L (2006) Plan B 2.0: rescuing a planet under stress and a civilisation in trouble. Earth Policy Institute, Washington
13. Brown L (2008) Plan B 3.0 mobilizing to save civilisation. Earth Policy Institute, Washington
14. Buch-Hansen H (2014) Capitalist diversity and de-growth trajectories to steady-state economies. Ecol Econ 106:167–173
15. Burnside C, Dollar D (2000) Aid, policies, and growth. Am Econ Rev 90(4):847–868
16. Carlin W, Hassel A, Martin A, Soskice D (2014) The transformation of the German social model. In: European social models from crisis to crisis: employment and inequality in the era of monetary integration, vol 49. Oxford University Press, New York
17. Chambers N, Simmons C, Wackernagel M (2014) Sharing nature's interest: ecological footprints as an indicator of sustainability. Routledge, London
18. Chesbrough H (2006) Open business models: how to thrive in new innovation landscape. Harvard Business Press, Boston
18. Chouinard Y, Ellison J, Ridgeway R (2011) The sustainable economy. Harv Bus Rev 89(10):52–62
20. Confino J (2014) 97 % of companies fail to provide data on key sustainability indicators. Available via The Guardian com. http://www.theguardian.com/sustainable-business/2014/oct/13/97-companies-fail-to-provide-data-key-sustainability-indicators-stocck-exchange-report. Accessed 20 Dec 2014
21. Daly EH (2009) A steady-state economy. University of Maryland, College Park
22. Das MC et al (2012) Decision making under conflicting environment: a new MCDM method. Int J Appl Decis Sci 5(2):142–162
23. Diener E, Suh E (1997) Measuring quality of life: economic, social, and subjective indicators. Soc Indic Res 40(1–2):189–216
24. Drucker P (1985) Innovation and entrepreneurship: practice and principles. HarperBusiness, New York
25. Drucker P (1998) From capitalism to knowledge society. In: Neef D (ed) The knowledge economy. Butterworth, Woburn, pp 15–34
26. Eicher T, Turnovski S (2003) Inequality and growth: theory and policy implications. The MIT Press, Cambridge, MA
27. Eidelman JG (1995) Z-scores a guide to failure prediction. CPA J online. Retrieved December 2014 from: http://www.cpajournal.com/old/16641866.htm
28. EIU (2015) Gauging growth in 2015. Available via Economist.com. http://www.economist.com/blogs/graphicdetail/2015/01/daily-chart. Accessed 11 Jan 2015
29. Euroactiv (2014) Available at: http://www.euractiv.com/sections/social-europe-jobs/uber-taxi-app-banned-germany-after-frankfurt-ruling-308187. Accessed 12 Dec 2014
30. Featherstone M (ed) (1990) Global culture: nationalism, globalization and modernity, vol 2. SAGE, London
31. Feder G (1983) On exports and economic growth. J Dev Econ 12(1):59–73
32. Ferguson J, Huysman M, Soekijad M (2013) Knowledge management in practice: pitfalls and potentials for development. World Dev 38(12):1797–1810
33. Fischer-Kowalski M, Swilling M (2011) Decoupling: natural resource use and environmental impacts from economic growth. United Nations Environment Programme UNEP, Geneva
34. Florida R (2010) Internet connectivity and economic development. Available via: http://www.creativeclass.com/creative_class/2010/07/30/internet-connectivity-and-economic-development/. Accessed 11 Dec 2014

35. Frey BC, Osborne AM (2013) The future of employment: how Susceptible are jobs to computerisation? Available via Oxford University. http://www.futuretech.ox.ac.uk/future-employment-how-susceptible-are-jobs-computerisation-oms-working-paper-dr-carl-benedikt-frey-m. Accessed 11 Jan 2014
36. Frey BS, Stutzer A (2002) What can economists learn from happiness research? J Econ Lit 20:402–435
37. Frey BS, Stutzer A (2010) Happiness and economics: how the economy and institutions affect human well-being. Princeton University Press, Princeton
38. Frey BS, Stutzer A (2012) The use of happiness research for public policy. Soc Choice Welf 38(4):659–674
39. Galli A, Wiedmann T, Ercin E, Knoblauch D, Ewing B, Giljum S (2012) Integrating ecological, carbon and water footprint into a "footprint family" of indicators: definition and role in tracking human pressure on the planet. Ecol Indic 16:100–112
40. Garetti M, Taisch M (2012) Sustainable manufacturing: trends and research challenges. Prod Plan Control 23(2–3):83–104
41. Ganderton PT (2005) Benefit-cost analysis of disaster mitigation: application as a policy and decision-making tool. Springer, London
42. Gibson B, Hassan S, Tansey J (2013) Sustainability assessment: criteria and processes. Routledge, London
43. Goudie AS (2013) The human impact on the natural environment: past, present, and future, 7th edn. Wiley-Blackwell, Hoboken
44. Hackett SC (2011) Environmental and natural resources economics: theory, policy, and the sustainable society. ME Sharpe, New York
45. Hannigan J (2014) Environmental sociology. Routledge, London
46. Hastie R, Dawes RM (2001) Rational choice in an uncertain world: the psychology of judgement and decision making. Sage, Thousand Oaks
47. Harkiolakis N (2013) Ecological footprint. In: Encyclopedia of corporate social responsibility. Springer, Berlin
48. Hausmann R, Pritchett L, Rodrick D (2005) Growth accelerations. J Econ Growth 10(4):303–329
49. Heinberg R (2011) The end of growth: adapting to our new economic reality. New Society Publishers, Gabriola Island
50. Inglehart R (1997) Modernization and postmodernization: cultural, economic, and political change in 43 societies. Princeton University Press, Princeton
51. Isaac NJ, Carbone C, McGill BJ (2012) Population and community ecology. In: Metabolic ecology: a scaling approach. Wiley, West Sussex, pp 77–85
52. Jackson T (2011) Prosperity without growth: economics for a finite planet. Earthscan, London
53. Kaplan AM, Haenlein M (2010) Users of the world, unite! The challenges and opportunities of social media. Bus Horiz 53(1):59–68
54. Kaplan SR, Mikes A (2012) Managing risks: a new framework. Harv Bus Rev 90(6):49–60
55. Klasen S (2008) Economic growth and poverty reduction: measurement issues using income and non income indicators. World Dev 36(3):420–445
56. Knight KW, Rosa EA, Schor JB (2013) Could working less reduce pressures on the environment? A cross-national panel analysis of OECD countries, 1970–2007. Glob Environ Chang 23(4):691–700
57. Knox P, Agnew J, McCarthy L (2014) The geography of the world economy. Routledge, London
58. Kotnik T (2013) Obeti in grožnje nemške minimalne place (Prospects and threats of the German minimum wage). Available via Finance.si. http://www.finance.si/8352591/Obeti-in-gro%C5%BEnje-nem%C5%A1keminimalnepla%C4%8De?metered=yes&sid=120753021. Accessed 9 Jan 2015
59. Kradin NN (2006) Cultural complexity of pastoral nomads. World Cult 15(2):2–21

60. Krausmann F, Gingrich S, Eisenmenger N, Erb KH, Haberl H, Fischer-Kowalski M (2009) Growth in global materials use, GDP and population during the 20th century. Ecol Econ 68(10):2696–2705
61. Krugman P (1999) Balance sheets, the transfer problem and financial crises. In: Isard P, Razin A, Rose A (eds) International finance and financial crises: essays in honor of Robert Flood. Kluwer Academic Publishers, Boston, pp 3–44
62. Kuznets S (1973) Modern economic growth: findings and reflections. Am Econ Rev 63(3):247–258
63. Maddison A (2007) Contours of the world economy 1-2030 AD: essays in macro – economic history. Oxford University Press, New York
64. Marshall TH (1964) Class, citizenship and social development. Doubleday, New York
65. Moe T, Alfsen KH, Greaker M (2013) Sustaining welfare for future generations: a review note on the capital approach to the measurement of sustainable development. Chall Sustain 1(1):16–26
66. Moldan B, Janoušková S, Hák T (2012) How to understand and measure environmental sustainability: indicators and targets. Ecol Indic 17:4–13
67. Montobbio F, Sterzi V (2013) The globalization of technology in emerging markets: a gravity model on the determinants of international patent collaborations. World Dev 44:281–299
68. Mulej M, Potocan V (2007) Requisite holism – precondition of reliable business information. Kybernetes 36(3–4):319–332
69. Murtaza N (2011) Pursuing self-interest or self-actualization? From capitalism to a steady-state, wisdom economy. Ecol Econ 70(4):577–584
70. Nadakavukaren SK (2007) Chilling the protection of human rights: what the Kimberley process waiver can tell us about the WTO's effect on international law. Working paper no 2007/03. NCCR trade working papers. Retrieved in January 2015 from: http://phase1.nccr-trade.org/images/stories/publications/IP4/chilling%20effect%20revised.pdf
71. Ny H, Hallstedt S, Ericson A (2012) A strategic approach for sustainable product service system development. In: Chakrabarti A (ed) CIRP design 2012. Sustainable product development. Springer, Heidelberg, pp 427–436
72. Nooraie M (2011) Decision's familiarity and strategic decision-making process output: the mediating impact of rationality of the decision-making process. Int J Appl Decis Sci 4(4):385–400
73. Piketty T (2014) Capital in the twenty-first century. Belknap Press of Harvard University Press, Cambridge
74. Rifkin J (2014) The zero marginal cost society: the internet of things, the collaborative commons, and the eclipse of capitalism. Macmillan, New York
75. Roblek V, Meško M, Pejić Bach M, Bertoncelj A (2014) Impact of knowledge management on sustainable development in the innovative economy. In: Ferri MA, Basile G (eds) Proceedings of Business Systems Laboratory, 2nd international symposium. Universitas Mercatorum, Rome
76. Roblek V, Pejić-Bach M, Meško M, Bertoncelj A (2013) The impact of social media to value added in knowledge – based industry. Kybernetes 42(4):554–568
77. Roseland M (2000) Sustainable community development: integrating environmental, economic, and social objectives. Prog Plan 54(2):73–132
78. Rose Krasnor L (1997) The nature of social competence: a theoretical review. Soc Dev 6(1):111–135
79. Schaffer HR (1996) Social development. Blackwell Publishing, Oxford
80. Starik M, Kanashiro P (2013) Toward a theory of sustainability management: uncovering and integrating the nearly obvious. Organ Environ 26(1):7–30. doi:10.1177/1086026612474958
81. Slatter S (1994) Corporate recovery. A guide to turnaround management. Penguin Book, London
82. Stiglitz J, Sen A, Fitoussi JP (2009) Report by the commission on the measurement of economic performance and social progress. Available via http://www.stiglitz-sen-fitoussi.fr/documents/rapport_anglais.pdf. Accessed 12 Dec 2014

83. Szirmai A (2005) The dynamics of socio-economic development: an introduction. Cambridge University Press, Cambridge
84. The Economist (2015) The future of jobs: the onrushing wave. Available via http://www.economist.com/news/briefing/21594264-previous-technological-innovation-has-always-delivered-more-long-run-employment-not-less. Accessed 9 Jan 2015
85. Toffler A (1984) The third wave. Bentlam, Oak Park
86. Tyner K (2014) Literacy in a digital world: teaching and learning in the age of information. Routledge, London
87. Von Hippel E (1988) Sources of innovation. Oxford University Press, New York
88. Wackernagel M, Kitzes J, Moran D, Goldfinger S, Thomas M (2006) The ecological footprint of cities and regions: comparing resource availability with resource demand. Environ Urban 18(1):103–112
89. Werding M, Konrad KA (2012) Reforming the European welfare state. In: Buettner T, Ochel W (eds) The continuing evolution of Europe. Massachusetts Institute of Technology, Boston, pp 71–118
90. WCED, World Commission on Environment and Development (1987) Our common future. Oxford University Press, New York
91. World Bank, World Development Indicators (2014) Available via: http://data.worldbank.org/data-catalog/world-development-indicators. Accessed 4 Dec 2014
92. Wright LA, Kemp S, Williams I (2011) Carbon footprinting: towards a universally accepted definition. Carbon Manage 2(1):61–72

Chapter 5
Emerging Leadership Strategies for Complex Times

Gaynell Vanderslice

Abstract This article will provide a brief introduction of emerging twenty-first century leadership theories based on various literature reviews. Earlier, industrial era, leadership theories are centered on the perspective of a single heroic leader who provides vision and guidance on behalf of the masses. These theories leverage a command-and-control leadership style to optimize productivity and increase organizational sustainability. However, with the onset of rapidly changing economic conditions, unpredictable financial markets and the short fall of Red Ocean and Blue Ocean strategies, the single heroic leader phenomenon has failed to optimize knowledge—the largest human capital investment. Several researchers have stated that the value of knowledge is in its capacity to generate innovative products and service and thereby create competitive advantages for the firm. Thereby, leadership strategies that encourage knowledge sharing through social interaction may enhance organizational sustainability. Several scholars have argued that it is this social dimension of knowledge sharing that calls for new leadership strategies. The article highlights several knowledge era leadership theories.

5.1 Introduction

The market place is global, the economic system highly connected, technological advancements disruptive, changes at a rapid rate and opportunities to exploit Red Oceans diminishing [1, 13, 15]. By the same token, most leadership strategies and practices that emerged during the industrial era fail to optimize the competitive advantages that exist within a complex knowledge-based economy [15]. One key omission is how to leverage knowledge; a new dimension engrossed in economically, technological, complex systems that requires the enactment of counterintuitive strategies [3].

G. Vanderslice (✉)
Management, Alliant University, San Diego, CA, USA
e-mail: gvanderslice@alliant.edu

Ş.Ş. Erçetin (ed.), *Chaos, Complexity and Leadership 2014*,
Springer Proceedings in Complexity, DOI 10.1007/978-3-319-18693-1_5

5.2 Economic Complexity

Today's social-economic system is connected in a multitude of ways, resulting in dynamically complex systems [11]. As such, unexpected changes within financial markets, the housing industry, oil industry, and automotive industry, can trigger a systemic event across the many with the added capability of spiraling out of control [9]. In like manner, technological advancement can cause a systemic shift across various markets as well. As a result of these, and other changes, organizations are hard pressed to remain sovereign. Beer et al. [3] summed it up by stating that, "terms such as 'discontinuous change', 'disruptive technologies' and 'age of revolution' describe the uncertainty and complexity that pervades the competitive environment" (p. 446). Furthermore, ongoing turbulence in our social-economic system gives rise to a complex array of challenges for leaders to navigate. This state of disequilibrium is evident in many complex systems and the application of conventional methods often times produces unexpected results. Hence, leaders would be well served to apply unconventional strategies to leading within complex organizations [9]. That being said, this ongoing state of disequilibrium can also expose opportunities for growth when combined with emerging strategic leadership practices [11].

5.3 Red Ocean and Blue Ocean Strategies

According to Drucker [6] the purpose of the firm is to create a customer. To this end, organizations leverage various strategies to create new customers and/or retain existing ones. Using a broad stroke, market strategies are built on a foundation of competition or value [4, 13]. Markets that lend themselves to competition are commonly referred to as existing in a Red Ocean. In contrast, organizations that enter markets where few have traversed are referred to as Blue Oceans [16]. Red Oceans and Blue Oceans have been juxtaposed for years; hence neither of which are new. Albeit, the difference between the two leadership strategies and is that supporting theories for Red Ocean strategies abound while theories supporting Blue Ocean strategies are still being developed [13].

Blue Ocean strategies do not equate to a state of nirvana and in some cases may be difficult to put into practice [1]. Hence, it must be noted that without barriers to competition, inroads into new markets and expanded profit margins will eventually draw the attention of future competitors, thereby increasing the competitive landscape over time ultimately resulting in a Red Ocean and shrinking market shares [4, 16]. It has also been argued that successful implementation of past structures and strategies can impede future opportunities to remain competitive [15].

Nonetheless, "far from encouraging managers to take an either/or position between Blue Ocean strategy and Red Ocean strategy, the optimal strategic approach appears to involve adopting a blend of both strategies with different temporal emphasis" ([4]: 23). However, if twenty-first century strategic positioning starts and ends with Blue Ocean and Red Ocean strategies, then one of the most valuable strategic opportunities will remain underutilized—human capital [1].

5.4 Knowledge Productivity

Unlike, the industrial era, aligned with assembly line workers who performed repetitive task [29] and bosses who used command-and-control tactics to rule the masses, the largest capital investment in a knowledge-based economy is the knowledge worker [11]. The knowledge worker's capabilities, skills, experience and social engagement of cognitive thought represents an ongoing stream of innovative ideas for the firm [29]. It is through social interaction that the value of knowledge within the organizations expands. As knowledge workers develop relationships and interact in a social context, they share ideas, resolve complex issue, and create new forms of advantages through the collective use of organizational knowledge [21] resulting in an increase in knowledge productivity.

Nissen [20] further argued that knowledge sharing or knowledge flow between individuals and groups oftentimes represents a slow process. It must be noted that, although the knowledge worker can share some of what she/he knows in verbal or written form, much of what is known will remain hidden in the subconscious mind [20, 25]. Nonetheless, knowledge productivity or creating new knowledge begins with knowledge sharing. According to Nonaka et al. [22], the knowledge creation process consists of four elements: *socialization, externalization, combination* and *internalization* that represent the flow between tacit (personalized) knowledge and explicit (codified) knowledge; yet despite its potency to create sustainable competitive advantages, individuals, groups and organizational barriers routinely hinder attempts by leaders to strategically leverage individual or collective knowledge [29].

Thus, the ongoing need to increase knowledge productivity places many organizations on a new trajectory that most leaders are not equipped to navigate [6]. Therefore, organizations need leadership strategies and practices that effectively influence the knowledge sharing process of knowledge workers [10].

5.5 Industrial Era Leadership

An array of definitions exists that define leadership theories born out of the industrial era, but at their core, the most common understanding has been expressed as "a process whereby an individual influences a group of individuals to achieve a common goal" ([23]: 3). Close examination of what has been accepted as the heroic leader phenomena opened the door to a plethora of leadership theories that are still practiced today. For example, Northhouse [23] provided the following overview of leadership theories (see Table 5.1) prior to the twenty-first century that emerged and promoted the image of the heroic leader and down played the role of the follower to the determent of the organizational effectiveness [7]. Continued reliance on the influence of leadership practices and theories founded on the principles of leader-follower leave twenty-first century leaders in a quagmire [5].

The tension between organizational leaders and workers was laced with mutual distrust and battles over working conditions and wages [14] amongst other things.

Table 5.1 Single heoric leader theories

Theory	Characteristics
Situational leadership theory	The situational leadership view expresses the need for the leader to adjust his/her approach based on current situation or circumstances and demonstrate a leadership style that is appropriate for the occasion.
Contingency leadership theory	Is about right fit or alignment of the leader's style within the current context or setting. Based on the context leaders were said to be task motivated or relationship motivated. The former being concerned with goal achievement and the latter with building relationships with followers.
Path-goal theory	The leader's ability to motivate followers to complete specific goals.
Leader-member exchange theory	Interaction between the leader and the follower becomes important.
Transformational leadership theory	Emerged during the early 1980s. The leader influences followers by supporting their intrinsic motivations. It is this level of motivation that encourages followers to extend themselves beyond their basic aspirations.

These tensions gave rise to unions as outcries for fair treatment of labor and had little to do with leveraging the employee's intellectual knowledge [2]. The irony of this complex and turbulent state is that the study and practice of leadership, after many years of building on a foundation of leader-follower influence is now in a state of flux, which calls for the development of new foundational truths. Stated differently, "we're in a knowledge era, but our managerial and governance systems are stuck in the industrial era" ([19]: 48).

In the same vein, Rost [27] argued that predominant leadership studies align with the various aspects of the industrial era. He also calls for the furtherance of leadership theories that may run counter to what we know today, but highlight challenges associated with a postindustrial era. Similarly, Drath et al. [5] asserted that as the context for leadership continues to shift the ontology of leader-followers and common goals becomes more of a hindrance to the development of leadership theory.

5.6 Strategic Opportunities in Emerging Leadership Theories

Understandably, leadership theories of the industrial era, at best, provide a simplistic view of what occurs in the realm of leadership such that the model of the single heroic leader has not stood the test of time [18]. The emergence of a knowledge-base economy creates a shift whereby the leader-follower relationship becomes more peer-like and collaborative in form that is highly relational with adaptive qualities that can optimize the new competitive advantage of the firm; in other words, its human capital [5, 19].

Unfortunately, there is no agreed upon definition for these emerging leadership theories [17], thereby, overlaps exist between concepts and meaning [8]. Nonetheless, leadership studies are poised to provide new insights into the meaning of leadership for the twenty-first century. Lichtenstein et al. [18] provided a definition of adaptive leadership to portray a contrary view of what leadership represents. Accordingly, "adaptive leadership does not mean getting followers to follow the leader's wishes; rather, leadership occurs when interacting agents generate adaptive outcomes" (p. 4). This definition can be used to examine emerging leadership theories like *complexity leadership, relational leadership,* and *distributed leadership theory.*

- **Complexity Leadership Theory** is well established in earlier theories of complexity science and complex adaptive systems (CAS) that respond to dynamic, interdependent, and interconnected part [31]. As such, leadership is constantly evolving because leaders both influence and are influenced by their environments, creating and responding to a network of interactions and outcomes among agents. In the same vein, it embraces the innovative, contextual interactions that occur across a social system [18].
- **Relational Leadership Theory**, not to be confused with relationship-based approaches, is a central framework that examines the relational process that facilitates and produces leadership. Within this context, the lines between who is leading and who is following can become blurred. Moreover, studies in relational leadership extend beyond the role of the leader/manager, and examine the social dynamics that contribute to the rise of leadership [30].
- **Distributed Leadership Theory** highlights the interactions between members in formal and informal settings and takes into account the contributions of all individuals regardless of organizational status [8]. In support of this view, "a distributed perspective presses us to consider the enactment of leadership task as potentially stretched over the practice of two or more leader and followers" ([28]: 16). Distributed leadership can also be viewed as a diagnostic tool or a framework for further development of other collective, interactive, theories [8]. For instance, emerging leadership theories, like *collective leadership*, *shared leadersh*ip, and *collaborative leadership*, to name a few, have similar qualities and overlap (see Table 5.2).

Each theory examines and embraces the social interconnectivity and interdependence between its members. Extending beyond these simple realizations is the understanding that as the organizational terrain changes new leaders may emerge, temporarily or permanently, at various levels within the organization. Stated differently, the interactions of one individual, or a select few, cannot fully leverage all of the interdependent relationships and subsequent interactions of the firm required to navigate complex environments [18]. Although challenging, emerging leadership theories co-exist in the mist of chaos and complexity, not apart from it.

Table 5.2 Emerging leadership theories

Theory	Characteristics
Collective leadership ([7], p. 933)	A dynamic leadership process in which a defined leader, or set of leaders, selectively utilize skills and expertise within a network, effectively distributing elements of the leadership role as the situation or problem at hand requires.
Shared leadership ([24], p. 1)	A dynamic, interactive influence process among individuals in groups for which the objective is to lead one another to the achievement of group or organizational goals or both.
Collaborative leadership ([26], p. 134)	Collaborative leadership means everyone is in control of and can speak for the entire team.

5.7 Conclusion

Chaos, complexity and leadership are realities of the twenty-first century economic environment and former leadership strategies may limit opportunities to compete and remain viable. As long as organizational leaders continue to buy into the single heroic leader phenomenon progress the adaptation of new strategies will be thwart. Nonetheless, the adaptation of new forms of leadership that seeks to engage members throughout the organization is gaining awareness. Yet, Harris [8] cautions that various forms of distributed leadership extend beyond the scope of creating a flatter organizational structure or assigning various task to others; it's a leadership practice.

As with any new phenomenon, there is a process of emergence during which ideas, theories, and concepts with overlapping definitions may abound [12]. In the interim, theoretical models that inform practice and the implementation of distributed leadership strategies have yet to emerge. Unfortunately, organizational leaders have to address today's challenges today meaning they may have to employ ad hoc solutions due to the lack of empirical research needed to guide decisions regarding these emerging leadership theories. Be that as it may, even in its nascent state, researchers and practitioners would benefit from exploring these emerging strategic opportunities more deeply—together.

References

1. Anwar A (2012) Benefits of management research to industries. Excel J Eng Technol Manag Sci 1(1):106–111
2. Beard MR (1920) A short history of the American Labor Movement, vol 3. Harcourt, Brace and Howe, New York
3. Beer M, Voelpel SC, Leibold M, Tekie EB (2005) Strategic management as organizational learning: developing fit and alignment through a disciplined process. Long Range Plann 38:445–465

4. Burke A, Van Stel A, Thurik R (2009) Blue Ocean versus competitive strategy: theory and evidence. Retreived from ERIM report series: http://www.ssrn.com
5. Drath WH, McCauley CD, Palus CJ, Van Velsor E, O'Conner PM, McGuire JB (2009) Direction, alignment, commitment: toward a more integrative ontology of leadership. Leadersh Q 19:635–653
6. Drucker PF (1996) The leader of the future. The Drucker Foundation, New York
7. Freidrich TL, Vessey WB, Schuelke MJ, Ruark GA, Mumford MD (2009) A framework for understanding collective leadership: the selective utilization of leader and team expertise within networks. Leadersh Q 20:933–958
8. Harris A (2008) Distributed leadership through the looking glass. Manag Educ 22(1):31–34
9. Helbing, D. (2012). Systemic risks in society and economics. In *Social Self-Organization*. Springer, Berlin/Heidelberg, pp. 261–284
10. Huysman M, Wulf V (2006) IT to support knowledge sharing in communities, towards a social capital analysis. J Inf Technol 21(1):40–51
11. Ireland RD, Hitt MA (2005) Achieving and maintaining strategic competitiveness in the 21st century: the role of strategic leadership. Acad Manag Exec 19(4):63–77
12. Khun T (1996) The structure of scientific revolutions, 3rd edn. The University of Chicago Press, LTD, Chicago
13. Kim WC, Mauborgne R (2004) Blue ocean strategy. Harv Bus Rev 82(10):76–84
14. Kimelodorf H (1999) Battling for American labor: wobblies, craft workers, and the making of the union movement. University of California Press, Berkley
15. Kotter, JP (2012) Accelerate. Harv Bus Rev 90(11):44–58
16. Kumar R, Gupta SK (2010) Blue Ocean marketing strategy. APJRBM 1(2):60–71
17. Li HC, Mirmirani S, Llacqua JA (2009) Confucius Institutes: distributed leadership and knowledge sharing in a worldwide network. Learn Organ 16(6):469–482
18. Lichtensteiin BB, Uhl-Bien M, Marion R, Seers A, Orton JD, Schreiber C (2006) Complexity leadership theory: an interactive perspective on leading in complex adaptive systems. Emerg Complex Organ 8(4):2–12
19. Manville B, Ober J (2003) Beyond empowerment: building a city of citizens. Harv Bus Rev 81(1):48–53
20. Nissen ME (2002) An extended model of knowledge-flow dynamics. Commun Assoc Inf Syst 8:251–266
21. Nonaka I (1994) A dynamic theory of organizational knowledge creation. Organ Sci 15(1):14–37
22. Nonaka I, Toyama R, Konno N (2000) SECI, Ba and leadership: a unified model of dynamic knowledge creation. Long Range Plann 33:5–34
23. Northhouse PG (2010) Leadership: theory and practice, 5th edn. Sage, Thousand Oaks
24. Pearce CL, Conger JA (2003) Shared leadership: reframing the hows and whys of leadership. Sage, Thousand Oaks
25. Polanyi M (1966) The tacit dimension. University of Chicago Press, Chicago
26. Raelin JA (2004) Don't bother putting leadership into people. Acad Manag Exec 18(3):131–135
27. Rost JC (1993) Leadership for the twenty-first century. Praeger Publishers, Westport
28. Spillane JP, Halverson R, Diamond JB (2004) Towards a theory of leadership practices: a distributed perspective. J Curric Stud 36(1):3–34
29. Swart J, Kinnie N (2003) Sharing knowledge in knowledge-intensive firms. Hum Resour Manag J 13(2):60–75
30. Uhl-Bien M (2006) Relational leadership theory: exploring the social process of leadership and organizing. Leadersh Q 17(6):654–676
31. Uhl-Bien M, Marion RB, McKelvey BC (2007) Complexity leadership theory: shifting leadership from the industrial age to the knowledge era. Leadersh Q 18:298–318

Chapter 6
Does Everyone in Turkey Benefit from Health Services with General Health Insurance?

Emine Orhaner

Abstract General health insurance was instituted with the aim of reducing the burden of health financing which fell upon the state, yet has ended up increasing it. The system was constructed with the aim of creating additional financial resources, but caused an important responsibility for the collection of unpaid premiums. Around 4.5 million people who do not work with general health insurance and who have not had means tests have received a high premium rate and indebted themselves, and since they cannot pay these, they cannot benefit from health services. On top of this, these people's debt increases with interest for non-payment. On the other hand, many citizens who reach 18 have to enter the general health insurance system, and yet they increase their premiums since they do not have the information to make informed choices; they only learn of their debts when they visit their doctor or a hospital. General health insurance has not been explained well to the public, and it has been implemented without enough information being distributed, so it does not seem that meeting its aim of allowing those without enough money to access health services will be possible. This paper sought to answer the question of whether everyone in Turkey benefits from health services with the general health insurance.

6.1 Introduction

Health insurance is a type of risk insurance in which a person's costs felt necessary in cases of sickness, accidents at work, occupational illnesses and general accidents are met for the examination, diagnosis, treatment, analysis and medicine prescribed by a doctor, according to the conditions of the social security institution, the level of subscription to a social security institution or the general conditions of the insurance policy in question. With health insurance, people pay premiums or contributions whilst they are healthy and thus participate in paying for the costs and losses of

E. Orhaner (✉)
Faculty of Economics and Administrative Sciences, Health Care Management Department, Gazi University, Ankara, Turkey
e-mail: eorhaner@gazi.edu.tr

Ş.Ş. Erçetin (ed.), *Chaos, Complexity and Leadership 2014*,
Springer Proceedings in Complexity, DOI 10.1007/978-3-319-18693-1_6

income associated with their illnesses, accidents, or incidents which leave them unable to work. For this reason, one of the ways in which fees for health services are funded other than taxes and direct payment is through premiums taken from those who are insured.

There are two ways in which health insurance in our country functions:

1. General health insurance. A branch of insurance that ensures the financing of costs encountered through health risks and in the protection of primary health. With general health insurance the aim is to provide an insurance system financing a state-guaranteed suite of health services which are equal, protective, diagnostic, quality, fair and easy-to-access for the whole population of the nation and resident foreigners.
2. Private health insurance. This is optional health insurance in which agreements are made according to the terms and conditions of the policies of private insurance companies. Private health insurance provides security in addition to that given by general health insurance.

Developing countries such as Turkey have an important social and economic responsibility for health insurance aimed at ensuring financial provision for health services and so the Social Insurance and General Health Insurance Law number 5510, which went into operation in 2008, brought in a new model of health organization. In the general health insurance model, which is generally in place in developed western countries, a large portion of the financing of health services is ensured through premiums taken no matter whether the individual benefits from the health services or not.

6.2 Health Insurance as a Method of Health Financing

In order to deliver health services, it is impossible to avoid increases in health costs. Increases in health costs are inevitable in Turkey, just as it is all around the world. In Turkey, increase in health costs can be attributed to the following causes:

- As in other countries we have an aging population
- Technology in the health sector is expensive and develops fast
- Patients and health providers wish to benefit from the latest technology.
- Chronic illnesses are on the rise
- Services are easy to access
- Sources of information have become easier to access
- People needing health services gradually want more from their health services

The trend towards increasing health expenditures create a problem with financing the cost of healthcare. Ensuring that all individuals benefit, that sources of income are created in order to ensure the continued provision of health services, that different financial methods are used and these sources are made efficient use of, are all important considerations in the provision of health services. The sources of

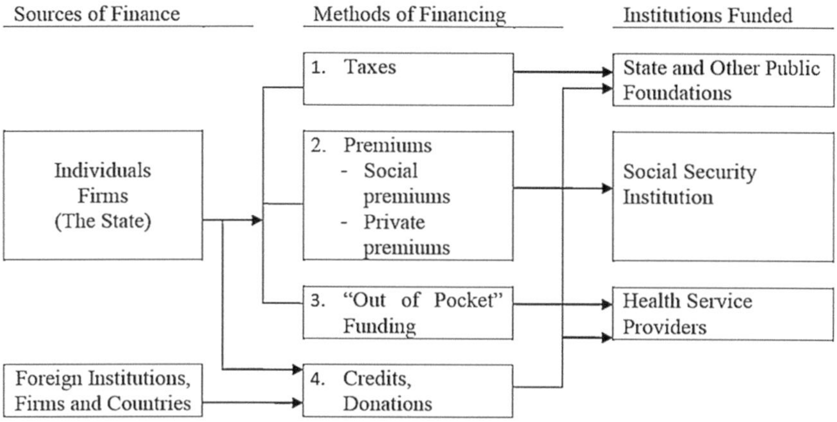

Fig. 6.1 Methods of health financing (Source: Adapted from Çelik [2, p. 170])

financing for health services are, as is often repeated in economics, the economic decision makers, who we can name individuals, firms and the state. When individuals wish to access health services the most important sources of the financial means are payments from their own pockets, the taxes they pay to the state or premiums paid for health insurance to the Social Security Institution. Firms are also a source of financing through the taxes and insurance premiums they pay and even the donations and other help they provide to health institutions. The state is a source of financing for health services through its contribution to insurance premiums and its creation of additional resources for health services outside of taxes. Aside from the above, there are also international organizations as well as foreign countries and institutions which provide credit or donations as a source of funding for health services.

These methods for health financing in Turkey may be divided into four main groups:

- Financing through taxation
- Financing through Insurance premiums
- Financing through "out-of-pocket" payments
- Financing through other methods (Fig. 6.1)

6.2.1 *Financing Through Taxation*

One of the most important sources of financing for health expenditure is tax revenue. The state collects direct and indirect taxes in order to provide the financial means for public services. In other words, some degree of purchasing power flows from the private to the public sector. The state uses these resources collected through taxation to finance public spending. It is possible for the state to use taxes collected from the people through legal compulsion to pay for public services. The state procures the following from its health expenses budget, of which the most important source is tax revenue:

- Premium expenses for poor, temporary village guards, and olympic and European champion sportspersons supported by the state,
- Expenses for preventative health services,
- Expenses for the training, accommodation, equipping and educational support of doctors and other health workers,
- Hospital expenses,
- Expenses for personnel, buildings and equipment,
- Investment expenses, etc.

6.2.2 Financing Through Insurance Premiums

An insurer is an institution which prevents and distributes the impact of socio-economic disasters through pooling together many people threatened with the same risk and reimbursing individuals or institutions up to an agreed monetary value for the material damage arising from chance risks, in return for a defined premium.[1]

From the perspective of social security, insurance can be divided into two groups: social insurance and private insurance. Types of social insurance are put into practice with a material contribution from those benefitting from the insurance. This material contribution is called a premium, and social insurance is also known as the premium-based social security system.

Social insurance is an expression of the principle of compensation for losses as a form of aid against the risks encountered within a society or group.[2] Social insurance is an insurance system which relies on the power of the state to cover the needs of those who professional, physiological and socio-economic risks have left with lowered income or earnings. To put it another way, social insurance is an institutional, compulsory form of insurance which aims to protect the labor of economically weak and working people, make their future more secure, and thus to ensure social security within social life.[3] On the topic of health, social insurance has been put in place for the entire population through General Health Insurance by the Social Security Institution. This General Health Insurance encompasses the diagnosis, treatment, medicine and healthcare costs, etc. resulting from sickness, occupational illnesses and accidents at work.

Private insurance compliments the activities of social insurance. Private insurers help to provide social security based on the idea of helping one another. Private

[1]Güvel, E. A. and Güvel, A. Ö., Sigortacılık, Ankara, 2004, p. 26.

[2]Talas, C., Sosyal Ekonomi, Ankara, 1979, p. 12.

[3]Çubuk, A., Sosyal Politika ve Sosyal Güvenlik, G.Ü. Yayınları, 31-4, Ankara, 1986, p. 186.

insurers have economic objectives in providing insurance and measure their profit in monetary terms. Private health insurers, provide health insurance for illness, health and travel health, as well as life and non-life plans.[4] Premiums collected from social health insurance or private health insurance are all financial resources used for the health expenses of the Social Security Institution.

6.2.3 "Out-of-Pocket" Financing

Individuals can directly make payments for the use of health services from their own pockets. These payments can be grouped into direct payments, user contributions, and unregistered payments.

Direct payments are made for services outside the scope of social security, such as plastic surgery, private doctor appointments, etc. Patients can benefit from health services outside the scope of social security as well as services which are difficult to access or which have long waiting lists through a direct payment.

A contribution is demanded from the user for health services in order to create more resources. The contribution creates additional income, prevents the extreme or unnecessary use of services and substitutes for sources of public financing. Typical unregistered payments are when the patient or his/her friends or family give closed-envelope payments give by the patient or patients' friend or family was a "knife Money" or gifts for surgeons, gifts or etc. in order to affect those providing the services or persuade them to carry out services provided within the scope of social security in a certain way.

6.2.4 Other Methods of Financing

Among the sources of finance used for the nation's health expenses it is possible to include foreign firms, organizations, international health organizations and foreign countries. In particular organizations such as the European Union and NAFTA and members of the G-8 give credit to projects aiming to develop health services. The World Bank and the European Investment Bank contribute to the health expenses through grants which designate credit for education in developed countries in order to allow more effective provision of public services such as health services. The citizens, firms and religious organizations of foreign countries donate health materials or the necessary buildings, plots and land for the provision of health services.

[4]Orhaner, E. Sigortacılık. Ankara, 2014, p. 319.

Table 6.1 Basic indicators of health expenditure

	2009	2010	2011	2012
Total health expenditure				
Million TL	57,911	61,678	68,897	76,278
Million US	37,493	41,067	41,091	42,332
Health expenditure per person				
TL	804	845	932	1019
US	521	563	556	566
Total health spending/GDP (%)	6.1	5.6	5.3	5.4
State health expenditure/Total health expenditure	81	78.6	79.6	76.8
Private sector health expenditure/Total health expenditure	19	21.4	20.4	23.2
Household health expenditure/Total health expenditure	14.1	16.3	15.4	15.4

Source: Adapted from TÜİK [10]

6.3 Indicators of Health Expenses and Sources of Financing in Turkey

Public sector health expenditures represent the total of the health spending of central government, social security institutions and their local administrations and other public bodies. As seen in Table 6.1, the public sector can also be considered as the state's health expenditures in general. In 2012, it made up 76.8 % of total health expenditures.

Health expenditures financed by the private sector, in contrast, increased to 23.2 % in 2012. The health expenditures of the private sector represent the total of household health expenditure plus other health expenses. Household expenditures here equal 'out-of-pocket' payments. Out-of-pocket payments made up 15.4 % in 2012.

6.4 General Health Insurance

General Health Insurance is a form of insurance which covers financial expenses incurred due to health risks and the protection of primary health for all people in the country including foreigners. The law which put this into place was No. 5510, the Social Insurance and General Health Insurance Law, passed through the Turkish Grand National Assembly on 13.05.2006, but with some articles postponed by the Constitutional Court.

General health insurance is an important type of health organization known worldwide. This model largely takes health premiums directly and indirectly from those who benefit from the health services. With General Health Insurance, which was put into practice in Turkey on 01.01.2012, health was for the first time dealt with in a comprehensive fashion, and important changes were made to the social security system.

The following were the reasons why a general health system was necessary in Turkey:

(a) Preventative health expenses were being overlooked and large share of health spending was related to treatment. It was possible to reduce spending on treatment expenditure by increasing the importance of preventative health services.
(b) The inequalities observed in the provision of health services. There were not enough hospitals and doctors for rural areas. Medical equipment, technology, medical information and skills, and hospital beds were concentrated in large cities.
(c) The services and service levels available at different social insurance institutions differed according to who was benefitting from them. The treatment given to those who had worked in the public sector for many years was different to that given to ordinary social insurance beneficiaries. In the same way, the insurance for small businessmen (Bağ-Kur) did not provide the same health services as the insurance for general service users (SSK) and retirees.
(d) In Turkey, the proportion of the budget devoted to health is very low. The share of the public sector of health expenditure has decreased and the share of the private sector has increased.
(e) The health system was composed of many health service units which had no communication or relation to another, causing much confusion.
(f) It was difficult for citizens in rural areas to access specialist treatment, since the doctors, hospitals, assistant health personnel, health personnel and equipment was at specialist hospitals in the cities.
(g) The proportion of people needing to make 'out-of-pocket' payments increased due to the implementation of new capital circulation rules in public hospitals.
(h) Due to the lack of a working referrals system, university hospitals were providing diagnosis and treatment instead of education and research.

6.4.1 The Scope of the General Health Insurance

Every resident in Turkey and foreigners with residential permission and who had resided for more than 1 year became part of the compulsory general health insurance system as of 01.01.2012. Those covered by the existing social security system – the SSK, Bağ-Kur and Emekli Sandığı – continued to be able to benefit from their existing insurance schemes without paying new premiums or taking a means test.

With General Health Insurance, those with (state-issued, means-tested) Green Cards or no health insurance at all became part of the health insurance system. Those with Green Cards or no health insurance could apply to be means tested by their regional Social Aid and Solidarity Foundations, and according to their income situation[5] they would either benefit from premium-free or premium-based health insurance.

The implementation of general health insurance encompasses some vulnerable groups without condition[6]:

- All children up to the age of 18
- Those requiring nursing by others for health reasons
- Emergency cases
- Cases of work accidents and occupational injuries
- Contagious diseases requiring notification
- Health services required for personal protection or the prevention of substance dependency dangerous to human health
- All types of medical care and treatment related to motherhood, whether in-patient or out-patient
- Assistance relating to precautions to prevent disaster which would affect ordinary life, and health services for those needing to look after those with general health insurance in a time of war

With the implementation of general health insurance, 1.7 million people not covered by any form of social insurance, 9.1 million people with Green Cards, 650,000 privates, NCOs and reserve officers (as of 2013), and those with private health insurance schemes were all able to benefit from general health insurance.[7]

With general health insurance, all the members of society ought to benefit from health services in a universal and effective way against the risks of illness which might arise in the future, without worrying about their costs or whether the services are available. For this reason, those covered by the insurance can seek help from any health institution, whether public or private.

[5]If a family's per person monthly income is less than a third of the minimum wage, all the members of that family can benefit from the health system without making any payments. Indeed, the state will pay premiums for each individual separately. Premiums for 2014 are the following:

- If a family's per person monthly income is less than a third (357 TL) of the minimum wage (1071 TL), then the state will pay its premiums.
- If a family's per person monthly income is between 357 TL and 1071 TL the monthly premium per person is 42.84 TL
- If a family's per person monthly income is between 1071 TL and 2142 TL, the monthly premium per person is 128.52 TL
- If a family's per person monthly income is over double 1071 TL, then the monthly premium per person is 257 TL

Those who do not have their means tested will be assumed to be earning over double minimum wage per person and will be assessed for 257 TL.

[6]See Karakaş, İ., Genel Sağlık Sigortası Uygulama Rehberi, Ankara, 2008, p. 32.

[7]Uras, G. 74 Milyona Sağlık Sigortası Başarıdır. Milliyet (27.12.2011) p. 8.

6.4.2 *Conditions for Use of General Health Insurance*

To be able to use General Health Insurance, these conditions must be in place:

(a) – Those considered insured under Law No. 5510 clauses 4/a, 4/b and 4/c
  – Those who choose to insure themselves
  – The citizens of foreign countries who have obtained residence permits and who are not legally insured by a foreign country (in a reciprocal fashion)
  – Those receiving unemployment payments under Law No. 4447 and similar laws on temporary employment
  – Citizens who do not have the right to health insurance in another country and those looking after members of the above groups who have paid their general health insurance premiums and do not have more than 30 days outstanding from the year preceding their request for health services.

(b) Those considered insured under clause 4/b of the general health insurance Law No. 5510 and who do not have the right to health insurance in another country, and those who have to look after them, in addition to the conditions of (a), must not have more than 60 days of debt relating to premiums outstanding when they apply for health services.

(c) When those who have become insured by their own choice, those foreign citizens who have residence permissions, those who are not covered by insurance in a foreign country or those who are their careers apply for health services, they must not have any debt relating to premiums on the date on which they apply.

For those with General Health Insurance and those who are full-time careers to be able to benefit from health services, they show either their identification (population) cards, a driver's license, marriage certificate, passport or Social Security Institution official health card. Those in emergency situations may show their documents after the emergency has ended.

6.5 The Evaluation of General Health Insurance

The implementation of general health insurance is expected to bring certain improvements. These are the expected outcomes of the measure with the general insurance:

(a) Revenues for health spending will increase
(b) Sources of revenue will be centralized
(c) The financing of the health system will be primarily covered from insurance premiums, and secondarily through a public contribution. For this reason the most important means of increasing health revenue will be increasing insurance premiums
(d) With general health insurance, the whole population plus foreigners living in the country will form an insurance arm which finances the provision of preventative, medicinal, quality and easy to access health services. This insurance will also be secured by the state

(e) Unorganised distribution of funding sources will be replaced by an efficient funding source.
(f) Referrals will be put more effectively in place and will no longer be a source of waste
(g) A competitive environment will be created between healthcare providers, who will endeavour to improve the quality of their services
(h) There'll be a reduction in the number of negative basic health indicators, and a reduction in the difference between Turkey and foreign countries which have better health care systems.

It is however important to note that having everyone benefit from the health insurance in a country is a toll order and tough asking given that the program presents a complex terrain. This complexity is imbedded within people's perceptions of the ideal health insurance package compared to what the policy makers have put in place. The complexity is then worsened by the internal contradictions within the program that may end up leaving some citizens excluded. The existing positives and negatives of the general health insurance measure are listed below:

6.5.1 Positive Outcomes of General Health Insurance

(a) From 2012 with the general health insurance measure, 85.3 % of the population and in 2014 all people will come under the scope of health insurance. Healthcare for 18 year olds and under without condition has already begun.
(b) Health service providers are working to increase their health service capacity and quality.
(c) The private sector has seen an increase in health investment.
(d) All residents in Turkey have the same rights to health services.
(e) The introduction of a General Practitioner system for entry has made access to the healthcare system easier.
(f) As public insurance is now the only buyer of medicine in the market, medication spending has become more efficient and the number of middle-men has decreased.
(g) The Social Security Institution has increased the number of private hospitals through signing agreements for them to provide in-patient and out-patient services. In this way, the burden on public services has been shared with the private sector, and the number of alternatives available to the consumer has increased.
(h) The general health insurance system has reduced the complexity of the health insurance sector, making unified norms and standards in provision possible.
(i) While the treatment costs in the Social Security Institution has increased, medication costs have decreased.
(j) The implementation of general health insurance together with state incentives has encouraged some hospitals to open new branches in different places and become chain hospitals.

(k) Following the implementation of general health insurance, professional business methods were introduced into public health institutions in order to prevent a loss of institutional income.

6.5.2 Negative Outcomes of General Health Insurance

(a) Individuals have to become part of general health insurance whether they want or not. The Social Security Institution must pay premiums every month according to income for all those apart from those insured under 4/a, 4/b and 4/c or who are not entitled to coverage. When premiums remain unpaid they gain interest, and the state imposes an interest rate higher than the market rate.[8] In this case, those who do not pay their premiums can only access their basic human right to health in cases of emergency.
(b) The burden of health expenses on the budget will increase due to the wide scope of general health insurance. Where it is impossible to limit expenses, the debt on premiums will not be paid by people with insurance; the financing for the health expenses will transfer to the state.
(c) The public have still not been effectively informed about issues such as referral networks, receipts and the share of each contributor. Mistakes might occur due to their lack of knowledge, such as paying too much to the wrong institution.[9] Primary and secondary health services have not been introduced properly by health service providers.
(d) With the introduction of general practitioners, general practitioners will collect statistics about prescriptions, referrals and advice services. However, general practitioners will put emphasis on primary health care treatment.
(e) Financial costs for health services have increased. Increased health costs have led to an increase in the costs of health insurance. As private hospitals have increased their prices due to increased investment for competitiveness has caused an increase in premiums.
(f) Precautions need to be put in place to ensure that the government does not treat patients unnecessarily or spend excessively. However, a performance-related pricing system increases health expenses.
(g) In Turkey, there are around one doctor for every 3,500–4,000 people (2013 figures). This number outside Turkey is doctor per 1,200 people.[10] The caseload of doctors can be a reason for ineffective referrals, leading to a preference for sending people directly to secondary and tertiary healthcare at hospitals, which increases costs.

---

[8] Bayer, Y. Genel Sağlık Sigortası Çöküyor. Hürriyet Gazetesi (05.02.2014).

[9] Kavuncubaşı, Ş-Yıldırım, S., Hastane ve Sağlık Kurumları Yönetimi, Ankara, 2013, p. 41.

[10] Accessed from http://www.ailehekimligi.gov.tr address in 20.06.214.

(h) The implementation of a performance-related pricing system has created an inconsistency between doctors, patients and health institutions. Patients increase their demands for health services due to their lack of knowledge, or else institutions and doctors may create artificial demand in order to make profit.
(i) The problems with the actuarial balance which some believed general health insurance would solve nonetheless continue and it may be difficult to continue general health insurance from a financial perspective.
(j) Not enough attention is being paid to payment methods such as global budgeting which consider the scope and quality of health services.

6.6 Conclusion

General health insurance was instituted with the aim of reducing the burden of health financing which fell upon the state, yet has ended up increasing it. The system was constructed with the aim of creating additional financial resources, but created an important responsibility for the collection of unpaid premiums. Around 4.5 million people who are not working with general health insurance and who have not had means tests have received a high premium rate and indebted themselves, and since they cannot pay these, they cannot benefit from health services. On top of this, these people's debt increases with interest for non-payment. On the other hand, many citizens who reach 18 have to enter the general health insurance system, and yet they increase their premiums since they do not have the information 0HL8-BJ9C-QVY3-BCXL to make informed choices; they only learn of their debts when they visit their doctor or a hospital. General health insurance has not been explained well to the public, and it has been implemented without enough information being distributed, so it does not seem that meeting its aim of allowing those without enough money to access health services will be possible. Before general health insurance was put into place, they used to find the money for treatment in one way or another, but now they cannot access hospitals or general practitioners without first paying their debts, and so their right to benefit from health services is effectively removed. As one of the characteristics of social states, government should provide the right of benefiting from health services for all citizens.

References

1. Bayer Y (2014) Genel sağlık sigortası çöküyor. Hürriyet Gazetesi (05.02.2014)
2. Çelik Y (2013) Sağlık ekonomisi, 2nd edn. Siyasal Kitabevi, Ankara
3. Çubuk A (1986) Sosyal politika ve sosyal güvenlik. G.Ü. Yayınları, Ankara
4. Güvel EA, Güvel AÖ (2004) Sigortacılık, 2nd edn. Seçkin Yayınları, Ankara
5. http://www.ailehekimligi.gov.tr
6. Karakaş İ (2008) Genel sağlık sigortası uygulama rehberi. Ankara

7. Kavuncubaşı Ş, Yıldırım S (2013) Hastane ve sağlık kurumları yönetimi. Ankara
8. Orhaner E (2014) Sigortacılık. Siyasal Kitabevi, Ankara
9. Talas C (1979) Sosyal ekonomi, Ankara
10. TÜİK, Haber Bülteni (2013, 10 October) No: 15871
11. Uras G (2011, 27.12) 74 Milyona sağlık sigortası başarıdır. Milliyet, p. 8

Chapter 7
A Classical Approach to Modeling of Coal Mine Data

Mehmet Yılmaz, Nihan Potas, and Buse Buyum

Abstract Data sets such as the occurrence time of random events or the lifetime of a certain product (or a system) are modelled by compound or mixture distributions especially in the last years. This situation is led to encounter proposal of more complex distribution models in the literature. One of the data set made a model proposal by in this way is coal mine data set. In this study, Two Component Mixed Exponential Distribution (2MED) model had more easier interpretation on this data set is used and compared with the other study results. Also, the extended coal mine data set with 191 observations is modelled by 2MED and the results are given.

7.1 Introduction

Mining is an important source of foreign exchange for many developing countries. But this important source of foreign exchange is also one of the sectors in which it occurs the most accidents all over the world. These coal mine accidents is still under investigation by many different disciplines in today as in the past. Coal mine accidents are often used as a real data set in studies in the field of statistics as in other fields. In these studies, some researchers have made trend analysis by taking the occurrence time of coal mine accidents, some of them have tried to model of accident occurrence time. Purpose of modeling studies is obtained the model that can best forecast (or model) these processes.

M. Yılmaz (✉)
Faculty of Science, Department of Statistics, Ankara University, 06100, Ankara, Turkey
e-mail: yilmazm@science.ankara.edu.tr

N. Potas
Faculty of Economics and Administrative Sciences, Department of Health Care Management, Gazi University, 06100, Ankara, Turkey
e-mail: nihanp@gazi.edu.tr

B. Buyum
Department of Statistics, Graduate School of Natural and Applied Science, Ankara University, 06100, Ankara, Turkey
e-mail: busebuyum@gmail.com

Ş.Ş. Erçetin (ed.), *Chaos, Complexity and Leadership 2014*,
Springer Proceedings in Complexity, DOI 10.1007/978-3-319-18693-1_7

Because of estimate of the occurrence time of these accidents is of vital importance to prevent them addition to the measures that can be taken.

In the current study, the data set, which is one of the most widely used in the literature, is obtained firstly by Maguire et al. [8] is firstly analyzed by Cox and Lewis [2], is handled. The data set obtains the time intervals (in days) between coal mine accidents concluded death of 10 or more men. In later, this data set is arranged by Jarrett [6] and is extended to 191 observations. Some researchers who try to model the data set with 109 observations are Adamidis and Loukas [1], Kus [7], Mirhossaini and Dolati [10], and Rodriguesa et al. [11]. Some of them use non-mixture distributions such as Exponential, Gamma and Weibull and the others use mixture distributions such as Exponential-Poisson (EP), Exponential-Gamma (EG) and Exponential Conway-Maxwell Poisson (ECOMP).

In this study, Two Component Mixed Exponential Distribution (2MED) is used to model this famous data set. Aim of this study, propose 2MED as a new distribution (model) in addition to distributions used in modeling study. First of all, properties of 2MED, then parameter estimations of Maximum Likelihood (MLE) and the Least Squares (LSE) will be introduced. In here, MLE is obtained by Expectation-Maximization (EM) algorithm which is one of the numeric way. The results that is obtained from other studies and from this study will be compared with Kolmogorov Smirnov Test Statistic (KS) and it is tried to indicate that how 2MED is successful about modeling this data set. Parameter estimations, KS values and p values (p) are obtained by MATLAB.

7.2 Parameter Estimations Methods for 2MED

In this section, some basic properties of 2MED are introduced and then the methods of maximum likelihood and the least squares will be given.

7.2.1 *Mixed Exponential Distribution with Two-Component (2MED)*

Probability density function (p.d.f) of 2MED is given below.

$$\begin{aligned} f(x;\alpha,\theta_1,\theta_2) &= \alpha f_1(x;\theta_1) + (1-\alpha) f_2(x;\theta_2) \\ &= \alpha\frac{1}{\theta_1}\exp(-x/\theta_1) + (1-\alpha)\frac{1}{\theta_2}\exp(-x/\theta_2) \end{aligned}$$

where $\alpha \in (0,1)$, $\theta_i > 0$ $(i = 1,2)$, $x > 0$. Similarly the cumulative distribution function (c.d.f) is as follows.

$$\begin{aligned} F(x;\alpha,\theta_1,\theta_2) &= \alpha F_1(x;\theta_1) + (1-\alpha)F_2(x;\theta_2) \\ &= \alpha(1-\exp(-x/\theta_1)) + (1-\alpha)(1-\exp(-x/\theta_2)) \end{aligned}$$

Survival function,

$$\begin{aligned} S(x;\alpha,\theta_1,\theta_2) &= \alpha S_1(x;\theta_1) + (1-\alpha)S_2(x;\theta_2) \\ &= \alpha\exp(-x/\theta_1) + (1-\alpha)\exp(-x/\theta_2) \end{aligned}$$

and the hazard function,

$$\begin{aligned} h(x;\alpha,\theta_1,\theta_2) &= \frac{\alpha h_1(x)S_1(x;\theta_1) + (1-\alpha)h_2(x)S_2(x;\theta_2)}{\alpha S_1(x;\theta_1) + (1-\alpha)S_2(x;\theta_2)} \\ &= h_1(x)\frac{\alpha S_1(x;\theta_1)}{\alpha S_1(x;\theta_1) + (1-\alpha)S_2(x;\theta_2)} \\ &\quad + h_2(x)\frac{(1-\alpha)S_2(x;\theta_2)}{\alpha S_1(x;\theta_1) + (1-\alpha)S_2(x;\theta_2)} \\ &= h_1(x)w_1(x;\alpha,\theta_1,\theta_2) + h_2(x)w_2(x;\alpha,\theta_1,\theta_2) \end{aligned}$$

where $h_i(x) = \frac{1}{\theta_i}$ and $w_1(.) + w_2(.) = 1$.

7.2.2 *Maximum Likelihood Method for 2MED*

Let $\underline{X} = \{X_1, X_{2,\dots,}X_n\}$ be a random sampling with independent and identically distributed as 2MED having a p.d.f $f(\underline{x};\underline{\Phi})$ where $\underline{\Phi} = (\alpha,\theta_1,\theta_2)$ is a parameter vector. The likelihood function and the logarithmic form of the likelihood function of $\underline{\Phi}$ are respectively given as below:

$$L(\underline{\Phi};\underline{x}) = \prod_{j=1}^{n}\left[\sum_{i=1}^{2}\alpha_i\frac{1}{\theta_i}\exp(-x_j/\theta_i)\right]$$

$$\log L = \sum_{j=1}^{n}\log\left[\sum_{i=1}^{2}\alpha_i\frac{1}{\theta_i}\exp(-x_j/\theta_i)\right] - \lambda\left(\sum_{i=1}^{2}\alpha_i - 1\right)$$

where $\sum_{i=1}^{2}\alpha_i = 1$. If the derivative of this function respect to $\alpha_i, i = 1, 2$ is equalized to zero,

$$\frac{\partial \log L}{\partial \alpha_i} = \sum_{j=1}^{n} \frac{\frac{1}{\theta_i}\exp(-x_j/\theta_i)}{\sum_{i=1}^{2} \alpha_i \frac{1}{\theta_i}\exp(-x_j/\theta_i)} - \lambda = 0$$

then

$$\sum_{j=1}^{n} \frac{\frac{1}{\theta_i}\exp(-x_j/\theta_i)}{\sum_{i=1}^{2} \alpha_i \frac{1}{\theta_i}\exp(-x_j/\theta_i)} = \lambda \tag{7.1}$$

Multiplying the both side of (7.1) by α_i and taking the sum over index i:

$$\sum_{j=1}^{n}\sum_{i=1}^{2} \frac{\alpha_i \frac{1}{\theta_i}\exp(-x_j/\theta_i)}{\sum_{i=1}^{2} \alpha_i \frac{1}{\theta_i}\exp(-x_j/\theta_i)} = \lambda \alpha_i$$

then $n = \lambda$. Based on Bayes' rule, the probability that x_j belongs to ith component when $X_j = x_j$ is observed is as follows:

$$P(i \mid x_j) = \frac{\alpha_i \frac{1}{\theta_i}\exp(-x_j/\theta_i)}{\sum_{i=1}^{2} \alpha_i \frac{1}{\theta_i}\exp(-x_j/\theta_i)}$$

Thus,

$$\hat{\alpha}_i = \frac{\sum_{j=1}^{n} P(i \mid x_j)}{n}$$

where $i = 1, 2$. If the derivative of $\log L$ with respect to θ_i is equalized to zero,

$$\frac{\partial \log L}{\partial \theta_i} = \sum_{j=1}^{n} \frac{\frac{\alpha_i}{\theta_i}\left(\frac{x_j}{\theta_i^2}\right)\exp\left(\frac{-x_j}{\theta_i}\right) - \frac{\alpha_i}{\theta_i^2}\exp\left(\frac{-x_j}{\theta_i}\right)}{\sum_{i=1}^{2} \alpha_i \frac{1}{\theta_i}\exp\left(\frac{-x_j}{\theta_i}\right)} = 0$$

$$\hat{\theta}_i = \frac{\sum_{j=1}^{n} x_j P(i \mid x_j)}{\sum_{j=1}^{n} P(i \mid x_j)}$$

where $i = 1, 2$. $\hat{\theta}_i$ is obtained and reminded that $P(2 \mid x_j) = 1 - P(1 \mid x_j)$, then the solutions will be

$$\hat{\theta}_1 = \frac{1}{n\hat{\alpha}_i} \sum_{j=1}^{n} x_j P(i \mid x_j)$$

$$\hat{\theta}_2 = \frac{1}{n(1-\hat{\alpha}_i)} \sum_{j=1}^{n} x_j (1 - P(i \mid x_j)).$$

As seen in the above, the parameter estimations can not been obtained directly from derivative equations. Therefore numeric ways are preferred for solving of these equations. In this study, EM algorithm which is one of the numeric way is taken into account [3, 4, 9]. These are step solutions obtained by EM which steps are given:

1. Input the initial values $(\alpha_i^{(0)}, \theta_i^{(0)})$, $i = 1, 2$.
2. Calculate the $P(i \mid x_j)$.
3. Calculate $\hat{\alpha}_i^{(k)}, \hat{\theta}_i^{(k)}$
4. After calculations of $\hat{\alpha}_i$ and $\hat{\theta}_i$, the values replace in $logL$ and get the value of function. For $\epsilon > 0$ selected small enough $\log L^{(k)} - \log L^{(k-1)} \leq \epsilon$ is provided then the values on the kth step will be used for parameter estimations. Steps 2–4 are repeated until converge is accomplished.

7.2.3 The Least Squares Method for 2MED

This method is based on the idea that there is a regression relationship between empirical $\hat{F}$ and parametric F distributions. Considering ordered observations $x_{(1)} \leq x_{(2)} \leq \ldots \leq x_{(n)}$ versus empirical distribution $\hat{F}(x_{(i)}) \equiv i/(n+1)$, the vector $\underline{\Phi}$ which minimizes the following expression is tried to determine. Detailed study was given in Gupta and Kundu [5] for non-mixture Generalized Exponential Distribution. System of equations that is occurred for the solutions for this optimization problem is as follows.

$$Q(\underline{\Phi}) = \sum_{i=1}^{n} \left(\hat{F}\left(x_{(i)}\right) - F\left(x_{(i)}; \underline{\Phi}\right) \right)^2$$

For solving of this optimization problem, since the expressions after derivative are related to parameters, it is difficult to obtain the solutions. Therefore it is necessary to use numerical ways. The values minimized $Q(\underline{\Phi})$ function are calculated numerically by current command in MATLAB. The stopping rule can be based on absolute value of the difference between the function values in the previous iteration and next iteration. So, when the measured absolute difference becomes less than $10^{(-21)}$ the search can be stopped.

Table 7.1 The time intervals (in days) between coal mine accidents

378	96	59	108	54	275	498	228	217	19	156
36	124	61	188	217	78	49	271	120	329	47
15	50	1	233	113	17	131	208	275	330	129
31	120	13	28	32	1,205	182	517	20	312	1,630
215	203	189	22	23	644	255	1,613	66	171	29
11	176	345	61	151	467	195	54	291	145	217
137	55	20	78	361	871	224	326	4	75	7
4	93	81	99	312	48	566	1,312	369	364	18
15	59	286	326	354	123	390	348	338	37	1,357
72	315	114	275	58	457	72	745	336	19	

Table 7.2 Parameter estimations, KS and p-values for 2MED

LSE			MLE		
$\hat{\alpha}$	$\hat{\theta}_1$	$\hat{\theta}_2$	$\hat{\alpha}$	$\hat{\theta}_1$	$\hat{\theta}_2$
0.9162	238.5660	27.8652	0.1757	592.0210	166.1348
KS Stat.		p-value	KS Stat.		p-value
0.0594		0.8138	0.0578		0.8386

7.3 Suggested and Current Models for Coal Mine Data

The data set, which is obtained firstly by Maguire et al. [8] and obtains the time intervals (in days) between coal mine accidents concluded death of 10 or more men is given in Table 7.1. In this section, the results obtained from the studies in the literature and from the current study will be compared.

First of all, in terms of providing comparison and ease of comment the parameter estimations, KS and p-values obtained by modeling with 2MED is given in Table 7.2.

In Adamidis and Loukas [1], firstly they are suggested Weibull and Gamma which are used frequently as non-mixture distributions and then they are tried to model the data set with EG. After modeling, KS value of EG is found as **0.076** and they said that the EG distribution fits the data set at least as good as the two popular alternatives. When this value and the KS value for 2MED according to two methods, it can be said that 2MED is more successful than EG distribution about modeling the data set.

In Kus [7], EP distribution is used in addition to the distributions used in Adamidis and Loukas [1]. The KS and p-value of EP is given in.

When the KS values for 2MED and EP given in Table 7.3 are compared, it can be seen that 2MED values are smaller than EP values. Therefore 2MED is the best amongst four distributions handled so far according to KS criteria.

Table 7.3 KS and p-values for Kus [7]

Distribution	KS value	p-value
EP	0.0625	0.7876
EG	0.0761	0.5524
WEIBULL	0.0773	0.5325
GAMMA	0.0852	0.4076

Table 7.4 KS values for Mirhossaini and Dolati [10]

Distribution	KS value
EXPONENTIAL	0.0776
ME	0.0667
GAMMA	0.0796
WEIBULL	0.2965

Table 7.5 A* values for Rodriguesa et al. [11]

Distribution	A*
ECOMP	0.432
EG	0.439
EP	0.480
EXPONENTIAL	0.658

Exponential distribution model is discussed in addition to the above non-mixture distributions in Mirhossaini and Dolati [10]. Besides non-mixture distributions, ME is used in modelling study and the results is given in the Table 7.4.

Considering the proposed model in the above, it is thinkable that ME is the closest model to 2MED. Even in this thought, it is clear that the KS values for 2MED is lower than the KS values for ME. The results is same for the other three distributions. The comment made on the results of other studies is also applied here.

The ECOMP distribution is used as well as commonly used for modeling this data set in Rodriguesa et al. [11]. In their study, the modified Cramer-von Mises (W*) and Anderson-Darling (A*) test statistics are taken into account but it is decided that which distribution is more successful according to the value. Therefore the A* value is calculated for 2MED while comparing with distributions in Rodriguesa et al. [11]. Computational code is taken from the first author Josemar Rodriguesa. The A* value is given in the Table 7.5.

The A* values calculated according to MLE and LSE methods for 2MED are found **0,426** and **0,722** respectively. A* value found according to MLE seems to be smaller than the value calculated for distribution in the above table. Accordingly, 2MED is more suitable for this data set.

Coal mine data set is arranged and is extended to 191 observations by Jarrett [6]. However any modelling study for 191 observations has not reached in the literature search. Here we try to show how this extended data set (given in Table 7.6) is modelled by 2MED in addition to the studies above.

When the results given in Table 7.7 are examined, modelling of this extended data set with 2MED for two methods is also successful according to KS criteria.

Table 7.6 The extended coal mine data set

157	123	2	124	12	4	10	216	25	19	33
66	232	826	40	12	29	190	53	17	186	23
92	197	431	16	154	95	250	80	78	202	36
110	276	16	88	225	24	91	538	187	34	101
41	139	42	1	112	43	3	324	56	31	96
70	41	93	2	0	143	16	27	144	45	6
208	29	15	72	193	134	420	95	125	34	127
218	59	315	378	36	15	31	215	11	137	4
286	114	96	124	50	120	203	176	55	93	326
275	59	61	1	13	189	345	20	81	354	307
108	188	233	28	22	61	78	99	123	456	54
217	113	32	388	151	361	312	462	228	275	78
17	1,205	644	467	871	48	217	120	498	49	131
182	255	194	224	566	19	329	806	517	1,643	54
326	1,312	348	745	156	47	275	20	66	292	4
368	307	336	952	632	330	312	536	145	75	364
129	1,630	29	217	7	18	1,358	2,366	65	17	19
37	19	12								

Table 7.7 Parameter estimations, KS and p-values for 2MED of extended data set

LSE			MLE		
$\hat{\alpha}$	$\hat{\theta}_1$	$\hat{\theta}_2$	$\hat{\alpha}$	$\hat{\theta}_1$	$\hat{\theta}_2$
0.1940	39.9254	224.5533	0.8210	134.7353	574.3638
KS Stat.		p-value	KS Stat.		p-value
0.0304		0.9926	0.0443		0.8334

7.4 Conclusion

In this study is handled seven different distributions used modelling of the time intervals (in days) between coal mine accidents. It can be seen that the data set is modelled mostly by Weibull, Gamma and Exponential as non-mixture distributions and by EG, EP as mixture distributions. Except for these distributions, ME and ECOMP are used in modelling study.

The results of KS statistic and the A* test statistic are used as a measure to compare three of these modelling studies. When the results obtained from both LSE and MLE methods for 2MED and the results of these distributions are compared, 2MED seems to be the best model according to KS and A* criteria (except for LSE). 2MED is the best model between mixture and non-mixture distributions used in Adamidis and Loukas [1], Kus [7], Mirhossaini and Dolati [10], and Rodriguesa et al. [11]. In addition to comparison, the extended data set is also analyzed. But studies for this data set is generally based on analyzing as a stochastic process. Therefore this extended data set is modelled only by 2MED and the results is given Table 7.7.

Finally, we can say that as an uncomplicated model 2MED can be recommended for the coal mine data set which is studied by many researches.

References

1. Adamidis K, Loukas S (1998) A lifetime distribution with decreasing failure rate. Stat Probab Lett 39:35–42
2. Cox DR, Lewis PAW (1966) The statistical analysis of series of events. Chapman and Hall, London/Methuen
3. Dempster AP, Laird NM, Rubin DB (1977) Maximum likelihood from incomplete data via the EM algorithm (with discussion). J R Stat Soc Ser B 39:1–38
4. Everitt ES, Hand DJ (1981) Finite mixture distributions. Chapman and Hall, London
5. Gupta RD, Kundu D (2000) Generalized exponential distribution: different method of estimations. J Stat Comput Simul 00:1–22
6. Jarrett RG (1979) A note on the intervals between coal mining disasters. Biometrika 66:191–193
7. Kus C (2007) A new lifetime distribution. Comput Stat Data Anal 51:4497–4509
8. Maguire BA, Pearson ES, Wynn AHA (1952) The time intervals between industrial accidents. Biometrika 39:168–180
9. McLachlan GJ, Krishnan T (2008) The EM algorithm and extensions. Wiley, Hoboken
10. Mirhossaini SM, Dolati A (2008) On a new generalization of the exponential distribution. J Math Ext 3(1):27–42
11. Rodriguesa J, Balakrishnan N, Cordeiro GM (2011) A unified view on lifetime distributions. Comput Stat Data Anal 55(12):3311–3319

Chapter 8
A Unique Classification on Adults: "Adult-Child, Adult-Adolescence, Adult-Adult" Different Perspective on Leadership

Şefika Şule Erçetin, Şuay Nilhan Açıkalın, and Halime Güngör

Abstract Because of wideness in adults' age range and differences in all ages' characteristics, there have been different definitions and classifications on adults in literature, and also making different classifications are possible. Adults have been classified in three groups as adult-child, adult-adolescence and adult-adult in this study. Age range has not been considered while classifying. Leaders and also political leaders has been focus of attention throughout history not only in written and visual media but also in academic researches. In this study, primarily adult-child, adult-adolescence and adult-adult properties have been defined in feeling, thought and behavior dimensions. Profiles were taken as two: feeling, thought and behavior dimension related to characteristics of individual in profile analysis; attitude frequency which contains adult-child, adult-adolescence and adult-adult characteristics of individual. According to results, a unique model of classification of adults, adult-child, adult-adolescence and adult-adult, has been formed. Although dimensions of leaders' interventions in chaotic situations have been differ in challenging, moderate, destructive, positive circle, this leadership attitudes have been placed even more chaotic to chaotic situations or have been transformed chaos to cosmos. In this context, finally, leaders' coping strategies in chaotic ambiance have been evaluated on the base of adult-child, adult-adolescence and adult-adult classification model.

Ş.Ş. Erçetin (✉)
International Science Association (ISCASS), Ankara, Turkey

Faculty of Education, Hacettepe University, Ankara, Turkey
e-mail: sefikasule@gmail.com

Ş.N. Açıkalın
Department of International Relations, Middle East Technical University, Ankara, Turkey
e-mail: suaynilhan@gmail.com

H. Güngör
Department of Educational Management, Planning, Supervision and Economics, Hacettepe University, Ankara, Turkey
e-mail: halime_gngr@yahoo.com

Ş.Ş. Erçetin (ed.), *Chaos, Complexity and Leadership 2014*,
Springer Proceedings in Complexity, DOI 10.1007/978-3-319-18693-1_8

8.1 Introduction

The word of "adult" was derived from the past participle tense of "grow up" verb (adolescere) in Latin, so an individual who is an adult has been assumed that an individual who is grown up and mature physically and psychologically [5]. Because of wideness in adults' age range and differences in all ages' characteristics, and difficulty to measure psychological necessities of being an adult, there have been different definitions and classifications on adults in literature according to age range generally, and also making different classifications are possible. In this conceptual study, we made an adulthood classification. We made group discussions, brainstorming, focus group studies, and observations to conceptualize a unique classification on adulthood. We did not handle adulthood according to physical characteristics of adults. We handle the subject on emotional, behavioral and thought dimensions. Aim of the study is that initiate a discussion on the subject; draw attention on the point of being an adult does not mean that growing up physically. So in this study, adults have been classified in three groups as;

- Adult-children
- Adult-adolescents
- Adult-adults

Adult-Children Adult children are adult as physical but they show child characteristics in cognitive, spiritual, emotional and psychological dimensions in one of them, more or all of them

Adult-Adolescents Adult adolescents are adult as physical but they show adolescence characteristics in cognitive, spiritual, emotional and psychological dimensions in one of them, more or all of them

Adult-Adults Adult adults show adult characteristics in physical, cognitive, spiritual, emotional and psychological dimensions. Adult-child, adult-adolescence and adult-adult individuals' characteristics are shown in Tables 8.1–8.3 and these individuals' characteristics are given comparatively.

8.2 Adult-Children, Adult-Adolescence and Adult-Adult Profiles

Peoples can carry one or more of the features in Table 8.1, this does not mean that individual is adult-children, adult-adolescence or adult-adult. For example;

"They can say everything come into their mind. There are no bad intentions. They believe what they hear without questioning." (Adult-children)
"They like hurt people around them. They object everything." (Adult-adolescence)
"They do not say empty promises; they do not hurt people around them." (Adult-adult)

Individuals generally have "do not like hurting people around them" but sometimes can hurt; it does not mean that he/she is an adult-adolescence. An

Table 8.1 Characteristics in feeling dimension

Ch. No	Adult-children	Adult-adolescence	Adult-adult
1	Physical development is completed.	Physical development is completed.	Physical development is completed.
2	Generally positive. When they are pessimistic, they quickly forget this.	Generally pessimistic	They are positive and look life positive
3	Inner peace of them is not stable	Inner peace of them is problematic	Inner peace of them is good
4	They are sincere about their feelings, behaviors but their thoughts are unconscious.	They are not sincere about their feelings, behaviors, thoughts	They are not consciously sincere about their feelings, behaviors, thoughts
5	Insights of them are not good.	Insights of them are not good.	Insights of them are good.
6	They are generally in peace with themselves.	They are not generally in peace with themselves. They have problems with themselves.	They are in peace with themselves.
7	Self-confidence and self-esteem are under the influence of their surroundings.	They have fake self-confidence and self-esteem	They have consciously self-confidence and self-esteem
8	They are not intuitive	Their intuitions are not powerful but they assume that they are intuitive.	They are intuitive
9	They are fragile	They are fragile and aggressive	They are not fragile and aggressive.

established finding of research Robinson et al. [6] is that adults prefer to guess before rather than after a chance event has happened like children. Human can show characteristics from all categories time to time. Furthermore, to define an individual as adult-children, adult-adolescence or adult-adult, is not necessary that individual always to carry related category's all features. Individual's profile has indicated according to how often to elicit behavior. In this sense, first of all, features of individuals divided into three dimension as feeling, thought and behavior into three categories; adult-children, adult-adolescence and adult-adult (Tables 8.1–8.3).

8.3 "Adult-Child, Adult-Adolescence, Adult-Adult": Different Perspective on Leadership and Chaos

Leaders are backbones of organizations. Leadership and personality are related [2]. Personal characteristics of leaders and effective management of organizations are interrelated concepts. According to Erçetin et al. [2], social constructs like societies,

Table 8.2 Characteristics in thought dimension

Ch. No	Adult-children	Adult-adolescence	Adult-adult
1	They have "other" concept in their minds. They believe in love	They have "other" concept in their minds. They do not believe in unconditional love. Their love has short and provisional	They have not "other" concept in their minds. They believe in unconditional love
2	They do not decide alone, they are dependent to others.	They are unstable.	They are clean in their decisions.
3	Their friends and not to be alone is so important.	Their friends are on the center of their life.	Their friends are in life as necessary.
4	They are not aware of responsibilities on their family, work and special life	They are aware of responsibilities on their family, work and special life	They are aware of responsibilities on their family, work and special life
5	Their behaviors and thoughts are unconscious.	Their behaviors and thoughts are semi-conscious.	Their behaviors and thoughts are conscious.
6	They are not respectful to people's beliefs, thoughts; they have not been developed conscious on this subject.	They are not respectful to people's beliefs, thoughts	They are respectful to people's beliefs, thoughts.
7	They are not respectful to human rights.	They are not respectful to human rights.	They are respectful to human rights.
8	Their thoughts and life are completely under influence	Their thoughts and life are completely under influence but they never accept this situation.	Their thoughts and life are not under influence
9	They sometimes use people or other things to reach their goals but they do not this consciously.	They use people or other things to reach their goals consciously	They are in conscious about not using people or other things to reach their goals
10	Their ego would not change much when nice developments realize in their life.	Their ego swells when nice developments realize in their life.	Their ego does not arise more than enough when nice developments realize in their life.
11	Their consciousness level is restricted.	Their consciousness level is restricted.	Their consciousness level is universal
12	They do not know people much. They cannot understand well others' aim, perspectives on life and paradigms.	They assume that know people well. Making judgment without restraint. They cannot understand well others' aim, perspectives on life and paradigms but they never accept this situation	They know people well. They can understand quickly others' aim, perspectives on life and paradigms.

13	They cannot know events which they live is their own choices and these events are stemmed from themselves.	They cannot know events which they live is their own choices and these events are stemmed from themselves.	They know events which they live is their own choices and these events are stemmed from themselves.
14	They have conditionings and obsessions	They have conditionings and obsessions but they generally do not accept this situation.	They have not conditionings and obsessions
15	They judge people but they do not behave consciously like this. They behave like as seen from the environment.	They judge people and they would like to oppress people	They do not judge people
16	Awareness level of them is low but they do not aware of this situation.	Awareness level of them is low but they think high.	Awareness level of them is high
17	They do not tolerate criticism; their heart is broken, is offended and is hurt.	They never accept criticism. They are always correct.	They can tolerate criticism.
18	They wait others' approval on their behaviors.	The approvals of other on their behaviors appears that not important but in fact it is important for them	The approvals of other on their behaviors are not important.
19	They sat that they do not wait provisions from others but in fact they wait in deep.	Everything is reciprocal	They do not wait provisions from others.
20	Others are more important than themselves. Other are more valuable	The most important person is himself/herself in life for them	They care about both themselves and others.
21	Their thought system is "yours" (First you)	Their thought system is "selfish" (First me)	Their thought system is "ours" (First us)
22	They have daily thought systematic.	They assume that have scientific thought systematic	They have scientific thought systematic.

Table 8.3 Characteristics in behavior dimension

Ch. No	Adult-children	Adult-adolescence	Adult-adult
1	They deal with persons, development of themselves on any area is not important at all.	They deal with persons, development of themselves on any area is quasi important. They assume to know everything.	They do not deal with persons. Development of themselves on any area is important.
2	They have not competition concept.	They compete with others.	They compete with themselves.
3	They do behaviors that attack on the rights of person such as mobbing, stigmatization, slander unconsciously by under the guidance of environment.	They do behaviors that attack on the rights of person such as mobbing, stigmatization, slander unconsciously or semi-consciously.	They do not do behaviors that attack on the rights of person such as mobbing, stigmatization, slander
4	Wearing and behaving appropriate to environment is not important for them.	They do not wear and behave appropriate to environment. They have problems on their appearance, they struggle with themselves much. Some of them are constantly on diet. They like draw attention on them	They can wear and behave appropriate to environment
5	They can say everything come into their mind. There are no bad intentions. They believe what they hear without questioning	They like hurt people around them. They object everything.	They do not say empty promises; they do not hurt people around them.
6	They cannot behave people as they deserve, they are passive and aggressive.	Behaviors of them are immoderate	They can behave people as they deserve
7	Conversations of them are sincere	Conversations of them are immoderate	Conversations of them are explicit and clear
8	IQ and EQ of them are not in balance	IQ and EQ of them are not in balance	IQ and EQ of them are in balance
9	They cannot see consequences of their behaviors before they do but they have not bad intentions, they see themselves as victim.	They cannot see consequences of their behaviors before they do and they are uncontrolled	They can see consequences of their behaviors before they do
10	They know and behave according their responsibilities often	Behave according their responsibilities is difficult for them	They know and behave according their responsibilities
11	They are externally controlled and motivated.	They are externally controlled and internally motivated.	They are internally controlled and motivated.
12	They cannot use effective communication skills and body language	They can use effective communication skills and body language for their own benefit.	They can use effective communication skills and body language well

organizations and leadership are chaotic, inter-dependent, non-linear systems that are closely tied to initial conditions and leadership and chaos can be considered fundamentally interconnected, topics of chaos and leadership have been classified, unitized and graded in a variety of ways [2]. In this mean, leadership has been evaluated in terms of chaos in the context of adult-children, adult-adolescence and adult-adult classification;

In the organizations managed by adult-children leaders;

Because of adult-children leaders cannot take effective and right decisions alone, management of organization is not expected to be in the management of leaders. A leader should be able to find different solutions to different problems [1] but when any problems has been living in organization, these leaders' management these problem situations effectively and strategically is very difficult. Creating organizational communication, organizational culture by leaders is also difficult. Therefore, leaders cannot manage organization effectively, frequency of experiencing chaotic situations may be increased. In addition, coping strategies with chaos is developed and applied by leaders are also difficult.

In the organizations managed by adult-adolescence leaders;

Taking appropriate decisions by adult adolescence leaders is difficult because adult-adolescence leaders generally behave emotionally. Because of their level of tolerate criticism is low, when they take wrong decision in organization, it is difficult to decide to change the rotation for them. When any problems have been living in organization, leaders' toleration to take decision with other members of the organization and apply together is low but they assume that they manage the organization effectively about themselves. Organizational communication and organizational culture is leader-centered in these organizations. They cannot manage organization effectively and organization-centered, so neglecting chaotic situations, blaming others for this chaotic situations and/or not making self-questioning by leaders is probable. In fact, coping with chaotic situations are teamwork for organization, so in the organizations managed by adult-adolescence leaders, coping with chaos is difficult.

In the organizations managed by adult-adult leaders;

Adult-adult leaders can take right and effective decisions and can manage the organization effectively, so when any problems have been living in organization, leaders' toleration to take decision with other members of the organization and apply together is high. Organizational communication is healthy and organizational culture is powerful in these organizations. Chaos is a source of life and also a reality for organizations and people as living and self-organizing systems [3] that order is a transient state that can lead to problems if it persists [2]. So adult-adult leaders by the characteristics of them can manage chaos effectively so organization overcome chaotic situations with the least damage.

When the organization envisions the future and possible probabilities, it will challenge change and uncertainty, and will make the chaos an opportunity to success [4]. Chaos can be lived in all organizations in everywhere, but if chaos can be managed by adult-adult leaders, that organizational system can function in an healthy way in future.

8.4 Future Trends

This study is a conceptualization study about adults. In the following study, political leaders' leadership profiles will be analyzed based on this different adult classification. Although dimensions of leaders' interventions in chaotic situations have been differ in challenging, moderate, destructive, positive circle, political leadership attitudes have been placed even more chaotic to chaotic situations or have been transformed chaos to cosmos. In this context, finally, political leaders' profiles and coping strategies in chaotic ambiance will be evaluated on the base of adult-child, adult-adolescence and adult-adult classification model.

References

1. Erçetin ŞŞ (2000) Lider sarmalında vizyon. Nobel Yayın Dağıtım Şirketi, Ankara
2. Erçetin ŞŞ, Açıkalın ŞN, Bülbül MŞ (2013) A multi-dimensional approach to leadership in chaotic environments. In: Banerjee S (ed) Chaos and complexity theory for management: nonlinear dynamics, vol 5. IGI Global, Hershey, pp 89–104
3. Erçetin ŞŞ, Hamedoğlu MA, Açıkalın SN (2014) Can managerial narcissism which likely leads to a chaotic process be measured? Managerial Narcissism Scale (MNS). In: Erçetin ŞŞ, Banerjee S (eds) Chaos and complexity in world politics, vol 7. IGI Global, Hershey, pp 89–104
4. Erçetin ŞŞ, Kayman EA (2014) How to be a quantum leader in an intelligent organization? In: Erçetin ŞŞ, Banerjee S (eds) Chaos, complexity and leadership 2012. Springer, Dordrecht
5. Onur B (1995) Gelişim psikolojisi: Yetişkinlik – yaşlılık – ölüm, 3rd edn. İmge Kitabevi, Ankara
6. Robinson EJ, Pendle JEC, Rowley MG, Beck SR, McColgan KLT (2009) Guessing imagined and live chance events: adults behave like children with live events. Br J Psychol 100:645–659. doi:10.1348/000712608X386810

Chapter 9
A Complex Adaptive System Framework for Management and Marketing Studies

Gianpaolo Basile and Gandolfo Dominici

Abstract This theoretical work aims to analyze the choice of strategic management activities, taking into account a complex systems perspective. Following this approach, we represent the firm as a complex adaptive system, in which the management must be able to develop and implement different behaviors in order to dynamically ensure the viability of the firm or system. This implies that the management governing the firm or system is capable of choosing, from among a number of heterogeneous entities, the relevant stakeholders within the competitive context and of creating and maintaining significant relationships with them, which are considered to be relevant in a turbulent environment.

As an expression of the relational dynamics between the direct and indirect stakeholders, the firm exchanges energy and information with the reference contexts, in order to survive. As part of these exchanges and adaptations, both the firm and its stakeholders disperse energy, producing a dissipative phenomenon (Prigogine I, Order out of chaos. In: Livingston P (ed) Disorder and order: proceedings of the Stanford international symposium (Sept. 14–16, 1981). Anma Libri, Saratoga, pp 41–60, 1984), which we describe—in an analogy with the second law of thermodynamics and complexity theory—as entropy.

Our research question is thus:

> Can complex adaptive systems theory help decision makers to deal with the dynamism of organizations and brand in turbulent environments?

Through a theoretical and descriptive framework, we will draw our conclusions to this research question.

G. Basile (✉)
Sociology and Political Science Department, Business Systems Laboratory, Roma, Italy

Faculty of Arts and Philosophy, University of Salerno, Salerno, Italy
e-mail: gianpaolo.basile@bslaboratory.net

G. Dominici
Business Systems Laboratory, Roma, Italy

Department S.E.A.S., Polytechnic School, University of Palermo, Palermo, Italy
e-mail: gandolfo.dominici@libero.it

Ş.Ş. Erçetin (ed.), *Chaos, Complexity and Leadership 2014*,
Springer Proceedings in Complexity, DOI 10.1007/978-3-319-18693-1_9

9.1 Introduction

In the last 40 years, several management scholars have argued that we need new paradigms and practices to make sense of the changes that have come about through the global restructuring of the economy and advances in social, political, and cultural sphere [26, 36, 37, 51]. The systemic approach, together with the concept of entropy, in the fields of economics, strategic management, and marketing, has been discussed since the mid-1970s by many scholars, including Kangun [33], Monieson [46], Bass [3], Georgescu-Roegen [23], Boulding [6], Reidenbach & Oliva [52], and Layton [40]. These studies contextualized and validated well-established principles in the field of market dynamics.

The firm can be depicted as a partially open system which is in constant change and adaptation induced by the influences of the context, by the endogenous organizational dynamics, and by the effect of exchanges of energy, information, materials, symbols, and ideologies with heterogeneous and relevant stakeholders, all with the aim of obtaining sustainable advantages [2, 4, 13, 15, 18, 21, 24, 39].

In order to gain its final objective of sustainable viability through the achievement of sustainable advantages, the firm must be able to effectively acquire and manage its dynamic capabilities. Therefore the firm needs dynamic capabilities; these are the capacities of the organization to purposefully create, extend, or modify the resource base [25].

The continuous exchange of energy with the context and the relevant stakeholders constitutes the state of entropy. Entropy is given by an *informative chaos* requiring the continuous ability to decode and adapt in terms of cognitive, behavioral adaptive dynamics [49].

In this regard, the parallel developments in the thermodynamic theory of irreversible phenomena, in the theory of dynamical systems, and dynamic capabilities theory [50] have converged to show, in a compelling way, that the gap between "simple" and "complex," between "disorder" and "order," is much narrower than previously thought.

9.2 Theoretical Framework

On the basis of the above, we are conducting a scientific effort with an analogical approach to consider complexity theory against the background of the laws of thermodynamics. We aim to show how the survival of the firm is based on its abilities to create and maintain relationships with different and heterogeneous stakeholders (the complexity condition), transforming energy in interactions by means of self-organizing activities, so as to create a dynamic order.

The second laws of thermodynamic assert that, in a system without environmental exchanges, entropic disorder will unavoidably increase. These laws, in fact, point out the relevance for any living system (including the firm) of continuously exchanging information with relevant stakeholders [47].

Total Entropy

=

Se (entropy from environmental exchanges)

+

Si (entropy caused by internal irreversible changes)

Fig. 9.1 Total entropy

Indeed, if isolated, all such systems undergo a reduction of transformational activity. This tendency towards exhaustion or dissipation of energy results in the state of disorder, entropy, or chaos—culminating in the implosion of the closed system.

Interaction with other systems which lack energy or information exchange processes is the cause of the dissolution of the system over time, due to growing disorganization. Therefore, any isolated system naturally evolves towards the state of maximum disorder.

At this point, we point out that entropy is a term that evokes a dynamic condition of survival—namely, the continuous transformation of energy. Hence, we can assert that the entropy in open systems is due to the disorder or dynamism caused by the combination of so many inputs, some coming from the contexts and some from natural business organizational processes which involve a loss of energy (see Fig. 9.1).

For several years, scholars of thermodynamics held that a system reaches a point of equilibrium in the absence of an external input. Prigogine [47] applied this principle to open systems and argued that they could reach a state of dynamic equilibrium in the event of minimal production of entropy. This implies that, to reach this condition, the firm needs to be able to stay synchronized with stakeholders' changes by growing its dynamic capabilities [50].

9.3 The Partially Open Adaptive Systems

The 'Opening' condition (Se) is fundamental to avoiding the negative entropic effect. Therefore, the system survives if it is open or partially open, and thus able to transform the energy possessed in interactions with other contexts.

Contextualizing these concepts in the firm, we can say that the firm survives if it is able to establish, develop, and/or modify its appropriate informative, cognitive, and adaptive behaviors and skills, allowing it to receive and provide answers in the area of systemic exchange [12, 35].

In this sense, the firm could covey a different approach to create dynamic capabilities that could be summarized as *inside-out* and *outside-in*.

In the *inside-out* approach, the management plans the behavior of the firm on the basis of the resources or capabilities that will be modified through the market–consumer relationships. The *outside-in* approach is based on a continuous dialogue between firm or brand and the consumer, through the medium of some spaces (virtual or actual) in which the firm can observe and meet the consumer's needs. (Here we can think about Amazon offering the consumer some access to their cloud computing network in order to meet their web service needs).

These adaptive conditions represent the state of systemic balance on which survival is founded and which, given continuous adaptation or anticipation, we could define as the condition of "dynamic order" in dissipative systems.

Prigogine's work is pivotal in arguing that it is not that the lack of informative exchange allows a steady state or orderly condition to be achieved, but rather that this lack contributes to an increase in entropy. Indeed, the absence of informative exchanges represents the isolation of a system which, if it is to achieve the condition of order, should become self-organizing, generating the conditions of its survival independently based on deterministic behavior ([1, 5], [20]: 241–257).

In a management studies, a similar situation might be that of a firm operating in a static condition (such as in a monopolistic industry) with managerial approach based on deterministic, linear, exact, cause-and-effect planning, and not on probabilistic planning, considering performance feedback on the basis of inside-out, routinized processes.

Under current markets conditions, the firm is a partially open system; it is somewhat isolated, but must be able to anticipate and respond to environmental turbulence and changes in order to survive, exchanging information and energy with the appropriate language-behavior (in branding, marketing, and communication) with heterogeneous stakeholders (the conditions of complexity). This "language" develops based on the partners' needs and on common coenetic variables (such as variables, input, stimuli, influences, interference, standards, and rules), inasmuch as they belong both to the cognitive schema of the system under analysis and to those stakeholders/partners considered relevant by decision makers. The relevance is expressed by the stakeholders' possession of resources that are essential for the survival of the system [4, 16, 24].

The system's ability to survive requires facilitators in reading and in dialogue (e.g., partner–consultants and managers) and arises from the moment in which it can:

– use resources to understand the inputs (the opening phase of the system),
– achieve output–behavior–language (the closing phase of a self-organizing system),
– propose responses to changes in the contexts (the opening phase of the system)
– acquire and adapt behavior in response to the contexts.

In fact, this response is intended to create a new condition of equilibrium with respect to the context, which has been altered by causal interference or disruption

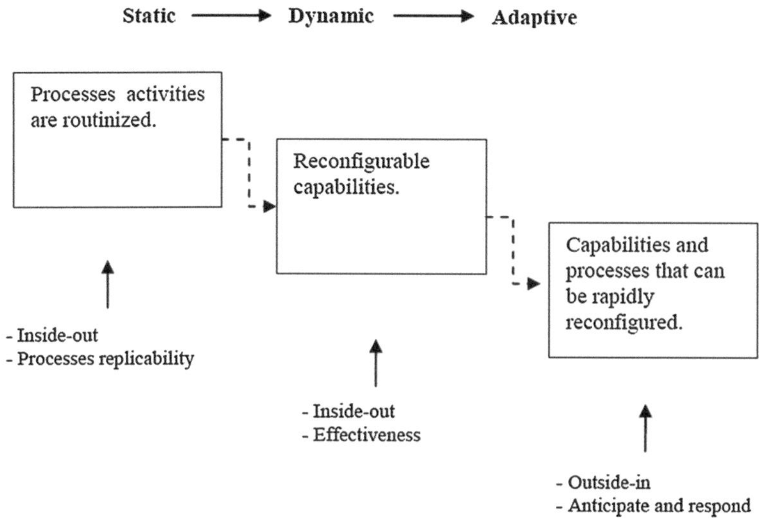

Fig. 9.2 The firm as adaptive system

caused by the contexts. This dynamic adaptation is the phenomenon of homeostasis and occurs by means of processes of self-organization [4, 44, 45].

Self-organizing phenomena are thus defined as processes in which a new dynamic order spontaneously emerges out of a chaotic state [8, 34].

The control of company dynamics, in any form, is the typical process supporting the homeostatic phenomenon (control of management, cost, finance, production, etc.), together with the processes needed to adapt to changes in the standards (laws and rules of conduct) that govern the dynamics of the system–market. These forms of control act as regulators that aim to correct a disruption in the context of systemic relations.

Hence, we can see control as a process directed towards achieving, in a certain period amount of time, a dynamic order—namely, the reduction of entropy caused by informative chaos; where entropy is a way of measuring both the state of equilibrium of the system and the degree of disorder and, possibly, the dissipation of the system itself (see Fig. 9.2).

Fantappiè [21, 22] argued in this regard that such negentropic processes are characterized by the principles of finality—that is, they are based on future results and not on past conditions, differentiation, order, or organization. Systemic survival is, therefore, supported by a vision of the future of the context and by the ability to create a dynamic order—that is, the coincidence between the variety of information that allows sharing of interpretive information and of influences coming from the contexts and the needs of relevant stakeholders.

Clearly, the ability to create dynamic order must be greater than the production of entropy, in order to ensure survival conditions ([32]: 161]).

9.4 The Adaptive Systems Approach in Marketing

Taking the traditional definition by Kotler [38], which describes direct marketing as a facilitator for meeting human wishes and needs through the purchase of products, services, or (even better) through brands, it could be argued that marketing is designed to meet the social needs of the individual, for himself and for his social identity, and the economic needs of the enterprise, for competitiveness and creation of value.

The continuous interactions between the agents thus create the business environment on the basis of the two major dimensions of complexity and turbulence. The complexity shows the richness of the heterogeneous and different stakeholders that compose the business environment, and with whom the firm must continuously create and maintain relationships [43].

The turbulence, on the other hand, conveys the level of competitiveness and change in the business environment [53].

The definition of complexity expresses three important characteristics. First, complex organizations stimulate outputs that cannot necessarily be predicted simply by understanding all of the inputs. The second characteristic is that complex organizations create behaviors that are neither predictable nor unpredictable (they lie on the edge of chaos). Third, the system's history is irreversible [17].

For these reasons, marketing objectives are reached by means of ever-more dynamic strategic and operational planning that takes into account, or anticipates, the many coenetic variables that make the context chaotic and dynamic.

In fact planning, in the field of marketing studies, tends more and more to take the capacity of stochastic analysis into consideration, due to the variability of rules and norms, and thus the behavior, of the components of the market and the contexts.

As a result of this planning, the social actor, the firm, and the individual consumers or stakeholders all tend to decrease their informative asymmetries in the dynamics of creating and maintaining relationships with partners. This effort is clearly intended to limit the dissipation of energy (entropy), due to the lack of correspondence both in the variety of information (in the input phase) and in the semantics between the meanings of language and behavior in the context of the creation and maintenance of relationships between social partners (in the phase of creation of relationships with stakeholders) [35].

In this regard, Wiener [55] argues that

> [...] *learning is a form of feedback in which the behavior of the model is modified, also, by past experience* [...]

Thus feedback, as a form of behavior, represents a learning process in which comparison is made between the conduct and the result to be achieved, so that success or failure alters future behavior.

Any planning process—in particular that of marketing and communications—is the result of the creative abductive, inductive, and deductive cognitive ability used by the governing body, management, or decision makers, based on the level of knowledge and learning (informative and cognitive variety) gained from impulses originating in the contexts [2].

In fact, the aim of planning and control, when applied according to a systemic methodology, is to detect or anticipate any gaps between what is expected from the applied model and what actually happens (a loop which represents the first homeostatic phenomenon), with the purpose of both observing the inability on the part of the model and the supporting variety in the contexts, and, finally, to reformulate the model for next time.

According to Ashby's Law of Requisite Variety "*only variety can control variety*" [1]; therefore, the variety of contexts can be addressed by management through equally informative varieties that act as an attenuating factor on complexity (Emery 1973).

At this stage, the following questions naturally arise:

> Does corporate management, at the decision stage, act in accordance with a deterministic or a complex model?
>
> As an effect of the complex dynamics, do different types of marketing models exist?
>
> How can these adaptive capabilities be built?

The firm, as a complex adaptive system, has always needed to consider the development of various kinds of relationships with a number of heterogeneous stakeholders—investors, institutions, employees, customers, partners, competitors, and others. Market globalization has produced an exponential growth in the number and heterogeneity of the subjects with which each company develops its relations.

This proliferation of relationships, made possible by new tools and new ways to communicate, is also accompanied by an increase in their speed or, to put it better, a different way of exploiting the time taken in business processes.

This scenario represents both the market and the firm as transient systems that increasingly express a kind of a probabilistic nature.

In addition to this is the time variable which, as an exogenous element compared to the life of the enterprise, and due to the globalization of relationships and dissemination of modern digital communication technology, has become a critical variable in competition. Each firm chooses whether and how to speed up their processes, when to activate them, and how to define their relative duration from a competitive perspective (time-based competition).

In the global context, businesses must therefore adopt a competitive approach to the market (market-driven management) or their relationship with components of the market (relation/complexity-driven management). This orientation demands not only the ability to know the market, the operators who work in it, their key characteristics, and their products, but also that the firm be able to constantly seek the opportunity to create and maintain "lasting" relationships with the relevant actors.

The focus of the relationship differs, therefore, from traditional approaches, because it brings out a form of company management in which, as a continuous process, the acquisition of cognitive input from the actors, along with the relationships between them, is placed before understanding the demand. The market is made up of these variables, and understanding them supports planning for determining both the present and the future of the enterprise system.

However, from the perspective of relationship/complexity-driven management, if marketing management states that knowledge of demand is the prerequisite from which all competitive developments in the market derive, then proposing a product that meets the defined expectations and ensures the firm or brand a competitive advantage causes the orientation of the relationship to become cooperative and competitive.

In global markets, extremely heterogeneous realities coexist within relationship systems that are composed of a number of different actors, both near and far away in terms of physical and competitive distance. In contrast to what was simulated in classical and neoclassical economic models, the diversity of competitive conditions is not based on the number of bidding businesses, but on the cooperative and competitive intensity that develops between businesses—namely the system of relations (competitive and partnership) established between the firms in the competitive market space.

In all markets characterized by scarcity of supply relative to demand, or by low turbulence levels (a stable environment), the bidding business not only governs demand by determining the quantities produced and then sold, but also has all the necessary knowledge (the variety of information) to set up future activities (e.g., Coca-Cola, Nutella, and all those companies and brands that operate as leaders in their markets in the mature stage of their life cycle).

When the entire production has been sold at a price defined by the manufacturer, who does not normally accumulate stocks of the finished product, the company information system tends to coincide with the system of internal records management, according to the inside-in type of information management model; this is characterized by the collection and processing of predominantly internal information, and an internal projection of the results of these calculations.

Under these conditions, business phenomena are especially significant, and are closely monitored and governed in order to continue to foster a company-system that focuses on itself, seeking continuous improvements in internal performance parameters. (For the authors, these situations characterize a static, predictable, and implosive system).

Certainly, sectors of the economy that are experiencing conditions of cooperative and competitive intensity are much more numerous in global markets. This situation of excess supply is characterized by the presence of saturated demand, with no possibility of increasing purchases and consumption, and encouraging diversification policies, research and innovation, and respect for market participants. In these markets, the only truly predictable phenomenon is the continuity of change operated by the actors (e.g., information technology market and social networking companies and products, such as Facebook and Google+).

The collection of information is therefore aimed at producing trends or offering profiles that are able to intercept and aggregate the preferences of a variety of buyers (conditions that characterize highly dynamic situations). This scenario represents an irreversible and unstable condition, in which the result of planning is not a certainty, but only a possibility, that an event or condition may occur.

Planning system behavior, therefore, aims to supply the dynamic result that we can define above as dynamic order or consonance–resonance [2, 15]. These conditions express the temporary ability to meet the needs of the stakeholders, who formerly communicated impulses and created disorder through their behavior (products and services, social impact, environmental impact, and so on).

9.5 Determinism vs. Complexity

The imperfect circularity represented in the previous process expresses the condition in which the firm and its management seek survival. This condition is no longer the expression of a purely deterministic approach, corresponding to cause and effect, but is rather a complex scenario in which the cause of a given future effect is absolutely indeterminable.

It is relevant to consider that, in contrast to the numerous studies that tend to blame the deterministic approach as being outmoded or inadequate in relation to the evolution of the discipline, the firm and therefore the management and decision makers find themselves working in market conditions that could require both a deterministic and a complexity approach [28].

Such a scenario also emerges, by analogy, from Prigogine's [43] studies of dissipative structures; he argued that relatively isolated systems could evolve into qualitatively different states, thanks to the continuous exchange of energy and information with the context (i.e., the stakeholders).

This consideration leads us to represent the firm as a system that can find itself in temporary conditions of balance, order, and reversibility, or experiencing conditions of nonequilibrium, chaos, and disorder (see Fig. 9.3). The point that separates the deterministic from the complex conditions can be named the discontinuity or the edge of chaos (see [7, 27, 29, 30, 41, 54]). This area is characterized by a limited condition between the government of a static, predictable, and implosive system and one in which order may be upset by minor changes and which, therefore, requires the capacity for stochastic analysis to support the creation of new features [56] (Fig. 9.4).

In this regard, Prigogine and Stengers [48] point out that the historical and temporal path that the system or firm goes through (market information, market share, total invoice, and others) is characterized by two kinds of conditions:

The first is a stable condition, in which deterministic approaches are applied. Here, we can see by analogy the management working in a dynamic and competitive environment, but with a regular relationship with the stakeholders. This situation is illustrated by a firm with a top-down organizational model, a strong brand, and a strategic approach based strongly on the imitative strategies and behaviors of

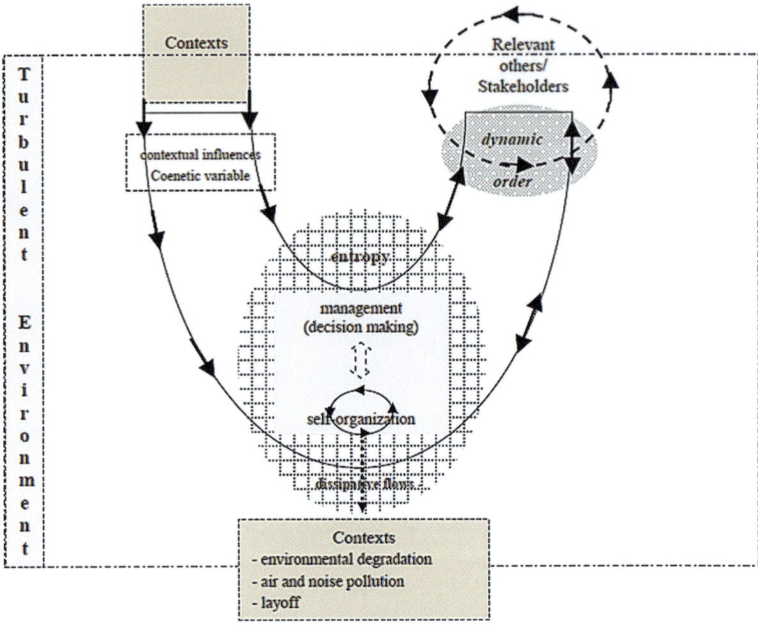

Fig. 9.3 Adaptive systems

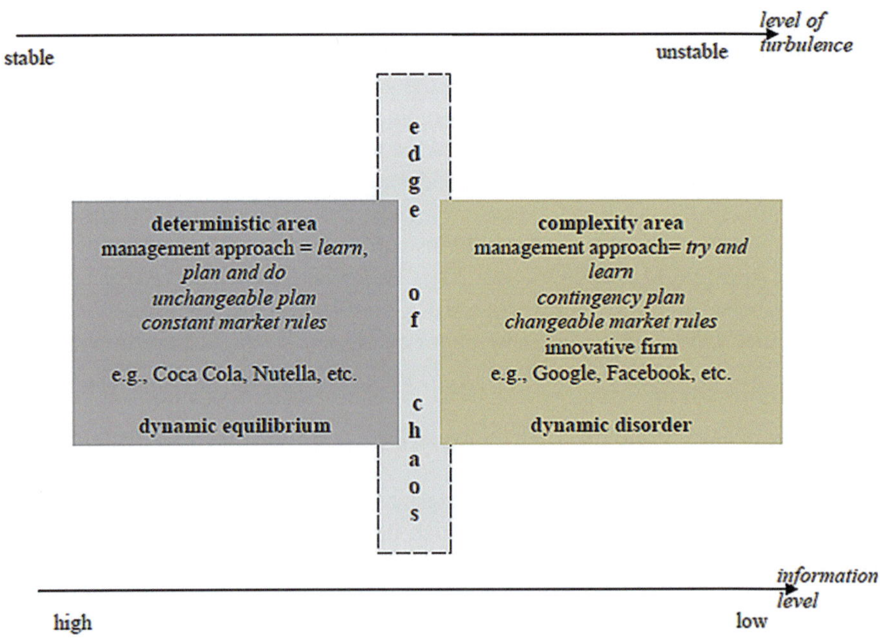

Fig. 9.4 The edge of chaos

competitors—the "me too" approach (e.g., the competition between Coca Cola and Pepsi Cola, whose strategies are strongly imitative and whose organizational models are both top-down).

The second is an unstable condition, near the edge of chaos, in which the firm chooses different strategies and behaviors on the basis of the level of information about the stakeholders. In this case, the firm presents a bottom-up organizational model that shows a networking approach with the stakeholders and competitive strategies that we could describe as "with you or between us" (e.g., Apple, which planned a connective strategy and behavior adding to its historical business as a producer of hardware and software; consider also its move into mobile phone production and computer-based entertainment). In this case, the firm works to go towards the edge of chaos, strengthening the relationships and the brand.

The difference between firms that work in the deterministic area and those that operate in the complexity area is that the former represent conditions in which the dynamic order is more enduring and the structural or organizational changes occur only in less frequent turbulence; while the latter represents the area of action, where the firm works in circumstances that are far from orderly. So, in the second case, the firms and brands operate in different industries and are frequently characterized by multifaced brands.

9.6 Entropy, Sustainability, and Curative Marketing Management

If the reduction of entropy in terms of the relationship between the firm's system and the stakeholders is not taken properly into consideration, there is a tendency for environmental entropy (whether social, economic, or natural) to arise. In this regard, the concepts of systemic entropy and dissipation are often associated with unethical, illegal, and immoral behavior.

At the planning phase, the behavior of the firm or system is aimed at competitive survival and the reduction of entropy, in order to be permanently sustainable. The consequences of such behavior, not only for the consumer or for market performance, but also on the environment suprasystem, must be taken into consideration [52].

In the last 20 years, the mainstream of marketing studies has focused on the process by which demand is satisfied by means of a set of products, and where environmental and social impact are the remit of consumer choice [9]. According to this branch of research, environmental and social entropy is a sacrifice—a price to be paid by the company—because the individual consumer achieves social progress by expressing greater economic well-being.

In Reidenbach and Oliva's [52], words

> [...] the marketing function, while extending our human existence, is reducing the ability of our environment to support our continued existence.

In recent years, some marketing scholars have begun to address behaviors aimed at reducing environmental and social entropy and supporting the survival of business systems or macromarket systems, from an environmental, economic, and social perspective [11, 12, 42]. On the basis of these developments, both the laws of thermodynamics—understood as irrefutable laws of nature to which all activities of social actors are subject—and the application of these laws have been brought into economic and marketing studies. This latter refers mainly to branches of sustainability marketing, medicinal marketing, and the bioeconomy.

In this regard, Nicholas Georgescu-Roegen [23], founder of the bioeconomic field of studies, has developed an economic theory that calls into question the "fundamentals" of decline in each production process, by applying the second law of thermodynamics to economics, and particularly to the economics of production. The author argues that this phenomenon does not decrease entropy on the planet, but increases it irreversibly, or at least leaves it equal—that is to say, the more energy is transformed into an "unavailable" state, the more energy will be subtracted from what is available for future generations, and therefore the more entropy (proportional disorder) is returned to the environment that surrounds us. The author, therefore, argues that it is not the quantity of production that adversely affects the environment, but the planning of this production in terms of the compliance and use of materials, in the very long-term view.

Relative to the condition of entropy and the role of marketing managers, Czinkota et al. [11, 12] introduced *curative international marketing*, asserting that restoring and developing international economic health may be the next marketing direction. By 'Restoring,' the authors mean "something lost which once was there;" 'Developing' refers to new issues addressed with new tools and frames of reference, and 'Health' underlines how the issue is essential to overall welfare, which marketing needs to address, resolve, and improve. Marketeers must deliver joy, pleasure, fulfillment, safety, and personal growth, while advancing towards a better society, and to do so across borders [10].

In line with these considerations, we may assert that marketing, both on the micro and macro levels—is beginning to overcome the past approaches of the mainstream widespread convictions, where it was regarded as a facilitator of individual economic well-being at the expense of the survival of future generations [31].

9.7 Conclusion

We have shown how complexity theory can be useful in providing a new lens through which to look at and analyze organizations and their ability to adapt their behavior by creating dynamic relationships with an ever-increasing number of heterogeneous stakeholders.

In this sense, the paper has presented a scientific and methodological framework to stimulate empirical research. Our study presents several ideas for further research aimed at overcoming the perceived abstraction of theory of complexity. We believe that managers and decision makers could benefit by considering complexity theory as a new way of reading dynamics and planning behaviors.

Empirically, this condition would be met by developing the propensity to identify potential entropic performance represented by a number of contextual variables, and to create interactive plans to continually adapt the behavior of the firm and brand to stakeholders needs.

Variables such as average income, consumption capacity, changes in turnover of distribution channels, and others, if treated stochastically, will allow managers to identify possible dissipative phenomena. Given that management and decision making are increasingly nonlinear activities and processes, we highlighted how a complexity management approach would allow marketeers to identify marketing and communications behaviors and conduct necessary for the company to contribute to the adaptation of the market system and, therefore, in reducing the instability that would otherwise be generated.

This study does not aim to provide marketeers with a contribution in terms of specific models or new planning practices, but rather aims to raise their awareness to a new concept of management and marketing, understood as a process conducted in chaotic/unstable environments that no longer requires a rational or linear approach from businesses/managers, which is designed to maximize results, but uses greater analytical capabilities to increase the level of information so as to better understand the dynamic environment governed primarily by stochastic influences.

We are confident that the disruptive shock to the system from the internet era will spawn a new generation of insights into how markets work and how organizations can anticipate and respond to fast-moving market signals.

Acknowledgements To cope with regulation on Academic publishing we declare that although this study is based on a collaborative effort by all authors, Sects. 9.1 (Introduction), 9.2 (Theoretical Framework) and 9.7 (Conclusions) can be attributed to Gandolfo Dominici; while Sects. 9.3 (The Partially Open Adaptive Systems), 9.4 (The Adaptive Systems Approach in Marketing), 9.5 (Determinism vs. Complexity) and 9.6 (Entropy, Sustainability, and Curative Marketing Management) can be attributed to Gianpaolo Basile.

References

1. Ashby WR (1956) An introduction to cybernetics. Wiley, New York
2. Barile S (2009) Management sistemico vitale. Decidere in contesti complessi. Giappichelli Editore, Turin
3. Bass FM (1974) Theory of stochastic preference and brand switching. J Mark Res 2:1–20
4. Beer S (1966) Decision and control. Wiley, Chichester
5. Boltzmann L (1964) Lectures on gas theory (Dover Books on Physics). The Regents of the University of California
6. Boulding KE (1966) The economic of the coming spaceship Earth. In: Daly HE (ed) Toward in steady state economy. WH Freeman and Co, San Francisco

7. Capra F (1997) The web of life. Anchor Books, New York
8. Coveney P, Highfield R (1995) Frontiers of complexity: the search for order in a chaotic world. Ballantine Books, New York
9. Crane A, Desmond J (2002) Societal marketing and morality. Eur J Mark 36(5/6):548–569
10. Czinkota M, Skuba C (2011) Global-the two faces of international marketing-effective marketing and ethical practices must exist together. Mark Manag 20(4):14–25
11. Czinkota M, Skuba C (2013) Trade policy and international marketing. Marketing News, pp 22–25, May 2013
12. Czinkota M, Kaufmann HR, Basile G (2014) The relationship between legitimacy, reputation, sustainability and branding for companies and their supply chains. Ind Mark Manag 43(1):91–101
13. Dixon DF (1984) Makromarketing: a social system perspective. J Macromark 4(fall):1–17
14. Dolan P (2002) The sustainability of sustainable consumption. J Macromark 22(2):170–181
15. Dominici G, Basile G, Palumbo F (2013) Viable systems approach and consumer culture theory: a conceptual framework. J Organ Transform Soc Chang 10(3):262–285
16. Dominici G, Palumbo F (2013) Decoding the Japanese lean production system according to a viable systems perspective. Syst Pract Action Res 26(2):153–171
17. Dooley K (1996) A complex adaptive systems model of organizational change. Nonlinear Dynamics Psychol Life Sci 1(1):69–97
18. Dowling GR (1983) The application of general systems theory to an analysis of macromarketing systems. J Macromark 3:22–32
19. Emery AR (1973) Preliminary comparisons of day and night habits of freshwater fish in Ontario lakes. J Fish Res Bd Can 30:761
20. Emery FE, Trist EL (1969) The causal texture of organizational environment. In: Emery FE (ed) Systems thinking. Penguin, Harmondsworth, pp 241–257
21. Fantappié L (1942) Teoria Unitaria del Mondo Fisico e Biologico. Di Renzo Editore, Rome
22. Fantappié L (2011) Che cos'è la Sintropia. Principi di una teoria unitaria del mondo fisico e biologico e conferenze scelte. Di Renzo Editore, Roma
23. Georgescu-Roegen N (1971) The entropy law and the economic problem. In: Daly HE (ed) Toward in steady state economy. WH Freeman and Co, San Francisco
24. Golinelli GM (2010) Viable systems approach. Governing business dynamics. CEDAM, Padua
25. Helfat CE (2007) Stylized facts, empirical research and theory development in management. Strateg Organ 5(2):185–192
26. Hermiter JD (1974) A comparison of the entropy model and the Hendry model. J Mark Res 2:21–29
27. Hibbert B, Wilkinson IF (1994) Chaos in the dynamics of markets. J Acad Mark Sci 22(3):218–233
28. Holbrook M (2003) Adventures in complexity: an essay on dynamic open complex adaptive systems, butterfly effects, self-organizing order, coevolution, the ecological perspective, fitness landscapes, market spaces, emergent beauty at the edge of chaos, and all that jazz. Acad Market Sci Rev 1–5
29. Holland J (1999) Emergence: from chaos to order. Helix Books, Cambridge, Massachusetts
30. Holland J (1999) Emergence: from chaos to order. Helix Books, Perseus Publishing, Reading, Cambridge, Massachusetts
31. Kadirov D (2011) Macro-systems role of marketing: do we trade environment for welfare? J Macromark 31(4):359–375
32. Kadirov D, Varey RJ (2011) Symbolism in marketing system. J Macromark 31(2):160–171
33. Kangun N (1981) Marketing and entropy process in a new world view. In: Proceedings of the 1981 of American marketing association educators conference, n.47, Chicago
34. Kauffman S (1994) The origins of order: self-organization and selection in evolution. Oxford University Press, New York
35. Kaufmann HR, Loureiro SMC, Basile G, Vrontis D (2012) The increasing dynamics between consumers, social groups and brands. Qual Mark Res Int J 15(4):404–419

36. Kelly K (1994) Out of control: the new biology of machines, social systems, and the economic world. Addison-Wesley, New York
37. Kelly K (1998) New rules for the new economy. Viking, New York
38. Kotler P (1980) Strategic planning and the marketing process. Business 30(3):2–9
39. Yolles M (1999) Management systems. A viable approach. Financial Times, London
40. Layton RA (1989) Measures of structural change in macromarketing systems. J Macromark 9:5–15
41. Langton CG (1992) Artificial life II. Addison-Wesley, Redwood City
42. Martin D, Schouten J (2012) Sustainable marketing. Pearson Prentice Hall, Upper Saddle River
43. Mason RB (2007) The external environment's effect on management and strategy. A complexity theory approach. Manag Decis 45(1):10–28
44. Maturana HR, Varela FJ (1980) Autopoiesis and cognition. The realization of the living. D. Reidel Publishing Company, Dordrecht, Netherlands
45. Maturana HR (1975) The organization of the living. A theory of the living organization. Int J Man Mach Stud 7:313–332
46. Monieson (1981) Marketing and the theory of dissipative structures. In: Proceedings of the 1981 of American marketing association educators conference, n.47, Chicago
47. Prigogine I (1955) Introduction to thermodynamics of irreversible processes. Charles C.Thomas, Springfields
48. Prigogine I, Stengers I (1979) *La nouvelle alliance: Les métamorphoses de la science* [The new alliance: the metamorphosis of science]. Gallimard, Paris
49. Prigogine I (1984) Order out of chaos. In: Livingston P (ed) Disorder and order: proceedings of the stanford international symposium (Sept. 14–16, 1981). Anma Libri, Saratoga, pp 41–60
50. Teece DJ (2009) The (new) nature and essence of the firm (with Christos N. Pitelis). Eur Manage Rev 6(1):5–15
51. Tetenbaum T (1998) Shifting paradigms: from Newton to chaos. Organ Dyn 26(4):21–32
52. Reidenbach RE, Oliva TA (1983) Toward a theory of the macro systemic effects of the marketing function. J Macromark 3(2):33–40
53. Vorhies DW (1998) An investigation of the factors leading to development of marketing capabilities and organizational effectiveness. J Strateg Mark 6:3–23
54. Waldrop M (1992) Complexity. Penguin, London
55. Wiener N (1954) The human use of human beings. Houghton Mifflin Company, Boston
56. Winsor RD (1995) Marketing under conditions of chaos. Percol Metaph Models J Bus Res 34:181–189

Chapter 10
Females Complex Leading Roles in the *Game of Thrones* Through Semiotics

Mina Abbasiyannejad and Nadya Supian

Abstract Women in *Game of Thrones*, a fantasy book by George R.R. Martin, have leading roles in several ways. Despite being represented as being subordinated and dominated by men, they show that when it is necessary they are able to reverse roles and show their ability to lead and have the men follow them. The complexity of their roles is presented in many different ways – as mature women, as the young and even as children. In this paper, it is shown that power conflict is the most important conflict in the novel. The women are seen in commanding positions and who attempt to develop their power despite the chaotic circumstances. This paper looks at the positions of two women in the novel using semiotics to shed light on words used, their signs and connotations. Interpreting signs will enable us to understand the hidden layers of the text.

10.1 Introduction

Fiction is one of the key sources of education and learning and fantasy writing, as a form of fiction mostly occurs in a timeless, imaginary world with fantastic characters and even animals with special powers. In fantasy stories conflicts are usually resolved by the great deeds of others. Despite being in the shadow of male authority in the majority of cases the roles of women are shown in all facets of the stories. While women can be in different positions, as a mother, daughter, wife, maid among others, all show their power and influence in their respective and individual ways. They are not shown just in relation to men, but are shown to have the ability to make their own decisions and to be in control of such decisions.

In the context of fantasy writing, women are often shown as malicious and jealous. Ironically in fantasy writing even though the female characters may remind

M. Abbasiyannejad (✉)
Faculty of Creative Industries, Universiti Tunku Abdul Rahman, Petaling Jaya, Malaysia

Faculty of Educational Studies, Universiti Putra Malaysia, Seri Kembangan, Malaysia
e-mail: minaabbasiyannejad@gmail.com

N. Supian
Faculty of Creative Industries, Universiti Tunku Abdul Rahman, Petaling Jaya, Malaysia

Ş.Ş. Erçetin (ed.), *Chaos, Complexity and Leadership 2014*,
Springer Proceedings in Complexity, DOI 10.1007/978-3-319-18693-1_10

us of their submission to the male characters, it is not always the case. Despite being shown as marginalized, and abused, the female characters in fantasy fiction show that under different circumstances they can be strong enough to beat their male opponents.

Game of Thrones is a series of fantasy written by George Martin which represents women in many different aspects. It is a saga of loyalty and treachery, love and hate, power and powerless, winners and losers. In the masculine world of the novel, women try to show their abilities in many different ways. Being in a leading position has always troubled women living in a masculine world. It is of significance to understand and differentiate women's leadership strategies in the novel. For example, both Catelyn and Cersei are married but their love relationships diverge in two different directions. Catleyn is a faithful, understanding wife to Lord Stark, while Cersei has an affair and shows no commitment to her husband. They are powerful in their own ways with different leading styles. The roles of these two women, Catelyn Stark and Cersei Lannister will be discussed in this paper.

The author locates the characters in the fantasy world and engages them in complicated conflicts, both public and personal, to present a holistic view of the environment in which people live, rather than only focusing on the conflicts between individuals. Power conflicts are shown in various ways, between friends, among family members, strangers and enemies.

10.2 Literature Review

In general, "Internationally, women are still under-represented in positions of power, responsibility and leadership, despite the dramatic increase in their formal employment over the last five decades" [8, p. 1]. Jones [10] comments on the game of Genders and women's role in her article "A Game of Genders: Comparing Depictions of Empowered Women between *A Game of Thrones* Novel and Television Series"

> Martin has already established strong and empowered women in his novel, yet in the show adaptation these women have often gained more strength, as 15 years later he is now catering to a society whose standards for television have changed and are now able to have leading women who are strong, without objectifying them. These five women ... allow the audience to view a range of pressures, trials and situations that modern viewers deal with. Cersei, Catelyn ... all exhibit different archetypal traits and all face unique trials and obstacles and threats to their positions and power .... In the end, he[Martin] shows that while our culture has changed over the past 15 years, there is still plenty of progress to be made before the Aryas of the world can wield their swords without needing to become men. (p. 20)

Adi [3] writes that Martin makes special innovations in the role of women. "The women dominate the whole points of view". Adi goes on to say that the roles and characteristics of women could be found in the depiction of women in narratives of the action genre and also emphasizes the point that despite the feminine appearance of women in *Game of Thrones* they have "male characteristics such as (being) brave, domineering, powerful and free" (p. 310). Gardner [9] in the Review of *Game of Thrones* pinpoints how families function in the novels. The game of power within the families is a central theme of the series (p. 177).

10.3 The Relationship Between Semiotics and Literature

Studying semiotics enables us to be more aware of the role of signs and their importance in our lives [1, p. 1298]. In the view of Chandler [5], semiotics encompasses both signs in everyday speech and also anything, which represents something else. It demonstrates the construction of meaning and how reality is depicted; therefore signs can be in the form of words, images, sounds, gestures and objects (p. 2). Semiotics is differentiated from other literary criticism. It is concerned with how "language and literature convey meanings" while other literary criticism, "asks what texts mean" [4, p. 61].

Generally, semiotics maintains that studying literature must involve "all investigation of the ways and means of literary signification". For some theorists the capability of interpreting individual works relies on "mastery of the system" and strategy that semiotics attempts to explain. Critics believe that compelling interpretation of a novel is only possible by "thorough understanding of the nature and conventions of narrative, the relations between story and discourse, and possibilities of thematic structure." On the other hand, the semiotician stresses that the interpretations of readers and critics are themselves considered part of the work they study. Investigation of literary signification involved the scrutiny of how works communicate to readers [6, p. viii].

A crucial element to be considered in semiotics is contexts. The words, gestures, and generally signs become meaningful in the context. It explains the relationship of meaning in semiotics and restores literature to its many contexts. Consequently, semiotics explores the inner structures, questioning meaning as expressed, within specific categories and how they have been organized [7, p. 61]. Connotations as well as context play an important role in understanding semiotics. Beyond what words mean, literally they have various other meanings, within every single societal and cultural context. It is appropriate to refer to Sebeok [13] who maintains the elation to usage and situation, for to a large extent human signs have the capacity to encode two primary kinds of referents, that are, denotative and connotative (p. 6).

Culler [6] in *The Pursuit of Signs: Semiotics, Literature, Deconstruction* writes that a methodology that is introduced by semiotics is lucid and clear and it is often disregarded when studying literature and culture (pp. xiv & xv). Robert Scholes [12] explains the relationship between semiotics and the study of literature in *Toward the Semiotics of Literature*. In this regard, he writes that generally, studying literature must involve semiotics and in particular in production and interpretation of the most important kinds of literature that are ruled by codes of semiotics (p. 15). Semiotics is an essential tool that will assist in understanding the hidden layers of the meanings.

It not only demonstrates the importance of understanding the signs but also, the ways these signs are interpreted could provide extra and tremendous meaning of the literary texts.

10.4 Discussion

Developing the right foundations and relationships could be one of the ways for women to progress toward being capable leaders. Both the women in the novel are connected to strong leaders such as King Robert and Ned Stark. Cersei, introduced as "The Lannister woman" is actually representative of Lannister's power. Almost everyone in the kingdom knows that, despite her charming face, she is a fearsome character. The king may be the leader but there are times when he inevitably follows what his lady orders. The scene in which they are Stark's guests is a good example of this. After Prince Joffrey hurts his arm in the fight with a direwolf, the queen plays a leading role in insisting on punishment as shown below,

> "The girl is as wild as that filthy animal of hers," Cersei Lannister said. "*Robert, I want her punished.*" "Seven hells," Robert swore. "Cersei, look at her. She's a child. What would you have me do, whip her through the streets? Damn it, children fight. It's over. No lasting harm was done." *The queen was furious.* … Robert started to walk away, *but the queen was not done.* "And what of the direwolf?" she called after him. "What of the beast that savaged your son?" The king stopped, turned back, frowned. "I'd forgotten about the damned wolf."… "We found no trace of the direwolf, Your Grace." Robert did not look unhappy. "No? So be it." *The queen raised her voice.* "A hundred golden dragons to the man who brings me its skin!" "A costly pelt," Robert grumbled. "*I want no part of this, woman.* You can damn well buy your furs with Lannister gold." The queen regarded him coolly. "I had not thought you so niggardly. The king I'd thought to wed would have laid a wolfskin across my bed before the sun went down." Robert's face darkened with anger. "That would be a fine trick, without a wolf." "*We have a wolf,*" Cersei Lannister said. *Her voice was very quiet, but her green eyes shone with triumph.* It took them all a moment to comprehend her words, but when they did, the king shrugged irritably. "As you will. Have Ser Ilyn see to it". [11, p. 130]

The King's reaction is comprehensively described with all the possible details. He is leaving the scene when, something makes him move back, and he frowns, showing that despite his will he needs to take action, and follow the queen's order to satisfy her. The Queen raises her voice after her second attempt fails. She raised her voice, wants the direwolf punished. The quite voice and shinning green eyes can predict she has already won the battle. In the end, the King agrees to what she wants, and this demonstrates her strong leading power despite living in a masculine world. Semantically speaking, the combination, eye contacts, rising and falling voice, ignoring what the king says, all confirm the hidden power of femininity which, instead of being shown in apparent violence, is masked by a womanish tranquillity.

Cersei's hand in killing the King is hinted through some signs. The King was killed by a boar while hunting in the forest. "I wonder, Ser Barristan," asks Varys, so quietly, "who gave the King this wine?" The King missed the point because he was overloaded with wine. The young lad, his squire who served him generously. "I hope the dear sweet lad does not blame himself. Children are so vulnerable in the innocence of their youth, how well do I remember." [11, p. 410] The word innocence is ironically used here, since in a later scene the affair between the lad and the queen could be a reason for the treachery of killing the King. There is an indication that the queen has plotted the death of the King very subtly, to be free and to have full authority. This is another sign of the queen's ambition to achieve what she desires, and take over as ruler. Although she cannot sit on the throne, because of her son, everybody knows that she is the real commander.

Catelyn Stark is another strong female character that shows her strong will and abilities in the story. She is introduced as such. "Catelyn had been anointed with the seven oils and named in the rainbow of light that filled the Sept of Riverrun. She is of the Faith, like her father and grandfather and his father before him" [11, p. 22]. Opposite to Cersei Lannister, Catelyn is known to be honest, and trustable. She is embrocated with high attention. In "Sept of Riverrun" Riverrun is the ancestral castle of the House of Tully. The main characteristic of this castle is that the gates are built in a way that in times of danger they can be opened and leave the castle surrounded on all three sides by water, turning Riverrun into an island and making it completely unconquerable. The whole generation of Catelyn was mingled with faith. The connotation is that no one could affect Catelyn and shake her faith and her beliefs.

When Ned is appointed as a king's hand, Catelyn is the only person who persuades him to take a position.

> "I will refuse him," Ned said as he turned back to her. His eyes were haunted, his voice thick with doubt. Catelyn sat up in the bed. "You cannot. You *must* not." "My duties are here in the north. I have no wish to be Robert's Hand." "He will not understand that. He is a king now, and kings are not like other men. If you refuse to serve him, he will wonder why, and sooner or later he will begin to suspect that you oppose him. Can't you see the danger that would put us in?". [11, p. 51]

Ned's reaction to the King's request is shown through his haunted eyes, and voice. Catelyn takes the leading position and orders him, reminding him he cannot refuse the King's order. She uses the words "must" which is strong enough for Ned to change his decisions. In the end Ned accepts to be the King's hand and goes to the South.

10.5 Conclusion

"Analyzing signs, or a semiotic reading, potentially offers a more complete understanding of the novel under consideration and allows for an exploration of the characters and their actions as implicit signs of conflict" [2, p. 2]. In the *Game of Thrones* women's role as leaders are demonstrated in diverse ways. Their power in the story is shown through their voice and eye contacts, choice of words, and the way they act. In the male-dominated world of the novel, when swords and weapons are the basic ways of communication, women find their way to command successfully despite the barriers. Reading between the lines reveals the hidden points below the surface. Therefore semiotics could be an excellent tool to assist us understanding the novel.

References

1. Abbasiyannejad M, Talif R (2013) The cultural conflict in the Ian McEwan's on Chesil Beach through semiotics. J Soc Sci Human Pertanika (JSSH) 21(4):1297–1308
2. Abbasiyannejad M, Talif R (2014) The exploration of political conflicts and personal relationships in Ian McEwan's the innocent. SAGE Open 4(1):1–9
3. Adi IR (2012) Popularizing epic narrative in George Rr Martin's a game of thrones. Journal Humaniora 24(3):303–314
4. Atkins GD, Morrow L (1989) Contemporary literary theory. University of Massachusetts Press, Amherst
5. Chandler D (2007) Semiotics: the basics. Routledge, London
6. Culler J (2001) The pursuit of signs: semiotics, literature, deconstruction. Routledge, London
7. Davis RC, Schleifer R (1989) Contemporary literary criticism: literary and cultural studies. Longman, New York
8. de la Rey C (2005) Gender, women and leadership. Agenda 19(65):4–11
9. Gardner L (2014) Review of game of thrones. Int J Jungian Stud 6(2):171–177
10. Jones R (2012) A game of genders: comparing depictions of empowered women between a game of thrones novel and television series. J Stud Res 1(3):14–21
11. Martin GR (1996) A game of thrones. Harper voyager, London
12. Scholes R (1977) Toward a semiotics of literature. Crit Inq 4(1):105–120
13. Sebeok TA (2001) Signs: an introduction to semiotics. University of Toronto Press, Toronto

Chapter 11
The Effect of Social Media and Social Networking on Perceptions of Leadership and Leaders

Şefika Şule Erçetin and Ssali Muhammadi Bisaso

Abstract Leadership, one of today's most important public undertakings, is brimming with problems in which all the challenges of complex social life and the globalizing world are directly felt. Together with the changing world and the person, perception of leadership and interaction systems among people are also changing. This paper argues that social media and social network analysis provide a useful theoretical framework for understanding the dynamic evolution of leadership and the complex interactions among leadership actors. It is argued that leadership can be conceptualized and modeled as a complex, dynamic system, which exhibits both unpredictability and underlying order. The relevance of social media and social networking for strategy and applicability is discussed, and a number of managerial implications are suggested. To illustrate the application of social media and social network analysis, a comparative view is presented that depicts the interactions between individuals and their perception of leadership. The views presented in this paper demonstrate how managers might underestimate the effect of social media and social networking on the perceptions of leadership. The paper concludes that, by understanding leadership as a complex system, leaders can improve decision making and adopt innovative approaches to dealing with social media and social networking in the organization. The paper thus is a jewel in the crown of the elusive body of knowledge within the social media and social networking realm.

Ş.Ş. Erçetin (✉)
International Science Association (ISCASS), Ankara, Turkey

Faculty of Education, Hacettepe University, Ankara, Turkey
e-mail: sefikasule@gmail.com

S.M. Bisaso
Islamic University in Uganda, Mbale, Uganda

Department of Educational Management and Supervision, Hacettepe University, Ankara, Turkey
e-mail: bisasossali@yahoo.com

Ş.Ş. Erçetin (ed.), *Chaos, Complexity and Leadership 2014*,
Springer Proceedings in Complexity, DOI 10.1007/978-3-319-18693-1_11

11.1 The Conceptual View of Organizations and Organizational Success

Organizations are considered to be social entities where a position can be measured by social-psychological and demographic data [1, 2, 45]. Location and connection to others provides information about members and group identification. A person's perceived position is essential in determining his or her beliefs, interests and motivation for action [8, 9, 40]. Additionally, people with close ties tend to maintain similar interpretations of the organizational environment [7] and tend to act similarly. This results in flows of social information and the tendency for people to seek out similar others.

Formal organizations are social groups that distribute tasks for a collective goal. That is why network research on organizations may focus on either intra-organizational or inter-organizational ties in terms of formal or informal relationships. Intra-organizational networks themselves often contain multiple levels of analysis, especially in larger organizations with multiple branches, franchises or semi-autonomous departments.

Research has shown that organizational success largely depends on employees who are considered as one of the most important assets of any organization because they are capable of creating value and enable organizations have a sustainable competitive advantage [17, 19, 29, 31, 45]. Indeed organizations that learn and adapt faster to social networking trends will have a clear advantage in the market place and ultimately increase their bottom lines. But this is preceded by the nature of leadership in the said organization which equally responds and reacts to the way it is perceived. Interestingly this leadership perception is shaped by social media and social networking.

Over the past 15 years, the world as we know it has been taken by storm through the onset of social media. According to Comscore [11] about 90 % of Internet users visit a social media site each month. Because we live in such a largely global-society, creating and maintaining an online presence has become most relevant in promoting your brand and expanding your social network.

By their very nature, educational institutions have a complex structure and exhibit a non-linear situation. Therefore, for an educational institution to be successful, it must define the education phenomenon as a chaotic phenomenon [23]. In Radford's opinion [37], schools are organizations under the influence of different effects of multiple implicit or explicit elements and also intending to create different effects. In these organizations there is no single and unchanging formula for success. So there are many factors affecting the success of those organizations. According to Garmston and Wellman [21], schools are also under the influence of their own feedbacks. As a child, these feedbacks increase resonantly and exponentially. Thus they create radical changes. So in order to get the desired result, it is necessary to note that the effect on the system shall be the right effect for the long term [10, 28].

Reilly refers to this as follows: [39, p. 424] "*the linear proportional relationship can not be established in the education system*" and he gives the following example:

"in order to assess whether educational investments are successful, looking the ratios of inputs and outputs is the result of a linear thought and this is a faulty assessment." Sullivan [42] also emphasizes this chaotic order in educational institutions. He also implies that the administrators can get the opportunity to bring a new dynamic to the school using the environment of educational institution and the network of relationships in chaotic environment. In other words, it is possible for the administrator to turn the chaos into an opportunity by taking more advantage of environment.

11.2 Conceptual Description of Social Media and Social Networking

The prevalence and continued growth of social media has changed virtually every facet of how we socialize and engage for both personal and professional activities [33, 44]. Social media is not only extremely pervasive; it is an activity in which you are expected to participate. Not all social media is Facebook and Instagram. Think LinkedIn, the new virtual business profile quickly replacing the traditional printed resume. This phenomenon is a tangible version of Rogers's concept of the ideal self. We have a general persona we construct and put out to the cyber universe based on the person we want to be, and more important, based on the person we want to be seen as.

While there are a number of positive aspects to social media, it is important to take a look at what effect (negative) it has presented on the perception of leadership and some alternatives to adopt a successful social media policy in various workplaces.

Social networks have also had a measurable impact on workplace activities as human resource policies are adjusted to keep pace with the new social norms. Social networks on the other hand are self-organizing, emergent, and complex, such that a pattern appears from the interaction of the elements that make up the system [34]. These patterns become more apparent as network size increases.

According to Drew Hendricks [24], a tech, social and environmental addict, "there are tremendous varieties of social networking sites, and many of them can be linked to allow cross posting. This creates an environment where users can reach the maximum number of people without sacrificing the intimacy of person-to-person communication. We can only speculate about what the future of social networking may look in next decade or even 100 years from now, but it seems clear that it will exist in some form for long as human are alive". In the world today communication and getting timely information is vital in all aspects.

Most larger social networks display features of social complexity, which involves substantial non-trivial features of network topology, with patterns of complex connections between elements that are neither purely regular nor purely random (see, complexity science, dynamical system and chaos theory), as do biological, and technological networks. Such complex network features include a heavy tail in the degree distribution, a high clustering coefficient, assortativity or disassortativity among vertices, community structure, and hierarchical structure. In the case of agency-directed networks these features also include reciprocity, triad significance profile and other features. In contrast, many of the mathematical models of networks that have been studied in the past, such as lattices and random graphs, do not show these features [41].

Social network theory is the study of how the social structure of relationships around a person, group, or organization affects beliefs or behaviors. Network analysis is a set of methods for detecting and measuring the patterns. The axiom of every network approach is that reality should be primarily conceived and investigated from the view of the properties of relations between and within units instead of the properties of these units themselves. It is a relational approach. In social and communication science these units are social units: individuals, groups/organizations and societies.

Social network analysis is a collection of concepts, measures, and techniques for relational analysis. It is an approach that is specifically designed to grasp the most important features of social structures and it is unrivalled in this task. It can be used to explore social relations themselves and also the cultural structures of norms and ideas that help to organize those relations in conjunction with material circumstances.

Social network analysis is focused on uncovering the patterning of people's interaction. Network analysis is based on these patterns, which is important features of the lives of the individuals. Choices of individual depend in large part on how that individual is tied into the larger social network. Social network approach is guided by formal theory organized in mathematical terms, and grounded in the systematic analysis of empirical data. It has found important applications in organizational behavior, inter-organizational relations, and the spread of contagious diseases, mental health, social support, the diffusion of information and animal social organization.

Social network analysis is a suitable method for investigating relations and interactions. Social network analysis is an approach rooted in anthropology, sociology and social psychology for assessing social structures [5]. The social network perspective illustrates social systems as networks of various relationships [6, 27]. Social network analysis is concerned with the structure and patterning of relationships and seeks to identify both their causes and consequences [5, 9]. Social networks are analyzed at the scale relevant to the researcher's theoretical question. Although levels of analysis are not necessarily mutually exclusive, there are three general levels into which networks may fall: micro-level, meso-level, and macro-level.

In general, network analysis focuses on the relationships between people, instead of on characteristics of people. Analysis of these relationships, it helps to uncover the emergent and informal communication patterns present in an organization, which may then be compared to the formal communication structures. These emergent patterns can be used to explain several organizational phenomena.

11.3 Conceptual View of People's Perceptions

Where a person lives in the world, the circumstances of life there affect the way that person views life. There are influences within the culture in which a person lives that affect his or her views on life. The world as a whole values life, yet it is the influence of a certain area's organizations and culture that influence the life system and how people see it.

Beliefs are usually conceptualized as a tacit set of often unconsciously held assumptions regarding educational issues and processes such as teaching, learning, curriculum, schooling, and knowledge [18]. Beliefs can be inferred from what people say, intend and do [36], and thus they can give insight into the reasons teachers act the way they do. In most cases the beliefs are spelled by attitude and perception held by an individual over something. Attitude permeates everything we do. It is an important component in all aspects of human endeavor. Attitude and perception influence whether we begin or continue with certain activities and whether we achieve in certain areas [32].

Perceptions have been explored by a host of researchers in varying degree [4, 13, 14, 25, 35]. A perception is a belief, theory, hypothesis, feeling, appearance, opinion, observation, insight, awareness, or sensitivity. It may or may not constitute reality, and initial perceptions often change with the passing of time, the changing of circumstances, or the receipt of additional information. The most important revelation here is that no two people experience and interpret sensations, situations, or their own feelings the same way [35]. According to Jackson [25], perception is reality. No matter how you view yourself, your profile as a leader will be constructed for you by those with which you work and interact. Based on that, knowing oneself is imperative; however, it may be that your ability to understand how others perceive you that will be a more powerful tool to help you be a better leader.

Green [20] writes that perception is everything; especially in the world of social media. In terms of perception, we all have an ideal self. We all wish to maximize our careers, our profession, and aspire to be like those who we find most successful.

Otara [35] contends that having the right perception is not only about becoming competent, polyvalent and productive but also about nurturing diversity and being able to live with all employees. Accordingly, there are very important things to consider in understanding the vital role of perception.

1. *Objective perception is difficult if not impossible. Each individual selects, organizes and interprets information gained from their senses and internal awareness differently.*
2. *In addition to mechanics of perception, it is also important to recognize that every person has a unique frame of reference that includes all of the internal and external factors that affect behavior. Interpretation of their experiences is dependent on many factors not directly related to their immediate situation.*
3. *It is important to recognize that values, beliefs and attitudes are important factors affecting perceptual process.*

4. *The amount of energy we expend accomplishing a task is directly related to our perception of its importance. This is significant in work environments. Unless employees perceive their job assignments as interesting and worth, while their level of effort is likely to be relatively low.*

Such views trickle down to the leadership process, as leaders are influenced by many outside issues; implicit and explicit understandings, ideologies, philosophies, theories, psychology, sociology among others [3, 15, 16]. So when comparing two or more people and their thinking, there are both common and differing perceptions about the idea of the leadership process.

This perception of leadership will directly influence the person's role in the organization [14, 25]. In organizations, perceptions of leaders, managers and employees shape the climate and effectiveness of the working environment. Perception is the way interpretation of experiences is made [35]. Most of the time people believe that they are effective and efficient leaders using their perception but their supposed followers may have a very different perception. Having the right perception is significant skill for any effective leadership. It is important to understand that perception is often portrayed through communication in any organization be it big or small and therefore it is a pertinent tool in leadership.

While in the research and practice of leadership we often focus on the leader as a person or his/her behavior, the role of followers is often neglected. However, followers constitute the context in which leaders operate and there is a long-standing argument, going back to Weber's charisma theory, that leaders can only be leaders if they have followers. So, one of the obvious questions to ask is why do people follow and how are leaders 'granted' influence [13].

Recent research is looking at followers' (and others') images or stereotypes of leaders (commonly known as implicit leadership theories) and how they influence how we see leaders. For example, if someone thinks of leaders as charismatic, almost any behavior shown by a person labeled 'leader' could be interpreted as a sign of charisma [4, 13]. This notion shifts the focus from the leader to the follower, in the sense that a leader's behavior is less important for the leadership process than the perception of that leader by his/her followers. The consequences for leadership development can be quite severe in that we train leaders to behave in certain ways but if ultimately followers do not perceive this behavior as 'leader-like', the leader will find it difficult to exert influence over his/her followers.

Equally, many recent studies have also come in handy to explain the perception and beliefs of employees about leadership and their influence on practice [14, 25]. Understanding of leadership therefore, must include a grasp of interaction with the environment around a leader. No matter what we view of our personal abilities or profile authenticity is measured when others working with you are open and willing to stand behind or with you [14, 25]. The way in which we interact with others and the task is the perception (and reality) of who we are as leaders. Not only should we be aware of this social construction, but you need to be a vigilant self-evaluator of this perception so that you can adapt and adjust as needed.

While you perceive yourself to be an effective leader even under stress, do your colleagues share this perception of you? Your perception of effective leadership may be shared by others who work with you. People in leadership may see a relationship between "leaders in title" and "leaders in action" from their own experiences, but this is not guaranteed. Understanding yourself is imperative [14, 25]. Embracing your strengths and appreciating others' perceptions of you help to be a better leader.

To ensure a strong study, however, a researcher draws on a multiplicity of paradigms to zero on one that is congruent with his or her beliefs about nature of reality [30]. Given ontological and epistemological beliefs, there are several points of departure along a spiral of methodological spectrum. Theoretically therefore, this paper builds on three major assumptions:

- People's beliefs and perceptions come from a variety of experiences, including their upbringing, life experiences, or schooling processes, yet the exact sources are still unclear [38]. Beliefs are tentative constructions and thus subject to revision.
- A person's view of the leadership process can present a major barrier to implementation of certain practices [22].
- Changing a paradigm is a complex matter. As Kuhn [26], indicated, paradigms control the methods, questions, and standards of a community, as well as the broader constellation of its cherished beliefs, values, and techniques. Changing educational views is therefore a gradual process and multiple conceptions co-exist in the transitional stage.

11.4 Social Media and Social Networking and Perceptions of Leadership

Education is an important means of interaction since humans use educational institutions to teach individuals mechanisms of dealing with needs in this world. However educational institutions cannot survive in a leadership vacuum. The leadership process therefore is regarded as one of the most critical undertakings of the education system and it is largely inextricably interwoven with the competence of the leaders and its effectiveness requires the comprehensive knowledge of the management underpinnings thereto. Therefore, the leadership process requires leaders' immense effort and followers' effective behavior. This however largely depends on what the followers consider viable for the institution which is synonymous with the way they perceive the leadership process. If the perceptions of employees about the leadership process are not worthwhile and thus incompatible with the general thrusts of the institution and education system in general, then effectiveness of leadership outcomes will be a myth than a reality since these perceptions will be translated into the actual task execution and behavioral practices.

Nevertheless the perceptions of the employees towards leadership are shaped by a number of critical factors ranging from the natural to the man made. However, none of the factors plays a defining role in contemporary leadership structures like social media and social networking. Indeed as the use of social media continues to evolve; the concept of presenting our ideal selves versus our real selves has become more and more prevalent on social media platforms such as Facebook, Twitter, Instagram, Google+ and even LinkedIn.

Sunstrum [43] writes that social media puts an interesting lens on the creation of the self, and how this construction affects our mental well-being. The ideal self is the self we aspire to be. According to Carl Rogers's theory of personality, every human has the basic instinct to improve herself and realize her full potential. Like Abraham Maslow, he called this achievement self-actualization. He believed this state was attained when the ideal self and the person's self-image were in line with each other. This person would be deemed a fully functioning person.

The patterns of relationships bring actors' similarities with the attitudes and behaviors of other organizational members; social network analysis may explain why members develop certain attitudes toward the organization. Distributed leadership and social network analysis have equally emerged as promising strategies for assessing the presence of and effectiveness of leadership in schools, examining the social and relational aspects of school actors, and identifying leverage within a school's networks to build capacity in leadership and instruction. This kind of analysis has the potential to provide schools with important insights about leadership and instructional capacity for improvement/reform.

Dowding [15] while exploring the aspect of perceptions of leadership contends that it is important to establish how leadership is perceived by people and why some might be thought to be 'strong' or 'weak' given their actions in the context in which those actions take place. Dowding [15] thus makes the following conclusions:

- Patterns of relationships can lead observers to perceive some leaders as weak and others strong.
- How we view individual traits in terms of leadership characteristics is, in part, dependent upon the context in which those traits emerge.
- Furthermore, once those traits are perceived by people, including the subject themselves, they further develop as traits.
- Perceptions of someone as a strong leader can make that person a strong leader or, perhaps more pertinently, once someone is labeled as a weak leader there is little that they can do to lose that reputation or not be a weak leader.
- Actions that would have appeared strong if carried out by someone with a reputation for strength will be viewed as the actions of a weak person, perhaps trying to be strong.
- Reactions of others to those actions will depend upon how they view the leader.
- Someone who is believed to be strong and shouts at a subordinate might lead that subordinate to quail and obey. The shouting leader the subordinate believes is weak might be laughed at, further weakening him.

Conery [12] meanwhile examined the relationship between social networks and leadership. He thus contends that social networks help to account for the distribution of leadership and colleagues' access to potential resources and expertise. Accordingly, the following conclusions came up:

- Access to potential resources is related to each colleague's social ties.
- Access to the expertise of colleagues through the school's social networks is limited.
- The pattern of advice-seeking and communication is largely one-directional
- There is variance between social networks based on subject matter and proximity.

Social media and social networking influence the communication structure of an organization. The use of network analysis techniques distinguishes structural features such as the (formal and informal) communication patterns in an organization or the identification of groups within an organization (cliques or functional groups). Thus information flow between members can be determined. Characteristics that can be investigated using network analysis techniques are the following:

- Communication load as perceived by employees
- The communication styles used and
- The effectiveness of the information flows

Realistically therefore, the perceptions of leadership premised on social network and social networking are influenced by how either party views, values or partakes in the process of the media. When an individual is part of social media and social networking systems, he or she bases all subsequent views on that and the other way round. Accordingly, perceptions of leaders based on social media and social networking systems can be viewed from the following perspective:

- Leaders who value social media and social network will always allow and install social media-laden equipment within their institutions or organizations for people thereof to use at all times. This can turn them into close allies of the members in the organization as a result.
- Leaders who see less value in social media and social network will not only reject installation of the social media but may even limit freedom of utilization of the same by others. Such leaders may be perceived as archaic by those around them.
- Leaders who utilize social media and social network understand the implication of the same and would even go an extra mile to promote it amongst their followers. Such leaders may be perceived as modern in the eyes of their followers.
- Leaders who do not utilize social media and social network cannot know for sure what they mean to others. They may actually criticize those who utilize the same and end up being labeled autocratic or dictatorial by their followers.

This however does not only reflect the leaders and their attachment to social media and social networking systems. Sometimes the perceptions are a making of the employees or followers' attachment to social media and social networking systems. In this case the way followers or employees value or perceive their leaders at all times.

- Followers who value social media and social network would wish to have a leader who approves of the same and actually hold them in high esteem
- Followers who do not value social media and social network always criticize those who love social media and social network. If it is leaders then they would always undermine them as weak or useless. They may doubt their ability to steer the organization to greater heights
- Followers who utilize social media and social network would want freedom and acceptance from their bosses and would feel enthralled by leaders who also utilize social media and social networking sites. They find them more of family indeed.
- Followers who do not utilize social media and social network may never understand why certain people behave in a certain way and act in a certain way. They may thus command unrealistic perceptions of such people. If it is leaders then the trust and respect they accord them may also be less or inappropriate.

Important to recall and note too is the fact that the nature and mode of utilization of the social media and social networking will also have an influence on the way a person perceives a leader. In such cases, issues of being too lousy, modern, responsible etc. may crop up based on which social media and for what reason, a leader utilizes social media and social networking systems.

- The type of social media and social network sites utilized by the leader would also affect or influence the way people perceive the leader and therefore, leadership itself. Some of the social media and social network sites are very demanding and would not be ideal for leaders but if the leader adopts them, then he or she may be perceived differently by the people around
- The frequency of utilization of social media and social network by the leader also influences the way people around the leader may perceive him or her. If it is extreme then such a leader may be perceived as flamboyant
- The forms of utilization of social media and social network i.e. for what is social media utilized? If a leader utilizes the social media and social network sites in professional or formal forms, then such a leader would be perceived as responsible. However if it is informal, then perceptions might take a weird direction indeed.
- The leader's involvement of followers in the utilization of social media and social networking. Sometimes a leader may utilize social media and social network sites by him or herself but other times could involve the people he or she leads. Any decision would influence how the leader is perceived by the followers.

The leadership styles as we know them ranging from the customary ones (Democratic, dictatorial and laisez-faire) to the more contemporary ones (Transactional, transformational and charismatic) as well as certain specificities in leadership (people leadership, service leadership, action leadership, goal oriented leadership etc.) cannot stand on their own and indeed may not be existent *per se*. What exists, therefore, is how people perceive the leader. It is therefore not surprising that a

Table 11.1 Effect of attachment to social media on the classification of leadership traits by followers in an organization

Attachment to social media and social networking systems	Plausible perception of leadership style
Attaches value to social media and social networking systems	Modern, democratic, transformational
Does not value social media and social networking systems	Archaic, dictatorial
Utilizes social media and social networking systems	Modern, transformational, transactional (Interactive)
Does not utilize social media and social networking systems	Archaic, dictatorial

leader who seems to be democratic in his dealings may end up being perceived as dictatorial based on how the followers view his or her activities in the realm of leadership.

The table below shows how attachment to social media can influence the classification of leadership traits by followers in an organization (Table 11.1).

Important to note is the fact that these perceptions may not necessarily reflect the real leadership trait of an individual leader but they may persist and remain the basis of understanding and conceptualizing the leader within the organization at all times.

11.5 Conclusion

Perceptions of the employees towards leadership are shaped by a number of critical factors ranging from the natural to the man made. However, none of the factors plays a defining role in contemporary leadership structures like social media and social networking. Indeed as the use of social media continues to evolve; the concept of presenting our ideal selves versus our real selves has become more and more prevalent on social media platforms such as Facebook, Twitter, Instagram, Google+ and even LinkedIn.

Social media puts an interesting lens on the creation of the self, and how this construction affects our mental well-being. The ideal self is the self we aspire to be. Leaders should thus be steadfast when it comes to how they conduct themselves and relate with others in the organization since how those around them view leaders is critical to the effectiveness of the leadership setting in the system. Crucially though, the way the leaders undertake to exploit the social media and social network systems is very profound given that these shape the way people perceive their leaders at all times. There is no gain saying therefore, that leaders and leadership may turn out to be victims of social media and social network, yet they may ride on the back of the same once properly and appropriately understood, conceptualized and utilized.

References

1. Barnat R (2013) Strategic management: formulation and implementation. http://www.introduction-to-management.24xls.com/en220
2. Bedeian A, Zamnuto R (1991) Organizations: theory and design, vol 9. Dryden, Chicago
3. Binmore, K (2007) Playing for real: a text on game theory. Oxford University Press, Oxford
4. Birgit S (2012) Leadership: why perceptions matter. http://www.trainingzone.co.uk/topic/perceptions-leadership/168409
5. Borgatti SP, Foster P C (2003) The network paradigm in organizational research: a review and typology. J Manag, 29:991–1013
6. Brass DJ (1992) Power in organizations: a social network perspective. Res Polit Soc 4:295–323
7. Burt RS (1976) Positions in networks. Soc Forces 55:93–122
8. Burt RS (1982) Toward a structural theory of action. Academic Press, New York
9. Burt RS (1992) Structural holes: the social structure of competition. Harvard University Press, Cambridge
10. Cahill C, Rios-Moore I, Threatts T (2008) Open eyes-different eyes: PAR as a process of personal and social transformation. In: Cammarota J, Fine M (eds) Revolutionizing education: youth participatory action research in motion. Routledge, New York
11. Comscore (2011) http://www.brafton.com/news/comscore-nine-in-10-americans-visit-social-sites-each-month-800393454
12. Conery MJ (2012) Mapping the terrain: a social network analysis of elementary teachers' perceptions of school leadership practices. Northern Arizona University, 2012, 378; 3548511
13. DeRue DS, Ashford SJ (2010) Who will lead and who will follow? A social process of leadership identity construction in organizations. Acad Manage Rev 35:627–647
14. Donna V (2011) Perception is reality: your strengths matter. J Leadersh Educ 10:1
15. Dowding K (2007) Perceptions of leadership. Public leadership in Australia and beyond' conference, 29–30 November, ANU
16. Dowding K (2008) The civil service. Routledge, London
17. Efron L (2014) The secret to self-management and organizational success. http://www.forbes.com/sites/louisefron/2014/08/29/the-secret-to-self-management-and-organizational-success/
18. Elen J, Lowyck J (1999) Meta-cognitive instructional knowledge: cognitive mediation and instructional design. J Struct Learn Intell Syst 13:145–169
19. Fong K (2010) The five elements of organizational success. A framework to analyze organizational systems, cultures and leadership styles within a context of change. http://elementalpartners.net/downloads/5_Elements.pdf
20. Green RK (2013) The social media effect: are you really who you portray online? http://www.huffingtonpost.com/r-kay-green/the-social-media-effect-a_b_3721029.html
21. Garmston R, Wellman B (1995) Adaptive schools in a quantum universe. Educ Leadersh 52(7):6–12
22. Gunstone RF (1994) The importance of specific science content in the enhancement of meta-cognition. In: Fensham P, Gunstone R, White R (eds) The content of science. Falmer, London, pp 131–146
23. Gunter H (1995) Jurassic management: chaos and management development in educational institutions. J Educ Adm 33(4):5–30
24. Hendricks D (2013) The complete history of social media: then and now. Retrieved 20 August 2013, from http://Smallbiztrends.Com/2013/05/thecompletehistory-of-socialmedia-infographic.html, 8 May
25. Jackson D (2011) Leadership: the only reality we have is perception. http://www.jordantinney.org/leadership-the-only-reality-we-have-is-perception/
26. Kuhn T (1970) The structures of scientific revolutions, 2nd edn. Chicago University Press, Chicago
27. Lincoln JR (1982) Intra- (and inter-) organizational networks. Res Sociol Org 1:1–38
28. Lortie D (1975) School teacher: a sociological study. University of Chicago Press, Chicago

29. Maltz' AC et al (2001) Defining and measuring organizational success: toward a dynamic, multi-dimensional model. In: Proceedings of the management of engineering and technology, 2001. PICMET'01. Portland international conference, 1
30. Mills JBA, Francis K (2006) The development of constructivist grounded theory. Int J Qual Methods 5(1)
31. Mckinney P (2013) Organizational success: factors, definition. http://study.com/academy/lesson/organizational-success-factors-definition-quiz.html
32. Morgan P (2008) Teacher perceptions of physical education in the primary school: attitudes, values and curriculum preferences. J Phys Educ 65(1):46–56
33. Nations D (2013) What is social media? And what are social media sites? http://webtrends.about.com/od/web20/a/social-media.htm
34. Newman MEJ, Barabási A-L, Watts DJ (eds) (2006) The structure and dynamics of networks. Princeton University Press, Princeton/Oxford
35. Otara A (2011) Perception: a guide for managers and leaders. Kigali Inst Educ. 2 (3):21–24 doi:10.5430/jms.v2n3p21
36. Pajares MF (1992) Teachers' beliefs and educational research: cleaning up a messy construct. Rev Educ Res 62(3):307–332
37. Radford M (2006) Researching classrooms: complexity and chaos. Br Educ Res J 32(2):177–190
38. Raths J (2001) Teachers' beliefs and teaching beliefs. Early Child Res Pract 3(1).
39. Reilly DH (1999) Non-linear systems and educational development in Europe. J Educ Adm 37(5):424–440
40. Reis HT, Collins WA (2004) Relationships, human behavior and psychological science. Curr Dir Psychol Sci 13:233
41. Strogatz SH (2001) Exploring complex networks. Nature 410:268–276
42. Sullivan TJ (1999) Leading people in a chaotic world. J Educ Adm 37(5):408–423
43. Sunstrum K (2009) How social media affects our self-perception. http://psychcentral.com/blog/archives/2014/03/14/how-social-media-affects-our-self-perception/
44. Walker L (2013) Social media definition: what are social media? http://personalweb.about.com/od/socialmedia101/a/Social-Media-Definition.htm
45. Warigon SD (2014) Managing people for organizational success. https://www.itmpi.org/Portals/10/PDF/WarigonPeople.pdf

Chapter 12
Political Violence in Plural Democracies: A Comparative Study of Uganda and Kenya

Sultan Juma Kakuba

Abstract This paper examines the nature, roots and implications of political violence in particular and the political system in general, in both Uganda and Kenya. It also looks into how these vexed issues are addressed within the framework of a democratic practice in Uganda and Kenya. Thus, the paper analyses the issues in a comparative perspective, while recognizing the exclusivity and specificity of the challenges involved in a specific national context. It also underscores the significance of creating a conducive and just socio-economic and political environment for promoting Political trust and tolerance. Besides, it is known that a violence-laden situation is one that drives society to the edge of chaos (which in chaos theory is the right place to be) where a system engages mechanisms of self-adaptation and transformation. Once properly negotiated, this chaos or violence may create a basis for effectiveness in all spheres of society. There is no gain saying therefore, that the political violence in Kenya and Uganda can be a basis for effective political transformation in the said countries if appropriate measures are taken to arrest the situation. Realistically therefore, understanding political violence in plural democracies becomes a sine-qua-non in contemporary political science studies.

12.1 Introduction

Political violence is common throughout the world-over, and Kenya and Uganda are not an exception. One of the major problems of plural democracies today particularly in Africa is how to come up with an appropriate framework to promote Political coexistence with the opposition. Both Uganda and Kenya share this need. In Uganda, since the early years at independence, the country has witnessed several political conflicts as a result of struggle for power. These conflicts have resulted in loss of thousands of lives, mass destruction of property and posed a threat to political pluralism and harmony. The political violence in fact posed the greatest threat to the unity and corporate existence of the country. Tackling these and related violent

S.J. Kakuba (✉)
Department of Political Science, Islamic University, Mbale, Uganda
e-mail: sultan_juma@yahoo.co.uk

Ş.Ş. Erçetin (ed.), *Chaos, Complexity and Leadership 2014*,
Springer Proceedings in Complexity, DOI 10.1007/978-3-319-18693-1_12

eruptions especially political violence that ensue during elections in the country should obviously be a major area of concern not only for policy makers but social scientists as well.

On the other hand, Kenya after experiencing mature and peaceful politics since its independence in 1963, the political landscape was threatened with similar problems like that of Uganda immediately after 2007 general election. Over the years and especially in recent times, the secular fabric of Kenya has come under severe stress. The Kenyan nation has witnessed hostile political experience to political harmony. These ranged from the very severe tribal crises between Kikuyu and the Luo after disputed general election where it is believed Raila Odinga won but was deprived of his victory through manipulation. The uglier outbreak of this political violence unfolded in Kisumu, the western part of Kenya.

Many considerations come to the fore in explaining political violence and tensions in both Uganda and Kenya. In all cases, lust and the urge to cling on power are a major cause. The socio-economic and political factors must also be taken into account. Nnoli [5] observes that conflicts can realistically be explained in terms of sharp economic and political competition among different religious, ethnic or communal groups.

Furthermore, crises clearly represent the failure of the state to live up to expectations. The failure of the state to create the required atmosphere that is conducive to create political trust and stability is clearly apparent. Given the complex and sensitive nature of communal conflicts as well as their serious negative implications for society, there is a pressing need to evaluate the strategies that need to be adopted to promote political stability and development. Given the fact that both countries Uganda and Kenya are multi ethnic and religious population, they both need to evolve long term citizen centered measures to promote lasting peace and stability

This paper examines the nature, roots and implications of political violence in particular and the political system in general, in both Uganda and Kenya. It would also look into how these vexed issues are addressed within the framework of a democratic practice in Uganda and Kenya. Thus, the paper analyses the issues in a comparative perspective, while recognizing the exclusivity and specificity of the challenges involved in a specific national context. It also underscores the significance of creating a conducive and just socio-economic and political environment for promoting Political trust and tolerance.

12.2 Uganda and Kenya: Basis for Comparison

The interest to undertake this comparative study takes up from the fact that both Uganda and Kenya share so many things in common and can therefore benefit from each other's experiences. Both Uganda and Kenya were colonized by Britain and only attained independence in 1962 and 1963 respectively. The long period of colonization had left an indelible mark in the socio-economic and political developments of the two countries, not the least of which is the sowing of the roots of ethnic tensions in the two countries. As third world countries, the two countries are facing the same challenge of fashioning out stable and economically developed polities. Both are democratic politics, though in Uganda, the military had ruled the country for some time unlike in Kenya which has never experienced military rule.

No doubt Uganda and Kenya are pluralistic democratic societies. But interesting to know is that these countries are important in East Africa Community because they have both played a major role in the socio-economic and political affairs affecting the East African region. It is thus the case that both Uganda and Kenya share a number of things in common, including the fact of the significant roles they played in International organizations such as the East African Community. In spite of whatever differences there may be, there is a convenient basis for comparison between the two countries especially that both countries stand to benefit from each other's unique experiences.

12.3 Democracy and Pluralism in Kenya

Since the attainment of independence from Britain, Kenya has operated a democratic system of government with KANU dominating political power for many years. Its political system was modeled on political establishment by British system its former colonial master. Unlike in Uganda, Kenya has never been ruled by the military. Kenya, as will be made clear in due course, is a plural democracy and in the last few decades, has witnessed the resurgence of post-election violence. Democracy as a system of government has been viewed by successive Kenyan leaders to provide the best framework for addressing the problems of a plural society. The logic is clear; democracy is premised upon the assumption of majority rule and minority rights. Minority rights are supposed to give them not only a sense of belonging but also the desired guarantees to live in peace and harmony with the majority. The constitution of Kenya has provided for an array of rights to be enjoyed by all citizens. They ranged from right to life, property, liberty and so on. Specifically, minority rights are provided and guaranteed for. The constitution also provides for the equality of all before the law, the prohibition of discrimination on grounds of religion, race, tribe, sex or place of birth and equality of opportunity in matters of public employment. This involves the freedom of conscience and free profession, practice and worship. Importantly, the Kenyan also constitution provided for cultural and educational rights.

Kenya has long made itself a multi-ethnic secular country accepting all religions. But secularism in Kenya does not follow the western root of the concept which imply that the state should have nothing to do with religion or that 'things' religious should not necessarily be brought into the domain of what is political. Secularism in Kenya is premised upon the non-adoption of any religion as a state religion, instead the government regards all religions with equal treatment and of their followers as well as the provision of the right to freedom of conscience and religion. But the democratic secular arrangement in Kenya is facing serious challenges. The greatest of these is manifested in the post-election violence, the very problem which broke the long political stability and directed democracy in this country supposedly needs appropriate measures to address similar occurrence.

12.4 Kenya Post-election Violence: Nature, Causes and Consequences

Unlike in Uganda, the 2007 post-election violence is the major violence Kenya experienced since it attained independence. This violence has impacted upon the socio-economic and political landscape of modern Kenya. This affected perfect harmony existed before hand among the different communities of Kenya. Scholars have described this violence as a remarkable blow to the politics in Kenya. Since then, incidents have continued to occur that suggest Kenya will never be the same as before. The number of people killed and injured in the post-election violence was alarming.

But the post-election violence in the Kenya is the latest and most complex in Kenya's recent politics. Though Kenyans have been known for practicing peaceful politics, post-election violence was immediately sparked up by the Kikuyu carnage in which the Luo were the target. It was a gruesome episode. Cheeseman [2] reveals that before the 2007 election, government security forces killed a total of 500 members of Mungiki around Nairobi as well as an estimate of 600 persons died during election campaigning. The consequences were disastrous. Officially, over 1000 have been reckoned dead, most of whom were Luos and Kikuyu. The killings were also brutal and cruel as many were burnt alive. Women were also assaulted and gang-raped. Houses, business premises, and other properties were destroyed. Economic activities were brought to a halt. Mutual fear and hatred intensified as communities affected became more polarized into rival combatant groups.

Not unexpectedly, the Kenya post-election violence attracted the attention of members of the international community. Foreign missions in Kenya principally drew similitude of this violence with what happened. It saw it as pre-planned and the pattern suggested that the attempt was to purge kikuyu and the Nandi from the centre of politics in Kenya. It expressed concern that the scale of violence and deaths may have been greater than documented. So serious was the post-election violence on the image of Kenya that the government reached by asking foreign missions to intervene in providing humanitarian assistance. The key actors in the post-election violence were mainly those from strongly-hold powerbroker based on ethnic animosity [1].

Two major developments kept the post-election violence alive 2 months after their outbreak. One is the continued violence itself though to a much lesser degree. Minor incidents were more often than not reported. The second issue is related to the debate within and outside of government on ethnic ground. The Luo consistently criticized government's dominance in handling of the "national cake"; they were therefore struggling for change in the status quo.

A number of factors have come to the fore in explaining the occurrence of post-election violence in Kenya in 2007 after experiencing a long spell of stable politics it is arguably the keen competition among the Kenyan elites lead communities, into aggravated demands of political change. This competition took place at different levels. One is political and the other being economic. In the former, the concern is with who wields which political power while the latter is mainly over control of the resources of the state. The two are closely interconnected. Poverty at the level of the lower classes made them easy prey for mobilization towards post-election violence.

The post-election violence in Kenya started simply as demonstration stage-managed by opposition and their allies but it later spread to different parts in the country and turned out to be like ethnic struggle between the Luo and Kalenjin supporters of Odinga on one side and the Kikuyu mainly the supporters of Mwai Kibaki. This made security forces to respond with maximum force in an effort to quell the violence. According to Kriegler report [4] stress that use of hatred language during election campaigns contributed greatly to the outbreak of this violence. The report emphasizes that the election was characterized by harassment of supporters of different political parties. Earlier in the 1990s, Moi was strongly opposed to multi-party politics, arguing that it was a major threat to unity among citizens in the country. In this regard, the Moi government encouraged citizens to arm themselves to deal with decisively with multi-party supporters [3]. In addition, the government used violent dehumanizing language against the opposition which openly called for violence. Political issues that would occupy election campaign such as health, education and generally development of the country were hardly raised.

The media, which coincidently could play an important role in combating post-election violence, had unwittingly compounded the problem. The daily coverage of violence events by both the electronic and print media often revealed the gruesome aftermath of the events to people hundreds or thousands of kilometers away in the country particularly the western part. According to the United Nations Development Program report [8], it argued that the media was instrumental in sharpening ethnic divisions which whetted ethnic violence. This did not only create awareness of the happenings in areas with violence but also raised passions and generated chain reactions. This reinforced the identity crisis. So important is this question that often community or ethnic identity (kikuyu or Luo or Nandi etc.) tended to overshadow national or regional identity. This problem became widespread in areas affected with violence and individuals preferred to live in areas where they felt secure [1].

The secular fabric of Kenya came under severe stress arising from this disastrous violence. Constitutionally, there could have been no problem with this is there was guarantee for power sharing among the different ethnic groups in the country.

Post-election tensions had arisen because of the failure of the state to manage the elections. If the state was not seen to effectively and fairly deal with post-election violence, people would take the law into their own hands. This led to the rise of local dissatisfaction. Secondly, the failure to seriously tackle the different socio-economic problems of society such as unemployment and poverty to create a reservoir of people that could easily be mobilized with no pay to cause violence. Thus, it is not surprising that many poor people particularly the peasants, petty traders and other lower social classes actively got involved in the post-election violence. In that regard, people from different communities lost their property and lives and many were left in perpetual fear.

What is apparent from the foregoing is that post-election violence in Kenya did not necessarily occur solely on account of poor election management but over issues of creed, norms or practices by the government of the day in Kenya at the time exploiting to fit the vote bank politics of the politicians.

12.5 Democracy and Pluralism in Uganda

Uganda attained independence on October 9th 1962. Since that time, one of the major problems that have threatened the existence of the country has been the rising tide struggle for political power. The more recent ones, since the country's 5 year NRA guerrilla war (1980–1985) include the Lord Resistance Army war. This war proved complex in scale and disastrous in consequences. Handling it by governments called for a proper understanding of not only the nature, basis, intensity and implications of the war but also of the handling by the authorities of the situations.

Uganda has been ruled in its political history by both the military and elected civilian democratic leaders. Uganda's estimated population of 33 million is made up of over 60 ethnic groups with the major linguistic/ethnic group being the Baganda. This major group predominates in the central region of the country.

The core aspect of Uganda's political system is the formal division of powers among levels of government i.e. national and local government; each performing some exclusive functions in some sphere while cooperating with others in carrying out others. The significance of this is that it allowed the local government units to deal with their unique issues delegated to it by the central government, while the central government remains the overall in handling most of the state affairs. The Ugandan constitution 1995 provides for central government powers to legislate on national matter, while local government councils are mandated by the constitution to make bye-laws on issues they deem necessary.

Second, to further create a sense of belonging, the 1995 constitution makes provisions for representative democracy so that the government at the center has representatives from different constituencies in the country. There is also a provision under the principles and directives of state policy which states that neither the state nor any of its constituent parts is to adopt any religion as state religion. This is the section upon which Uganda is said to be secular; otherwise no mention of the word has ever occurred in the constitution. Other relevant sections describe the kinds of rights and freedoms which Ugandans are entitled to. These include the right to freedom of belief and conscience, the right to profess, practice and propagate any religion and of the non-discrimination of persons solely on account of religion, race, and tribe among others.

Within the above framework, the Ugandan has sought to manage the problems arising from pluralism. A pertinent question is why there has been civil war in spite of the consensus by many writers that democracy provides the best frameworks for dealing with plural societies? The next part is devoted to this phenomenon of conflicts and consequences.

12.6 Violence/Demonstrations in Uganda: Nature, Bases and Consequences

Two kinds of conflicts that have taken place in Uganda are addressed here. They include civil war (Lord Resistance Army – LRA, war) and political demonstrations. The LRA war is that civil war which was waged by LRA for over 20 years led by Joseph Kony against the NRM government, while political demonstrations took place mainly after election most notably the "walk to work" launched after 2011 elections. The immediate causes of these kinds of political conflicts include the sharp disagreements over doctrines of political beliefs and power sharing. The LRA war has dragged on for over 20 years posing serious security threat in Uganda. In recent times, the most critical one, the "walk to work" violence took place in the Kampala the capital city and business center district. Included in this category are chains of demonstration that raked different parts of the country in protest for government failure to arrest the increasing price of commodities in the country.

In addition, a series of strikes by different groups in the country such as teachers demanding for an increase in their pay; strike by business community (traders) upon the decision by the central bank to increase lending interest rates; sit down strike stage managed by lawyers over police brutality; tax drivers' strike against Kampala City Capital Authority (KCCA) new policy demands and strike by academic staff at the oldest institution of higher learning (Makerere University) pressing for improvement in their pay, all have undermined NRM re-election into power for the fourth term.

Important to acknowledge is that when Uganda transited into multi-party election in 2006 since 1980 after the referendum endorsed its re-introduction, the ever-first multi-party election held under National Resistance Movement (NRM) witnessed violence which claimed 17 lives [7]. In terms both of complexity and the enormous destruction that was caused to life and property, by both the LRA people were attacked and killed on account of their failure to give information about government forces or for being government collaborators. On the other hand, those people involved in the walk to work protest, beaten, tortured and arrested by the police; vehicles, shops, houses, were set on fire, lives lost and the scale of destruction unimaginable. This occurred after the February 18th 2011 elections.

Polarisation along the pro-and against re-election of NRM for the fourth term in office assumed a very dangerous proportion which became the most weapons for the opposition to mobilize and attract the attention of the public. On the whole, most violence took place in the central business district, Kampala. All the demonstrations took place in urban areas with the urban poor being the worst victims. The foregoing demonstrates the fact of confined political violence in Uganda. The most critical issue is how could the high incidence of political violence in both Uganda and Kenya be lessened or solved? The next section addresses this.

12.7 Towards Promoting Communal Harmony

Understanding of why political violence occurs and how it develops and facilitate the devising of strategies and procedures whereby their negative effects could be mitigated. The most important issue in addressing political violence is devising an appropriate policy that will constructively deal with political justice within the framework of democratic governance, justice and respect for human rights. Given that political violence of recent, it has continuously shaken the political foundations of both Uganda and Kenya; this should be accorded significant attention. These countries have adopted policies to deal with the pluralistic nature of their respective countries. These include constitutional provisions relating political pluralism, to equality of all and of rights to freedom of belief and conscience as well a number of socio-economic and political measures designed to give a sense of belonging to all.

A major question is whether the recurrence of political violence could be termed as not only signaling the failure of the measures adopted, including the democratic framework for resolving problems of plural societies. The matter is that this is not necessarily so. Without the measures adopted, including the potentials offered by the democratic framework, many countries including Kenya and Uganda would have had to grapple with political violence that are of greater complexity and intensity. What perhaps is more important is the failure of respective governments to effectively handle pluralistic issues and political violence within the stipulated frameworks and in ways that are consistent with justice and fairness. In other words, the major problem has been with the practitioners of politics-civilians or military.

One of the most important measures that could be taken to arrest political violence during election period are preventive. A number of issues or considerations are involved here. First is the need to forecast and prevent political violence. Political violence did not just emerge full-blown; they undergo period of incubation. It is imperative for early warning mechanisms to be developed so that potential political violence especially during elections is nipped in the bud. The second preventive measure involves addressing those issues that makes violence feasible. The most important of this is poverty, a feature that is endemic in both Uganda and Kenya. Statistically most Ugandans and Kenyans are poor or poverty stricken.

Interesting to know is that many of the political violence that took place in the two countries, it is evident that most of those affected could not meet the lowest of psychologist Abraham Maslow's hierarchy of human needs. These ranged from food, shelter and security, through self-esteem to self-satisfaction. Yet basic needs must be satisfied either way. Not only do political violence arise from unmet needs but that hunger, poverty and despair can provide a fertile ground for mobilizing people around narrow vested interests and act as a major source of conflict. Related to this is the resolution of the crises generated by urbanization and modern life.

The next set of issues relevant for promoting harmony has to do with the attitude of the state in respect to management of political violence when it occurs. States or governments need to deal with political violence in a neutral manner. Of course, in general, states rarely are innocent or neutral in political violence either in their causation or in the way such political violence are handled. Nonetheless, it is significant that any charge of government bias could worsen the matter. Post-election violence that took place in Kenya, accusation of bias by security agencies and government in the arrests, prosecution and punishment of accused persons played a major role in the emergence of political violence that took place. The accusation of bias against the Kibaki government in its handling of the opposition ethnic clashes had also worsened.

The same problem is created by politicization of conflicts. While life and property were being wasted in the Kenya in 2007 election, the government was trading blames on this post-election violence to opposition political parties. Kriegler correctly summarized that post-election violence in Kenya reflects the keen competition by so called leaders of ethnic groups to control the power and resources of the state. It is thus the case that political violence partly is the attempt by dominant elites, politicians or others, who impose, preserve or extend their hegemony. The conflict among rival claimants to the power and resources of the state is translated into confrontations between this and that ethnic or political group, that majority and minority and that dominant and subordinated marginalized group, among others. As noted by Stravenhagen [6] and Nnoli ([5]: 1), conflicts that appear as confrontations between ethnic groups are in fact mostly confrontations between politically mobilized ethnic groups and an existing state. The point being made is that governments must stir away from politicization of conflicts and handle them with all the seriousness they deserve.

Ultimately, states face a political moral question which is directly related to their legitimacy. Protection of life and property (some will add liberty) is the most important role of the state. If such a state could not guarantee life and property, the question can as well be asked as to whether it has any legitimacy to demand obedience from the citizenry. States therefore faces the moral obligation to address the issue of conflicts just like they do when the sovereignty of the nation is threatened.

The foregoing shows some of the measures that governments could adopt to promote harmony. Beyond the above, however, the question of stemming the tide of conflicts is a cooperative one involving not only governments but also the civil society and members of the international community. The most important role citizens could play is to avoid being used in communal violence. There is already a sign of this in both Kenya and Uganda.

The media could also play a vital role by not only objectively reporting on conflictual situations but also helping with ways through which issues could be resolved. The media could also point to issues or problems capable of leading to conflicts. The point being made is that the task of combating conflicts most be seen as a cooperative one involving not just states or governments but the wider society as a whole.

We can conclude the discourse on communal harmony by looking at the international dimension to the issue. Today, more than ever before, issues of conflicts, human rights and so on have ceased to be matters solely of domestic attention. The international community has increasingly become concerned with such matters as well. A decade ago (1992 to be precise), the General Assembly of the United Nations adopted the declaration on the Rights of Persons Belonging to National or Ethnic, Religious and linguistic minorities. This signifies the continued concern of the international community over the rights of culturally distinct peoples who live in independent states but who do not fully identify with the dominant cultural model of the nation-state.

12.8 Concluding Remarks

The last few decades have been characterized by so many political conflicts in both Uganda and Kenya. In Uganda, the more serious ones include the LRA, ADF, Walk-to-Work etc. and the main violence in Kenya include 2007 post-election and mungiki as well as threats of Al-Shabab controversy and the eventual bombings and attacks. In both Uganda and Kenya, different ethnic groups have competed with one another for recognition, equality or autonomy (like the attempted overthrow of government by LRA from Uganda) within the existing territorial state. Ethnic groups in both countries have been an important feature of the conflicts. Notwithstanding that most of the conflicts examined in this paper assumed ethnic and socio-economic and political form, it is evident that economic considerations are key in the emergence of such conflicts. Most participants/victims are urban marginalized poor and unemployed youth with the middle or upper class only watching and or directing dastardly act from a far.

The state could not be exonerated in the rising tide of conflicts. Conflicts in the two countries have largely been manifestations of elite struggle for the control of the power and resources of the state. The same elites, as actual or potential rulers of the state have used conflicts to serve their vested interests by peddling their personal influence. This is further demonstrated by the laxity with which conflicts are handled, compared to, for instance, threat of external aggression. Yet, the greatest challenge to security in Uganda and Kenya today appears to be within. The consequences for allowing conflicts to take root have been enormous. Several thousand lives have been lost in the post-election violence and Mungiki in Kenya, LRA, ADF, Walk-to-work etc. in Uganda respectively to mention a few. More terrible has been the gruesome and bizarre nature of the killings of innocent people, women being gang raped and then burnt, for instance. Massive destruction of properties have also been witnessed. Conflicts also made sustainable and balanced development impossible especially that resource for development has been diverted for reconstruction and investment declined in the absence of peace. Pluralistic Democracy has remained pertinent in providing the appropriate framework for addressing political violence highlighted in this paper.

References

1. Branch D, Cheeseman N (2009) Democratization, sequencing and state failure in Africa: lessons from Kenya. Afr Aff 108(430):1–26
2. Cheeseman N (2008) The Kenyan elections of 2007: an introduction. J East Afr Stud 2(2):166–184
3. Haugerud A (1997) The culture of politics and modern Kenya. Cambridge University Press, Cambridge
4. Kriegler Commission, Report of the independent review commission, Sept 2008. http://www.hackenya.org/index. Accessed 29 Dec 2014
5. Nnoli O (1997) Ethnic politics in Nigeria. Fourth Dimension, Enugu
6. Stavenhagen R (1996) Ethnic conflicts and the nation state. MacMillan, London
7. Stremlau N, Price EM (2009) Media, elections and political violence in Eastern Africa: towards a comparative framework: an Annenberg-oxford occasional paper in communications policy research. University of Oxford Centre for Global Communication Studies, London
8. UNDP (2008) Media monitoring: the experience and future. Strategic Public Relations Research Ltd, Nairobi

Chapter 13
Assuming an Epistemology of Emergence: Classrooms as Complex Adaptive Systems

Chase Young

Abstract This review of literature was conducted to identify practical implications of complex adaptive systems in the classroom. The article uses complexity thinking to analyze conditions for emergence. Emergence is understood in this context as a "teachable moment", and in order for students to capitalize on these many moments, the conditions for emergence must be set by the teacher and experienced by the students.

13.1 Introduction

I think cliff-diving is great. It may not be my passion, but it is a good time when the opportunity presents itself. And yes, it presented itself. First, consider the context. My teacher was not coaching me on the edge, nor were my friends—in fact, they simply yelled intelligibly. How did I learn this? Was I born with an innate ability to jump off high places? What a strange gift, but no, it was a moment of emergence. All my prior experiences from other situations came together to create a unique one—cliff-diving. Educationally speaking, it resembled a "teachable moment". Fortunately, I possessed the agency to capitalize on the moment and execute the jump. The intense learning occurred at the precise moment I shifted my weight, bent my knees, prepared my mind, cast away my fears, swung my arms, and pushed off the rocky edge. Retrospectively, it was a complex process, especially having never done it before. Although the first jump was simply a matter of survival, subsequent jumps used related information to increase my proficiency. A nonlinear, dynamic process seemed to emerge from nothing—a new world was revealed. It is this reason, and the intense learning that occurs in the "teachable moment" that I have assumed an "emergentist" epistemology. I believe that "knowledge emerges from our transactions with our environment and feeds back into this same environment, such that our environment becomes increasingly meaningful for us (p. 223, [14])."

C. Young (✉)
Department of Educational Leadership, Curriculum and Instruction, Texas A&M University – Corpus Christi, Corpus Christi, TX, USA
e-mail: chase.young@tamucc.edu

Ş.Ş. Erçetin (ed.), *Chaos, Complexity and Leadership 2014*,
Springer Proceedings in Complexity, DOI 10.1007/978-3-319-18693-1_13

13.2 The Purpose

It is nearly impossible to predict and control educational outcomes. Learning is dependent on teaching, but what the students learn is unpredictable [4]. Teachers cannot delineate individual causes of the student's learning [3]. Classrooms are unpredictable because of their ecological multidimensionality, simultaneity [7], and nonlinear dynamics. Linear models fail to capture the complexity of learning communities. Educational researchers are now utilizing complexity thinking when observing classrooms as complex adaptive systems [4]. So, how does knowledge of complex adaptive systems inform teacher practice? Answering this question is important because learning environments can suffer from reductionist beliefs resulting in linear teaching models and frameworks. Teachers need to be aware of the underlying nonlinear dynamics in the seemingly chaotic classroom [18]. In fact, teachers should strive to inhabit the zone between order and chaos—this is where the intense learning occurs [1, 13]. Teachers should provide conditions for emergence, and prepare their students to capitalize on it.

This review of literature provides a conceptual framework for understanding the role of complexity in education, and how teaching should be understood as a complex adaptive system. The review will begin by discussing complexity thinking. The discussion will lead to the concept of Complex Adaptive Systems (CAS). Next, the CAS is related to the classroom. After building the foundation, the impact on educational research is analyzed. Finally, implications for teachers are explored. The primary purpose of this paper is to move complexity and education from the theoretical to the practical.

13.3 Complexity Thinking

The term complexity is often followed by the words science, theory, or thinking. For the purpose of this paper, complexity thinking is used. Scholars who ascribe to this thinking will be referred to as "complexivists." The purpose of complexity thinking is not to establish fact (science), nor explain what seems to be (theory), but to establish practical means for thinking about what already is, and what it might be. Complexity thinking can be used heuristically, as tools for interpretation, problem solving, and, ultimately, guide action [5].

Complexity is a lens with which to examine nonlinear, dynamical, emergent systems, or CASs. CASs are characterized by complex interactions of adaptive agents. Complexity thinking is particularly useful when explaining social groups. The dynamics of a group are similar to (and work with) the intrapersonal dynamics of individuals which are both subject to the complex dynamical systems of external influences. Thus, it makes complexity a necessary lens when observing social constructs.

Nonlinear systems have a fundamental order, but often appear to be chaotic [15]. These systems can only be determined from one point to another. This seems to be the only tangible aspect of complexity. A current state forms the foundation for next state of the system. When studying one point in time, it is rather straightforward. However, the direction in which the system will take is still unpredictable. Essentially, nonlinear systems undergo "successive iterations" that feed back into the system.

Emergence is also referred to as self-organization [14]. Emergence is the ability to organize into existence. There is no leader, or creator [5]. Although systems seem to appear from thin-air, the new form is actually the result of the system organizing its sufficient disorganization [8]. Thus, the system remains on the edge of chaos—never completely stabilizing, but never reaching disintegration.

The concept of emergence is particularly important in this review. Emergence in the classroom resembles the "teachable moment", or the "Ah-hah" experience [4]. This is when knowledge organizes itself into a unique understanding. Teachers covet these moments, and examining emergence might help bridge the theoretical to the practical.

Complex systems cannot be represented by anything less complex. According to complexity science, systems cannot be compressed or reduced. Compression results in losing parts of the system [5, 12, 14]. The reductionist method of complexity, however, is a matter of debate among complexity scientists. Complexity thinking recognizes the limitations of the human mind, and allows for some reduction when studying complex systems. The shift is primarily caused by the need to use knowledge of complexity in a pragmatic way. Complete understanding of a system that no human can analyze requires reduction. Complexity thinking lends itself practically in order to promote appropriate actions in complex systems [3, 5, 4, 12].

Complexity thinking moves beyond the epistemologies of structuralists and post-structuralists [4]. Structuralists are primarily concerned with how the mind creates knowledge through relations [2]. Essentially, knowledge does not have to relate to the real world, but is created in the mind based on countless influences. Structuralists focused on the cognitive organization of knowledge. The post-structuralists hold similar beliefs, but are interested in culture and power's influence in the structure of learning. They agree that learning cannot be decontextualized, and power relationships in culture and language are no exception [5]. Complexity thinking goes beyond explanation, and contends to turn the nebulous of complexity into practical implications for interested parties—in this case, teachers. So, complexity thinking is used reflexively with CASs. This is important because this author will use complexity thinking to focus on functionalism and pragmatism [14].

13.4 Complex Adaptive Systems

CASs differ from complex systems in that CASs evolve, aggregate, and anticipate. CASs evolve constantly in a competitive fashion to become more fit. The system works in a realm that is ever-changing [10]. It is in a constant state of "becoming" without ever reaching an ultimate goal [14]. However, the CAS becomes better fit for the temporal environment [13]. Holland [10] uses the concept of antibodies to explore this evolution as adaptation. Antibodies cannot ascribe to a set list of antagonists, but work together to fight a constant barrage of new and different attacks. Therefore, to keep a human system healthy, the immune system must constantly evolve to win ongoing battles of varying circumstances. An immune system's inability to adapt would result in the destruction of the human. This is where the competitive evolutionary concept is essential in CASs.

Aggregation is the ability to categorize separate systems. CASs, although not without overlap, attempt to delineate behaviors within. The rules for aggregation are also always developing and changing. Thus understanding the concept linearly can provide a model for understanding the truly nonlinear process. Though the system categorizes, the categories adapt with evolution [10].

The third characteristic of the CAS is anticipation. This ability is the least understood of the three. Of course, it makes sense because of the discussion earlier regarding the inability to predict outcomes in CASs. However, the system itself can consistently anticipate outcomes [10]. This author, heretofore, believes the system can only anticipate immediate outcomes. The belief stems from knowing CASs are endless, and only exist to improve the temporal fitness [14].

Within the CAS, simultaneity occurs. Many parts are active at once. Also, there are many aspects at work in the system, that is, they are multidimensional. All of these factors in this section contribute to the unpredictability of these systems.

The CAS is often understood through the use of metaphors. One such metaphor is "The Butterfly Effect." This is in direct opposition to Newton's belief that small causes result in small effects as well as large causes render large effects. The Butterfly Effect illustrates how effects cannot be predicted by the magnitude of the cause. The wind emitted from a butterfly's wing in Korea could possibly cause a tornado in Texas {{Robinson 1993}}.

13.5 The Classroom as a Complex Adaptive System

The classroom is a CAS. Doyle [7] conducted an ethnographic ecological analysis of secondary classrooms in which university supervisors observed ten groups of four to six student teachers and their classrooms. The study analyzed instructional episodes and student/teacher interactions. The student teacher, cooperating teacher, and the university supervisor completed the necessary triangulation of qualitative data. The researcher noted the impossibility of representing all the descriptive data, but noted the most salient characteristics. The most salient features of the interactions and classroom phenomena were simultaneity, multidimensionality, and unpredictability.

Reactions (both students and teacher) to the three salient characteristics observed, like a CAS, could not be determined before the occurrence. One thing is certain: after each interaction, the classroom collective was never the same again. What was before evolved into something new, which will then immediately and indeterminately evolve again. The classroom is, also like a CAS, always "becoming" and never "arriving." Linearly, one would have assumed some sort of arrival, but truly, classrooms never "get there", they only become increasingly fit for emerging situations [10].

The brain is a complex adaptive system [1]. Essentially, then, there are complex adaptive systems within the complex adaptive classroom. This phenomenon is understood through fractal geometry and the concept of self-similarity. Each level is similar to the underlying level [5]. A study on academic motivation will be used to illustrate the complexity of individual students.

Dowson et al. [6] conducted a large scale, longitudinal, mixed method study of variables influencing student academic motivation. The experimental group consisted of 107 high school seniors and eight teachers from a boy's school in Sydney, Australia. The control was approximately 800 students from previous cohorts. Results of the study indicated that motivation is nonlinear. One might expect student motivation to continue to rise throughout the year, and of course, this would be optimal. However, motivation at particular intervals did not reflect linear growth/decline. Interviews, surveys, observations, and academic records were used to determine reasons for the fluctuating motivation.

The researchers found a constant interplay between personal, interpersonal, institutional, teacher motivation, and other various factors influencing student motivation [6]. They conclude that one cannot separate the ecological factors from the intrapersonal when assessing student motivation. The system must be studied from a holistic perspective. No single variable can be entered into a cause and effect equation to determine an outcome. Instead, classrooms, and the students within, reflect nonlinearity and adaptation (for better or for worse).

13.6 Implications for Educational Research

The lack of empirical research on this topic made this review difficult. Only two research studies were found to examine the interpersonal perspective of CASs. The remaining three empirical studies focused on cognitive processes of reading as a CAS. There is an obvious need for further inquiry. However, there was no shortage of theoretical papers. Many authors overlapped in their beliefs, regardless of differing vocabulary. Most of the researchers agreed that education cannot be viewed in a linear model; the learning environment is too complex and nonlinear.

How can research focus on the initial conditions of emergence? Teachers are very adept at dealing with nonlinearity. In fact, some researchers claim teachers are already experts in the field, and should assume leadership roles promoting CASs [13]. Would this experiential knowledge contribute to reading research methods? Teachers possess a tacit understanding of CASs because of the daily art/science of teaching. Whether they know it or not, they are completely immersed in complexity.

The "teachable moment" is highly desirable, but almost indescribable [5]. Assuming the epistemology of emergence will help researchers focus on the initial conditions present immediately prior to the teachable moment [14]. It is this author's goal to understand this phenomenon.

What is the teachable moment? It must first be easily identifiable. It occurs in transitional stage between order and chaos. Perfectly disorganized information reorganizes to create a new thought [8]. This, naturally, might be difficult to see with the naked eye, or even a highly trained one. Though some teachable moments are recognized, it is this author's contention that many go unnoticed. This may be the result of a threatening environment or the lack of an established collective. It is unknown whether a lack of, or inability to recognize, teachable moments is the case. However, the problem of identifying the moment still exists. One must understand either the conditions, or the identification of emergence to study it successfully. The previous seems more viable.

Knowledge emerges from every transaction with the 'the world' [14]. If this is true, teachable moments are everywhere, and initiated by everything. Researchers may have to identify a classroom with a strong collective and decentralized teacher, observe the classroom, and conduct a path analysis. Of course, the generalizability would be zero assuming that learning is particular to a specific collective engaging in a unique construction of knowledge. The existing conditions of the collective classroom could be analyzed against classrooms with centralized teachers. From that, scale scores for level and predictability of learning can be compared.

Although conditions of emergence are not completely clear, one thing is—these conditions cannot be examined linearly. This might be the reason for the absent of clarity in emergence. Despite complexivists' warnings, empirical research is still being conducted linearly [15]. A need for extensive ethnographic and naturalist research paired with discourse and path analysis is apparent.

13.7 Impact on Teacher Practice

Human behavior is inherently undeterministic [3], and learning is a human behavior. Learning is not determined by teaching. Nevertheless, learning is dependent on teaching [4, 14]. Of course, this statement is assuming that anyone, thing, or experience can act as a teacher. The question in mind may resemble, "Why teach?" However, a better question might be, "How should we teach?" Alternatively, a more specific question would be, "What is the role of the teacher?"

In complexity thinking, the role of the teacher is to create opportunities for emergence. The teacher is required to assume a decentralized role in learning. The complex adaptive system centers on the most abundant flow of information. This, historically, is the teacher. However, if the teacher assumes the central role, the student is limited by the knowledge of the teacher. Likewise, if any one student gains more power, the group is limited by their level of knowledge. If the classroom collective is characterized by emergence with a primary goal of exploring what is not-yet-imaginable, then the teacher's control should be limited. Otherwise, the teacher acts as an impediment to optimal learning. The teacher, of course, should not step back completely and rely on learning to happen by chance, but should carefully control the container of the complex adaptive system [8]. Unfortunately, for the teacher, the container is also part of the CAS; therefore, the boundary is constantly shifting. This boundary, like the rest of system, should be allowed to evolve.

Teaching reading is a complex process. Roehler [16] studied teachers who embrace these complexities. His subjects included a first grade teacher and 24 students, a third grade teacher and 27 students, and a sixth grade teacher and 24 students. As Davis and Sumara [4] suggest the quality of teaching was based on student's learning, not on teaching methods. However, the teacher beliefs and understanding were particularly important in the study. Once the study was complete, researchers conducted a correlational analysis of teacher beliefs and students' strategic reading [16].

The researchers interviewed the three teachers (pre-, mid-, and post-year) to determine the degree in which they embraced the complexities of reading. Researchers also tested the students using a modified GORP test adapted from the Qualitative Reading Inventory (pre-, mid-, and post-year). Teacher A demonstrated the highest understanding of reading's complexities. Similarly, the students' strategic reading abilities increased the most. Teacher B was described as reducing the complexities of reading. The students demonstrated a lower understanding of strategic reading than students of Teacher A. Lastly, Teacher C decreased in understanding of reading's complexities, and the students made little or no growth in the reading categories assessed [16].

The previous study correlates teacher understanding of complexities of reading with higher strategic reading among students (Roehler & Michigan State Univ, East Lansing Inst for Research on Teaching, 1991). In this study, Teacher A was also a learner. It would have been interesting to study how the increased knowledge of complexity affected the role of the teacher. Did Teacher A create more opportunities for emergence in the literacy classroom?

Sumara and Davis (2006) argue that emergence is much like the desirable "teachable moment". The teachable moment is not a product of the teacher. Nor is it a manifestation from a single student or idea. In fact, the teachable moment does not even have to be addressed by the teacher. Other students or objects can instigate learning. Teachable moments, or emergence, are the direct result of creating a classroom collective—a place where student learning occurs in relation to other students, ideas, and experiences. Arguably, this is the place where optimal learning occurs. It is sometimes referred to as "the edge of chaos." On the edge of chaos, the system must self-organize to avoid disintegration. This is when something new emerges. This is the key to learning and teaching. Of course, no one obtains the key without a price. The price is decentralizing the role of the teacher. It can be hard to give up such power. Losing control in a classroom is like executing a first cliff-diving back flip. The height is merely estimated; the depth of the water is unclear. Yet both affect the necessary speed of rotation. The power required for lift-off is based on quasi-related experiences. All of which do not precisely predict the real outcome. Rest assured, one can always execute multiple attempts. Fortunately, the second attempt is based directly on experience. With each jump, new information is reorganized. Remember, there is no such thing as a perfect back flip, but some are more aesthetically pleasing. The cliff and surrounding factors, created a situation for emergence. Although the back-flipper had no formal education in this area, the jump was a success, and got progressively better with increased experience.

The indication that learning occurs with or without a certified teacher, prompts researchers to question teaching methods. Throughout this research, the author has assumed an epistemology of emergence, that is, classroom ecology that promotes maximum opportunities for emergence encourages learning [4]. Most educators would agree that learning is essential in classrooms. Mason [12] describes learning as "a process of emergence and co-evolution of the individual, the social group and the wider society (p. 25)."

Learning to read is not easy. Reading, itself, is complex and nonlinear [9]. Teaching reading is also complex. Teachers need to create classroom collectives that co-evolve adaptively toward something greater (or different) than before. The goal is to create strategic lifelong readers. When looking at a young reader, it may be difficult to see them as such. In fact, it is a clear representation of what is not-yet-imaginable [4]. Obviously one mind cannot conjure incalculable outcomes. Therefore, the students, as well as the teacher, need to depend on the classroom as a CAS. Specifically, information must flow freely in the network collective. Theoretically, it sounds perfect. The teacher has to create good lines of communication—easier said than done.

Students will not speak up if they believe it does not matter. Each student needs a sense of agency. They must view themselves an important factor in literacy learning. For a student to be agentive, they must acquire a literate identity. This requires action from the classroom collective, including a teacher with knowledge on the subject. Essentially, the first identity that needs to change is the teacher's. Building a collective of readers and authors begins with the teacher releasing control to the students (decentralized role of the teacher). Eliminate the standard authoritarian-type relationship to prepare the room for serious identity overhauls. Balancing teacher and student power is more conducive to a learning organization. As the concept "we" breaks through the power differential in classrooms, students begin to take responsibility for learning. Not only are they responsible, but students become the key agents in the construction of knowledge [11]. Thus, it creates more opportunities for emergence [14]. Note, the teacher provides this opportunity; it is not orchestrated or managed.

Sometimes a simple twist of a common phrase makes all the difference. A student's role is often the teacher-pleaser; however, this does not conjure the most independent and intrinsically motivated learner. These students long for the phrase "I'm so proud of you" to spring from their teacher's lips. What if, for example, the phrase was slightly altered in a way that actually builds pride in the students rather than the teacher? "You should be proud of yourself (p. 25)" does just that; it effectively gives permission for the student to accomplish tasks for him. It eliminates teacher-pleaser mentalities and instills internal motivation [11].

Now that the students are not simply "doing school", a new identity must replace "student". This slot can be filled with whatever is necessary to collectively complete a task. For example, when writing, "author" is used; when reading, "reader"; if students are conducting research, replace with "researcher". Affixing of labels, of course, will not eliminate the need for instruction; students will need guidance as authors, readers, and researchers. In the case of creating a literacy classroom collective, equal weight is given to student input [11].

Now that students have acquired a literate identity, they can become agentive in the classroom. Johnston [11], eerily similar to research in CASs, describes agency as "the perception that the environment is responsive to our actions (p. 29)." In fact, students, through CAS, can understand how massively correct they are once assuming agency. This is a great example of The Butterfly Effect [15]. The student, unknowingly, has changed the dynamic of the classroom and learning forever. The student's contribution was unfathomable. Amazingly, it is the first step in taking literacy learning to places unimagined. The teacher alone could not create this trajectory; therefore, building agency in students is imperative. Teachers must insist on students adopting agency. The teacher can facilitate agentive thinking by helping students realize they are responsible for their learning. The students actively plan, reflect, and modify literate experiences [11]. Sharing these experiences will invite the collective to learn, process, and assimilate them.

Teachers are lucky that learning is a CAS. It helps define the teacher's role. The classroom is a direct response to interactions of the students within. The teacher is responsible for creating the responsive environment desperately needed for learning. Teachers need to create safe environments as threat often hinders learning. Conversely, a challenging classroom enhances learning [1]. Once safe, students are free to develop their literate identity and assume agency. At this point, the collective is formed. The ride begins. The destination is unknown, but it is supposed to be. If it were known, a teacher could simply state it, or show it. To learn, knowledge must emerge from the minds of the collective. The students create knowledge in a way that has never been done before. The students did not discover it. The teacher did not uncover it. The collective learning community created it.

The collective's power cannot be measured by the sum of its parts. Once it comes together, no one can describe it conventionally. Of course, the teacher has a responsibility in this—they dissipate into the collective becoming just another node in the CAS, and join the journey to the undreamed of learning on the edge of chaos—an endless source for what is not-yet-imagined.

Teachers should focus on the conditions of emergence, and build a classroom around it. Although, in the past, education has representational epistemology, where learning is decontextualized, but represented as "real life" in the classroom, a movement towards an 'emergentists' epistemology acknowledges the boundaries of classroom, but uses its contents to extend learning to unknown destinations. The learner and 'the world' are in the same complex system [14].

13.8 Conclusion

Outcomes in learning are unpredictable [4]. I hope that this claim is no longer threatening. In fact, it should act as relief, for learning is about constructing the unimaginable. Teaching is not about ultimate control. It is about affecting learners with the least possible influence, and most positive outcomes. These outcomes, of course, are left to the devices of the classroom collective. To completely understand and buy into such a claim, one must shift beyond the umbrella of post-structuralism, and assume an 'emergentist' epistemology. My students and I are a part of the greater CAS. Through schooling, students create meaning that helps us understand 'the world' a little more. Even though the world never reveals itself completely, students continue to re-negotiate prior conceptions [14]. This act of re-negotiations happens through interaction in everything they do. A teacher's ultimate control is detrimental in the construction of knowledge. Step back; let the collective experiences, and innate need to create order drive learning [17]. Do not let cliff-diving be a student's only opportunity for emergence.

Acknowledgements Many thanks to Dr. Leslie Patterson who helped me assume a new epistemological stance.

References

1. Caine RN, Geoffrey C (1997) Education on the edge of possibility. Association for Supervision and Curriculum Development, Alexandria
2. Cunningham JW, Fitzgerald J (1996) Epistemology and reading. Read Res Q 31(1, 01/01): 36–60
3. Cziko GA (1989) Unpredictability and indeterminism in human behavior: arguments and implications for educational research. Educ Res 18(3, 04/01):17–25
4. Davis B, Sumara D (2007) Complexity science and education: reconceptualizing the teacher's role in learning. Interchang Q Rev Educ 38(1, 03/01):53–67
5. Davis B, Sumara DJ (2006) Complexity and education: inquiries into learning, teaching, and research. Lawrence Erlbaum Associates, Mahwah
6. Dowson M, Cunneen T, Irwin A (1999) A chaotic look at students' motivation: exploring the interface between chaos theory and goal theory. American Educational Research Association, New York
7. Doyle W (1977) Learning the classroom environment: an ecological analysis of induction into teaching. American Educational Research Association, New York
8. Eoyang GH (1997) Coping with chaos: seven simple tools. Lagumo, Circle Pines
9. Francis N (2004) Nonlinear processing as a comprehension strategy: a proposed typology for the study of bilingual children's self-corrections of oral reading miscues. Lang Aware 13(1, 01):17–33
10. Holland JH (1992) Complex adaptive systems. Daedalus 121(1, Winter92):17–30
11. Johnston PH (2004) Choice words: how our language affects children's learning. Stenhouse Publishers, Portland
12. Mason M (2008) Complexity theory and the philosophy of education. Educ Philos Theory 40(1, 02):4–18

13. Oekerman C (1997) Facilitating and learning at the edge of chaos: expanding the context of experiential education. Annual AEE, North Carolina
14. Osberg D, Biesta G, Cilliers P (2008) From representation to emergence: complexity's challenge to the epistemology of schooling. Educ Philos Theory 40(1, 02/01):213–227
15. Robinson R, Yaden DB (1993) Chaos or nonlinear dynamics: implications for reading research. Read Res Instruct 32(4, 06/01):15–23
16. Roehler LR (1991) Embracing the instructional complexities of reading instruction, Research series no. 208. Michigan State University, Institute for Research on Teaching, East Lansing
17. Southerland C (2008) Capturing complexity: a complementary methods study of early readers' fluency and comprehension development. Ph.D., Texas Woman's University
18. Trygestad J (1997) Chaos in the classroom: an application of chaos theory. Unpublished PhD Thesis, Texas, Chicago

Chapter 14
Part I: The Headscarf Ban in Turkey

Sema Bölükbaş

Abstract Official ideology suppressed the headscarved women as it saw them as a symbol of political Islam. But it is clear that headscarf is a different face of new social classes emerging with the changing economic policies. So, it is a clear demonstration of the fight between different social classes as evidenced by legal and practical ebbs and flaws about the ban. This fluctuation in the practical and legal situation sometimes crippled all the system, causing conflict between the different organs of the state. It is said that the ban neither related to secularism nor the 'sacredness' of public sphere but to the tension between military and civil bureaucracy at the one side and society at a whole at the other.

14.1 Introduction

The headscarf ban implemented in universities and official organizations in Turkey has caused serious social problems, particularly during the last 20 years. This ban revealed the perspectives of the state and its institutions and as well as of the society on human, women and citizens rights. The obstacles and restrictions for women to entry to study and work in institutions of higher education and public sector institutions, rested on a mentality that views women's choices to wear headscarves as an attack on the secular regime, instead of their choices as a manifestation of their freedom of religion and belief [21]. And consequently, these unjust practices gradually deteriorated to create further violations of rights to education, to employment, equal access to public life and basic rights and liberties [21].

Although there are plenty of analysis, dialogue, debate and research on the matter, most of them are restricted to whether the headscarf ban in universities is justified or not. The impact of the ban in the public sector on the professional lives of higher educated headscarved women was not included in many studies. As many

S. Bölükbaş (✉)
Social Policy Department, Faculty of Political Science, Yıldırım Beyazıt University, Ankara, Turkey
e-mail: semabolukbas@gmail.com

Ş.Ş. Erçetin (ed.), *Chaos, Complexity and Leadership 2014*,
Springer Proceedings in Complexity, DOI 10.1007/978-3-319-18693-1_14

researches show this ban also hindered or restricted professional headscarf women as they apply for or work in private sector jobs.

Headscarf ban was a serious obstacle with regard to democratization process in Turkey. This antidemocratic application undermined the trust in the future of Turkish society which took the steps day by day to complete its democratization process (Hazar Eğitim Kültür ve Dayanışma Derneği HAZAR [13]). Practicing ones religion, which was guaranteed in many international conventions and also in our constitution, became a reason for social tension and polarization.

The problem was not only about headscarved women but mainly about individual rights and freedoms and the state's approach to these rights. So, it was related to the democratization process.

In the last few years, some efforts were observed to ease the restriction. It was said that this relaxation was due to the attitude of the administration. But again it was said that there should have been sound institutional and legal arrangements to eradicate the problem completely.

In this study, the issue will be examined in two parts as it is of multidimensional feature and a long history. In the first part, acknowledging that this is also very controversial and highly debated issue and caused a lot of tension in the society for a long time, the causes and the consequences of the problem as well as the historical, political, social, economic and legal dimensions will be assessed after the definition of the problem. While doing so, alternative approaches, the definition of some vital concepts such as secularism and public sphere, some figures about the implementation and consequences of the ban, current international situation will be focused on.

In the second part, the effects of the ban on the professional headscarved women will be studied. Through highly debated questions, the groundless of the ban will be tried to explain. The questions are as follows:

Is it a political symbol?
Was compromise indispensible for the solution of the problem?
Whose problem was the ban?
Do the headscarved women cover as a result of oppression?

These questions and their answers are of vital importance to grasp the problem completely. Because these arguments were used to legalize the ban by the proponents of it. After looking at the different approaches to these questions, the significance of settling the problem will be focused on.

After that, the process of the solution of the problem will be examined. Finally if there are any chaos in the universities or the public sphere upon lifting the ban will be discussed.

References for the study will be presented at the end of the second part.

14.2 What was the Headscarf Problem?

Because arguments on covering had a central place in politically polarized discussions on the issue, they remained restricted by the parameters applicable to political splits. Academic studies on headscarf revolved around the question 'Veiling is objection to what?' There are two trends identifiable in the debates [20].

One approach argues that veiling is an objection to the secular modernization and westernization project of the Republic. This is an approach based on expectations generated by theories of modernization: Urbanization, education, women's increasing participation in labor force will subdue the traditional and Islamic structures and identities in society, and as a result women will be liberated. In this perspective, veiling prevents women from being liberated and is considered a symbol of pressure exerted on women on the basis of religion [2].

Contrary to modernization project theory, other approach points to the possibility of imagination of an alternative modernity that builds on non-Western codes [20]. This approach suppose that veiling is a modern phenomenon in which it reflects an Islamic way of life in contemporary times, and claims that covered women represent a protest against loading the concept of 'civilization' with the Western based content ([20] cited from [22]), as well as against the West's domination of 'lifestyle' [11]. Göle argues that the turban and veiling both keep women in a private space while she is in public and thereby facilitate her presence there, and serve as a sign of protest against the image of docile and repressed Muslim women [11]. İlyasoğlu argues that headscarved women turn inside out the definition of modernity by appearing in educational and professional life, domains otherwise coded as a 'modern' ([20] cited from [22]).

Bayramoğlu states that women's appearing in public sphere thanks to the veiling, with the experiences they gain there, results in the creating an 'Islamic modernization' effect in both the private and public spheres [4].

Official ideology did not consider citizen's choice of dressing as a matter of preference; rather it thought that it was a way of westernization and modernization. Headscarf was banned due to the fact that it did not seem western and 'modern'. Official ideology expected to transform the society in line with its own ideology. But it was not a sound attitude for official ideology to consider some kind of dressing illegitimate and primitive because of their conventional and religious style of dressing [17].

Official ideology and state elites thought that headscarf was a symbol of political Islam. But Bostancı stated that, headscarf and its being used by more and more women day by day was a result of new social classes emerging after economic transformation policies applied during Ozal term. Headscarf was face of these new classes. So, in fact it was related to the changes in social class structure in Turkey [6]. Bostancı also emphasized that turban was a symbol of new elites who moved upward thorough the class structure but it was not a symbol of political Islam and this revealed that the real problem was an objection to 'circulation of elites', and concerns about secularism was a pretext for it [6].

Covering was perceived as an state of 'not becoming docile' and so it was opposed. There were several splits about any kind of diversity for example being Kurd, non-Muslim, Alevism besides headscarf that risked the modern national state's claim, intention and representation of homogeneity. We can say that different 'others' as well as headscarf are subject to the mechanisms of exclusion. Despite these exclusionary and control mechanisms, it doesn't seem possible to control life and homogenize social segments [15]. Sambur also points to this matter. According to him, official ideology brings about a problem: To accept the lifestyles which are compatible to official ideology as legitimate and to consider others illegitimate and primitive. So, through the headscarf ban it has revealed that most important problem of Turkey is official ideology [17]. Bumin underlines this future of republic, and she defines the women with turban as ultramodern while defining women not veiled as modern and headscarved women as conventional (According to Bumin, headscarved women are those who cover traditionally. The one used by educated and urbanized women is turban). She uses term 'ultramodern' for the women with turban because they refuse to adopt not just tradition but also the lifestyle imposed by the Republic and they demand their well deserved place in the public sphere. They don't cover like their mother and refuse to stay at home and also they refuse to adopt a lifestyle the Republic wants them to. So, they are ultramodern. Bumin says that turban is individualization of religion and is a kind of expression of individualism. Turban does not imply returning of religion but instead, it means emerging of religion as an identity chosen by individuals. And this identity does not confront with modern state; on the contrary it is a result of modern state [7].

14.3 The History of the Ban

As Kentel says, Turkish modernization did not appear as a result of long struggles as in Europe. It arose as a result of the direct implementation of the process of European modernization without making any adjustment [15]. This direct assemble is realized by state elites and majority of people who felt the effects of it in their daily life remained passive in this process. Turkish modernists embraced French central modernization model and jacobanism as a model instead of Anglo-Sakson liberalism [12]. Therefore there did not emerge a harmony between these elites and the majority of the people as in the case in Europe. On the contrary this process brought about a permanent conflict between these sections and the state elites [3].

The headscarf issue arose as a repercussion of this conflict between the passive part of the society and those who were trying to build up this modernization process. This distinctive nature of Turkish modernization necessarily brought about some contradictions. The most concrete example of this was the bureaucratic elites trying to include these people into the process on the one hand, and embracing some methods which contradict with these efforts on the other hand [3]. Furthermore emerging of a new capitalist class and the handover of capital starting with Democrat Party's taking office in 1950s, continuing with the Ozal term in 1983 and with the AK Party government in 2000s help us to understand the socioeconomic roots of headscarf issue.

During the foundation years of The Republic, a considerable amount of population was living in rural areas. As a result of the long migration process from rural areas to urbans, today majority of the population live in the cities. This transformation created new cities where the people who newly migrated came together with the urbanized elites. These elites were consisting of İstanbul bourgeoisie who became rich through the privileges implemented within the 'project of creating a national bourgeoisie' during the one party era and 'bureaucratic oligarchy'. These elites lost their privileged position with the foundation of DP (Democrat Party) and its coming into power. They defined any political party and its supporters as 'reactionary, counterrevolutionist, regime opponents'. Behind this behavior it was the motivation to preserve their privileged position. In this sense, it was possible to see their support to the military coups as a reflex created to secure their situation [3].

As a consequence of this process, headscarf became a symbol of 'not being one of them' in the eyes of the elites. In other words, headscarf symbolized 'other classes' which became visible in every field of city life and broke down the monopoly of elites in these fields [3].

In sum, it was evident that regime debates going on due to headscarf ban were created because of the loss of hegemony on the part of city elites consisting of bureaucratic oligarchy and old capitalist class. It was the matter of hegemony but not regime [3].

To be able to comprehend the headscarf issue truly, being aware of these facts are of vital importance.

14.4 What is Secularism and Its Relation with the Ban?

Before talking about why headscarf freedom isn't contradictory with secularism, we should define secularism first.

Secularism is an understanding based on political and legal principle that requires state to be neutral in the face of various religious and non-religious beliefs, and not to favor one of them against the other [23]. A secularist state can't adopt a hostile approach to preclude a certain religion from social arena as well as it can't try to legalize itself through a certain religion [23]. Because of all these traits, secularism is very important, useful and needed principle for building up democracy, and freedom of belief and worship in Turkey. However if it doesn't based on democracy and freedom and isn't applied properly it can serve to an oppressive and totalitarian state mechanism, and can become a threat to freedom of religion and conscious [23].

In Turkey, we see that state mechanism can't remain neutral between non-homogeneous and different beliefs [3]. As Prof Dr. Mustafa Erdoğan has stated, freedom of religion and conscience not just secures individual's beliefs but at the same time contains the freedom of behavior in accordance with their individual belief in both individual-special and social realms [8].

In Turkey state couldn't realize this kind of secularism. In the public sphere a women who was atheist or was not covered could practice her philosophy of life freely but a covered women was obliged to put aside her own philosophy of life. That meant that state could not remain neutral in the sense of building secularism and favored one side. Furthermore, the reason of why the sense of secularism could not gain functionality was not freedom of headscarf in public sphere but on the contrary prohibition of it [23].

What should be done by a state which claims to adopt the principle of secularism is to secure the individual's freedom of religion and conscience, to prevent any lifestyle from oppressing another one, to make it possible for every citizen to live as they want to without restricting other's lifestyle [3]. Therefore someone claiming that freedom of headscarf contradict with secularism either does not know what it is or wants to use it as a means to realize their totalitarian ideals [3].

Falk explain this point of view as follows:

> The hostility to covered women on the part of old secularists are part genuine, part cynical manipulation. The anxiety of many old secularists has to do with their genuine attachment to a frozen image of modern Turkey as shaped by a number of societal forces including the resolve at the founding of the republic to render religion as invisible as possible in public spheres. The cynical manipulation of this issue is associated with increasingly desperate efforts by the old secularists under current conditions to find ways to overcome their loss of power, influence, and popular support in Turkey. This efforts depend on a portrayal of wearing the headscarf, especially in its turban form, as expressive of a hidden agenda by the newly elected political leadership to ensure that sharia becomes the law of the land for all Turks.... [9].

There is no doubt that a political party's efforts to lift the ban is a matter of expanding the freedoms and this effort does not have anything to do with building a state taking religion as a reference. There is nothing more natural than the fact that political parties are concerned about citizen's demands of freedom in the political arena [3].

The Chairman of Constitutional Court Haşim Kılıç, who cast negative vote on the case of AK Party's shutting down, said that the arguments made by AK Party to lift the ban on headscarf couldn't oppose to secularism because the main capital of political parties is social demands and problems [3].

Richard Falk also states that the headscarf problem is mainly about Turkish democracy. It is not really about the interplay of religion and politics or the old secularist's allegation that AKP is a vehicle for political Islam [9].

Falk draws a distinction between old secularists of pure Kemalist persuasion and new secularists that favor a democracy of inclusive rights. According to Falk, the old secularists are adherents of the approach associated with French laicize emphasizing freedom from religion, whereas the new secularists whether consciously or not, are adherents of the American preoccupation with freedom of religion. Both attitudes toward the proper relation of religion, politics and the state were grounded in the distinctive national experience of the two countries. The main goal of the French Revolution was to free the French State from the Catholic Church, whereas the founders of the United States were devoted to achievement of a governing process that would tolerate a plurality of religions without their persecuting worshippers or showing favor to one over the other religious identities. The American ideal was to keep the state neutral, but to be active in protecting a diversity of religious practices, and ensuring an equality of rights of he religiously observant as well as for the non-observant [9].

Kavakçı states that the secularist state in Turkey devalues the headscarved women and this contradicts with its modernization project [14]. Maybe French style secularism can best be demonstrated in her sentences.

> The Republican system of representations devalues the religious women (who in order to meet the ideals of the Turkish Republic, needed to be liberated in the 1920s) and privileges the 'modern' Turkish women whom the republic intended to be its full citizens. The oppositional binary created between these two groupings of women poses a serious obstacle in The Turkish Republic's attempt to implement its progressive female agenda and attests to the inconsistencies of the Turkish modernization Project [14].

As Kırık said the position of the state against basic rights and freedoms is strictly indexed to the sustainability of secularism unconditionally and without making any concession [16].

14.5 What is Public Sphere and Its Relation with the Ban?

To be able to understand what public sphere is and its relation with the ban firstly we should look at the definition of 'public sphere'. As this study mainly tries to focus on the implementation of the ban in the public sphere, the meaning of the term 'public sphere' and changes in this meaning, and the challenges it faces today are very important for us. So, first of all, in the first subtitle the term will be tried to be defined in a detailed way with its background and then in the second subtitle its relation with the ban will be discussed.

14.6 Public Sphere

The birth of public sphere in which new classes and power relations came out, correspond to bourgeoisie's search of freedom for itself. The high interest of the bourgeoisie was espoused as the interest of all classes. People 'discussed' and contributed to operation of the capitalist system. The rationality of bourgeoisie, having the control and administration of rising industry and modernity, made its mark on this field of discussion. An understanding of science that believed human could find out everything and he was above everything accompanied with it. Having the control of knowledge became the means of power. Discussion became the means of this knowledge but this discussion has never been the one between the equals. This debate is realized under the dominant ideology and others. The powerful ones controlled and banned them if necessary or eradicated when they were too dangerous for he system. That is, 'inequality' has become the unchanging feature of public sphere. Public space required individuals who gain autonomous against private sphere. The individuals of modernity were those who were autonomous against anything in the private space for example, community, belonging to religion or ethnicity. In order to complete consolidation of its power this ideology needed private space to limit itself. As a result public sphere was the space of dominant classes and was a field where the powerful becomes apparent with its ideology and symbols and where religion, region and local languages were defeated and where the ideology of bourgeoisie were legalized. So, the public space corresponded to a concrete history and to a time period. But the universalization of this historical element toward time and society meant that the ideology of a specific class (bourgeoisie) was accepted as belonged to all regions and classes [15].

Turkish modernization and establishment of public sphere were in accordance with this story. But there was a significant difference: The modernization process faced by Turkish people came out as a result of adaptation of an outcome gained as a result of a lot of struggle in the European modernization process. Thus it brought about a lot of conflict and polarization. The most significant aim of Turkish modernization was creating a new, homogeny nation and making several changes in the field of religion. Many efforts were exerted to eliminate the power of religion as reference and to exclude it from public sphere which would be built up with a strict and authoritarian understanding of secularism [15].

The model, faced with society in a strict and severe way, didn't include the people who embraced traditional values. But at the same time these people refused the model. The public sphere, where modernity was learned, experienced and finally espoused, became the field of those who succeeded to 'internalize' the model. The differentiation of 'public sphere- private sphere', which had functionality with regard to modernity, became fields of two different lives [15].

But those people, who couldn't be tamed, have created their own modernity in time, and started to demand their place in public sphere. This disobedient alternative modernity made the realm which the elites tried to protect by attributing sacredness to it, fragile, and ruined the balance they suppose to have. The permanent reconstruction of this ruined balance have been tried to accomplish by military through coups or through the violent methods legalized by laws [15].

Today public sphere face with a great deal of challenge all over the world. As capital and capitalism can't remain limited within the national borders, public space is no longer national. Sovereignty has gone beyond national borders and public sphere has been globalized as 'free area of discussion'. Today societies can't generate themselves as the society of bourgeoisie in the sense of only 'nation' and 'national market'. To be stuck in such a narrow space strangle the modern person and they want to 'aggregate' dissociated parts [15].

In conclusion the public sphere of modernization could not fulfill its claim not only in the European countries who claim to be the center of modernity and civilization, but also in the countries who imported modernity and some have some 'special' conditions such as strict application of secularism and military coups. Modernity is subject to challenges by different kinds of modernities in different parts of the world especially by immigrants. Private sphere can't be eradicated or even be limited. The power relations included by public sphere have been revealed.

It is no longer possible to monopolize and have the power by insulting people and proclaiming them illegitimate. Because the elites no longer carry the modernity. Being modern is no longer realized in the public sphere defined and controlled by elites but in a public sphere which is being redefined [15].

14.7 The Headscarf Ban and Public Sphere

After 1980s the concept of public sphere was again high on the agenda. Because societies were in search of how diversity of people live together. And the answer to it came from public sphere. It was discussed within the sense of democracy. But still there is no certain definition scholars agreed on [10].

The concept is an issue being examined and debated in the field of political philosophy and legal arrangements that affect the life deeply can't be implemented with the help of these debated terms [23]. According to Gole the ban has revealed the limitations of public sphere [10]. There can be some definition about the term but with none of them the headscarf ban can be justified [23]. Yayla talks about each of these alternative definitions and emphasizes that the ban can't be based on a sound and meaningful ground in any situation. For example, we can take it as the field where a public authority takes place. But in social life there is public authority almost everywhere theoretically such as streets and parks. Then the ban should be expanded to the parks and streets. Furthermore in different situations there can be public authority even in our houses. If you mistreat your wife or children, state can use its authority in our houses. Then headscarf will have to banned in our houses [23].

If we say that any place a public service is being conducted is public sphere, then universities, tax offices, hospitals should be among the places where the ban should be implemented [23].

If we think that the places where state officials appear are public sphere then we face the question that when an official goes home in the evening, does it make his home public sphere? And should his wife put off her headscarf? It can be said that a public sphere is the place where an official works. But again definition isn't enough to solve the problem. For example a nurse working for the Health Ministry goes to a village for a vaccine campaign. This event makes the village public sphere. In this situation should both the nurse and the women taking that service put off their headscarf? [23].

As Yayla demonstrated by these examples explicitly, it is impossible to reach a meaningful conclusion through the public sphere debates. Yayla says that in a liberal country citizens have the rights to be in every kind of public sphere or any place claimed to be public sphere [23].

The ambiguity in the meaning of the public sphere has caused a lot of unjust practices by several state organs. For example, old president of High Education Council Erdoğan Teziç once said:

'A judge can't be headscarved in a market while she is uncovered in the court. She can't say 'it is my own special life and it is my belief' . . . When you are walking on the street, if a police can't recognize you, you should put off your headscarf. It is also true if you are at home. When police leaves, your home again becomes your special space' [1].

Teziç either doesn't know that Islam doesn't require a women to cover in her home, or he says all these intentionally to prevent headscarved women to be in the public sphere. Because as generally said, headscarf facilitates covered women to participate in public life.

According to AKDER, the unjust practices applied to covered women reached to a level that sometimes even private institutions refuse to serve to the covered women. For example a private factory didn't allow a headscarved mother of a student who was on the school trip to come into the factory. It is possible to enhance these examples [1].

As these examples demonstrate, public sphere should be redefined in Turkey. Sambur emphasized that official mentality which undermines the possibility of living together with the people who embrace different lifestyles and beliefs is against the pluralism and freedom [17]. The redefinition of public sphere requires also redefinition of secularism. In that point, it can be said that it will be useful to think about Anglo-Sakson style of secularism, as it is more pluralist and is trying to be neutral in the face of different beliefs. Kentel underlines this issue: 'The redefinition of public sphere requires the redefinition of the secularism. A sense of pluralist secularism is needed for a public sphere which is pluralist, inclusive and open to different religious identities' [15].

Kırık says that this problem is related to publicity and democracy can be achieved on the ground of basic rights and freedoms [16].

14.8 The Legal Situation About the Ban

If we consider some major turning points in the history of the ban, legal and practical ebbs and flows are generally seen to be the case, with day to day politics constantly seeking to shape the practice with which the headscarf ban put into effect [21].

The first ever regulations by The Ministry of National Education came out in 1991. 'The Regulations Concerning Dress and Attire Of Staff and Students in Schools Under The Ministry of National Education and Other Ministries' stipulated the following with regard to the dress code applicable to students attending postsecondary schools under the Ministry of National Education:

> A-Female students: Clean, neat, ironed, plain clothes; plain and no high heeled shoes and boots; head uncovered, hair combed flat with a topknot, no headscarves to be worn while in the building of the institution.
>
> B-male student: Clean, neat clothes and shoes, head to be covered while in the building of institution, no long whiskers on hair, no bear will be grown, moustache, if grown, will be clean, combed and cut so as not to run over the lip . . . (TESEV 2011 cited from Official Gazette 1981).

The first regulations stipulating rules for the dress and attire of headscarved employees working in public jobs was issued in 1982. The 'Regulations Concerning Dress and Attire of Employees in Public Institutions and Organizations' took effect in that year (TESEV 2011 cited from Official Gazette, 25 October 1982, ıssue 17849). The document contains provisions that female and male civil servants will their head uncovered. Still in the same year the regulations issued by The Ministry of National Education on 22 July 1981 was amended to remove the phrase 'postsecondary schools' found in the original text of the document. As a result of this amendment university students were moved outside of the scope of the Regulation [21].

In 1982, The Higher Education Council (HEC), issued its Circular Order concerning dress and attire, which stipulated "a modern dress". The Circular Order provided that:

> A-All staff and students of institutions of higher education are required to have dress and attire that accord with the revolutions and principles of Ataturk and are of a civilized and modest shape.
>
> B-All male and female students, including those with foreign nationalities, will wear clean, neat and plain dresses, have their head uncovered and will not wear a headscarf while in the building of the institution [21].

Headscarf ban was put into effect with these regulations in universities and public institutions and organizations. From that day on, there were great deal of changes in the regulations, sometimes to ease the restrictions but sometimes to make it more strict. For example, facing with increasing pressure from students, in 1984, HEC issued another Circular Order which provided that 'students may use a modern turban', and delivered it to the presidents of universities. But in 1987 HEC again prohibited using headscarf. 'Freedom for headscarf and turban' was granted by an article added to the law on the Higher Education Council in 1988. Turkish Grand National Assembly adopted this additional article 16 by way of law No: 3511. This amendment provided that hair and neck may be covered with headscarf or turban because of religious beliefs. But in 1999, upon an application by then-president Kenan Evren, The Constitutional Court reviewed the matter again. The court repealed the sentence that allows using headscarf. It found the section 16 violated the constitutional principle of secularism, equality before the law, and freedom of religion [5]. Interestingly this time The Court of State (DANIŞTAY) upheld the headscarf regulation based on basic individual liberties granted by the constitution. Thus a contradiction between the decisions of two different Turkish High Courts, resulting in 'a true impasse: neither a clear yes, nor a clear no but bifurcation and ambivalance' [5].

In the 1990, Turkish Grant National Assembly passed Law No: 3650 to add Article 17 to Law 2547 of HEC. Article 17 reads 'At institutions of higher educations, there is no dress code so long as students do not wear outfits that are not antithetical to the laws enacted'. The opposition (Republican People's Party) went to the Constitutional Court to repeal article 17. The court decided that article 17 is in accordance with the Turkish Constitution.

Section 17 of Higher Education Act formed another attempt to circumvent The Constitutional Court's holding. The sections vague wording allowed for an open interpretation and served to prevent any recognition or affirmation of one religion. In a subsequent ruling in 1981, the court upheld section 17, despite noting in dicta that 'in higher education institutions, it is contrary to the principles of secularism and equality for the neck and hair to be covered with a veil or headscarf on grounds of religious belief'. The court reasoned that because it had previously struck down a regulation (Article 16) that permitted headscarf as be worn in universities, and therefore section 17 didn't violate the Constitutional Court's judge-made law [5].

A 1999 Constitutional Court judgment stated 'The legislative and the executive are bound by both the operative provisions of the Court's judgments and the reasoning taken as a whole. Furthermore, legislative activity must be measured against and guided by these judgments'. The Constitutional Court thus essentially proclaimed that its own dictation bind other Turkish Courts when those courts review legislative and executive action. Therefore, because The Constitutional Court previously mentioned in a judgment that wearing headscarf in public universities is contrary to the constitutional principle of secularism, the legislative and executive are now prohibited from enacting any legislation that would contradict this conclusion [5].

The brief history of the legal aspect of the ban indicates that there were no actual bans in laws. (Section 17 of the Law of HEC). However this ban implemented through regulations by some institutions and by court judgments. In a country where rule of law prevails, basic rights and freedoms can't be restricted by regulations but only buy laws or constitution. As Yayla said, there was no ban actually; on the contrary there was freedom for headscarf [23].

This was also true for the public sector jobs. The section 70 of The Turkish Constitution reads: 'Every Turkish citizen has the right to enter the public service. In acceptance to the service, there can be no discrimination except for the qualifications needed for the service itself'. But in every level of public institutions, the condition of being uncovered is needed, without considering the qualifications required by that service. After 1999 headscarf was banned in the entrance exams of public institutions. Furthermore some women are expected to uncover even out of institutions they worked for. For example Council of The State (DANIŞTAY) decided that it is a reason for punishment covering even at the out of the workplace AKDER [1]. Especially after 28 February process the ban in higher education and public institutions was implemented more strictly. Because of the ban about 5000 women civil servant were fired and about 10,000 of them were forced to resign [1].

When we consider the discriminative applications in higher education institutions, we can't expect much about the basic rights and freedoms of headscarved women working in public sector or wanting to do so.

As many writers emphasized there was not any ban on the headscarf in laws or in the constitution. But the courts violated the laws and the constitution by creating a ban through their interpretations [3, 6, 15, 17]. This situation pointed out to the struggle between social classes as mentioned in previous chapters. We saw this struggle in the most apparent way on the headscarf debate. This fight between the emerging capitalist classes and those consisting of the old capitalist classes and the bureaucratic elites was the main cause of the ban. This could be understood from the volatility of the legal arrangements and their changing implementation. When we searched the people's approach toward this ban (fight), we could see that majority of the people did not support the ban and also they severely condemned using of headscarf as political material by politicians. 69 % of people doesn't support the ban in public institutions and 78 % of them wants the ban to be lifted in universities (Türkiye İş Kadınları Derneği [19]). Likewise, 95.9 % of people did not want to see politicians making use of the matter [13].

Headscarf problem was not only the problem of one section of the society, but it transformed into a challenge effecting the relation between judiciary and legislative. All efforts to lift the ban by legislative blocked by judiciary. On the other hand parties who support the freedom of headscarf paid a high price such as being shut down. The governing party who tried to lift the ban in the universities was about to be shut down. The Constitutional Court's decisions about AK Party and about the constitutional changes narrowed down the political sphere and curtailed the capacity of legislative to solve the social problems (Siyaset, Ekonomi ve Toplum Araştırmaları Vakfı (SETA) [18]).

14.9 International Approach Toward Headscarf and European Court of Human Right's Decision on the Ban

It seems significant to realize that it is American understanding of secularism that found its way into the development of international human rights law after World War II. The norms in the most relevant, authoritative international texts, Universal Declaration of Human Rights and The Covenant of Civil and Political Rights both incorporate freedom of religion as the only legal basis upon which religious freedom can be based [9]. Likewise Bumin states that European Union is a transnational organization and based on freedoms. Secularism is related to the national states. There is no place for secularism which is a state term and which implies the approach of state towards society in a transnational union. As a matter of fact, the international law covenants do not include secularism, but they cover the freedom of religion, belief and conscience [7].

The terrorist attack on the September 2011 caused some negative perception on the side of the Western States and institutions. One of the most significant repercussions of this event is maybe the decision of European Court of Human Rights (ECHR) on the headscarf in 2004 which ruled that headscarf ban was not contradictory with human rights and it is a requirement for secularism. Although it was a very controversial decision it is said that The Court may have been influenced by its own feelings about the headscarf and its post-September 11 fears of fundamentalist Islam [5].

Bleiberg criticize the ECHR's ruling and states that it is incorrect on several legal grounds. First, decision gives too much deference to The Turkish Constitutional Court. Second, The ECHR's review of Turkey's historical, social, and legal context is incomplete and unsophisticated. Third The ECHR incorrectly applies its own precedent in arguing that the headscarf violates the rights of others [5]. The Court's ruling is important in terms of human rights of headscarved women in Turkey and in the rest of Europe, and there is no doubt that it will have adverse effect on their lives [5].

In addition to the adverse affect of September 11, some countries faced with the fact of multiculturalism and they could not cope with it properly. One of these countries is France. In France, The Republic planned to create a homogeny nation (like Turkey) through mandatory education but now there are people who challenge this project by putting forward their own identities (again like Turkey). France has been facing with this meeting and she bans the headscarf in the schools and this is her defeat [7].

However, in many countries there is no headscarf problem in public sphere especially in those where the Anglo-Sakson understanding of secularism prevails.

The Convention on Elimination all Forms of Discrimination against Women (CEDAW) should also been mentioned about. Turkey signed this convention in 1985 and by doing so, Turkey pledged to take every needed steps aiming to realization of the political, social and legal rights mentioned in the convention and to end any kind of discrimination against women. In this sense, the obligation of uncovering for getting job in the public sector blatantly violates the provisions included in this convention [1].

References

1. Ayrımcılığa Karşı Kadın Hakları Derneği (AKDER) (2010) Toplumsal cinsiyet, eşitlik ve başörtüsü yasakları from http://www.ak-der.org/toplumsal-cinsiyet-esitlik-basortusu-yasaklari-2010.gbt
2. Arat Y (1999) Political Islam in Turkey and women's organization. İstanbul: TESEV Publications
3. Başkan B (2009) Başörtüsü sorunu ve meşrulaştırılamayan yasakçılık. Liberal Düşünce 55(14). Retrieved October 1, 2012, from http://www.ozgurtoplumundegerleri.com/res/Burak
4. Bayramoğlu A (2006) Çağdaşlık hurafe kaldırmaz. İstanbul: TESEV Publications
5. Bleiberg DB (2006) Unveiling the real issue: Evaluating The European Court of Human Rights decision to enforce the Turkish headscarf ban in Leyla Şahin Turkey. Retrieved October 26, 2012, from http://chinisejil.oxfordjournals.org/content/5/3/7/9.full.pdf
6. Bostancı N (2012) Başörtüsünün Arkasındaki Sosyoloji. Retrieved October 24, 2012, from http://www.nacibostanci.com/makaleler/item/247-başörtüsünün-arkas\T1\indaki-sosyoloji.html
7. Bumin T (2003) Türbanlı Kadınlar Ultramodern, interview with Neşe Düzel. Retrieved October 3, 2012, from http://www.radikal.com.tr/haber.php?haberno=10041
8. Erdogan M (2001) Sivil Özgürlük Olarak Din ve Vicdan Özgürlüğü. Liberal Düşünce (21):s.42
9. Falk R (2012) Decoding The Turkish Headscarf Debate. Retrieved October 15, 2012, from http://www.global.ucsb.edu/orfaleacenter/luce/luce08/documents/Falk_TurkishHeadscarfDebate.pdf

10. Göle N (2003) Nilüfer Göle ile Kamusal Alan ve Sivil Toplum Üzerine Söyleşi, Sivil Toplum Dergisi (2). Retrieved October 10, 2012, from http://www.siviltoplum.com.tr/?ynt=icerikdetay&icerik=47&id=133
11. Göle N (1991) Modern Mahrem Medeniyet ve Örtünme. İstanbul: Metis Publications
12. Göle N (2009) İçiçe Girişler İslam ve Avrupa. Metis Publications, İstanbul
13. Hazar Eğitim Kültür ve Dayanışma Derneği (2007) Covered head of truth in Turkey. İstanbul: Umut Publications
14. Kavakçı Mİ (2010) Headscraf politics in Turkey. Palgrave Macmillan, New York
15. Kentel F (2008) Teorinin Raconunu Bozan Başörtüsü: Bir Direnişin Anatomisi, Örtülemeyen Sorun Başörtüsü, Edisyon Book. İstanbul: AKDER Publications.
16. Kırık H (2007) Turban Sorunu? Kamusal Alan ve Demokrasi. İstanbul: Salyangoz Publications.
17. Sambur B (2012) Başörtüsü Yasağı, Özgürlük ve Çoğulculuk. Retrieved 3, 2012, from http://www.liberal.org.tr/incele.php?id=Nzkw
18. Siyaset, Ekonomi ve Toplum Araştırmaları Vakfı (2009) Başörtüsü AK Parti ve Laiklik: Anayasa Mahkemesinden İki Karar İki Gerekçe. Retrieved October 5, 2012, from www.seta.org
19. Türkiye İş Kadınları Derneği (2010) Başörtüsü ve Toplumsal Uzlaşma', Press Summary. Retrieved October 1, 2012, from http://gundem.bugun.com.tr/basortusu-anketinde-carpici-sonuc-129673-haberi.aspx
20. Türkiye Ekonomik ve Sosyal Etütler Vakfı (2012) Başörtüsü Yasağına İlişkin Değerlendirme ve Öneriler, TESEV Demokratikleşme Programı. İstanbul: TESEV Publications.
21. Türkiye Ekonomik ve Sosyal Etütler Vakfı (TESEV) (2011) Headscarf and discrimination: professional headscarved women ın the lobor market. TESEV Publications, İstanbul
22. İlyasoğlu A (1994) Örtülü Kimlik: Islamcı Kadın Kimliğinin Oluşum Öğeleri. İstanbul: Metis Publication.
23. Yayla A (2012) Bir Atilla Yayla Metni: Ahlak, Hukuk ve Başörtüsü Yasağı. Retrieved 25, 2012, from http://gundemmaddesi.blogcu.com/bir-atilla-yayla-metni-ahlak-huku-ve-basortusu-yasagi/2941461

Chapter 15
Part II: The Effects of Headscarf Ban on the Professional Headscarved Women

Sema Bölükbaş

Abstract The right of entering to public job was secured by the constitution. But the condition of uncovering was imposed at every stage of recruitment in the public sector. The ban had a spillover effect and influenced the practices in the private sector. The headscarf was not a political symbol and covered women did not cover due to the pressure from their families. On the other hand it was not only the problem of covered women but of society. As it was well established, there was not any need to reach a compromise to be able to get the basic rights and freedom. It seemed that society as well political parties realized these facts and at the end the problem that caused a lot of tension and human rights violations in the country.

15.1 The Ban's Effect on the Professional Headscarved Women

Section 70 of the constitution secures the citizen's rights of entering into public job. But at every stage of recruitment in the public institutions, the condition of uncovering is imposed. After 1999, headscarf was prohibited in the entrance exams for the public institutions [2].

Most people may think that headscarved women could not get a job only in the public sector. However when we look at the whole picture more closely we see that there was also discriminative applications in the private sector. TESEV, in its research named 'Headscarf Ban and Discrimination: Professional Headscarved Women in The Labor Market', showed that the headscarf ban in the public sector had direct and indirect effects over practices in the private sector. Working conditions were very different from those in blue-color and pink-color jobs. Professional employees do not relate to a limited number of individuals and organizations located in a certain place; they need to remain in contact with several individuals and organizations over the course of their work. Therefore the ban in the public sector

S. Bölükbaş (✉)
Social Policy Department, Faculty of Political Science, Yıldırım Beyazıt University, Ankara, Turkey
e-mail: semabolukbas@gmail.com

Ş.Ş. Erçetin (ed.), *Chaos, Complexity and Leadership 2014*,
Springer Proceedings in Complexity, DOI 10.1007/978-3-319-18693-1_15

prejudiced the public sector as well [10]. To illustrate the point, we can give the example of a journalist wearing headscarf: If she wants to pursue political news, the settings she will be allowed to access will be limited as a result of the ban. The ban could accordingly be said to have such a spillover effect of the ban [10].

Additionally, headscarved women working professional jobs were being asked to remain invisible at the workplace due to the spillover effect of the ban.

The findings of the study on the ban and its spillover effect lead to a number of conclusions at specific levels [10]:

At the level of women wearing headscarf, the headscarf ban is cause for trouble in terms of having a government job; performing one's job at work; getting a job, advancing through the ranks and being promoted at a private company. In cases where the ban in the public sector reflects onto the private sector through the spillover effect, fewer professional headscarved women participate in job markets. In a society where educated women are not allowed access to labor force, the creativity, productivity and solution-oriented thinking that come with the female labor force cannot be taken advantage of. Moreover, in a world where women are unable to join the labor force, there is no chance of shifting the balance in women's favor in terms of the traditional, domestic division of labor characterized by a male-dominant nature, where the role assigned to women is supporting the man by taking care of household chores while he provides for the family [10].

It is also obvious that private businesses keep headscarved women out of labor market and employ them in underpaid positions of an invisible nature, and thereby hold sway over wages to the disadvantage of women in general and of headscarved women in particular. Furthermore, potential social demands of the labor force- such as equal pay for equal work, half-time work, maternal leave-come to be easily disregarded due to women's decreasing participation in job markets. Because the social rights have more of a chance to proceed to the agenda in public or corporate companies, and given the fact that headscarved women are mostly employed in small-sized businesses or self-employed they have trouble accessing the rights that will keep them in job markets [10].

15.2 Is Headscarf a Political Symbol?

Most of those who oppose the freedom of headscarf claim that it is a political symbol and headscarved women serve to political interests of some politicians. In other words according to these claims the main motives behind using headscarf is completely political. Firstly, the main fallacy of this claim is that it accepts all headscarved women as a homogene group thinking in the same way and aiming the same things ([3] cited from Özipek 2007). It is almost impossible to find a group of people thinking in the same way and having the same ideals, and almost every social entity has its own variation in itself. Moreover it is very difficult to realize such homogeneity by the members of the group willingly and the realization of homogeneity requires an external intervention. But in the case of headscarf ban there is no such intervention to that entity. Accordingly the argument that ‘Headscarf is a political symbol’ doesn’t make any sense because of the fact that it is impossible to determine the purpose of every headscarved women for using it [3].

On the other hand the fact that the motives behind an action only can be known by the person doing it reveals the meaninglessness of the discussion. The meaning of a symbol only can be defined by those who use it. Besides, it is required that the meaning should be loaded on the symbol to make it meaningful. However, in the case of headscarf issue, the women using it don’t load that meaning on the headscarf. They are in a passive position. They don’t make any political definition about the covering. Those who state that it is a political symbol are those advocating the ban [3]. Moreover, a person can use headscarf just because she wants to do so, without any reason. No one has to adopt a determinist approach, and show reasons for their actions.

The results of the research conducted by TİKAD in 2010 are in accordance with this argument. According to the research, 93 % of the participants say that they cover only because of their religious beliefs [8]. So, the terms of ‘political symbol’ is the most debated point of the ıssue, but at the same time the most groundless one [3].

15.3 Do They Cover Due to the Pressure from Their Families?

The proponents of the ban claim that majority of headscarved women use it because of pressure from their family. This argument has two main flaws. Firstly this approach means that headscarved women constitute a homogeny group and they are unconscious. This people are at a certain age, most of them are university graduate and have the ability to show different reactions to the different situations with their free will. To accept that these women are so passive to use headscarf out of their will is a very shallow argument [3].

As Vahap Coşkun says, there are people having quitted their schools and jobs because of the ban, and their lives deeply affected by this [4].

On the other hand claiming that all headscarved women behave in the way because of their husbands or families’ desire means to underestimate them and the argument of pressure is therefore very problematic [3].

The TİKAD’s research suggests that, women start to use headscarf voluntarily (77 %). The findings of the research didn’t support the idea that they cover as a result of any kind of pressure. 96 % of the covered women says if they had the chance to go to back to the days when they covered, they would cover again. This figure also indicates their determination in using headscarf and their willingness [8]. So claiming there are pressure on these women to cover and therefore supporting the ban means to punish these women. Therefore the ban can’t be legalized by using this argument.

15.4 Whose Problem Is the Ban?

According to several researches, no less than 60 % of women with university degrees are wearing some type of headscarf in Turkey [10]. But as discussed in the previous chapters, headscarf is not only related to these sections of society. Because it is the symbol of conflict between the different classes in the society and this conflict has sometimes crippled the whole system including different organs of the state. Furthermore, when we look at the matter from the view of the public we see that, as studies and researches show, there is no conflict in the society over the headscarf issue. People are not prejudiced against headscarved women and vice versa. They have close relationship with each other [8]. For example there are covered women in the families of uncovered women. Even if they don't cover, the rate of having covered women in family circle is 70 %. This data is very important because it shows us the headscarf is very widespread in the society. 87 % of the uncovered women say that headscarved women's having career will not bother them. 88 % of them state that they see the covered women as equal citizens. 66 % of them say that they support covered women's working [8]. Additionally, a lot of politicians, writers, intellectuals and even some CHP politicians criticized the ban. 45 % of the CHP supporters doesn't think that headscarf is contradictory with secularism.

As a result, the problem belongs to the whole society and the state institutions. It is mainly about the basic rights and freedoms of the citizens and about democracy and rule of law.

15.5 Was Compromised Indispensable for the Solution?

As it is shown in many studies that while the majority of the people in Turkey wants the ban to be ended, relatively smaller and marginal group want to sustain the ban. This marginal group tries to make the society adopt their truths forcefully, and demand compromise without stepping back.

It will be useful for the course of the discussion to remember John Stuart Mill's thoughts on basic rights and freedoms: 'Even if all humanity adopts an idea and only one person think otherwise, humanity doesn't have the right to silence this person and vice versa' ([3] cited in [6]). These thoughts reveal explicitly that the individuals can't be violated by majority. In Turkey the supporters of the ban amount to 20 % of the people and they demand compromise with the majority of the society which condemn the ban. A state which respects to the human rights, is assigned to protect the citizen's rights and freedoms in spide of the majority's opposition [3]. The words of Özipek provides a satisfactory explanation about the issue: 'There are a lot of discrepancies in the suggestion that headscarf problem should be solved through social compromise rather than by legal and constitutional arrangements. First of all, human rights do not depend on approval of state or society, or the compromise between the different sections of the society. A political will which wants to build up democratic state of law, is obliged to protect to individuals rights against the society when needed' [7].

If basic rights and freedoms are the subject of the compromise and bargaining, this will cause to sacrifice the individual differences for the sake of society's general convictions. And this also cause the powerful ones to have the power of limiting basic rights and freedoms in a way that will benefit them [3].

If there will be any compromise, this can be realized by prohibitions' coming to the position required by human rights [7]. From this point of view the main reconciliation point in the debate is that the prohibitive mentality gives up its position and makes it possible for everyone to enter to the public sphere with their lifestyle and dressing.

15.6 The Policy of the Government on the Issue and Lifting of the Ban

As mentioned above, the headscarf ban is rather a matter of fight between the newly emerging socioeconomic classes and the old bureaucratic and capitalist elites. Again as mentioned before, there were severe debates, tensions between the institutions and state organs, especially between the judiciary and legislation-executive parts of the state. The Constitutional Court prevented by its judgments the executive and legislative to take any action toward lifting the ban. The governing party (AK Party) was almost to be shut down because of its efforts toward emancipation of the headscarved women. Even though these rulings and judgments of Constitutional Court are very controversial in terms of rule of law and democracy, it is an irrefutable fact that they suffocated the civilian political arena.

We could see some initiatives by the current government about the democratization process such as 12 September Referendum, some amendments of which can cause democratic changes in the membership structure of The Constitutional Court. It is claimed that it would be more difficult for the Constitutional Court to rule anti-democratic decisions with its new structure. According to Haber7.com (18.10.2012), the MHP and AKP were studying on a new constitutional package to lift the headscarf ban in universities and in the public sphere and to allow the headscarved women to be elected as deputy and mayor [1]. AK Party presented headscarved candidate in the Parliament elections in 2011. Even though she was ranked in a place that it was impossible for her to be elected, it could be seen as an indicator of the AK Party's intention. It should be also stated here that a parliamentary commission for the investigation into the coups experienced by Turkey so far was established. We know that a Merve İslam Kavakçı, who was fired from the Parliament because of her headscarf just before the 28 February process, gave testimony to the commission as one of the victims of the 28 February Postmodern Coup D'état.

Some think and say that the government was apparently wary of stirring another backlash not to cause tension in the country [5]. As mentioned above and in the definition of the problem the ban was deeply rooted in the power fight between the socioeconomic classes of the society, and the headscarf was the most apparent symbol of this fight. Hence, a complete eradication of the ban did not seem so easy.

In that situation, as the ban is not based upon any law or the constitution but based upon only a regulation and the unjust and illegal implementation of this regulation by the previous administrations, the government tried to change the implementation, and so, to create some relaxation for especially headscarved students in universities. But the ban was still in effect; it was being enforced in a limited fashion or fully in university campuses depending on the approach of the university administration in question. While this was in one way considered a 'relaxing' or 'softening' of the ban, young women aspiring to have university education see the uncertainty and arbitrariness as an ongoing cause of concern. The inability to be certain of one's own place and not knowing what treatment and regulation was awaiting one upon entering the campus were sources of tension and worry, to say the very least, on a daily basis. In public sector jobs, the continuing ban was incomparably stricter than on campus, and rested on different legal and political grounds [9].

In sum, the government tried to lift the ban through implementation, but on the other hand it tried to enact some legislation which would allow the headscarved women to work in public jobs, to be elected as mayor and deputy. But it was too cautious and wary to do so, and when we think about the previous experiences of it regarding the headscarf, we cannot say that there was not any reason to behave in this way.

On the other hand we could also observe softening of the approach of the CHP, most enthusiastic proponent of the ban within the political parties. CHP discharged its Kırklareli Youth Branch Chairmen as he insulted those who wanted to freedom for headscarved women in the public sphere, and he was also dismissed from the municipality he worked for. Hence no political party could stand against to the solution of the problem easily.

We saw at the same time some non-governmental organizations arranging some campaigns to condemn the ban in public sector jobs. For example MEMUR-SEN called for all the civil servants in the country to go to work with free dresses. The Chairman of the union said that they would continue the campaign until the ban was lifted.

Finally in 2013, seeing that there was no support for the ban on the side of the public and that CHP changed its approached towards to ban, the government issued a Prime Ministry Circular and changed the article 5 in the Regulations Concerning Dress and Attire of Employees in Public Institutions and Organizations. The phrases preventing headscarved women to work in public sector abated, thus an era characterized by bans and violations of the basic rights of the covered women closed officially. At the same time some women deputies joined the Grand National Assembly with their headscarves. As opposed to estimates of the proponents of the ban, neither there have been any signal of conflict in the public sphere nor has the secularism ended. And left were the women who lost years fighting for their rights.

References

1. AK Parti ve MHP'den Kamuda Başörtüsü İşbirliği! October 18, 2012, from http://www.haber7.com/partiler/haber/941556-ak-parti-ve-mhpden-kamuda-basortusu-isbirligi
2. Ayrımcılığa Karşı Kadın Hakları Derneği (2010) Toplumsal cinsiyet, eşitlik ve başörtüsü yasakları from http://www.ak-der.org/toplumsal-cinsiyet-esitlik-basortusu-yasaklari-2010.gbt
3. Başkan B (2009) Başörtüsü sorunu ve meşrulaştırılamayan yasakçılık. Liberal Düşünce 55(14). Retrieved October 1, 2012, from http://www.ozgurtoplumundegerleri.com/res/Burak
4. Coşkun V (2007) Anayasa Mahkemesi Başörtüsü ve Sorunu ve Siyaset. Retrieved October 7, 2012, from http://liberal-dt.org.tr/index.php?lang=tr&message=article&ort=340
5. Falk R (2012) Decoding The Turkish Headscarf Debate, Retrieved October 15, 2012, from http://www.global.ucsb.edu/orfaleacenter/luce/luce08/documents/Falk_TurkishHeadscarfDebate.pdf
6. Mill JS (2004) Hürriyet Üstüne (trans: Orman M). Dostel LDT Publications, Ankara
7. Özipek BB Mahalle Baskısına Hayır, Devlet Baskısına Evet mi?' Zaman Gazetesi, October 1, 2007
8. Türkiye İş Kadınları Derneği (2010) Başörtüsü ve Toplumsal Uzlaşma', Press Summary. Retrieved October 1, 2012, from http://gundem.bugun.com.tr/basortusu-anketinde-carpici-sonuc-129673-haberi.aspx
9. Türkiye Ekonomik ve Sosyal Etütler Vakfı (2012) Başörtüsü Yasağına İlişkin Değerlendirme ve Öneriler, Tesev Demokratikleşme Programı. Istanbul: TESEV Publications
10. Türkiye Ekonomik ve Sosyal Etütler Vakfı (2011) Headscarf and discrimination: professional headscarved women in the labor market. Istanbul: TESEV Publications

Chapter 16
Social Network Analysis: A Brief Introduction to the Theory

Şefika Şule Erçetin and Nilay Başar Neyişci

Abstract Social networks are self-organizing, emergent, and complex, such that a pattern appears from the interaction of the elements that make up the system (Newman M, Barabási AL, Watts DJ. The structure and dynamics of networks. Princeton studies in complexity. Princeton University Press, Oxford, 2006). These patterns become more apparent as network size increases. Social network analysis is a collection of concepts, measures, and techniques for relational analysis. It is an approach that is specifically designed to grasp the most important features of social structures and it is unrivalled in this task. It can be used to explore social relations themselves and also the cultural structures of norms and ideas that help to organize those relations in conjunction with material circumstances. The study discusses the ways in which relational and cultural structures can be investigated with a few simple network concepts. Theories of social structure inform and sustain the methods of social network analysis.

16.1 Introduction

Social network analysis is an applicable method for investigating relations and interactions. Social network analysis (SNA) is an approach rooted in anthropology, sociology and social psychology for assessing social structures [3, 16]. The social network perspective illustrates social systems as networks of various relationships [2, 12]. SNA centers upon the structure and design of relationships and explores to describe both their causes and consequences [3, 7, 17].

Ş.Ş. Erçetin (✉)
International Science Association (ISCASS), Ankara, Turkey

Faculty of Education, Hacettepe University, Ankara, Turkey
e-mail: sefikasule@gmail.com

N.B. Neyişci
Department of Educational Management, Planning, Supervision and Economics, Faculty of Education, Hacettepe University, Ankara, Turkey
e-mail: nilbasar@hacettepe.edu.tr

Ş.Ş. Erçetin (ed.), *Chaos, Complexity and Leadership 2014*,
Springer Proceedings in Complexity, DOI 10.1007/978-3-319-18693-1_16

It is over than 50 years that theory of social networks and the assumptions of sociometry and sociograms were developed. Barnes [1] is credited with the invention of the social network idea, in the early 1950s about a Norwegian island parish.

Organizations are considered to be social entities where a position can be measured by social-psychological and demographic data. Location and connection to others provides information about members and group identification. A person's perceived position is essential in determining his or her beliefs, interests and motivation for action [6, 7, 13]. Additionally, people with close ties tend to maintain similar interpretations of the organizational environment [5] and tend to act similarly. This result flows of social information and the tendency for people to seek out similar others.

Social network theory examines how the social structure of relationships encompassing people, group, or organization influences beliefs or behaviors. SNA addresses a set of methods for distinguishing and determining the patterns. The proposition of network approach is about the conception of reality that fundamentally perceived and investigated from the aspect of the attributes of relations between and within components instead of the attributes of these components themselves. In this relational approach, these components are social components: individuals, organizations and societies.

16.2 Outlook

SNA predominantly concentrates on the relationships between people, instead of on typical features of people. Analysis of relationships will support to discover the informal communication diagrams in an organization, which may then be declared to the formal communication structures. These diagrams can be used to define certain organizational experiences. The relationship diagrams contribute to actors with some analogies of the attitudes and behaviors of other organizational members; SNA may explain why members develop certain attitudes toward organization.

16.3 Utilization

SNA techniques spotlight the communication structure of an organization. The use of network analysis techniques distinguishes structural features such as the (formal and informal) communication patterns and the description of groups within an organization like cliques or functional groups. Also information flow between members can be determined. By using SNA techniques, further characteristics could be investigated which are the communication task perceived by employees, the communication styles and the effectiveness of the information flows. Social networks are examined at the extent to the theoretical question of researcher. Even though levels of analysis are not necessarily reciprocally restricted, there are three network levels: micro-level, meso-level, and macro-level.

16.4 Micro Level

At the micro-level, social network research usually founded with an individual, escalating with social relationships, or beginning with a small group of individuals in a particular social context.

Dyadic level: Dyad means a set social relationship between two individuals. Social network investigation about dyads focuses on framework of the relationship (e.g. complexity, effectiveness), social equality, and trends toward reciprocity/mutuality.

Triadic level: "Add one individual to a dyad, and you have a triad". In this level, SNA may consider closely on factors such as balance and transitiveness, along with social equality and tendencies toward reciprocity/mutuality [11].

Actor level: In a SNA the fundamental unit is an individual in his/her social environment, like "actor" or a "ego". Ego-network analysis concentrates on network features such as intensity, connection strength, density, centrality, status, connections, and ties [10]. This level of analysis is generally used in psychology or social psychology, ethnographic kinship analysis or genetic studies of ties.

Subset level: In this level, issues may start at the micro-level of analysis, however may traverse into the meso-level. It brings out on distance and attainability, cliques, united subgroups, or other group actions [9].

16.5 Meso Level

Essentially, theories in this level may interest in demography between the micro- and macro-levels. Nonetheless, it refers to analyses that are particularly constructed to acknowledge connections between micro- and macro-levels. With its low density feature Meso-level networks may demonstrate causative processes differing from interpersonal micro-level networks.

Organizations: Formal organizations are social groups that arrange tasks for a collective goal [14]. Organizational SNA may concentrate both on intra-organizational and inter-organizational connections as regards formal or informal ties. Networks inside an organization generally comprehend multiple levels of analysis, notably with multiple or semi-autonomous divisions. In such organizations, SNA is managed at a division and organization level, focusing on the interaction between the two networks [14].

Randomly distributed networks: In the 1980s, exponential random graph models of networks emerged as brand new methods of SNA. This structure represents social-structural consequence witnessed mainly in many human social networks, besides the general degree-based structural consequences in any social networks including reciprocity and transformation, and at the node level and attribute-based activity and popularity effects, because of borrowed from explicit theories about reliance among network ties. Parameters are disposed with respect to the predominance of small sections of graph composition in the network. Description of these parameters may be as expressing the consolidations of social processes, which bring about emerging a network. These probability network models for a set of actors allow regulation beyond the definitive dyadic self-reliance acquisition of micro-networks, allowing designs to be raised from theoretical structural organizations of social behavior [8].

Scale-free networks: Their degree distribution at least asymptotically accompanies a power law. In SNA theory, these networks have the degree distribution that extricates social groups' size distribution [8]. The distinguishing aspects of scale-free networks diverge alongside the theories and analytical mechanisms accustomed to construct them. Nonetheless scale-free networks in general have some collective characteristics. The relative commonness of vertices with a degree that greatly overtake the average is one of their outstanding characteristics. The highest-degree nodes, often called "hubs", may give definite functions in their networks, even if depending mainly on the social context. The clustering coefficient distribution, which declines as the node degree inclines, is a further characteristic of scale-free networks. This distribution again accompanies a power law [4].

16.6 Macro Level

Macro-level SNA; commonly detect the reactions of communications such as economic or other resource transfer interactions over a huge population, instead of detecting interpersonal communications. "Macro-level" network is primarily used similar as large-scale network in social and behavioral sciences, in economics. The term was applied initially extensively in the computer sciences.

Macro-level social networks exhibit specialties of social complexity, including important significant characteristics of network topology, alongside patterns of complex connections between complex elements (chaos theory, dynamical system and complexity science). These characteristics involve a heavy tail in the degree distribution, a high clustering coefficient, hierarchical structure, and community structure, assortativity or disassortativity among vertices. These characteristics further involve reciprocity, triad significance profile in the matter of agency-directed networks. On the contrary, several mathematical networks models as lattices and random graphs, do not demonstrate such characteristics [15].

SNA is concentrated on revealing the design of interaction between people. It is based on this design, which is an important feature of individuals. Choices of individual depend noticeably on how s/he is connected with the larger network. Social network approach is regulated by formal theory organized in mathematical terms, and grounded in the systematic analysis of empirical data. It has found important applications in organizational behavior, inter-organizational relations, and the spread of contagious diseases, mental health, social support, the diffusion of information and animal social organization.

References

1. Barnes J (1954) Class and committees in a Norwegian Island Parish. Hum Relat 7:39–58
2. Brass DJ (1992) Power in organizations: a social network perspective. Res Polit Soc 4:295–323
3. Borgatti SP, Foster PC (2003) The network paradigm in organizational research: a review and typology. J Manag 29:991–1013
4. Barabási A-L (2003) Linked: how everything is connected to everything else and what it means for business, science, and everyday life. Plum, New York
5. Burt RS (1976) Positions in networks. Soc Forces 55:93–122
6. Burt RS (1982) Toward a structural theory of action. Academic, New York
7. Burt RS (1992) Structural holes: the social structure of competition. Harvard University Press, Cambridge
8. Cranmer SJ, Desmarais BA (2011) Inferential network analysis with exponential random graph models. Polit Anal 19(1):66–86
9. de Nooy W (2012) Graph theoretical approaches to social network analysis. In: Meyers RA (ed) Computational complexity: theory, techniques, and applications. Springer, New York, pp 2864–2877. doi:10.1007/978-1-4614-1800 9_176. ISBN 978-1-4614-1800-9
10. Jones C, Volpe EH (2011) Organizational identification: extending our understanding of social identities through social networks. J Organ Behav 32:413–434
11. Kadushin C (2012) Understanding social networks: theories, concepts, and findings. Oxford University Press, Oxford
12. Lincoln JR. (1982) Intra (and inter-) organizational networks. In: Bacharach SB (eds) Research in the sociology of organizations, vol 1. JAI Press, Greenwich, CT
13. Reis HT, Collins WA (2004) Relationships, human behavior and psychological science. Curr Dir Psychol Sci 13:233
14. Riketta M, Nienber S (2007) Multiple identities and work motivation: the role of perceived compatibility between nested organizational units. Br J Manag 18:61–77
15. Strogatz SH (2001) Exploring complex networks. Nature 410:268–276. doi:10.1038/35065725
16. Tichy N, Fombrun C (1979) Network analysis in organizational settings. Hum Relat 32:923–956
17. Tichy NM, Tushman ML, Fombrun C (1979) Social network analysis for organizations. Acad Manage Rev 4:507–519

Chapter 17
Students' Activity as a Pedagogical Clause in the Formation of Leadership Skills

Anar Alimbekova, Kadisha Shalgunbayeva, and Aida Meirkulova

Abstract The article considers the types of activities which serve as pedagogical clauses in the formation of students' leadership skills. The influence of leadership skills on achievement goals, working in a team, motivation, self-management and culture of communication in the future professional actions is largely discussed in view of activities the students are exposed to. The formation of leadership traits however remains a complex undertaking at all levels and times. This complexity indeed begins with understanding and conceptualizing what leadership skills formation is all about, to understanding the technical components in the process of formation of leadership skills. The paper thus makes a deliberate attempt to examine these components via students' activity as a pedagogical clause.

17.1 Introduction

The development of human society has been passing through several historical periods. In Latin the term – civilis is used to evaluate the level of civil society's culture. The civilian society aims at unity of people's material and internal development. In this case it is considered that man should understand what to do in his own business, craft, what kind of product they have to produce and the appropriate ways of using labour, ability, finance and time. This is all about man's consumption need. The consumption need plays an important role in developing of human society. Every person belongs to a particular social group, category and nationality. There are activists in economic, social, political, cultural life as well as leaders among the social groups. Nowadays management of any structure, organization, group etc. by a leader is appropriate and profitable in the point of social view.

A. Alimbekova (✉)
Faculty of Social Sciences, Department of Social Pedagogy and Self-Knowledge, Eurasia National University named after Gumilev L.N., Astana, Kazakhstan
e-mail: alimbekova_66@mail.ru

K. Shalgunbayeva • A. Meirkulova
L.N. Gumilyov Eurasian National University, Astana, Republic of Kazakhstan
e-mail: santik@mail.ru; Aida_meirkulova@mail.ru

Ş.Ş. Erçetin (ed.), *Chaos, Complexity and Leadership 2014*,
Springer Proceedings in Complexity, DOI 10.1007/978-3-319-18693-1_17

The university as an institute of professional society has a big impact on formation of competitive, competent and qualified leaders. Nowadays society requires the development of professional-pedagogical preparation's quality in a way of directing the subject of educative experience in the process of education. One such kind of requirement will be the guarantee that the development of the professional specialist's leadership skills leading to proper results of any creative action.

17.2 Main Text

The alterations in human society have a direct connection with the people's activities. It is known in history that the development of society occurs via different kinds of directions of leader's actions. The activity tells the state of the person in a process of action. As a way of existence, the development of a person's communication skills, cognition and self-breeding activities take place [1]. If the base of existence in the world depends on the movement and action, the rate of it's quickness or slowness depends on the limit of need and keenness; so the driving force of activity is germinated from need and keenness (Maslow's theory of needs) [2].

It is imperative to consider several kinds of activities in this analysis. These are elementary activity, biological activity, and social activity. These kinds of activities are searched differently in a way of specificity of areas. Activity in sociology is one of the signs of a person's character and actions. It is divided as political activity, social activity and psychological activity.

Political activity defines the level of intervention of person or social group with the political process. This activity influences the political system through the aim of the social group and tries to change this system. Social activity is the complex structure that is directed towards solving consciously the obligation of a social group in a union. It can be estimated with the person's cultural level, education level and the tendency to relate with an environment with the help of a social activity.

The social activity shows that the people get behind their common social activities, their education and decision. Social activity is person's social activity. The social activity has different kinds as industrial, social-political, cultural-cognitive, domestic etc. The administrative-command system in management of society and over control of actions etc. brings about bad influence on spreading and fortification of social activity. When new objective situations are formed in the transition period, the programme of their implementation will be defined and the social activity will have public quality.

The activity in psychology shows the relationship of the person with the environment. In this case the activity is established according to the actions. In psychology the activity is divided into a person's activity, activity that depends on situations, searching activity, business activity and overmuch activity. A person's activity is a grasp of creation, actions, society and relationships. It is a feature showing a person's intervention to social alterations.

Searching activity is one of the many types of attitude. Searching activity is a person's attitude that directed to change unknown situation of the future. The person's activity describes his strongest in his ideological position and own view, the active life position that is shown by the unity of the words and actions, superior situation activity means being high from the level of subject's situation requirements.

With the help of a superior situation activity, the subject wins over the external and internal (for example, indifference, nonchalance etc.) obstacles. This kind of activity is shown in creational situations, in competing at cognitive activities and in overmuch activity. According to the pedagogical aspect the activity is considered in the same realm as intelligence activity, intellectual activity and cognitive activity.

Intelligence activity is a real universal activity clause that starts from elementary kind of creative actions to the hardest ones. Intellectual activity is a generator of leader's ideas. If there is no ideological or intellectual activity in the leader, he cannot interest the performers and cannot lead them. Scientists have different opinions about the cognitive activity. Some consider it as the action while others consider it as the special feature of the person. The cognitive activity cannot be described with only one sign, it is clearly noticed in it's diversity of features. Its duty is very powerful. By developing the cognitive activity the leadership possibilities in an unconcerned student can be seen.

Mentioned activities play a significant role in the formation of leadership skills of the students. These students' activities give an opportunity to consider the formation of the leadership skills as a pedagogical clause.

The first clause: cognize the origin of students' leadership and activity, consider the importance of certain defining of their suitability for collective society according to the labour and moral activity in the teaching and educational process. For example, we can pay attention on psychological, social, biological factors that have influence on activity:

1. Psychological factor: defines the person's view, features, and the unity of psychological processes. The main one of these factors is psychological features (orientation, temperament, quality, ability). Movement of psychological processes and origin of mental states depend on them.
2. Social factor: influences due to social relations that occur with being a member of one of the social groups, nation, etc. according to the student.
3. Biological factor: can be referred to as one of the types of the nervous system, structure of analyzers, natural reflexes, feelings, physical force, all body features, the colour of the skin etc.

In the formation of the leadership skill we should pay attention to the following factors:

1. Biological level: temperament, sex, state of health.
2. Social-psychological level: emotions, will, memory, ability, intellect, character.
3. Social-pedagogical level: according person's need, human relations, views, position, attitude etc.
4. Political level: according to political sociology, political choice, making decision, political character types etc.

Second clause: Training of the leader to communicate with people, to make interpersonal contact, to make constructive contact, knowledge of the practical psychology and their application for the achievement of the objectives, and also abilities to bewitch and interest people, and to be able to do certain activities for cooperation. To develop activity with the help of the various leadership interests. It is a self-government process. The level of performance of any task by a student is defined by his requirement and necessity to do this business, and thus manifestation of the leadership skills by the student shows his interest and desire to work. This is his desire to be developed as a personality. The leader is a person who set goals and a way of their achievement; is always ready for self-development. Therefore interest is a driving force of leadership.

Third clause: development of the leader and business activities. Business activity is directed towards the formation of a strategy of the organization for the purposes and tasks. Leadership activity is directed towards the creation of the conditions for the achievement of the objectives and tasks. Leadership and business activities are two parts of the unique administrative process. There always has to be students with the leadership skills who apply leadership in a collective manner. Leadership skills define a leader's role. Some leaders are capable of leading a crowd, to convince and interest anyone, and some are absolutely on the contrary, very sluggish, not resolute and not authoritative.

It is possible to notice traits or characters of those ought to be leaders. For example, manifestation of the leadership skills. The real leader always knows what he wants and what he has to demand from his team. He is able to listen to the internal voice. He can't be overpersuaded if he would start any action. He can't almost be manipulated as he always knows what is necessary for it. By the words of the psychologists, a feckless person goes into a tide of life and strong people are those capable of changing this tide using their own discretion. Here is a pertinently national statement about life, it says that the life of each person is in his own hands. Many people don't find what is a real necessity for them for the whole of their lives. Many examples in the book "Bring up the leader in yourself" by John Maxwell are given for such cases of life. Being a leader is not a rank and is not a position. It is the way of life and actions of a person. Leadership can't be considered as an inherited quality because a person's leadership develops over time looking at his aspiration to victory [3].

Fourth clause: the rule that we should always remember is that the activity of the leader should be equal with patience; otherwise the activity with a rush can lead to vanity. If the leader is patient in a relevant way, he will have an opportunity to act precisely and shortly with no long time actions. That's why ability of using activities adequately and precisely and ability of accepting and waiting belongs to leader's art. These two abilities should be equal, that kind of equality can help to solve the problems easily.

The leader student is a member of the group and has a right to make a decision in the important situations for the group and the group admits it, thus the leader plays the main role in the organization of integrated works and bringing order to unmutual relations. The leader student is an authoritative person who can save the unity of the team and in integrated works can defend the team's interest by having honesty and resourcefulness. However we shouldn't think that the leader is a person who is always leading everyone. The leader is an organizer, a person who directs, who gives ideas, a supporter of the group's interest but he cannot be seen as one in the major position. He is able to influence others for unity to work according to collective interest.

The leader's ability to influence students and teams and organize actions to achieve one of the aims determines the rate of leadership:

- active aspiration for responsibility and finishing the act begun;
- the power of will in the way of achieving aims
- initiation
- ability to influence character and actions of surroundings
- ability to accept consequence of activities and decisions
- ability to resist group frustration and disintegration
- influence of authority of the person or group member's character.
- knowing ownself, feelings, emotions
- being ready for the risk
- being confident to change his life
- active life style
- initiativeness and incentiveness
- unity, standard, invariable, reliable.

One more feature of the leader among students is to be not normal, original, orthodox, give a standard opinions in a non example way, have an criticism point of view, give own opinions through another, more interesting activities. In students' groups leaders act in the following cases:

- Planning the ways and means to achieve the goal
- Defining the people who are responsible for group activities
- If there is an important goal for the group, in this case he defends the group's interest and make the suitable psychological area
- Improve the social activities of group members

17.3 Results

It is possible to consider the formation of the leadership skills of the students as a pedagogical clause according to humanity. Considering the leadership abilities of the students and the leadership skills remains one of the topical problems.

Developing students' activities for formation of leadership skills should be organized through the following activities:

- Organize the leadership courses
- Lead the trainings
- Define and improve the individual, intellectual and human leadership skill

- Provide the social competence and humanity of group members
- Give precise tasks and seek precise results
- Develop stronger self belief
- Develop responsibility
- Lift authority

17.4 Conclusion

Considering students' activity as a pedagogical clause for formation of leadership skills in the educational and informative process gives us an opportunity to make the following conclusions:

1. It is impossible to define leadership abilities and leadership skills
2. It teaches us how to improve the self-belief, to respect each other, self-management, and how to work in a team.
3. It provides ways of reaching achievements in the future professional work. Олардың болашақ кәсіби іс-әрекеттерінде жетістіктерге жетуін қамтамасыз етеді.

References

1. Abraham M (2008) Motivation and personality. Petersburg Publisher, St. Petersburg, p 352
2. John M (2006) Bring up the leader in yourself. Poppuri Publishers, Billerica, p 201
3. Pavlodar (2006) Explanatory dictionary: sociology and politology. Pavlodar, Moscow, p 569

Chapter 18
Role of the Leaders' in Mobbing Cases as a Chaotic Phenomenon

Halime Güngör and Şuay Nilhan Açıkalın

Abstract Mobbing is a phenomenon that led to people crisis and at times has been caused serious health problems which has been applied individuals by leaders, by people who are in an equal position with victim, by subordinates or by other employees/workers. Mobbing lead to people dead end of the chaos; primarily in terms of work efficiency, motivation in the business environment; secondly in terms of individuals' psychological-physical health, family peace and finally in terms of economic difficulties when individual fired from work. Therefore, mobbing is naturally chaotic phenomenon. In this study, the role of leaders in cases of mobbing was discussed in terms of leaders' reactions to applied mobbing by leaders and by others. Leaders' approach to mobbing in organizational environment has been discussed in terms of chaos theory, have been explained by strange attractors phenomenon. Interviews were conducted with six individual who experience mobbing in Turkey that selected by using criteria sampling technique from purposive sampling methods. Mobbing related experiences, reactions of individuals who live mobbing, reactions of their colleagues and leaders, their views on coping with mobbing by legal ways have been analyzed according to individuals' perceptions who exposed mobbing. Furthermore, ten mobbing cases that seen in social media during the last 10 years in Turkey and leaders' roles and reactions to experienced mobbing cases in organization environment have been examined. As a result, transmission of chaotic life process to normal take long time when individual prefer to cope with mobbing by legal ways.

H. Güngör (✉)
Department of Educational Management, Planning, Supervision and Economics, Hacettepe University, Ankara, Turkey
e-mail: halime_gngr@yahoo.com

Ş.N. Açıkalın
International Relations Department, Middle East Technical University, Ankara, Turkey
e-mail: suaynilhan@gmail.com

Ş.Ş. Erçetin (ed.), *Chaos, Complexity and Leadership 2014*,
Springer Proceedings in Complexity, DOI 10.1007/978-3-319-18693-1_18

18.1 Introduction

In this study, the role of leaders in cases of mobbing was discussed in terms of leaders' reactions to applied mobbing by leaders and by others. Leaders' approach to mobbing in organizational environment has been discussed in terms of chaos theory, have been explained by strange attractors phenomenon. Interviews were conducted with six individual who experience mobbing in Turkey that selected by using criteria sampling technique from purposive sampling methods. Mobbing related experiences, reactions of individuals who live mobbing, reactions of their colleagues and leaders, their views on coping with mobbing by legal ways have been analyzed according to individuals' perceptions who exposed mobbing. Furthermore, ten mobbing cases that seen in social media during the last 10 years in Turkey and leaders' roles and reactions to experienced mobbing cases in organization environment have been examined.

18.2 Mobbing Is . . .

Mobbing in the etymological meaning is Latin origin and is derived from the word of "mobile vulgus" [8]. Mobbing means "undecided crowd" as a word [5]. Mobbing was used by the late Lorenz [11] in describing animal group behavior, he called the attacks from a group of smaller animal threatening a single larger animal "mobbing" (cited in [10, p. 67]). In this sense, mobbing is that an angry crowd attack collectively to other people by shouting [9]. Mobbing is the misusage of power [13]. Leymann is the first person that using the concept of mobbing as "violence at workplace". So, mobbing has been used to define moral harassment at workplace after Leymann [4].

Psychological violence is a long lasting torture applied to an employee usually due to his knowledge, success and positive behaviours, which set threats to others, by one or more individuals [6]. Mobbing is attitudinal, behavioral and emotional attacks to an employee in an organization by that employee's superiors/other employees or subordinates for the purposes of giving psychological harm [14]. Mobbing is a phenomenon that led to people crisis and at times has been caused serious health problems that has been applied individuals by leaders, by people who are in an equal position with victim, by subordinates or by other employees/workers.

18.3 Mobbing as a Chaotic Phenomenon

Chaos dominated life before order. Greek philosopher Hesiodos who lived in eighth century B.C. said that "... There was chaos before everything ..." in cosmic-themed work of *Thegony* [1]. Accordingly, when an individual whose life is ongoing in a stable way start to live mobbing, there will be negative changes in his/her life. Because mobbing is a phenomenon that has negative effects. According to Leyman [10] mobbing activities divided into five categories depending on the effects that they have on the victim:

1. Effects on the victims' possibilities to communicate adequately (management gives you no possibility to communicate; you are silenced; verbal attack against you regarding work tasks; verbal threats; verbal activities in order to reject you; etc.).
2. Effects on the victims' possibilities to maintain social contacts (colleagues do not talk with you any longer or you are even forbidden to do so by management; you are isolated in a room far away from others; you are "sent to Coventry"; etc.).
3. Effects on the victims' possibilities to maintain their personal reputation (gossiping about you; others ridicule you; others make fun about your handicap, your ethnical heritage, or the way you move or talk; etc.).
4. Effects on the victims' occupational situation (you are not given any work tasks at all; you are given meaningless work tasks; etc.).
5. Effects on the victims' physical health (you are dangerous work tasks; others threaten you physically or you are attacked physically; you are sexually harassed in an active way; etc.).

Mobbing lead to people chaos; primarily in terms of work efficiency, motivation in the business environment; secondly in terms of individuals' psychological-physical health, family peace and finally in terms of economic difficulties when individual fired from work. So, we can say that mobbing is naturally chaotic phenomenon.

On the other hand, psychological harassment in the work place cannot be predicted before that living. Mobbing is an unpredictable phenomenon. Furthermore, possibilities of future in human life are non-linear, open to surprises. An element that entering from any point to human's life system can be caused to drag system to chaos. If it is exemplified from the point of mobbing, let us think on a woman who started to new job, she is married, has no children, happy in her daily life. After a while, she gets pregnant. Then she can live mobbing by organization leader because of this pregnancy. In this case, her life can be dragged to chaos and at the end of this process, she can be unemployed. It is famously defined that strange attractors. In this case baby is a strange attractor. Erçetin et al. ([7]: 94) explain "strange attractor" concept such as:

> In chaotic systems it is possible to speak of more than one influential element. These sources of influence repel and attract and interact with the other elements of the system. The elements that attract are at the same time ones that are attracted. According to Newton's laws of motion, each attraction or repulsion is accompanied by an equal and opposite reaction. If you are attracted by a force, you will also attract the source of that force with an equal force in the opposite direction. This two-way interaction is the source of motion. Taking this attraction and repulsion to be a single phenomenon, we can refer to "attractors," or in the terminology of chaos, "strange attractors." These attractors give us information about the critical turning points in models that show system motion in chaotic environments.

18.4 Role of the Leaders' in Mobbing as a Chaotic Phenomenon

Mobbing at work comes out in different ways [14]. These are Downwards Mobbing has been applied employees by leaders; Upwards Mobbing has been applied leaders by employees (rare) and Horizontal Mobbing has been applied employees by other colleagues [3]. In addition, in research indicates two kinds of mobbing as individual and group mobbing [16].

This study's aim is to investigate leaders' roles on mobbing experiences as a chaotic phenomenon. This study has adopted a qualitative research method. This method aims to look at the underlying features beyond the details of everyday life [2].

Interviews were conducted with six individual who experience mobbing in Turkey that selected by using criteria sampling technique from purposive sampling methods. Participants that meet the criteria was determined by voluntariness. Participants coded as K1, K2 . . . . . . K6.

Mobbing related experiences, reactions of individuals who live mobbing, reactions of their colleagues and leaders, their views on coping with mobbing by legal ways have been analyzed according to individuals' perceptions who exposed mobbing.

Firstly, interviews were written on paper and computer, and then the content analysis was conducted. Later, the categories were identified and the findings interpreted. In the analysis of qualitative data, interview records were examined by putting them down on paper. Content analysis was used to analyze the data. Content analysis method is often used in the analysis of the qualitative interview data and open-ended questions. In this research, 'categorical analysis', one of the types of content analysis, was used. While analyzing the data in the content analysis, first of all, the data is converted into units that can be analyzed [12]. Then, themes were coded and categorized based on the research questions. Quoting categories directly from the data and so data validity is provided [15]. To ensure the validity, direct quotations have been given. Research results are shown below:

18.4.1 Mobbing Related Experiences

Victims lived excessive workload, not to get permission, not promoted in mobbing experiences. *"They began to hold responsible me from all negative developments in school. I had been going to home unhappy each day."* (K1). *"They assign another person to instead of me to my earning position. I asked why? Manager said: merit"* (K2). *"I lived lots of things that me down, but I can't believe that there is any subject that need to explain someone."* (K3). *"Manager's last stroke was that: carrying my room to another place in the work place which is under the stairs, narrow, stuffy, ceiling flowing without my information in summer holiday like threatened me before."* (K4).

18.4.2 Reactions of Individuals Who Live Mobbing

Victims' reactions on mobbing are that unhappiness, depression, not to forget they live, burnout, humiliation, feeling worthless, negative feelings to person who applied mobbing, reduction in organizational loyalty. *"I feel unhappy while going to job"* (K1). *"I feel burnout."* (K2). *"I cannot understand why I live these experiences, because I am always a positive person"* (K3). *"I cannot forgive and I will never forgive these people who try to damage me."* (K4).

18.4.3 Reactions of Their Colleagues

High proportion of the participants expressed that colleagues stay being quiet, did not give support. *"Colleague in my team never and ever gives support me and he did not help me on any subject."* (K1). *"I heard one of my colleague's sayings to another colleague on me. He said that she is not your special servant, leaving your garbage in everywhere is rudeness."* (K3). *"Colleagues that are not follower prefer to be quiet"* (K4). *"Colleagues are quite and they prefer to do their jobs." (K5)*

18.4.4 Reactions of Their Leaders

High proportion of the participants expressed that leaders of their organization applied mobbing. Only one participant's leader support to him/her. Leaders' reactions are southing, do not give permission, which is legal right of victim, rumoring. *"He called me and immediately he started to southing, said that you must to come to work, why you are not here and ring off."* (K1). *"I call my leader; I express my wife's situation to him. My wife is on the hospital. I said him to take permission, give me permission for one week. I call him again after one week than he said me I dismiss you from the profession."* (K2). *"They give me documents late so I must take doings to."* (K3). *"They are rumoring me with clients. They provocate the clients."* (K4). *"One day in weekend they remedy workplace, and I poisoning. Manager did not care with me."* (K3).

18.4.5 Their Views on Coping with Mobbing by Legal Ways

Participants did not cope with mobbing by legal ways without two participant. *"My professional enthusiasm was broken because of those terrible experiences. Because of my fatalistic life view, I did not get depressed, but I lived burnout. I change my job by the way of transition between institutions and I have been working in education."* (K2). *"God give them their penalties . . . "* (K3). *"I cope with mobbing by legal ways, I win. I feel exhausted. I have been living disappointment; my victory is not enough to cope with this disappointment."* (K4). *"I am forced to changing my workplace."* (K5). *"I cope with legal ways and I won."* (K6).

18.4.6 Mobbing in Turkey in Terms of Law

First lawsuit on mobbing in Turkey was on 2006. The person sued was the first person who wins the mobbing case in Turkey. Supreme Court approved the decision in July 2008 then this decision has been an example for the other mobbing experiences (http://tr.wikipedia.org/wiki/Mobbing).

"Prevention of Psychological Harassment in Workplace Prime Ministry Circular" published in Legal Newspaper in 19 March 2011, this circular is a legal development for Turkey. Cases about mobbing in Turkey are compensation cases and there is no law on mobbing in Turkey. When we examine mobbing ten cases which have been on media we can say that; eight of these cases' decision were positive for victim, other two cases' result were negative because of inadequate evidence. Mobbing process resulted with firing of the victims in many of the cases. Discipline punishment contrary to law, forced victims to resign of job, fired, from job, forced victims to unnecessary doings, overload of victims were seen in mobbing cases. Leaders applied mobbing in the seven of these cases and leaders did not support victims in the other three cases.

18.5 Conclusion

According to research results, we can say that;

- Victims lived excessive workload, not to get permission, not promoted in mobbing experiences.
- Victims' reactions on mobbing is that unhappiness, depression, not to forget they live, burnout, humiliation, feeling worthless, negative feelings to person who applied mobbing, reduction in organizational loyalty.
- High proportion of the participants expressed that colleagues stay being quiet, did not give support.
- High proportion of the participants expressed that leaders of their organization applied mobbing. Only one participant's leader support to him/her. Leaders' reactions are southing, do not give permission, which is legal right of victim, rumoring.
- Participants did not cope with mobbing by legal ways without two participants.

On the other hand; cases about mobbing in Turkey are compensation cases and there is no law on mobbing in Turkey. Leaders often applied mobbing according to cases in Turkey. Therefore, when we examine mobbing according to chaos theory, leaders like strange attractors. Because, leaders drag to chaos victims'/groups' life to chaos as an attracted element. As a result, transmission of chaotic life process to normal take long time when individual prefer to cope with mobbing by legal ways.

References

1. Alyanak A, Gündoğmuş B (2008) Karmaşıklık, araştırmacılıkta yeni paradigma. XI. Araştırma Zirvesi, Türkiye Araştırmacılar Derneği
2. Balcı A (2010) Sosyal bilimlerde araştırma: Yöntemler, teknikler, ilkeler. PegemA Yayıncılık, Ankara
3. Branch S, Sheehan M, Barker M, Ramsay S (2004) Perceptions of upwards bullying: an interview study. The fourth international conference on bullying and harassment in the workplace, London, UK
4. Cengiz SA (2007) Kişilerarası iletişimde sapkın şiddet: Manevi taciz (Mobbing). Kriz Dergisi 15(2):1–14
5. Davenport N, Schwartz RD, Elliott GP (2003) İşyerinde duygusal taciz (çev: Önertoy OC). Sistem Yayınları, İstanbul
6. Erçetin ŞŞ, Hamedoğlu MA, Çelik S (2008) Mobbing in primary schools: a case study for Hendek County, Sakarya. World Appl Sci J 3(6):945–955
7. Erçetin ŞŞ, Açıkalın ŞN, Bülbül MŞ (2013) A multi-dimensional approach to leadership in chaotic environments. In: Banerjee S (ed) Chaos and complexity theory for management: nonlinear dynamics, vol 5. IGI Global, Hershey, pp 89–104
8. Gül H, Özcan N (2011) Mobbing ve örgütsel sessizlik arasındaki ilişkiler: Karaman İl Özel İdaresi'nde görgül bir çalışma. Kahramanmaraş Sütçü İmam Üniversitesi İBBF Dergisi 2:107–134
9. Gün H (2010) Çalışma ortamında psikolojik taciz, mobbing/bullying, işyeri sendromu, 2nd edn. Lazer Yayıncılık, Ankara
10. Leyman H (1996) The content and development of mobbing at work. Eur J Work Organ Psychol 5(2):165–184
11. Lorenz K (1991) Hier bin ich- wo bist Du? Ethologie der grauganz. Here I am- Where are you? The behaviour of geese. New Edition. Piper, München
12. Mayring P (2000) Qualitative content analysis. Forum Qual Soc Res 1(2):20
13. Yaman E (2012) A scale of mobbing impacts. Educ Sci Theory Pract 12(1):237–240
14. Yaman E (2009) Yönetim Psikolojisi Açısından İşyerinde Psikoşiddet-Mobbing-. Nobel Yayın Dağıtım, Ankara
15. Yan Z, Wildemuth BM (2009) Qualitative analysis of content. In: Applications of social research methods to questions in information and library science. Libraries Unlimited, Westport, pp 308–319
16. Yagil D (2004) Mobs and bullies: collective and individual negative interpersonal behaviors. The fourth international conference on bullying and harassment in the workplace, London, UK
17. http://tr.wikipedia.org/wiki/Mobbing. Accessed 24 May 2014

Chapter 19
Chaotic Situations in Information Management Process

Nursel Yardibi

Abstract Information management aims to reach organized information easily at the time of need and to use in accordance with the intended purpose. The most important point in information management process is that managing information within the organization in the most effective way by using modern information technologies. Today, because of information management process uses information technology and management information systems if stored information cannot protect or back up, the chaotic situations will be existed in the organization. Intellectual memory of the organization would not be protected. This situation has a direct negative impact on the success of the organization. Chaotic situations that may occur in the information management process will be discussed in this chapter.

19.1 Introduction

Today, it is clear that parallel to the technological development processing of information and access to information is extremely easy. Accessing information is not problem in today, this process has occurred due to enter computerized systems to our lives. So organizations have been forced to renew and to transform themselves in parallel with the developments. Available information is stored on computers instead of files, so the possibilities of losing or changing information are reduced to extremely low. Information communication technologies has been started to use in management, so that the work to be done for the development of organizations may be healthier.

N. Yardibi (✉)
Dr. Fazıl Küçük Faculty of Education, KKTC, European University of Lefke, Lefke, Cyprus
e-mail: nurselyrdb@gmail.com; dryardibi@gmail.com

Ş.Ş. Erçetin (ed.), *Chaos, Complexity and Leadership 2014*,
Springer Proceedings in Complexity, DOI 10.1007/978-3-319-18693-1_19

19.2 Information Management and Importance of Information Management in Terms of Educational Organizations

Information management is defined as "to decide correctly in the right form, to the right person, with the right cost, at the right time, in the right place, providing the right information" [15]. Townley [14, p. 44] defines the information management as activity to control information to realize the mission of an organization or to reach out the goals of an organization and in terms of producing and sharing to implement the decisions taken in the most effective way. Information management, shortly, aims organizing crucial information and reaching that information where – whenever needed aims can be achieved [8].

Information management aims to reach organized information easily at the time of need and to use in accordance with the intended purpose. The most important point in information management process is that managing information within the organization in the most effective way by using modern information technologies. There are essential tools such as intranets, databases, networks, newsletters and video conferences as technological to store and distribute information. In addition, trust based relationships must be established in the organization to share and manage information successfully, face to face interviews and meetings should be held and individuals should be encouraged [3].

In our century, educational organizations have been forced to change their traditional structure and restructured themselves according to the needs of information society. Information society that, first of all, values information and knows how to use it and who can use it, can produce information. With these features, the information society is shaped by information technology. Decisive aspect of information society for training programs is "continuous learning" principle. In the information society, people will have to learn continuously [11]. The importance of knowledge management for organizations begins the process of knowledge generation. The importance of information management for organizations is started with the process of production of information. Information management is an organizational activity that was attended almost all employees in an organization. So, information management is needed an organizational culture which encourages learning, sharing information and using information technologies. This is, there is an important linkage between the concept of learning organization and information management [7]. Educational organizations should be learning organizations; basic building block of this is realization of information management process in a healthy and productive way.

Organizations that continue to be successful when the markets changes, technological developments and competitors increased are organizations that creating new information in a stable manner, disseminating this information largely to all parties in the organization and rapidly use the new technologies and products in the organization [9]. Information management applications are so important for organizations that resist to competition and steady. Sustainable success and quality is a value that is desired and expected by all stakeholders in today's schools and realization of this process is possible with the institutionalization of educational organization. The outstanding aspect of this success of institutionalized organizations is that they have managed successfully information management processes.

Information management allow using and managing resources in an efficient and productive way in the sense that to provide a baseline based on accurate information for planning and managing educational services, to create a system for data evaluation and data analysis, to provide support in the sense that system for using resources effectively. Using information management is required to progress by taking healthy and conscious steps towards success in organizations.

19.3 Management Information Systems

It is possible to find different definitions when we look for the definition of management information system. According to Argyris [2] Management Information Systems are systems that using different formal procedures to create information which is necessary for planning, directing and controlling of activities that are managers that all levels of management and which is convenient to make decisions on time and effectively by based on data from internal and external sources. To sum up, Management Information Systems are set of functions and systems related to processing, using and managing of information which is an organizational resource. Aim of the management information system is that [13]:

- to provide to information an active role in the organization by allowing production, storage, sharing and using the organizational information.
- to provide the necessary support in problem solving and taking decision processes.
- to develop and create information using and technological infrastructure of the organization
- to coordinate and integrate subsystems.
- to set up a good financial control system
- to prevent losing, destruction and forgotten of information
- to teach change and learn
- to provide saving and productivity
- to provide the best practical methods
- to provide useful information for troubleshooting
- to hold decision support systems
- to provide process documentation
- to provide human resources management
- to recognize the opportunity to work with a common data set to all units
- to reduce the risk of data inconsistency

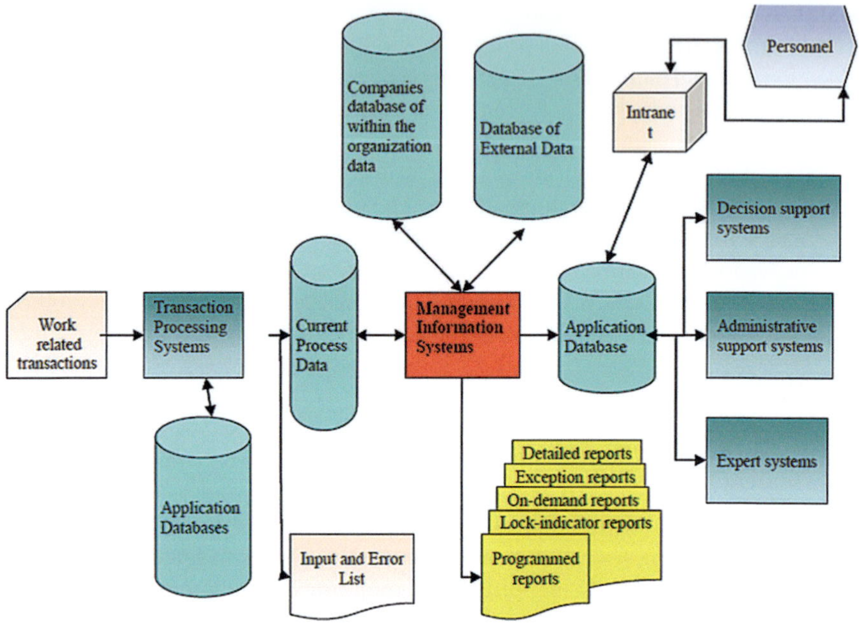

Fig. 19.1 Structure of management information systems, SGDB [13]

- to recognize the possibility to be recorded in a detailed manner to income and expenditures
- to contribute to the formation of organizational identity by providing to organizational works in more systematic and regular manner
- to provide different information for different users

The process that forms the basis of management information system concept is the process of converting data into information, obtained information at the end of transformation process should be capable of meeting these needs of managers at all levels when deciding. Structure of Management Information Systems is shown in Fig. 19.1.

19.4 Chaotic Situations in Information Management Process

Information management can is clear information that can be encoded, stored and transmitted [12]. Making organizational information registered information, forward information to the correct place in the organization and transfer it to the next the employees will prevent the recurrence of the worked steps and will also reduce the error rate [10]. Information management process will be damaged when any human error, faults or inaccuracies occur in information, which is transmitted, by an employee or manager during storage, recording or transmission of information. Social constructs like societies, organizations and leadership are chaotic, inter-dependent, non-linear systems that are closely tied to initial conditions [6]. According to the chaos theory, chaos has sensitive to initial point [1]. In this sense, in information management process chaotic situations will be experienced. So, necessary information that cannot be transmitted in time and correctly causes disruption of activities carried out in the organization and this situation will prevent the achievement of the organization's success and reaching goals. Because organizations' intellectual accumulation and information that play a decisive role in success or failure and in daily decisions have been transferred by information management [10].

Today, because of information management process uses information technology and management information systems if stored information cannot protect or back up, the chaotic situations will be existed in the organization. Intellectual memory of the organization would not be protected. This situation has a direct negative impact on the success of the organization. Chaos is a source of life and also a reality for organizations and people as living and self-organizing systems [4]. So, if principals cannot manage crisis situations and chaos in work environment, that organization should not function [5]. So, necessary precautions should be taken to prevent information loss and managing information management process in an healthy manner.

References

1. Açıkalın ŞN, Erçetin ŞŞ (2014) Great illusion in twenty-first century-chaos knocking door. In: Banerjee S, Erçetin ŞŞ (eds) Chaos, complexity and leadership 2012, Springer proceedings in complexity. Springer, Dordrecht. doi:10.1007/978-94-007-7362-2_1
2. Argyris C (1971) Management information systems: The challenge to rationality and emotionality. Manag Sci 17(6):275–292
3. Bayrakcı M (2007) Türkiye'de eğitim yönetimi bilgi sistemleri ve uygulamada karşılaşılan sorunlar. Kuram ve Uygulamada Eğitim Yönetimi 51:395–420
4. Erçetin ŞŞ, Hamedoğlu MA, Açıkalın ŞN (2014) Can managerial narcissism which likely leads to a chaotic process be measured? Managerial Narcissism Scale (MNS). In: Erçetin ŞŞ, Banerjee S (eds) Chaos and complexity in world politics, vol 7. IGI Global, Hershey
5. Erçetin ŞŞ, Kayman EA (2014) How to be a quantum leader in an intelligent organization? In: Erçetin ŞŞ, Banerjee S (eds) Chaos, complexity and leadership 2012. Springer, Dordrecht
6. Erçetin ŞŞ, Açıkalın ŞN, Bülbül MŞ (2013) A multi-dimensional approach to leadership in chaotic environments. In: Banerjee S (ed) Chaos and complexity theory for management: nonlinear dynamics, vol 5. IGI Global, Hershey, pp 89–104
7. Eren Gümüştekin G (2004) Bilgi yönetiminin stratejik önemi. Atatürk Üniversitesi İİBF Dergisi 18, 3(4):201–212, Erzurum
8. Maglitta J (1996) Know-How, Inc. Computerworld 30(1):15
9. Nonaka I (1999) Bilgi yaratan şirket. Bilgi Yönetimi İçinde, 29–50. Türkiye Metal Sanayiicileri Sendikası, Ankara
10. Odabaş H (2003) Kurumsal bilgi yönetimi. Türk Kütüphaneciliği 17(4):357–386
11. Özden Y (1998) Eğitimde dönüşüm. Pegem Yayınları, Ankara

12. Özmen F (2002) 21 Yüzyılda bilgi yönetimi ve eğitim örgütleri. Uluslararası Katılımlı Açık ve Uzaktan Eğitim Sempozyumu, Anadolu Üniversitesi, Eskişehir
13. Strateji Geliştirme Daire Başkanlığı (2010) www.hssgm.gov.tr/stratejikyonetim/bilgipaylasim
14. Townley CT (2001) Knowledge management and academic libraries. Coll Res Libr 62(1):44–55
15. Woodman L (1985) Information management in large organizations. In: Cronin B (ed) Information management: from strategies to action. Aslib, London, pp 97–114

Chapter 20
Complexity of Measuring Advertising Efficiency: An Application of DEA Method in Turkey

Yasemin Keskin Benli and Çağla Pınar Bozoklu

Abstract The aim of this research is to measure the efficiency of advertising expenditures on sales revenue of businesses that are traded on BIST in Food, Beverage and Tobacco Sector by data envelopment analysis. According to that, 22 businesses' 'Advertising Expenditure' and 'Net Sales Revenue' data for the years between 2009–2013 are gathered and they are respectively determined as 'input' and 'output' data. As a result, AEFES, BANVT and SKPLC coded businesses are ascertained as absolute efficient for all years. It is also found that CCOLA, DARDL, KENT, KERVT, TBORG and ULKER coded businesses' advertising budgets remains idle.

20.1 Introduction

Today, brand perception created by advertising strategies has become one of the most important factors because of the characteristic similarities of the rival products. It is observed that impact of advertising expenditures on sales revenue and sector advertising elasticity rises especially under conditions of emotion oriented purchase decisions and/or intensive brand competition.

Marketing, Selling and Distribution Expenses is defined as 'various expenses related to marketing, selling and distribution functions of a business' commodities and services [2, p. 188] and *Advertising Expenditures* are given as footnote information under this account. Therefore, detailed information about this expense cannot be reached and the measurement of direct impact of advertising expenditures on

Y. Keskin Benli (✉)
Faculty of Economics and Administrative Sciences, Department of International Trade, Gazi University, Ankara, Turkey
e-mail: ykeskin@gazi.edu.tr

Ç.P. Bozoklu
Faculty of Economics and Administrative Sciences, Department of Healthcare Management, Gazi University, Ankara, Turkey
e-mail: cputkutug@gazi.edu.tr

Ş.Ş. Erçetin (ed.), *Chaos, Complexity and Leadership 2014*,
Springer Proceedings in Complexity, DOI 10.1007/978-3-319-18693-1_20

sales revenue creates a chaotic and complex situation. However, indirect impact can be evaluated by analyzing the concept of advertising expenditures and advertising budget strategies carefully.

Within this context, the purpose of this research is to measure the performance of advertising expenditures on sales revenue of businesses that are traded on BIST in the Food, Beverage and Tobacco Sector by DEA. According to that, effects of advertising expenditures on sales revenue, conducted methods to measure this effect and theoretical base of data envelopment analysis are explained first, and then advertising performance within and between businesses of sector are analyzed.

20.2 Effects of Advertising Expenditure on Sales Revenue

Advertising expenditure identified under the Marketing, Selling and Distribution Expenses is one of the most significant factors that impacts sales revenue. However it does not appear in the footnote information by a common term in Turkey. It can be titled as *advertising expenditures, advertising and publicity expenditures, advertising or marketing expenditures* and etc. Since there is no commonly accepted word as an account term, measurement of the direct impact of advertising expenditures on sales revenue brings out significant complexity. However it seems possible to measure the advertising expenditures' indirect impact by means of analysis of advertising budget strategies of businesses.

Businesses can utilize various methods for setting advertising budget. Comparing with others, the competition and percentage of sales are the most preferred methods.

The first method enables businesses to avoid unnecessary competition of advertising and overwhelming expenses. But also, it may lead to miss the opportunities to increase their market share. In addition, there is no sufficient proof whether the method decreases the competition [7, 30].

Thomas Kamber [21] compared the change of sales, profits and market shares of businesses, which decrease, stabilize or increase the advertising budget in the recession period. As a result, he found significant growth on sales, profits and market shares of businesses, which stabilize and increase advertising budget in the prevailing and following periods.

On the other hand, the second method suggests determining the advertising budget depending on a certain percentage of sales revenue or profit. The data of sales revenue or profit can be generated from previous year's statements or prevailing year's pro forma statements. This method can make easier to find out the effects of advertising expenditures on sales, profit and market share [30]. However, the perception of sales revenue as a cause of advertising instead of a result might be dangerous. It can lead businesses to unnecessarily increase or decrease the advertising budget related to change of sales [7]. It may be more convenient to think that daily sale is a function of the past and prevailing advertising activities [30].

Kristian S. Palda [35], Russell S. Winer [48], Kamber [21], Tellis and Tellis [42] and many other researchers state that there is a strong and positive correlation between advertising expenditures and sales revenue including economic recession periods. Moreover, it is indicated that advertising expenditures have a strong impact on sales in the short term-for a year- and also decreasingly growing impact on sales in the long term-more than a year.

20.2.1 Methods Used for Measuring the Effects of Advertising Expenditures on Sales Revenue

The most widely used method is the historical analyses for measuring the effects of advertising on sales. This method based on an assumption that relation between previous years' advertising expenditures and sales revenue presents a similar tendency for following years. Within the context of this method, correlation and regression analyses can be conducted. Roland S. Vaile [46], Palda [35], Winer [48], Kamber [21] has found positive relation between advertising expenditure and sales revenue.

The other method is called as experimental method. In terms of this method, a series of markets, which have similar rate of advertising expenditure/sales, are chosen. Expenditure rates are reorganized as lower, same and higher for some specific markets in the testing period. At the end, changing of sales revenue is analysed for reorganized markets [7, 10, 30]. Different models also can be developed with the expansion of other related variables [30]. Clarke [14], Assmus et al. [5] and Bemmaor [8] also found positive impact of advertising expenditures on sales by conducted experimental models.

Previous researches, which have some limitations, present that consistent advertising expenditures affect sales revenue positively with a decreasingly growing rate in the long term [34]. *Cumulative Advertising Effects Theory* suggests that current advertising expenditures' additional impact on sales does not last only for the prevailing accounting term; it also continues for a longer time. On the other hand, there are some researches, which do not support the theory (e.g. [1, 14, 36]) although it is quite popular and widespread. For instance, *Current Advertising Effect Theory* defends that sales is a function of the current advertising expenditures and lasting effect of advertising cannot be totally loaded to previous years' advertising expenditures [47].

20.2.2 The Measurement of Advertising Expenditures' Performance by DEA

DEA method is recently used for analyzing the efficiency of advertising expenditures. DEA was introduced by Charnes et al. [12] and the main purpose of the model is the development of the best production curve without any limitations on production technology. DEA method is oriented to limits instead of central tendency.

In other words, it is focused on developing a linear rational plane, which involve marginal observed data rather than a regression plane, which adjust center of data.

DEA, which employs a kind of mathematical programming technique without dependence on parametric estimations, is accepted as a practical and simple method for measuring relative efficiency of a business. The main idea is the measuring efficiency of a decision-making unit relatively to another decision-making unit in the same industry. The limitation of the model is that all decision units should be over or under the efficiency limit. Efficiency limit of well-behaved observed points is set by means of linear programming and efficiency of all observations is measured relatively to that limit.

The units equal to one are valued as efficient whereas units lesser than one are valued as inefficient. The difference between one and the business's efficiency value represents that the same output can be maintained by lesser input according to the mentioned difference [49].

Another factor should be taken into consideration is related to the selection of the approach; whether input oriented or output oriented approach. Within the context of input oriented approach, minimum input is determined for producing specific quantity of output. On the other hand, output oriented approach is based on the idea of maximum output production with specific quantity of input. Solving of dual optimization problem results as the same efficient limit however some differences can be rarely occurred on inefficient units [49].

It should be decided whether returns to scale should be constant or variable. The constant returns to scale approach assumes that each of the decision-making units works at optimum scale and there is no inefficiency of scale. In the condition of some inefficient units' existence even though constant returns to scale is utilized, inefficiency of scale can only be viewed in the technical inefficiency and it is impossible to disaggregate. Briefly, efficiency under constant returns to scale is composed of both technical and scale efficiency. On the other hand, variable returns to scale approach assume that there is no inefficiency due to scale. According to that assumption, the source of the inefficiency is accepted as pure technical inefficiency and higher efficiency values are obtained comparatively to constant returns to scale approach. Efficiency values within both approaches are proportioned by the terms of measuring whether there is inefficiency due to scale [40, 43]. The mathematical method of DEA can be reached from many published researches such as the article of Tim J. Coelli [15].

Application of this method has been recently conducted in many researches ([3, 4, 6, 9, 11, 13, 16–20, 22–29, 31–33, 37–39, 41, 44, 45, 49, 50]. Most of them focus on the efficiency of production. Joachim Buschken [9], as one of the research on measuring the efficiency of advertising expenditures, assesses the advertising efficiency in German automobile industry by means of DEA. As a result, it is found that only 8 % of advertising expenditure remains idle. Similarly, Lou and Donthu [31] utilize DEA for analyzing the efficiency of 100 businesses, which expense higher advertising expenditures. According to the findings, it is stated that advertising expenditures are used efficiently and also sales revenue might be increased 20 %.

20.3 Method

The purpose of this research is to measure the technical efficiency values by DEA for analyzing the performance of 30 businesses that are traded on BIST in the Food, Beverage and Tobacco Sector between the years of 2009 and 2013. However 8 of 30 businesses' data is missing for several years since they are established later. Because of the nature of that research, panel data should be utilized and therefore all businesses should have the data for all years. Thus, the research sample is limited by 22 businesses. Annual *Advertising Expenditure* and *Net Sales Revenue* are respectively accepted as input and output variable. Accordingly, input and output oriented approaches are applied under the variable returns to scale assumption. In addition to that, advertising budgets of businesses are measured by *Annual Advertising Expenditure/Annual Sales Revenue* and it is used for finding out the advertising strategy of businesses. All gathered data is analysed by Win Deap 2.1 (developed by [15]) and IBM SPSS Statistic 21.

20.4 Research Findings

Research findings are presented under the *Technical Efficiency Values of Businesses, Percentages of Annual Advertising Budgets of Businesses and the Sector, Improvement of the Inefficient Businesses* titles.

20.4.1 Technical Efficiency Values of Businesses

Under the output oriented assumption, efficiency values of 22 businesses and the average technical efficiency values of Food, Beverage and Tobacco sector (between the years of 2009–2013) are given in the Table 20.1.

The mean technical efficiency value of Food, Beverage and Tobacco Sector is found as .530 for the year of 2009. According to the measured index of mean technical efficiency, inefficiency rate is equal to (1−.530) = .470 or 47 %. That rate can be also referred as the sector's inefficiency degree of source using. Moreover, it is seen that AEFES, BANVT, MERKO, SELGD, SKPLC and TATGD coded ones have the full efficiency (1.00) whereas the others are found inefficient. Sales revenues of these efficient businesses are respectively as 3,811,067,000 TL, 791,600,038 TL, 65,296,784 TL, 7,870,771 TL, 261,421,552 TL and 693,686,750 TL. Their advertising expenditures are also respectively 405,857,000 TL, 7,404,735 TL, 1,051 TL, 0 TL, 111,618 TL and 4,952,651 TL.

The mean technical efficiency value of Food, Beverage and Tobacco Sector is measured as .514 for the year of 2010. Within the context of mean technical efficiency index, inefficiency rate is equal to .486 or 48.6 %. Moreover, AEFES, BANVT, MERKO, SELGD and SKPLC coded businesses are valuated as full

Table 20.1 Output oriented technical efficiency values of businesses in the food, beverage and tobacco sector under the variable returns to scale assumption

Businesses	2009	2010	2011	2012	2013
AEFES	1.000	1.000	1.000	1.000	1.000
BANVT	1.000	1.000	1.000	1.000	1.000
CCOLA	.936	.953	.986	1.000	1.000
DARDL	.119	.058	.060	.097	.060
ERSU	.059	.104	1.000	1.000	.305
FRIGO	.057	.109	.199	.307	.277
KENT	.561	.383	.400	.376	.300
KERVT	.197	.163	.178	.181	.158
KNFRT	.251	.435	.183	.218	.719
KRSTL	.164	.099	.118	.114	.384
MERKO	1.000	1.000	.907	.578	1.000
MRTGG	.594	.483	.492	1.000	.087
PENGD	.276	.273	.268	.253	.838
PETUN	.376	.375	.329	.325	.280
PINSU	.111	.120	.071	.093	.087
PNSUT	.574	.550	.539	.512	.440
SELGD	1.000	1.000	1.000	.686	.137
SKPLC	1.000	1.000	1.000	1.000	1.000
TATGD	1.000	.970	.668	.573	.463
TUKAS	.365	.337	.183	.197	.207
TBORG	.228	.176	.176	.236	.259
ULKER	.797	.718	.791	.923	.822
Sector mean	.530	.514	.525	.530	.492

efficient (1.000) whereas the others are not. Sales revenues of these efficient businesses are respectively as 4,168,793,000 TL, 1,002,379,302 TL, 66,635,246 TL, 8,215,953 TL and 286,847,421 TL. Their advertising expenditures are also respectively 449,321,000 TL, 8,432,089 TL, 28,377 TL, 0 TL and 290,315 TL.

The mean technical efficiency value of the sector is measured as .525 for the year of 2011. Inefficiency rate of the sector is equal to .475 or 47.5 %. AEFES, BANVT, ERSU, SELGD and SKPLC coded businesses are valuated as full efficient (1.000). Sales revenues of these efficient businesses are respectively as 539.413,000 TL, 5,642,583 TL, 31,070,815 TL, 10,310,595 TL and 335,816,234 TL. Additionally, their advertising expenditures are also respectively 539,413,000 TL, 5,642,583 TL, 2,371 TL, 0 TL and 248,915 TL.

It is observed that SELGD coded business is efficient without any allocated advertising budget (0 TL) in 2009, 2010 and 2011. The reason of this efficiency may depend on model's measuring relative efficiency.

The mean technical efficiency value of Food, Beverage and Tobacco Sector is .530 in 2012. According to the measured index of mean technical efficiency, inefficiency rate is equal to (1−.530) = .470 or 47 %. AEFES, BANVT, CCOLA, ERSU, MRTGG and SKPLC coded businesses are evaluated full efficient (1.000) whereas the others are inefficient. Sales revenues of efficient businesses are respectively as 4,319,725,000 TL, 1,261,068,037 TL, 3,819,302,000 TL, 26,392,425 TL, 5,028,161 TL and 401,772,270 TL. Their advertising expenditures are also respectively 742,789,000 TL, 9,392,514 TL, 367,713,000 TL, 3,914 TL, 0 TL and 232,772 TL. MERTGG coded business is found full efficient without any allocated advertising budget like SELGD's efficiency values in 2009, 2010 and 2011. On the other hand, the efficiency value of SELGD is measured as 68.6 % in 2012 despite the allocated 533 TL advertising budget.

The mean technical efficiency value of sector is equal to .492 in 2013. According to the measured index of mean technical efficiency, inefficiency rate is equal to 49.2 %. AEFES, BANVT, CCOLA, MERKO and SKPLC coded businesses are found full efficient (1.000). Their sales revenues are 9,195,773,000 TL, 1,655,406,636 TL, 5,186,445,000 TL, 73,798,054TL and 62,748,470 TL respectively. The advertising expenditure of them is also 1,008,841,000 TL, 8,357,439 TL, 443,849,000 TL, 39,045 TL and 1,518 TL.

As a result, AEFES, BANVT and SKPLC coded businesses are measured as efficient between the years of 2009 and 2013. The highest mean technical efficiency values of sector are seen in the years of 2009 and 2012 whereas the lowest value is occurred in 2013. In addition to that, the mean technical efficiency value of Food, Beverage and Tobacco Sector is tending to decrease in 2010 and 2013. The mean technical efficiency value of Sector is lesser than 1.000. It signifies that advertising expenditures could not generate the highest sales revenue or prevailing sales revenues could be generated by lesser advertising expenditures. Briefly, it is referred advertising expenditures remain idle. Because of that result, input oriented relative technical efficiency values of businesses are also analyzed. They are presented in Table 20.2.

According to values in Table 20.2, MRTGG coded business is measured as full efficient in 2011 in addition to results of output oriented DEA analyze.

20.4.2 *Percentages of Annual Advertising Budgets of Businesses and the Sector*

Advertising budget strategy is thought as a significant indicator for assessing the effect of advertising expenditure on sales revenue. Within this context, the rate of *Annual Advertising Expenditure/Annual Sales Revenue* is measured depending on the *Percentage of Sales* budget method. The mean of annual advertising expenditures of businesses, the mean of annual sales revenue and the rate of Annual Advertising Expenditure/Annual Sales Revenue are given in Table 20.3.

According to the Table 20.3, the advertising expenditures of businesses equal to 7.7 % of sales revenues in 2009. This rate measured as 7.9 % in 2010, 8.2 % in 2011, 9.3 % in 2012 and 8 % in 2013. It is seen that the increase of advertising budget continues up to 2012 and it begins to fall in 2013. The rates of annual advertising expenditure/annual sales revenue of each business are presented in Table 20.4.

Table 20.2 Input oriented technical efficiency values of businesses in the food, beverage and tobacco sector under the variable returns to scale assumption

Businesses	2009	2010	2011	2012	2013
AEFES	1.000	1.000	1.000	1.000	1.000
BANVT	1.000	1.000	1.000	1.000	1.000
CCOLA	.910	.931	.980	1.000	1.000
DARDL	.002	.003	.005	.015	.019
ERSU	.000	.029	1.000	1.000	.101
FRIGO	.000	.031	.022	.122	.053
KENT	.130	.058	.037	.043	.061
KERVT	.006	.008	.009	.010	.093
KNFRT	.010	.302	.079	.169	.017
KRSTL	.006	.032	.010	.029	.005
MERKO	1.000	1.000	.834	.484	1.000
MRTGG	.014	.001	1.000	1.000	.607
PENGD	.019	.132	.191	.206	.533
PETUN	.057	.135	.045	.032	.140
PINSU	.000	.009	.007	.007	.036
PNSUT	.190	.239	.127	.117	.126
SELGD	1.000	1.000	1.000	.144	.029
SKPLC	1.000	1.000	1.000	1.000	1.000
TATGD	1.000	.956	.370	.256	.250
TUKAS	.029	.079	.033	.026	.119
TBORG	.007	.009	.007	.007	.087
ULKER	.674	.493	.603	.855	.661
Sector mean	.366	.384	.426	.387	.361

Table 20.3 The rate of annual advertising expenditures/annual sales revenue of businesses in food, beverage and tobacco sector

Year	The mean of annual advertising expenditures (TL)	The mean of annual sales revenues (TL)	Annual advertising expenditures/annual sales revenue (%)
2009	40,804,096.59	532,448,234.14	7.663
2010	46,148,441.68	581,698,166.73	7.933
2011	54,861,635.73	669,625,469.68	8.193
2012	66,695,648.00	715,925,853.91	9.316
2013	83,573,373.27	1041,394,944.00	8.025

According to Table 20.4, all businesses expense advertising expenditure lesser than the mean of sector advertising budget rate excluding AEFES, CCOLA and ULKER in 2009. CCOLA and ULKER coded businesses are not efficient though they allocate advertising budget over the sector trend (see Tables 20.1 and 20.2). In the year of 2010, all businesses set advertising budget below the sector trend with exception of AEFES, DARDL, KERVT and ULKER coded businesses. Only AEFES is found as efficient in that group. In the year of 2011, all businesses with the exception of AEFES, CCOLA, KENT, TBORG and ULKER set advertising

Table 20.4 The rates of annual advertising expenditure/annual sales revenue of each business in food, beverage and tobacco sector (%)

Businesses	Annual advertising expenditure/annual sales revenue (%)				
	2009	2010	2011	2012	2013
AEFES	10.65	10.78	11.33	17.2	10.97
BANVT	.94	.84	.51	.74	.50
CCOLA	10.07	9.84	10.22	9.63	8.56
DARDL	5.95	13.44	7.29	2.26	9.26
ERSU	3.74	1.21	.01	.01	.07
FRIGO	3.78	.78	.32	0.16	.15
KENT	4.34	8.84	8.69	8.63	7.66
KERVT	5.85	11.39	7.73	5.41	4.22
KNFRT	.16	.12	.58	0.26	.15
KRSTL	.21	.97	.58	0.48	.64
MERKO	.00	.04	.05	.09	.05
MRTGG	.00	.02	.00	.00	.05
PENGD	.74	.41	.27	.23	.22
PETUN	3.61	2.19	3.38	2.96	3.23
PINSU	4.97	5.17	7.38	7.41	5.73
PNSUT	2.82	2.61	2.92	4.34	3.83
SELGD	.00	.00	.00	.01	.51
SKPLC	.04	0.10	.07	.06	.00
TATGD	.71	.79	1.13	2.05	1.93
TUKAS	1.05	1.00	1.75	1.97	2.29
TBORG	5.43	9.79	9.45	7.88	5.17
ULKER	10.30	10.77	9.73	8.03	7.88

budget below the sector rate. Similar to the previous year, AEFES is found efficient but the others' advertising budgets remain idle. In the year of 2012, all businesses excluding AEFES and CCOLA allocate advertising budget below the sector rate and both businesses are evaluated as efficient by DEA. It is figured out that all businesses except AEFES, CCOLA and DARDL set budget lesser than the sector trend in the year of 2013. In this case, DARDL coded business is classified as having an idle budget due to inefficiency value in DEA.

20.4.3 Improvement of the Inefficient Businesses

According to analyses, it is realized that some businesses are inefficient for generating sales revenue depending on their advertising expenditure and they also do not follow the sector trend. Input and output oriented DEA analysis values are ascertained for developing suggestions to improve efficiency of allocating advertising budget of inefficient businesses. These values express the additional sales revenue that businesses should generate and the over-allocated advertising

Table 20.5 Additional sales revenues that businesses should generate for improving their efficiency values (TL)

Business	2009	2010	2011	2012	2013
AEFES	0	0	0	0	0
BANVT	0	0	0	0	0
CCOLA	165,531,482	134,879,168	47,420,146	0	0
DARDL	599,827,364	769,365,367	771,388,309	431,730,103	1,562,909,224
ERSU	294,750,924	263,271,502	0	0	46,360,972
FRIGO	793,398,386	134,653,582	105,367,091	65,905,538	51,143,473
KENT	395,891,595	767,761,847	847,659,985	969,941,402	1,366,110,648
KERVT	647,063,238	909,548,476	977,848,581	1,051,654,446	1,413,548,966
KNFRT	163,469,008	43,415,228	289,073,106	263,944,948	23,346,707
KRSTL	140,922,315	202,621,196	156,402,662	170,052,107	77,202,899
MERKO	0	0	5,936,528	46,358,472	0
MRTGG	4,700,241	5,403,509	5,240,119	0	57,581,277
PENGD	222,416,606	210,896,598	215,328,661	252,844,966	16,503,166
PETUN	511,787,161	587,729,404	782,694,019	865,486,193	1,233,681,858
PINSU	439,695,311	506,079,054	1,038,636,307	973,631,392	1,154,297,775
PNSUT	357,255,469	472,883,902	556,691,127	692,155,888	1,029,315,621
SELGD	0	0	0	2,489,901	65,990,481
SKPLC	0	0	0	0	0
TATGD	0	24,160,572	377,763,500	556,077,752	918,557,365
TUKAS	243,063,988	245,844,160	456,255,302	478,999,909	526,334,235
TBORG	626,666,312	885,736,024	1,001,642,864	1,051,782,989	1,319,671,199
ULKER	395,074,353	597,248,610	473,976,478	194,134,933	596,244,978

expenditure that businesses should reduce for improving their efficiency values. Tables 20.5 and 20.6 respectively summarize these values. Accordingly, relevant strategic suggestions are given for each business.

According to Table 20.4, it is observed that CCOLA and ULKER coded inefficient businesses set advertising budget over the sector average in 2009. Therefore reduction of advertising budget is recommended. Table 20.6 shows the reduction amount of advertising budget (21,843,721 TL for CCOLA and 52,134,457 TL for ULKER) in order to keep prevailing efficiency level of both businesses.

According to Table 20.4, CCOLA, DARDL, KENT, KERVT, TBORG and ULKER coded inefficient businesses set advertising budget over the sector average in the year of 2010. Similarly, in order to prevent idle advertising expenditures, these businesses should have respectively identified advertising budget as 18,780,468 TL, 6,296,709 TL, 39,633,300 TL, 28,979 TL, 18,346,134 TL and 83,160,419 TL being lesser than prevailing year's budgets.

In the year of 2011, CCOLA, KENT, TBORG and ULKER coded inefficient businesses are determined as having advertising budget over the sector average. According to Table 20.6, the reduction amounts of advertising budget are respectively 6,944,217 TL, 47,330,596 TL, 20,102,995 TL and 69,409,227 TL for keeping prevailing efficient level.

Table 20.6 Over-allocated advertising expenditures that businesses should reduce (TL)

Business	2009	2010	2011	2012	2013
AEFES	0	0	0	0	0
BANVT	0	0	0	0	0
CCOLA	21,843,721	18,780,468	6,944,217	0	0
DARDL	479,592,593	6,296,709	3,582,975	1,029,800	9,016,948
ERSU	69,196,750	359,479	0	0	13,470
FRIGO	667,283	124,873	81,899	40,181	27,166
KENT	19,080,346	39,633,300	47,330,596	48,161,596	42,008,959
KERVT	9,241,811	28,979	16,203,348	12,437,085	10,107,787
KNFRT	86,472	28,270	345,031	160,919	87,108
KRSTL	58,941	209,862	119,485	101,733	308,958
MERKO	0	0	4,803	28,263	0
MRTGG	68	1,083	0	0	982
PENGD	614,390	285,694	174,204	154,152	86,798
PETUN	10,475,529	6,687,557	12,394,465	11,925,350	13,285,883
PINSU	2,719,901	3,538,654	5,785,952	7,342,628	6,070,973
PNSUT	10,984,928	11,464,548	16,614,515	27,854,940	27,107,300
SELGD	0	0	0	456	51,722
SKPLC	0	0	0	0	0
TATGD	0	274,914	5,385,051	11,388,185	11,481,607
TUKAS	1,424,726	1,149,523	1,726,463	2,246,584	2,768,230
TBORG	9,970,134	18,346,134	20,102,995	25,421,694	21,779,545
ULKER	52,134,457	83,160,419	69,409,227	27,191,619	73,536,345

Surprisingly, none of the inefficient businesses defined advertising budget over the sector average in 2012. Thus, the insufficient sales revenues in spite of idle advertising expenditures cause inefficient values. It is recommended that they might consider revising strategies of advertising, media and public relations for developing their efficiency values. Table 20.5 states the amounts of sales revenues that should have been additionally generated by the inefficient businesses in order to be identified as full efficient.

In the year of 2013, it is seen that solely DARDL coded inefficient business set advertising budget over the sector average. According to Table 20.6, DARDL might conserve efficiency value in the condition of setting its advertising budget 9,016,948 TL lesser.

Briefly, it is advocated that inefficient businesses (2009–2013), which set their advertising budget over the sector average, should revise related strategies such as advertising budget, media and public relationship for improving their efficiencies.

On the other hand, in the year of 2009 DARDL, ERSU, FRIGO,KENT, KERVT, KNFRT, KRSTL, MERKO, MERTGG, PENGD, PETUN, PINSU, PNSUT, TUKAS and TBORG coded businesses; in the year of 2010 ERSU, FRIGO, KNFRT, KRSTL, MERTGG, PENGD, PETUN, PINSU, PNSUT, TATGD and TUKAS, in the year of 2011 DARDL, FRIGO, KERVT, KNFRT, KRSTL,

MERKO, PENGD, PETUN, PINSU, PNSUT, TATGD and TUKAS; in the year of 2012 DARDL, FRIGO, KERVT, KNFRT, KRSTL, MERKO, PENGD, PETUN, PINSU, PNSUT, SELGD, TATGD, TBORG, TUKAS and ULKER; in the year of 2013 all businesses except AEFES, BANVT, CCOLA, DARDL, MERKO and SKPLC found inefficient and they also set their advertising budget under the sector average. It is proposed to increase their sales revenues by means of improving sales promotion techniques and related sales strategies while reducing their advertising expenditures (see Table 20.5 for additional sales revenue that each business should generate for improving their efficiency values).

20.5 Conclusion and Suggestions

Within the context of this research, the efficiency of advertising expenditures on sales revenues of businesses that are traded on BIST in the Food, Beverage and Tobacco Sector is measured by data envelopment analysis (DEA). It has seen that the complexity of measuring direct impact of advertising expenditures on sales revenue can be eliminated by means of this method.

Research findings are organized in three dimensions. The first dimension includes values of output oriented DEA. According to results of this dimension, efficiency values of AEFES, BANVT and SKPLC coded businesses are found full efficient, with the value of (1.00) for all years. Amount of the generated sales revenue of SELGD in the years of 2009 and 2010 and MRTGG in the year of 2011 are also found efficient though no advertising budget is allocated. This result might be regarded quite surprising and interesting. In the sector perspective, it is determined that advertising budgets have remained idle with respect to the mean technical inefficiency values of Food, Beverage and Tobacco Sector for all years.

In the second dimension, input oriented DEA findings are given. In addition to output oriented DEA's results MRTGG coded business is also defined as efficient in 2011.

In the last dimension, advertising budgets are identified inconvenient according to the percentages of sales method by using annual sales revenue and annual advertising expenditure data. Therefore advertising budget strategies of inefficient businesses are analysed comparatively to the sector average. According to that CCOLA, DARDL, KENT, KERVT, TBORG and ULKER are the businesses which, set advertising budget over the sector average. It is observed that these businesses' advertising budgets remain idle. Rest of the businesses in the sector generates insufficient sales revenue while setting advertising budget under the sector average. Under these circumstances, reduction of the advertising budget according to values of input oriented DEA is recommended to the businesses such as CCOLA, DARDL, KENT, KERVT, TBORG and ULKER which have advertising expenditures over the sector average. Additionally, revising of sales promotion techniques and strategies related to advertising and media is also recommended to the businesses such as ERSU, FRIGO, KNFRT, KRSTL, PENGD, PETUN, PINSU, PINSUT and TUKAS, which have advertising expenditures under the sector average.

As a conclusion, it is suggested that the efficiency values of businesses in Food, Beverage and Tobacco Sector could be advanced. These results may also produce some useful insights to investors and practitioners who want to enter Turkey market. Especially, businesses that found efficient may attract more interest of hedge investors.

References

1. Aaker DA, Carman JM, Jacobson R (1982) Modeling advertising-sales relationships involving feedback: a time series analysis of six cereal brands. J Mark Res 19:116–125
2. Akdoğan N, Tenker N (2001) Finansal tablolar ve mali analiz teknikleri. Gazi Kitabevi, Ankara
3. Ali AI, Seiford LM (1993) The mathematical programming approach to efficiency analysis. In: Fried HO, Lovell CAK, Schmidt SS (eds) The measurement of productivity efficiency. Oxford University Press, New York, pp 120–159
4. Ar İM (2012) Türkiye'deki havalimanlarının etkinliklerindeki değişimin incelenmesi: 2007–2011 dönemi için Malmquist-TFV endeksi uygulaması. Atatürk Üniversitesi İktisadi ve İdari Bilimler Dergisi 26(3–4):143–160
5. Assmus G, Farley JU, Lehmann DR (1984) How advertising affects sales: meta-analysis of econometric results. J Mark Res 21:65–74
6. Atan M, Tanrıöven C, Kayacan M, Boztosun D (2004) Dünya borsalarının finansal göstergelere göre dönemlerarası karşılaştırılmalı analizi. Gazi Üniversitesi İİBF Dergisi 6(2):127–145
7. Babacan M (2008) Nedir bu reklam? Beta Yayınları, İstanbul
8. Bemmaor AC (1984) Testing alternative econometric models on the existence of advertising threshold effect. J Mark Res 21:298–308
9. Buschken J (2007) Determinants of brand advertising efficiency. J Advertising 36(3):51–73
10. Cemalcılar İ (1988) Pazarlama, kavramlar kararlar. Beta Yayınları, İstanbul
11. Charnes A, Cooper WW, Lewin AY, Seiford LM (1995) Data envelopment analysis: theory, methodology and applications. Kluwer Academic, Boston
12. Charnes A, Cooper WW, Rhodes E (1978) Measuring the efficiency of decision making units. Eur J Oper Res 2:429–444
13. Cingi S, Tarım AŞ (2000) Türk banka sisteminde performans ölçümü dea-malmquist tfv endeksi uygulaması. Türkiye Bankalar Birliği Araştırma Tebliğleri Serisi No.01
14. Clarke DG (1976) Econometric measurement of the duration of advertising effect on sales. J Mark Res 13:345–357
15. Coelli TJ (1996) A guide to frontier version 4.1: a computer program for stochastic frontier production and cost function estimation. CEPA working papers
16. Coelli TJ, Rao PDS, Battese GE (1998) An introduction to efficiency and productivity analysis. Kluwer Academic, Boston
17. Daneshvar RB, Erkan TE (2010) Ankara'da bulunan 4 yıldızlı otellerin, VZA-AHS sıralı hibrit yöntemiyle etkinlik değerlendirmesi. Gazi Üniversitesi İktisadi ve İdari Bilimler Fakültesi Dergisi 12(3):69–90
18. Deliktaş E (2002) Türkiye özel sektör imalât sanayiinde etkinlik ve toplam faktör verimliliği analizi. ODTÜ Gelişme Dergisi 29(3–4):247–284
19. Ertuğrul A, Zaim O (1999) Ekonomik kriz ve Türk bankacılık sisteminde etkinlik. METU Stud Dev 26(1–2):99–116
20. Güran MC, Cingi S (2002) Devletin ekonomik müdahalelerinin etkinliği. Akdeniz İİBF Dergisi 3:56–89

21. Kamber T (2002) The brand manager's dilemma: understanding how advertising expenditures affect sales growth during a recession. Brand Manag 10(2):106–120
22. Karacabey AA (2002) Türk bankalarındaki üretim değişiklikleri ve nedenleri. Iktisat Islet Finans 17(191):68–78
23. Karakaya A, Kurtaran A, Dağlı H (2014) Bireysel emeklilik şirketlerinin veri zarflama analizi ile etkinlik ölçümü: Türkiye örneği. Yönetim ve Ekonomi Araştırmaları Dergisi 22:1–23
24. Karsak E, İşcan F (2000) Çimento sektöründe göreli faaliyet performanslarının ağırlıklı kısıtlamaları ve çapraz etkinlik kullanılarak veri zarflama ile analizi. Endüstri Mühendisliği Dergisi 11(3):2–10
25. Kesbiç CY, Tokatlıoğlu İ, Ürüt S (2004) Rekabet gücü göstergeleri: Türk imalat sanayi için veri zarflama analizi-malmquist toplam faktör verimliliği endeksi uygulaması. Iktisat Islet Finans 19(220):63–75
26. Keskin Benli Y (2006) İstanbul Menkul kıymetler Borsası imalat sanayi için etkinlik ve toplam faktör verimliliği analizi. Seçkin Yayıncılık, Ankara
27. Keskin Benli Y (2012) Veri zarflama analizi (VZA) ve malmquist toplam faktör verimliliği (TFV): Konaklama işletmelerinde bir uygulama. Ege Akademik Bakış 12(3):369–382
28. Keskin Benli Y, Değirmen S (2013) The application of data envelopment analysis based malmquist total factor productivity index: empirical evidence in Turkish banking sector. Panoeconomicus 2(Special Issue):139–159
29. Koçoğlu D, Haşıloğlu SB (2008) Reklam harcamalarının işletmelerin etkinlik seviyesi üzerindeki etkisine yönelik bir araştırma. Afyon Kocatepe Üniversitesi İİBF Dergisi 10(1): 39–65
30. Kotler P (1997) Marketing management: analysis, planning, implementation and control. Prentice Hall International, Upper Saddle River
31. Lou X, Donthu N (2005) Assessing advertising media spending in efficiencies in generating sales. J Bus Res 25:28–36
32. Lovell CAK (1993) Production frontiers and productive efficiency. In: Fried HO, Lovell CAK, Schmidt SS (eds) The measurement of productivity efficiency. Oxford University Press, New York, pp 3–67
33. Orçun Ç, Çimen A, Şahin A (2014) Şirket etkinlikleri: IMKB 100 imalat sanayi şirketleri uygulaması. Dumlupınar Üniversitesi Sosyal Bilimler Dergisi 39:21–34
34. Özbek Y, Utkutuğ ÇP (2012) Cumulative advertising effects: presentation on financial statements of long term advertising effects on sales. Int J Arts Comm 1(4):237–247
35. Palda KS (1964) The measurement of cumulative advertising effects. Prentice-Hall, Englewood Cliffs
36. Parsons LJ, Schultz RL (1976) Marketing models and econometric research. North-Holland, New York
37. Salimi AM (2010) Türk sigortacılık sektöründe etkinlik: Veri zarflama analizi yöntemi ile bir uygulama. Gazi Üniversitesi İktisadi ve İdari Bilimler Fakültesi Dergisi 12(1):185–204
38. Seiford LM (1996) Data envelopment analysis, the evolution of the state of the art (1978–1995). J Prod Anal 7(2/3):97–137
39. Sezen B, İnce H, Aren S (2005) Türkiye'deki hayat dışı sigorta şirketlerinin veri zarflama analizi tekniği ile göreli etkinlik değerlendirmesi. Iktisat Işlet Finans 20(11):87–95
40. Stavárek D (2003) Banking efficiency in Visegrad countries before joining the European Union. Available at SSRN 671664
41. Tarım A (2001) Veri zarflama analizi matematiksel programlama tabanlı göreli etkinlik ölçüm yaklaşımı. Sayıştay Yayın İşleri Müdürlüğü Araştırma İnceleme Çeviri Dizisi No.15
42. Tellis GJ, Tellis K (2009) Research on advertising in a recession: a critical review and synthesis. J Advertising Res 49:9–40
43. Türker Kaya Y, Doğan E (2005) Dezenflasyon Sürecinde Türk Bankacılık Sektöründe Etkinliğin Gelişimi. Rapor No: 2005/10. Ankara: BDDK
44. Ulucan A (2000) Şirket performanslarının ölçülmesinde veri zarflama analizi yaklaşımı: Genel ve sektörel bazda değerlendirmeler. Hacettepe Üniversitesi İİBF Dergisi 18(1):405–418

45. Ulucan A (2002) İSO 500 şirketlerinin etkinliklerinin ölçülmesinde veri zarflama analizi yaklaşımı: Farklı girdi bileşenleri ve ölçeğe göre getiri yaklaşımları ile değerlendirmeler. A.Ü. Siyasal Bilgiler Fakültesi Dergisi 57(2):185–202
46. Vaile RS (1927) The use of advertising during depression. Harvard Bus Rev 3:323–330
47. Weiss DL, Windal PM (1980) Testing cumulative advertising effects: a comment on methodology. J Mark Res 17:371–378
48. Winer RS (1979) An analysis of the time-varying effects of advertising: the case of Lydia Pinkham. J Bus 52(4):563–576
49. Yayla M, Kaya Türker Y, Ekmen İ (2005) Bankacılık sektörüne yabancı girişi: küresel gelişmeler ve Türkiye. Bankacılık Düzenleme ve Denetleme Kurumu ARD Çalışma Raporları No.6
50. Zaim O, Taşkın F (1997) The comparative performance of public enterprise sector in Turkey: a malmquist productivity index approach. J Comp Econ 25:129–157

Chapter 21
Risk Society and Humanity

Rail M. Shamionov

Abstract Risk society is a society with nonlinear dynamics, where a person finds him/herself in a situation of constantly changing external conditions. This fact demands from a human being constant readiness for change, ability to make quick decisions, and evaluate potential risks and losses. That is why one of the major problems nowadays is that of psychology of a human being as a subject of social behavior under conditions of uncertainty, chaos, and dynamics. A special area of the study is the psychological image of a risking person and a risk-potential of his/her behavior, as well as determination of personal regulators of risk emergence and studying of the process of subjective risk evaluation. It is also dully argued by the paper that a situation of risk is that of chaos and complexity and a person who is at risk is equally chaotic and complex. Such a person may not even be able to understand themselves. A risk society equally is that where the normal balance has been lost and chaos can easily creep in. The paper thus explored humanity in the realm of a risk society while examining the possibility of dynamic balance thereof.

21.1 Is Personal Dynamic Balance Possible in the Risk Society?

Risk is a notion, which, as a rule, deals with a human being and his/her behavior. It is a phenomenon which is typical of the modern society. In other words, it characterizes both social phenomena and personality involved in them. However, this simplified scheme is more of an involuntary one, due to the fact that psychological knowledge is not enough to explain mechanisms and factors of risk in terms of the society as a whole, as well as any individual human being. It is still unclear what characterizes risk society and how it is interconnected with a personality.

R.M. Shamionov (✉)
Department of Social Psychology of Education and Development, Saratov State University, Saratov, Russia
e-mail: shamionov@info.sgu.ru

Ş.Ş. Erçetin (ed.), *Chaos, Complexity and Leadership 2014*,
Springer Proceedings in Complexity, DOI 10.1007/978-3-319-18693-1_21

Nevertheless, some peculiarities of the processes, which are taking place at the personality level in the risk society (or at least, a society, which is classified as such) could be identified through social and psychological analysis.

Investigations of risk, as a category of social human behavior, have recently increased [3, 4, 9, 25]. At the same time, researchers point out growth of manifestations of such behavior, this is based on internal acceptance of risk by a human being.

Beck characterized post-industrial society as a risk society. In his theory he claims that under new conditions people and communities are becoming more distracted from the social structure; industrialization strengthens as a consequence. Risk society is characterized by strengthening of the *threatening force of modernization and feelings of uncertainty and fear, which it breeds* [3].

Any social object gains psychological significance for a personality only in case when it (the object) involves personality. Therefore, social phenomena become phenomena for a personality due to the fact, that a person forms an attitude towards them, which is based on values, norms, and settings, which have been learned by a person through socialization.

Dynamics of life with its frequently unpredictable situational and qualitative resulting variables produces stimulating effect on the system of attitudes of a personality. Risk society in this case is an instance, which is stimulated and is stimulating different sorts of instability based on the objective tendencies of its development. This process is accelerating a variety of social and psychological personal phenomena, including formation of subjective uncertainty. Dialecticism of modern situation is hypothetical opportunities for satisfaction of personal needs, especially the basic ones, which are progressively improving (and of course, they are being stimulated externally), but they are accompanied by phenomena, which in reality either exclude or significantly complicate them. They are hiding certain dangers, which are caused by the society's actions aimed at luck (including the hope for lucky coincidences). These actions can be initiated by individuals, groups, or different social institutes. The existing unity of opposites can be easily "transferred" to a personal level, which, in its turn, is reflected on the system of its attitudes and emotions.

Under such conditions a person has to be in a constant state of readiness and experience a certain degree of intensity. It can be clearly seen in people, who are strongly stressed by the transition stage from subjective stability to instability, due to different circumstances. External conditions, which are perceived by any person individually, through his/her concepts, values, orientations have their impact on the system of personal experiences, therefore, we are faced with a situation, which requires balance between external and internal personal instances. It is quite obvious that depending on these strategies, which allow reaching the balance (and behavior to some extent); a person reacts to external instability. In terms of mundane consciousness and scientific knowledge, external instability is characterized as a negative state of the "personality" system. However, it is not so.

Firstly, not every personality experiences destruction when faced with instability. Secondly, when instability is reflected on personality, in a number of cases it creates conditions for activity, and, therefore, for subjective vision of a situation, which is no longer subjective. Finally, there is a sufficient number of mechanisms of psychological protection, which allow personality to preserve a certain balance between external instability and internal system of value and meaning formations. The external situation is perceived in such ways that its various elements are associated with corresponding "convenient" personal qualities. It can be turned into a frustrator only in cases when personality cannot reach the balance between external and internal instances with ordinary methods.

The most important factor for preservation of dynamic balance of a personality, in our view, is not objective instability, but rather internal instability of a personality, groups in various aspects of life, which is caused by stimulation of needs: starting from "average" advertising (rather controversial) created by producers of goods and services, and ending with significant changes in the sphere of ideology, which, notwithstanding various opinions about it, still exists and is transmitted by different socialization institutes. In this sense, the risk society (largely managed by joint movement of all its objects towards well-being), which is being filled with this kind of content, becomes an external circumstance for a personality, as well as a factor, which requires the release of certain energy and resources in order to preserve readiness for unexpected and unpredictable situation changes.

There seems to be a certain contradiction in the title of this paragraph, but, one has to understand that the notion of "risk" is in this case mutually applicable, i.e. from the point of view of external and internal events, which surround a person; a person is taking risks every time he/she is making a decision, which is aimed at preservation, creation (for oneself and for the lives of other people, due to the fact that the consequences of such decisions made "for oneself" can involve people around as well). Alongside with that, in our opinion, we cannot oppose the category of "dynamic balance", which is linked to desire for comfort, internal balance, and the category of "risk", especially, if it characterizes the society; objective external situation is not identical in the subject's understanding, its reflection can vary depending on different personal characteristics: individual-psychological and social-psychological. Apart from that, external instability can actualize mechanisms of adaptation, and therefore, stimulate a person to include certain phenomena into his/her internal instances. These phenomena are characterized by the notion of "risk" (internal risk norm), which can subsequently serve as a factor of subjective well-being. Psychologically, it can be referred to the phenomenon of "learned aggression", which is realized in such a way, that a person does not experience comfort without facing the necessity of attack [11].

Something similar is obviously and clearly applicable to a modern person in the risk society; his/her behavior could be described as a behavior of a person, who has been "taught to risk", moreover, personal risks are being externally stimulated, therefore, a *norm of taking risks* is being formed. Modern conditions often implement behavior, which is caused by a tendency to risk (for the sake of risk only), rather than the necessity to risk for making mundane, industrial, etc. decisions. But this tendency often acts as a motivator of behavior; presence and implementation of such behavior act as factors of subjective well-being [23]. For instance, in social professions (educators, doctors), where risk is more likely not to be approved of, there are individuals, who lace risk highly in their hierarchy of values. These people are characterized by higher level of satisfaction with life and work, although within the scope of other parameters (self-satisfaction, social relations) we have not discovered any significant differences [22]. Risk is an element of valuable-meaningful system for this category of people. Within modern socio-cultural situation desire for comfort and risk can be seen in various combinations, as well as level characteristics of both phenomena.

Over the last decades, psychologists and psycho-physiologists have been pointing out that a human being not only learns from the experience, but also forms his/her own experience in culture [32]. Here the attention has to be paid to psychology of a subject in order to find out how external and internal risks, which are being transformed by an individual personality, influence its subjective characteristics, which form the foundation of its behavior and emotions. Due to the fact that risk presupposes unpredictability of consequences of a certain decision, it can be conditioned by insufficient competence in a corresponding area, where a decision is being made. Apart from that, unpredictability has a certain "deviation step", i.e. the result can be approximate or "opposite from what is being expected". Therefore, risk is subjective ("more or less risk"), depending on the competence linked to conditions, situation, one's personal possibilities and other variables. In this sense, personal subjective position acts as a formation, which influences the behavior of a number of variables in the situations of risk; first of all, it is evaluation of risks, correlation between the significance of one's own values and meanings and the situational traits, etc.

The major problem, according to Grachev [8], is that instability of social situation creates conditions for uncertainty in one's own life path's conditions. It is extremely difficult for a modern person to find his/her lace in the dynamic and unpredictable system, where only short-term, or even situational stability is possible. In the work by Tugusheva [29], the youth's vision of social success includes such characteristics, as an ability to orient oneself under unstable conditions, take risks, etc.; while unsuccessfulness is linked to fear of difficulties, lack of self-confidence, the key discovery being that there is the connection with long-term life planning, strategies of planning for the future. Therefore, successfulness is understood as a quality, which can be achieved through personal activity, its subjective position. Formation of subjective positions is one of the essential points in the dynamics of subjective well-being. Subjectivity of a personality makes it less vulnerable in the risk society. Belief in one's luck objectifies personal uncertainty (lack of self-confidence) and personal belief and hope for successful outcome of the situation. Together with that, as a number of investigations show, people, who are oriented towards serious achievements, but are not quite confident in being capable of achieving them, achieve most [1, 2].

According to investigations of social psychologists [13, 28], the majority of people tend to avoid risk and make a choice, that can help them to avoid risk and gain some profit, therefore, they can accept risky offers only in cases when they are trying to avoid guaranteed losses; in other cases people are content with small-scale achievements or acquisitions. Therefore, the degree of "reasonable risk" exists every time when there is a situation of choice from various alternatives. Together with that, the "vision" of alternatives and possible "right" "choices" depends on the ability to "control the situation" and create multi-meaningful context. It is obvious that in the society, which is rich in different opinions and is tolerant, there are more opportunities for forming divergences. However, both groups and individuals tend to simplify and categorize different complex social objects, which promote the opposite: rigidity, intolerance, prejudices, and stereotypes. One has to understand that the risk society is a phenomenon, which is externally stable towards the groups and individuals; it is also formed due to their risk-oriented tendencies. Therefore, one cannot detach and set off personality against the society here; nowadays, one of the major tools for decreasing various risks, which have negative consequences for an individual and the society, is democratization, and, more importantly, development of institutional models of interaction and various communication schemes.

One more condition, which is in our opinion necessary for consideration is manipulation of mass consciousness, which is used for implementation of corporate interests at various levels of political (or any other) system. Here we mean speculation of instability proper, strategies of attribution of risk society as a consequence of impact of certain powers, which are "interested" in chaos, rather than attribution of risk society as an objective society, caused by natural transformations and dynamics within it. Presence of such processes, especially for a person who is not capable of analyzing events which are taking place, and who is stimulating the necessity for safety can act as an instability factor and is capable of causing a variety of negative emotions (the mass media and subculture promote it as well). It is obvious the society has to pay attention to critical thinking formation, because without it such tendencies will not decrease, but on the contrary, they act as additional factors of instability and lack of subjective well-being.

It is, however, rather contradictory to, on the one hand, stimulate individualism, and on the other hand, collectivist tradition, which is becoming stronger on the account of uncertainty, as personal safety, especially in "collectivist" societies is linked to building relationships, group efforts, which are aimed at achievement, desire to belong to a certain group. This correlation also becomes a factor of tensity and lack of well-being.

Therefore, we can claim that negative consequences are capable of increasing personal subjective well-being; they become an essential factor of the epoch; the decrease of possible risks for a personal state is linked to information provision for events, that are taking place at all stages, the increase of communicative culture of people, that influence decisions regarding many people (groups), competence regarding acceptance of culture, foreseeing their consequences and psychological culture as a whole, which allows to regulate one's personal life in the society, one's states and form the subjective positions of a personality effectively. It is important to point out that the notion of "risk" is as applicable to the variety of personal manifestations, as to those decisions, which it is making towards itself. In this case the same mechanisms are being implemented as in the situation with external risks.

21.2 Psychological Aspects of Social Risks

The problem of risk as a behavioral characteristic is becoming more topical in the modern era of humankind's development. Permanent changes in both external conditions of existence, personal realization, and (in case of their transformation) in the internal social-psychological and psycho-physiological characteristics are turning into a substantial foundation for consistent behavior of an individual. This behavior can be characterized as a risky one (risk promoting), as well as risk-prone in the sense of its implementation in the risky situation. To this extent, social risks should be perceived as a significant condition, because this kind of risks, which are conditioned by interactions between personalities and (or) are engendered by a personality (groups) and (or) characterize the relations between personality and society. Having pointed out the interdisciplinary nature of the problem, we would like to discuss its psychological aspects.

Personality in the process of its socialization learns the norms and values, which are forming the internal substance and are constantly "bumping into" the norms and values of others. On the one hand, they possess universality and stability (to a certain degree) and characterize personality as a bearer of general human norms, and on the other hand, they form individual profiles, which predefine specifics of social behavior of an individual personality. Socialization of a personality presupposes its social congruence (in this respect it acts as a part of society with all its peculiarities, including fulfillment of the requirement to correspond to a changing situation (situations) and undertake responsibility for risky actions, connected with social self-esteem) and personal congruence, which is connected with implementation of internal beliefs, acquisition of meaning of one's existence and realization of subject position. For that matter, actions of a socialized personality are determined by many variables and their analysis is impossible without understanding the whole history (biography) of formation and existence of a personality. What is meant here is not a certain epoch or political reconstruction of society with the process being implemented against their background, but primarily those interrelations, which are formed as a result of "collision" with another one, about those spheres, where socialization is being implemented.

Apparently, in the society with a high level of instability, these interrelations possess more complicated nature and at the same time "the price of expenses" is also higher: small temporally but extremely qualitative lag is possible between success and failure (as, for example, between poverty and wealth). In fact, because of personality socialization any personality is capable of actions, which are typical for manifestations of personal activity level, i.e. based on norms, values, attitudes, role, positions etc., learned in its process and thereby, personality is capable of actions, which are aimed at asserting oneself in the society. Obviously, the stronger the contradictions, differences between status groups personality is included into (economic, political, social, ethnic, confessional statuses etc.), the stronger this statement is fraught with risks (not only for a particular person but for his entourage as well). It is especially important to point out that in the process of socialization personality acquires an ability to make decisions on the basis of correlation of the process and results of actions (action) with the configuration of their value orientations. The role of subjective assessment as a decision-making situation is very important here as well as assessment of characteristics of decisions themselves, including possible expenses and risks, connected to their implementation.

Personality of a human inclined to risk is characterized by a number of properties. They are, for instance, decisiveness, proactiveness, thrill, tendency to negotiate social norms, envy, selfishness, external locus of control, etc., configurations of value orientations, where the value of risk has a high status, etc. [16–21]. At different times risk was associated with such personal characteristics as endurance [14], extraversion, neuroticism and psychoticisms [5], impetuosity [4, 6], riskiness [4, 6], egocentrism [12], self-effectiveness [30]. The issue of correlation between personal characteristics and riskiness, however, requires special empirical investigation. It is reasonable to suppose that individual characteristics proper are not determinants of riskiness, but their certain correlations could be linked to risky actions. It is true for social risks as well, which are actions of an individual or a group, which can violate their well-being or well-being of other individuals.

One of the significant reasons of social risks is social competition, which is a phenomenon characterizing competition between a personality and a group with other subjects in various spheres of social being. In this regard, actions, which are oriented towards possible success with a high obstacle influencing component (risky actions), turn into a tool for competition. At the same time social competition is seen as a socially approved form of competition for the best way (result) of personal (group) self-realization. An ability to compete, enter into competition with others is not only a form of subject positions, but also a condition for achieving social success, which is being programmed by the whole socialization process. However, in this case we do not mean doing harm to another (which is often demonstrated through envy towards another person's success), but rather an orientation towards one's own achievements, which is totally different due to the fact that the Other here acts as an instance of competition, as a sort of criteria of achievement (actual or potential); it is not an object of destruction. If we look at social risks from this angle, envy is one of the worst scenarios of the way the events unfold. As a result of studying acceptability of methods (tools) for achieving high economic standing, we obtained data testifying to clear tendencies of acceptability of unlawful means in envious personalities [24]. Thus, envious people have a tendency to use unlawful methods and consider them to be acceptable. At that, envious people tend to avoid paying the public transportation fare, as well as taxes. This data is supported by the correlation analysis results. Thus, acceptability of tax evasion is linked to envy towards the physical and material superiority of the other; non-payment of public transportation fare is linked to envy towards the social and material superiority; acceptability of bribe is linked to envy towards the social and material superiority [24].

In other words, among other effects of making behavioral decisions, envy is one of the destructive modi of human existence. Self-realization here is narrowed to realization at the account of another, his/her diminution, and destruction of things that he/she has created. Oftentimes there are cases when the essence of risk is that personal steps towards achieving a goal do not take into consideration possible consequences of these actions for others.

It should be mentioned that social risks in many respects are predetermined by cognitive abilities of a personality, and also by its emotional and conative sphere and orientation. However, the choice in favor of the strategy "all or nothing" or "better less, but of a higher quality" or "if it goes wrong, at least I tried" etc. depends on effects of personality socialization, mastering those behavior patterns that, in the opinion of the subject, can become "conductors" of success. It is clear that people focused on the rapid progress are at greater risk of failure, than those, whose steps to success are well considered, systematic and without high-level requirements. It is necessary to understand that many social institutes stimulate the individual's needs, broadcast the value of rapid success achieving, and the value of risk. This often leads to the opposite: having no other options, a personality finds itself in a situation of "convergence deadlock", i.e. makes a lot of attempts to achieve the goal or moves towards it using a single method; lack of variability, search for a wide range of solutions in certain kinds of situations creates a threat for claims, and thus the risk increases.

Social risks can be connected with inadequacy of personality in both cases of changing standard situation (situational social risk) and under social norms. While situational risk often has no strong public response, and has more to do with the fate of a particular person (and possibly his/her closest social circle), risks of inadequacy towards social norms have much stronger public response. This sphere of risks is particularly important today in terms of socio-psychological analysis of both factors and grounds in the form of effects of personality socialization. Going beyond social norms as a possible option of behavior implementation has recently become quite common in societies. This is happening due to many reasons, which have objective and subjective background (for example, changes of social arrangement and persistent legal collisions, problems of legal culture formation and legal nihilism, contradictions of subcultural norms, priority of subcultural norms over general cultural or traditional ones, etc.). But often such risks are associated with group behavior of a personality, and are mainly conditioned by its presence in the group. It is connected with adoption of group norms, expectations, and with the identification of personality and group in general.

Social destructive groups (stable or situational) unite individuals that have a specific value and semantic structure, backed with identity categories in which "insiders" and "outsiders" are quite clearly "separated". "Simple" categorization together with emotional and evaluative attitude towards the fact of belonging to a group may intensify (or in some cases, reduce) risk factors in a consistent social behavior of group members. Back in the early 1990s of the twentieth century L.M. Popov, who studied independent activity of youth formations, concluded that a group's exposure to romance is implemented through creating the effects of the empathic kind, where the sense of "us" is created against the emotional sympathy background [15]. Moreover, there are investigations where special aspects of behavior of group members under the circumstances of their life are specified. Thus, according to experiments of Sherif [27], competing groups can demonstrate antagonism, which intensifies on the account of common ill-being (for example, failure in activity, unattainable social status, economic well-being etc.). This is especially important when analyzing social risks, where the subjects are particular groups or their representatives. At the same time, the nature of personality and group behavior that forms a risky situation is connected with overcoming social norms; it is defined by certain conditions, among we have to outline emerging and related to the public mood, situational and personal preconditions.

One more psychological aspect of social risks is introducing such external cultural phenomena that do not presuppose (or have) any ready-made forms of personality and group adaptation. It often actualizes protective forms of response (including destructive ones). But there is equally high destructions probability in case of insufficient knowledge of the group culture, i.e. superficial socialization and lack of strong mental anchors leads to intolerance (which, by the way, has been shown, in the study by A.I. Dontsov, T.G. Stefanenko, Zh.T. Utalieva in 1997).

In the study of R. Shamionov [17] it is shown that there is a negative relationship between the level of the ethnic culture knowledge and ethnic fanaticism, between significance of preservation, development of ethnic culture and ethnic fanaticism. These findings show that cultural ignorance is a significant factor of absolute domination of ethnic interests and objectives, readiness to make any sacrifices and produce any actions for their sake, including aggression against representatives of other ethnic groups.

Meanwhile, many of the social risks are connected with political and economic decisions without any regard to cultural phenomena. In this respect the idea of Zhuravlev [31] that a significant social transformation can only be based on people's mentality is quite remarkable, while attempts to "change" mentality lead to rather sad consequences representing large-scale social risks. Frustration at a personal level can lead to a variety of negative consequences, but the same thing happens at the level of large groups. Since the behavior of large groups is governed by their social psychology, social representations of these groups possess important knowledge, which is opening the way to cognition and to some extent to the prediction of their behavior. It is important to take into account the fact that the risky action of a personality under conditions of objective social risks enhances the riskiness, which does not refer to a certain personality or sphere of two or more persons' interaction, but affects wider communities.

The individual proper experience is of particular importance within the context discussed in the present monograph. Experience of previous actions and their effect appear to be the prism which refracts everything that forms perceptual subject field; developing circumstances under which it is necessary to act, correlate with (sometimes unconsciously) the previous similar ones. It should also be noted that due to the fact that one of the outcomes of personality socialization is its inclination to use the already known schemes in decision-making, experience may determine risky behavior differently: in some cases it causes destructive, in other cases constructive behavior. In particular, a personality prone to stereotypical and conventional actions may resort to a well-known, but not constructive mode of action that leads to a programmed failure. However, actions related to risk avoidance can also contain a certain share of risk.

It is known that "the experience of victories" captures certain attitudes and related emotional states, which become motivators for seeking opportunities of achieving a "victory". Thereby a norm of risk is being formed, which usually accompanies every activity. This leads to the fact that many risks are either accepted as natural hazards on the way to achievement or their significance is reduced subjectively as something that is much less important in comparison with the expected successful result.

Thus, formation of personal determinants of risky behavior is connected with the process of socialization and its effects. However, for each specific case of riskiness manifestation they may have their own specifics. Moreover, let us point out the fact that many social risks (objective risks) are not qualified by their participants as such. This requires a study of value and motivation grounds, as well as the characteristics of public moods of the corresponding groups. It is known that group frustration of interacting individuals may lead to actions related to social risks with implications for larger communities. In this respect, one of the most important steps for prevention of social risks is a socio-psychological analysis of large social groups, as well as subcultures monitoring and socio-psychological states of groups tracking, psychological examination of decisions related to life of various communities.

Meanwhile, it is necessary to distinguish between constructive risks, which make advances in a particular activity possible, these advances lead to achievement of an objective and are not connected with possible damage to others, and destructive risks, which are based on the task of overcoming social norms, or possible disturbance of other people's well-being.

21.3 Risk-Potential of Social Activity

Social activity is recognized as an important and necessary part of social behavior of a personality and groups. However, despite all its desirability in terms of social development and personal self-realization, it is often limited by objective and subjective barriers. These barriers relate to all types of activity and its intensity. Long history of studying the psychology of social activity of a person allows us to discover a range of important parameters, which have been recorded in the course of time; among them are changeability, change of determinants and their correlations, risks.

Dissemination and realization of social activity of a person and group are connected to objective changes, which occur in our society, as well as in the world. These changes primarily relate to dissemination of informational field of mobility and emerging opportunities and prospects of activity dissemination. As social activity can be implemented at different levels of intensity and orientation, its personal regulation, which gains great significance, is based on acquisition of norms, values, settings, regulations of the "proper" and the "possible", representations of activity boundaries. These formations determine the system of subjective relations of a personality to the world and his/her place in it. Therefore, one of the most important circumstances of "application field" of social activity is the system of social images, which regulate both personality and group behavior. It is also necessary to note that the importance of solving problems linked to identification of physiological (social and psychological) determinants of social activity of a person and its constellations for determination of conditions of its normativity, overcoming of hindrance-influencing factors, including different types of risks, as well as the character of its application in the system of different relations and impact on the experience of subjective well-being.

Social activity can be connected to the risks that define both changes of social situation and personality. The problem of risk-potential of social activity has two main aspects: risks, which are given by realization of social activity, connected to uncertainty of its results, close and distant consequences, because its "social check", inclusion (integration) into the social system can be characterized by contradictions; risks that are given by personality instances at the level of their relationships, coherence and incoherence, character of self-regulation and decision making. The object of risks as in both cases can be a social situation, a group, and a personality itself together or separately.

Design and implementation of social activity of a personality are connected to the necessity of making strategic and efficient decisions under conditions of uncertainty, which is the most significant objective basis of risk. The multiplicity of options of the development of situation of social activity leads to the necessity of constant correlation and evaluation of a range of variables, including the return to initial conditions, that creates a zone of multiple uncertainty (therefore, the choice in favor of refusal of risk sometimes becomes "venture" in the view of realization of activity goals). In this case it is necessary to consider the idea of Kornilova [9] that in the situation of choice a person completes a definition of the situation of uncertainty as a subject (existence, activity, perception) that implements possibilities of its intellectual and personal potential, psychological picture of which is determined according to the axis of new formations actualized by the subject that allow to evaluate the potential of a person by means of analysis of the "height" of achievements based on the accomplished regulation of its choices. The level (the difficulty of tasks, pretensions) and intensity of social activity depends on the ability of a person to create zones of certainty based on self-regulation.

In accordance with objective circumstances, social (and not only social) activity of a person can appear at different degree of intensity and extensiveness. Subjective "necessities and possibilities" and "obligation" can come into contradiction. This is happening due to intensive external stimulation of social activity, its elevation to the rank of social norm upon the insufficiency of objective conditions of its implementation, therefore social frustration of a personality intensifies; its reverse effect (refusal from activity) appears quite often as a result. Besides, external stimulation of social activity (for instance, through the mass media) often presupposes only indication of a goal and results, which instigates the illusion of possibilities of the usage of different means, including the asocial ones. Growing social activity of a person and groups which is not supported by spiritual and moral regulators in a number of cases turns into a source of asocial and non-normative types of activity and sometimes possesses destructive character.

Depending on the configuration of dispositional formations, social activity can instigate different risks, which relate to a person her/himself and others. For that matter it is possible to talk about risk-potential of these formations for social activity of a person. According to the hypothesis, the more tension there is in the system of dispositional formations connected to their rivalry, the more the subjective uncertainty is expressed. Besides, their internal correlations can become the basis of risk behavior. So the high level of internal proneness to conflict (for instance, the sphere of values) "sets" the range of behavior effects, which are connected to (as it was showed by Krasilnikov [10]) the decrease of adaptational potential and the increase of aggression and irritancy.

For that matter the question regarding social development of a person (socialization of a personality) and coherence of his/her effects gains significant importance. First of all, it is referred to coherence and certain balance of social and personal congruence, because the primate of this or that leads to hazards or decreasing activity, which directs it into the negative channel. Social congruence at the expense of personal congruence intensifies conformity and practically reduces the activity to full submission to social circumstances, while personal congruence at the expense of social congruence intensifies activity that enters into conflict with public interests. In the meantime, coherence of socialization effects can be achieved through expressed subjective position of a person, its ability to build up the hierarchy of short-term and long-term goals and their reflections (as well as their realizations), to determine the most important of them, it identifies abilities for personal life creation, production of meanings and thus forms up internal entirety. Socialization of a personality also states "norms" of risk. For example, the studies show that depending on the conditions of ethnic socialization these norms can differ [21]. So in the sample of Russians the value of "appetite for risk" prevails over "caution" ($p < 0.001$); in the examined groups of Kazakhs, Armenians, Georgians such correlation is not found, but for example, the value of risk is more expressed in the Kazakhs unlike the Armenians ($p < 0.05$). However, there are spheres where representatives of different ethnic groups are not inclined to risk, such as the sphere of law, relations between friends, love and health.

One of the significant factors of personality's social activity is social comparison. Comparing the results of one's own activity to the achievements of the others, including the virtual ones (for instance, achievements presented with the help of depicted images) not in one's own favor, as a rule, causes a certain tension that acts as a motivational mechanism for activity. However, its direction depends on many other circumstances: value-meaning, moral, etc. For example, as the range of studies shows (V.A. Labunskaya, T.V. Beskova, R.M. Shamionov, and others) this activity may be directed towards the object of superiority of the other and may show an envious attitude. On the other hand, adequate assessment of one's own capabilities in achieving (material, non-material, status, etc.) success correlated with the results of social comparison, can act as the basis for display of certain types of social activity. With that, because of social comparison, some personal hazards connected with the changes in identity and penetration of components, which are preventing the activity, into one's self-consciousness are possible. Risk-potential of social activity also lies in solving the dissonance, which appears in the course of social comparison in somebody else's favor and can cause the occurrence of social dependency. We are not talking about the refusal from social activity of any form, but about the refusal connected with consumption on account of somebody else, i.e. decrease in labor activity or refusal from intellectual activity in case of the group activity situation, refusal from job search, decline of initiative in interpersonal relationship, and other areas as well. Meanwhile, in some cases, such behavior is not totally passive; it may be connected to a conscious decision and may represent a definite position.

We turn our special attention to the problem of risk-potential of social activity itself. Firstly, any form of activity possesses a definite risk-potential because its results are not always predictable or achievable both from the point of view of personality and group, and from the point of view of social consequences. Secondly, external stimulation of social activity often leads to its situational, spontaneous demonstration without consideration of circumstances and possible consequences. Thirdly, under conditions of social non-equilibrium and uncertainty there is always a risk of such a change in the situation that may lead to its temporary or permanent deterioration, destructivity having close or distant consequences. Finally, risk-potential of social activity may also involve inadequate application not meeting the conditions (or needs) of the others' or personality's interests; in the latter case the possible hazards may exceed both efforts made and psychological resources wasted. Besides, all these phenomena and parameters are characterized by a different level of risk defined by the extent of hazards: in particular cases we are talking about dissatisfaction and disappointment, in other cases – about serious harm done to health of a subject, in other cases – about the distress caused to the social groups (communities).

Significant place in the given aspect belongs to the time factor. Risk calculations at the level of different time spans may differ immensely. In particular, this has to do with decisions having distant results: quite an adequate assessment of risks at the moment of decision making with consideration of current life (social) situation can produce an unexpected effect as a consequence due to the changes in general social situation, biographic situation of a particular personality, etc. Besides, personality's social activity itself moves him/her to the new level of social experience and social situation with the changing set of risks. This calls for a great deal of competence in the area of estimation of the dynamics of social activity. For instance, realization of professional self-identification without consideration of risk-potential of the particular area of activity may corrupt the targeted result to the great extent.

At that, it is necessary to distinguish between the risks of objective and subjective hazards both at the level of subjects of activity and at the level of objects or social environment not having any direct links to activity of a personality or a group. As we are also talking about psychological risks, we would like to point out their special status residing in the fact that any unpredictable change in the situation without a certain readiness to it leads to de-adaptation of a personality. To overcome it means to change the whole social-psychological structure of personality because changes in value-meaning orientations, ideas, motivation and everything that serves as a regulator of social behavior, are inevitable in this case. It follows that absence of a certain readiness to risk may provoke changes in personal regulators of behavior (the deep ones as well). That is why, it is necessary to regard risk-potential of activity from the same angle as risk-readiness of personality. Accordingly, in case of optimal balance of these phenomena, psychological risks at the level of personality (or a group) are minimal.

Readiness for risk presupposes tendency to it and, primarily, dispositional readiness, because riskiness is a value. Our investigations [26] have shown that such values as "riskiness" and "prudence" are deeply interconnected, as they are characterized with orientation towards prudent risk. We discovered that riskiness and prudence are interconnected in the samples of youth and older generation ($r = 0.369$, $p < 0.001$ in the group of youth and $r = 0.663$, $p < 0.001$ in the group of older generation). It means that in the structure of values of both groups risk does not come into contradiction with prudence, but on the contrary risk presupposes it. Obviously, the position of "prudent risk" is typical for the cast majority. In the middle of the twentieth century Milton Friedman and Leonard J. Savage came to the conclusion that [7] the majority of people tend to avoid risk and make a choice that might cause them a certain advantage or profit, therefore, they can accept risky proposals only if they are making an attempt to avoid guaranteed loss. In other cases they are content with less. Therefore, the degree of "sensible risk" takes place every time there is a situation linked to choosing from a variety of alternatives.

Apart from that, according to the latest data [21], riskiness is interconnected with such personal characteristics as purposefulness, initiative, decisiveness, self-control, persistence, etc. In other words, readiness for risk acts as an integral social-psychological characteristic of a personality which possesses a unique combination of personal qualities, values, orientations, social representations regarding a certain area and subjective attitude of a personality to a certain object of reality, which act as regulators of social behavior. At the same time, lack of rashness and lack of competence, which are based on a certain subjective attitude to risk, where the danger margin could be slightly raised (which leads to non-objective evaluation of a situation); these qualities act as a stimulus of increased risky activity on a number of occasions. It leads to the increase of risk-potential of any activity, including the social one.

An important circumstance of the aspect under study is subjective evaluation of the risk area, i.e. its area, which can be seen as an optimal one from the point of view of identification of possible losses and gains. On the one hand, any problematic situation possesses an unknown component, where the change is difficult to calculate.

In this respect assessment of its status order seems rather important – in case of uni-ordinal status with known variables a compensation of its changes is possible due to them. In case of a higher order the risk potential increases, liquidation of which depends on possibilities of information receiving (fast access to it), emotional involvement, power of motivation and social communication, including immediate interaction with a support group. On the other hand, assessment of uni- and multi-ordinal variables correlation, which is characterized by different valency, predictability, defined by the level of personality competence in respect to a task and a situation, is rather important. In other words, the determinant type of risk here is physical, "relation" risk, property and financial risks, etc.

Riskiness as a personal characteristic, which is based on the notion of risk as a way of living, is an important regulator of behavior. Judging by the obtained results [24], we can state that risky activity is (inversely) correlated with orientation towards material wealth as a result of work ($p < 0.001$), which is indicative of a tendency for risky activities in people, who are rejecting the necessity of painstaking work for achieving material (monetary) well-being. In other words, *risky activity is linked to rejection of the necessity of work for achieving material wealth.*

Finally, there is evaluation of reversibility of irreversibility of a situation due to social activity and hazards in the case of the latter. Any choice can be accompanied by dissonance relations with the rejected alternatives (even if they are subjectively not rejected). It requires constant information processing and attributive decisions. Therefore, risk can involve cyclic circulation (jamming) of cognitions, which are used in the course of decision-making processes and in the state when a person has been "cornered". Together with that, every consecutive level (risk area) contains a nonlinear increasing risk-potential, which leads a personality to the level of psychological risks, i.e. such risks, which are inevitably reflected on the state and subjective well-being. The problem is that, on the one hand, personal subjective well-being as its dynamic characteristic, is interconnected with riskiness and utilization of risky strategies in behavior, on the other hand, implementation of risky activity can significantly destabilize the system of its relations and, therefore, the system of subjective well-being (may result in extreme states). Nevertheless, our investigations have shown that in a number of cases, for example, with close relationships, refusal from risk is linked to life satisfaction and other indicators of subjective well-being. This means, that subjective well-being is linked to "sensible", selective risky activity.

Therefore, social activity of personality may have different levels of risk-potential and presumes a certain extent of readiness to risk. Nevertheless, risk is a necessary characteristic of activity, because in a range of cases dynamics and contents of social activity acquire new quality due to acceptance of risk (and the corresponding decisions). Besides, it is necessary to mention the fact that veering away from the direct objective (and social norm as well) may be of multiple nature not necessarily conflicting with the social progress both at the micro- and macro-levels. For instance, deviation from the norm connected with producing or progress of values not acknowledged by the majority of the society at the present moment but shared by people as a consequence. Perhaps it will take much time so that these new values become important for many people. That is why it is significant to manage risks of personality's social activity by means of reflection and assessment of dispositional formations, development of critical thinking, and their correlation to the risk fields (spheres).

21.4 Conclusion

With that, today there is a need for research directed towards the problem of interconnection between the external and internal determinants of social activity, and especially towards the definition of their compensating role in case of high uncertainty at different stages of its realization and non-planned (unpredictable) result. The studies of the process of goal-setting in social activity both by the parameter of its intensity and by the level, broadness of comprehension, definition of subjective limits, are also efficient. In order to analyze social activity of personality with consideration of his/her risk-potential it is necessary to have such a methodological basis that would allow not only to understand the origin of risk and its psychological image, define its significant predictors, but also to help a person to organize his/her life in such a manner that he/she could control everyday life risks with the account of internal and external risks. Analysis of different instances of personality organized in different ways depending on objective and subjective configuration of social activity situation (and its dynamics), as its direct causes and boosters, is important. This presumes acceptance of the principle of instances diachrony [23]. In this case the opportunity to study social activity and its correlation to objective and subjective personality instances appears; it becomes possible to follow the multiple dynamics of personality instances where the integrator is the subject. By doing so we can point out their regulative role in realization of social activity, define the nature of internal and external risks and changes connected to facing uncertainty and the results of risky activity.

References

1. Atkinson JW (1964) An introduction to motivation. Van Nostrand, New York
2. Bandura A, Ozer EM (1990) Mechanisms governing empowerment effects: a self-efficacy analysis. J Pers Soc Psychol 58(3):472–486
3. Beck U (1992) Risk society. Toward a new modernity. Sage, London
4. Breakwell GM (2007) The psychology of risk. Cambridge University Press, New York
5. Eysenck HJ (1990) Biological dimension of personality. In: Pesvin LA (ed) Handbook of personality: theory and research. Guilford, New York, pp 244–276
6. Eysenck SBG, Eysenck HJ (1978) Impulsiveness and venturesomeness: their position in a dimensional system of personality description. Psychol Rep 43:1247–1255
7. Friedman M, Savage LJ (1948) The utility analysis of choices involving risk. J Polit Econ 4:279–304
8. Grachev GV (2004) Personality and society: information and psychological safety and psychological protection. Izdatel, Volgograd

9. Kornilova TV (2009) Uncertainty negotiation by the subject in the process of decision making. In: Zhuravlev AL, Znakov VV, Ryabikina ZI, Sergienko EA (eds) Subjective approach in psychology. Publishing house, Institute of Psychology of Russian Academy of Sciences, Moscow, pp 325–339
10. Krasilnikov IA (2006) Intrapersonal conflict and psychological adaptation. Publishing House of Saratov University, Saratov
11. Kudryavtseva NN, Smagin DA, Bondar NP (2011) Modeling fighting deprivation effect in mouse repeated aggression paradigm. Prog Neuropsychopharmacol Biol Psychiatry 35(6):1472–1478. doi:10.1016/j.pnpbp.2010.10.013
12. Lavery B, Siegel AW, Cousins JH, Rubovits DS (1993) Adolescent risk-taking: an analysis of problem behaviors in problem children. J Exp Child Psychol 55:277–294
13. McDaniels TL, Gregory RS, Fields D (1999) Democratizing risk management: successful public involvement in local water management decisions. Risk Anal 19(3):497–510
14. Nagy S, Nix CL (1989) Relations between preventative health behavior and hardiness. Psychol Rep 65:339–345
15. Popov LM (1992) Psychology of self-activity: abstract of doctoral thesis in psychology. Institute of Psychology of Russian Academy of Sciences, Moscow
16. Shamionov RM (2006) Subjective well-being and valuable and semantic formations of a person in his/her professional self-determination. Izvestiya Saratov Univ Ser Philos Psychol Pedagog 6(1–2):104–109
17. Shamionov RM (2008) Subjective well-being of personality: a psychological pattern and factors. Publishing House of Nauchnaya Kniga, Saratov
18. Shamionov RM (2009) Youth's characteristics of value orientations in correlation with the representations about Russia and Russians values. Sociol Educ 4:39–49
19. Shamionov RM (2010a) The link "subjective" personal properties and envy. Izvestiya Saratov Univ Ser Philos Psychol Pedagog 10(4):93–98
20. Shamionov RM (2011) Envy in the socio-economic attitudes and the individual activity. Theor Exp Psychol 4(4):46–52
21. Shamionov RM (2012) Strategies of behavior interrelation subjective well-being representatives contact ethnic groups. Izvestiya Saratov Univ Ser Philos Psychol Pedagog 12(1):79–83
22. Shamionov RM (2010b) On some transformations of personality's subjective well-being structure in different conditions of professional socialization. World Psychol 1:237–249
23. Shamionov RM (2013) Socialization of personality: systematic-diachronic approach. Psikhologicheskie Issledovaniya 6(27):8. http://psystudy.ru. Accessed 28 Feb 2014
24. Shamionov RM (2014) Envy, personality's riskiness and methods for achieving of high economic status. Open J Soc Sci 2:110–113. doi:10.4236/jss.2014.25022
25. Shamionov RM, Grigoryeva MV, Usova NV (2013) The subjective well-being of Russian migrants in Spain and of foreigners in Russia. Procedia Soc Behav Sci 86:498–504. doi:10.1016/j.sbspro.2013.08.604
26. Shamionov RM, Klenova MA (2009) Value and motivation of risk in different generations. Herald Kostroma State Univ 15(3):63–67
27. Sherif M (1966) A contribution predicament. Social psychology of intergroup conflict and cooperation. Houghton-Mifflin, Boston
28. Slovic P, Peters E (2006) Risk perception and affect. Curr Dir Psychol Sci 15(6):322–325
29. Tugusheva AR (2006) Special aspects of a socially successful person pattern in the view of the youth. Probl Soc Psychol Personal 3:87–93
30. Wiegman O, Gutteling JM (1995) Risk appraisal and risk communication: some empirical data from the Netherlands reviewed. Basic Appl Soc Psychol 16(1–2):227–249
31. Zhuravlev AL (2010) What modernization in Russia should be like? Psychological newspaper: We and the World 10(170):1
32. Znakov VV (2005) Psychology of a subject and psychology of human existence. In: Znakov VV, Ryabikina ZI (eds) Subject, personality and psychology of human existence. Publishing house, Institute of Psychology of Russian Academy of Sciences, Moscow, pp 9–44

Chapter 22
Investment Behaviors of Businesses on Ad and R&D According to Their Sizes: Food, Beverage and Tobacco Sector in Turkey

Çağla Pınar Bozoklu and Yasemin Keskin Benli

Abstract It is observed that the advertising and R&D expenditures of businesses, which are traded on Borsa Istanbul (BIST) in the Food, Beverage and Tobacco Sector, differ according to their business sizes. Within this context, two assumptions are developed. At first, the large businesses invest on the advertising and R&D expenditures regularly from sales revenues. At second, the relatively smaller businesses invest on these operations only in the periods which they generate over-expected sales revenues. In order to test these assumptions, the impacts of advertising and R&D expenditures on the sales revenues of large and middle-sized businesses in Food, Beverage and Tobacco Sector in BIST (2009–2013) are analyzed by panel regression after 22 businesses in the sector are classified as the large and middle-sized businesses according to their number of employees. As a result, it is proved that the large businesses invest on the advertising and R&D expenditures regularly. On the other hand, the middle-sized businesses set the R&D budget depending on their annual sales revenues whereas they act as the large businesses in terms of setting advertising budget.

22.1 Introduction

Advertising expenditures are sited as a footnote in Marketing, Selling and Distribution Expenses in the income statement. International Accounting Standards Committee (IASC) and Turkey Accounting Standards-15 (TMS-15) accept the term of research as an original and planned investigation and search effort for acquiring

Ç.P. Bozoklu (✉)
Faculty of Economics and Administrative Sciences, Department of Healthcare Management, Gazi University, Ankara, Turkey
e-mail: cputkutug@gazi.edu.tr

Y. Keskin Benli
Faculty of Economics and Administrative Sciences, Department of International Trade, Gazi University, Ankara, Turkey
e-mail: ykeskin@gazi.edu.tr

Ş.Ş. Erçetin (ed.), *Chaos, Complexity and Leadership 2014*,
Springer Proceedings in Complexity, DOI 10.1007/978-3-319-18693-1_22

new scientific, technical knowledge and comprehension. Development function is also considered as a transformation process of research findings to a plan or a design before production of new or developed materials, equipment, products, actions, systems or services. According to these definitions, R&D (research and development) operations are the expenses that businesses should face before launching new or developed products to market. Moreover, these expenses have an impact on sales revenue in the following periods. This impact provides some intangible and tangible benefits. Generally, the research activities generates intangible benefits whereas, development activities maintain tangible benefits such as increase of sales revenue and profit [35, p. 355].

After the R&D operations, it is expected that the advertising operations become more important while the developed products, services or ideas begin to commercialize. Therefore, the advertising expenditures are also seen as another interesting factor, which have an impact on the sales revenue in Marketing, Selling and Distribution Expenses. However, coherent and detailed data cannot be reached by reason of the fact that the advertising expenditures are referenced as footnote information without a common term in Turkey. The measurement of direct impact of the advertising expenditures on sales revenue causes complexity of developing effective strategies, but it seems possible to measure the advertising expenditures' indirect impact by means of analysis of advertising budget strategies of businesses.

Science of accounting accepts marketing expenses as a cost factor instead of an intangible investment [32]. According to that perspective, increment of the sales revenue generated by marketing expenses is disregarded. Actually, the marketing and advertising expenses have negative impact on sales revenue in the short-term; but in the long term they act as an investment instrument [13] and also add value to the businesses by means of increasing future sales, cash flows and profits. Moreover, the marketing expenses such as R&D and advertising expenditures, which have also a positive impact on the intangible assets such as brand value, may yield competitive advantage and an increase in market value [9].

The impacts of these expense accounts on the sales revenue are presented widely in the literature [25]. Vaile [31], Palda [23], Winer [34], Tellis and Tellis [30], McGraw-Hill research [20], Kamber [16] and many other researchers state that there is a strong and positive correlation between the advertising expenditures and the sales revenue including the economic recession periods. Moreover, it is indicated that the advertising expenditures have a strong impact on sales in the short term (shorter than a year) and also decreasingly growing impact in the long term. On the other hand, Clarke [8], Assmus et al. [3], Bemmaor [4] found a positive impact without a cumulative effect of advertising expenditures on sales revenue. There are also some researches, which could not identify any positive impact (e.g. [1, 8, 24]). Current advertising effect theory assumes that the sales are a function of advertising expenditures of current accounting period. According to that, continuing effect of advertising cannot be associated totally with the previous advertising expenditures [33].

The R&D expenditure is one of the most effective factors for developing new technological processes and productivity. The endogenous growth theory, which attracted wide interest at the late 1980s, admits R&D expenditures as a trigger force for growth. This perspective leads many researchers investigating the role of R&D expenditures on innovation, efficiency and especially economical growth [2, 15, 28]. But the notion of time has a critical significance for R&D projects [26]. Completing successfully of a project generally last more than an accounting period. Therefore, all expenses begin to generate income only after the completion of the project and the introduction of the developed product to the market. That long span of time challenges the measurement of R&D expenditures' impacts on the outputs such as the sales revenue, the profit and the market share.

Leonard [18] and Rosenberg [29] conducted two original researches on R&D expenditures and their effects on sales revenue. As a result, significant and positive relationship was found between R&D expenditures and sales revenue whereas; Doi [11] found a negative relationship [21, p. 61].

Within the context of this research, data of businesses that are traded on BIST in Food, Beverage and Tobacco sector between the years of 2009 and 2013 is reviewed. It is observed that the advertising and R&D expenditures of the large and the middle-sized businesses have different tracking patterns. This differentiation is especially obvious in terms of R&D expenditures. The reason of this is assumed as the large-sized businesses can generate satisfactory sales revenue, profit and profit reserves more than the relatively smaller businesses. Therefore, the large-sized businesses invest more regularly on the advertising and the R&D expenditures without depending on the change of annual sales revenue.

On the other hand, it is suggested that the relatively smaller businesses invest on the advertising and R&D expenditures only in the periods of over-excepted sales revenue. This possible diversity among large and relatively smaller businesses may lead a chaotic situation by terms of developing effective budget strategies.

According to that, the aim of this research is to determine the budget strategies for the advertising and the R&D operations whether they differ according to the business size criteria. The impacts of advertising and R&D expenditures on sales revenue is analyzed by means of data of 22 large and middle-sized businesses in Food, Beverage and Tobacco sector in Turkey using panel regression.

22.2 Method

By the year of 2014, there are 30 businesses traded on BIST in Food, Beverage and Tobacco sector. But the number of businesses is only 22, which traded in the sector between the years of 2009–2013. The research covers the data of these 22 businesses for maintaining validity of using panel data.

At first, the businesses are classified according to the number of employee criteria. In the literature, it is seen that there are a lot of classification criteria for business size concept. However, Republic of Turkey Ministry of Science, Industry and Technology changed 'The Regulation about the Definition, Qualifications and Classifications of Small and Middle-sized Businesses' in 9.10.2012 (www.resmigazete.gov.tr). According to that, the numbers of staff in micro, small and middle-sized businesses are as follows:

Table 22.1 Employee numbers and sizes of businesses between the years of 2009 and 2013

	Employee numbers of each year					
Businesses	2009	2010	2011	2012	2013	Business size
AEFES	6,422	6,348	6,227	5,918	19,852	Large
BANVT	2,643	3,076	3,238	3,414	3,877	Large
CCOLA	8,451	8,628	9,244	10,893	10,864	Large
DARDL	770	459	351	418	501	Large
ERSU	74	89	88	69	62	Middle
FRIGO	193	212	232	241	188	Middle
KENT	1,697	1,756	1,590	1,432	1,397	Large
KERVT	1,192	1,178	1,208	1,122	1,106	Large
KNFRT	119	129	131	141	125	Middle
KRSTL	66	78	79	76	75	Middle
MERKO	190	171	170	184	199	Middle
MRTGG	74	48	55	45	57	Middle
PENGD	147	150	150	192	181	Middle
PETUN	507	502	621	702	794	Large
PINSU	325	373	420	435	430	Large
PNSUT	725	719	856	957	1,067	Large
SELGD	201	160	128	112	118	Middle
SKPLC	1,311	1,479	1,504	1,599	680	Large
TATGD	1,013	1,081	1,096	1,127	1,044	Large
TUKAS	263	225	247	231	213	Middle
TBORG	515	548	555	574	630	Large
ULKER	1,280	1,254	7,226	8,627	9,218	Large

- Micro Businesses: 1–10 Employee
- Small Businesses: Lesser than 50 Employee
- Middle Businesses: Lesser than 250 Employee

It is also suggested that the large businesses employ over 250 staff. Table 22.1 presents number of employee and size of businesses in Food, Beverage and Tobacco sector between the years of 2009–2013. According to the table, there are 13 large and 9 middle businesses.

The research hypotheses are:

H_0: The change of the sales revenue ($SR_{sector_{it}}$) do not effect advertising (Ad_{it}) and R&D (RD_{it}) expenditures of businesses in Food, Beverage and Tobacco sector between the years of 2009 and 2013.

H_0: The change of the sales revenue $\left(SR_{large_{it}}\right)$ do not effect advertising (Ad_{it}) and R&D (RD_{it}) expenditures of *large* businesses in Food, Beverage and Tobacco sector between the years of 2009 and 2013.

H_0: The change of R&D expenditures ($RD_{middle_{it}}$) do not effect the sales revenue (SR_{it}) of *middle* businesses in Food, Beverage and Tobacco sector between the years of 2009 and 2013.

H_0: The change of advertising expenditures ($Ad_{middle_{it}}$) do not effect the sales revenue (SR_{it}) of *middle* businesses in Food, Beverage and Tobacco sector between the years of 2009 and 2013.

According to the hypotheses, regression models are:

$$SR_{sector_{it}} = \beta_0 it + \beta_1 (RD_{it}) + \beta_2 (Ad_{it}) + \varepsilon_{it} \quad (1)$$

$$SR_{large_{it}} = \alpha_0 + \alpha_1 (RD_{it}) + \alpha_2 (Ad_{it}) + \varepsilon_{it} \quad (2)$$

$$RD_{middle_{it}} = \Omega_0 + \Omega_1 (SR_{middle_{it}}) + \varepsilon_{it} \quad (3)$$

$$Ad_{middle_{it}} = \theta_0 + \theta_1 (SR_{it}) + \varepsilon_{it} \quad (4)$$

For all tests, e-Views 6.1 software is utilized.

22.3 Results

22.3.1 *The Results of Stationarity and Unit Root Tests*

Analyses of panel data maintain both time and horizontal section analyses simultaneously similar to time series analyses [12, 28]. Within the context of panel data analysis, variables must be also stationary by reason of not causing fake relationships between the variables ([17]: 100). According to that, Levin et al. [19], Fisher-ADF Chi Square (1979) and Fisher- PP (1988) unit root tests are done to analyze the stationarity of variables. For all three tests, hypotheses are:

H_0: There is unit root.
H_1: There is no unit root.

Table 22.2 shows the values of these tests. According to the probability values ($p < .05$), H_0 is rejected. So, there is no unit root in the series.

22.3.2 *The Results of Breusch-Pagan Tests*

There are two approaches such as fixed effects and random effects by terms of pooled data regression analysis. Before these approaches are applied, Breusch-Pagan (B-P) test is recommended to test whether there is a serial correlation [5]. The hypotheses of the test are:

H_0: There is no serial correlation
H_1: There is serial correlation.

Table 22.2 Statistical results of panel unit root tests

Variables	Levin, Lin & Chu T statistic				Fisher- ADF chi square test				Fisher- PP test			
	Level				Level				Level			
	Individual intercept		Individual intercept and trend		Individual intercept		Individual intercept and trend		Individual intercept		Individual intercept and trend	
	Stat.	Prob.	Stat.	Prob.	Stat.	Prob.	Stat.	Prob.	Stat.	Prob.	Stat.	Prob.
Sales revenue	−2.08	0.02	–	–	41.74	0.57	73.89	0.00	48.90	0.28	99.31	0.00
R&D expenditures	−2.49	0.01	–	–	21.76	0.35	32.59	0.04	29.90	0.07	42.02	0.00
Advertising expenditures	−1.47	0.07	−3.35	0.00	80.46	0.00	–	–	88.36	0.00	–	–

Table 22.3 Breusch-Godfrey serial correlation LM test

Variables	F-statistic	Obs*R-squared	Prob. F	Prob. chi square (2)	
Model 1	1.722910	3.495203	F(2,105)	0.1836	0.1742
Model 2	0.432674	0.924131	F(2,60)	0.6508	0.6300
Model 3	1.692151	3.431249	F(2,41)	0.1967	0.1799
Model 4	0.131007	0.285750	F(2,41)	0.8776	0.8669

Test results for all variables related to the developed models are given in the Table 22.3.

Prob. F(2, 105) and Prob. Chi-Square (2) values are found respectively .18 and .17 for model 1 $\left(X^2 > 0.5\right)$. Prob. F(2, 105) and Prob. Chi-Square (2) values are found respectively .65 and .63 for model 2 $\left(X^2 > .05\right)$. Prob. F(2, 105) and Prob. Chi-Square (2) values are found respectively .20 and .18 for model 3 $\left(X^2 > .05\right)$. At last, Prob. F(2, 105) and Prob. Chi-Square (2) values are found respectively .88 and .87 $\left(X^2 > .05\right)$ for model 4. According to these results, H_0 is accepted, as there is no serial correlation.

22.3.3 *The Results of Hausman Tests and Regression Analyses*

Hausman test is utilized to decide whether fixed effects or random effects model should be chosen [15]. In other words, Hausman test analyzes whether the difference between parameter estimators of fixed effect or random effect models is statistically meaningful [6, p. 717].

The distinction of fixed and random effects models is about the relationship status between the fixed-time effects and the descriptive variables. In the condition of random effects model's acceptance, fixed effects estimator continues to generate consistent parameter estimations. Fixed effects estimator should not be preferred over the random effects estimator unless all fixed-time factors, which are related to other descriptive variables, can be measured. The most important reason is that random effect estimator generates estimations over the real effect whereas fixed effects estimator generates estimations under the real effect [12, p. 403]. Within the context of Hausman test, random effects model is preferred under the null hypothesis due to higher efficiency, while under the alternative fixed effects is at least consistent.

22.3.3.1 The Results of Hausman Tests and Regression Analysis for Model 1

Table 22.4 presents the Hausman test results for model 1.

According to the probability value (p > .05), H_0 is accepted for model 1 and random effects model is utilized for estimation. The regression results are given in the Table 22.5.

Table 22.4 Correlated random effects – Hausman test

Test cross-section random effects			
Test summary	Chi-Sq. statistic	Chi-Sq. d.f.	Prob.
Cross-section random	0.064187	2	0.9684

Table 22.5 Model 1: the results of pooled regression according to random effects method

Dependent variable: sales revenue				
Method: Pooled EGLS (Cross-section random effects)				
Sample: 2009–2013				
Included observations: 5				
Cross-sections included: 22				
Total pool (balanced) observations: 110				
Variable	Coefficient	Std. error	t-statistic	Prob.
C	1.76E + 08	66,618,116	2.634474	0.0097
R&D expenditures	57.24588	19.22643	2.977458	0.0036
Advertising expenditures	8.452197	0.326520	25.88573	0.0000
R-squared	0.868350			
Adjusted R-squared	0.865889			
S.E. of regression	2.60E + 08			
F-statistic	352.8796			
Prob(F-statistic)	0.000000			

As it is seen in the Table 22.5, the R&D and the advertising expenditures can explain 87 % of the dependent variable, which is determined as sales revenue. Probability value and the F statistic indicate that the validity of model 1's equation is quite high. The constant values of independent variables are found positive and statistically meaningful. According to that, all businesses in Food, beverage and Tobacco sector set the R&D and advertising budget without depending on the annual sale revenues. Thus, H_0 of the research hypotheses is rejected and the equation of model 1 is:

$$SR_{sector_{it}} = 1.76E + 08 + 57.24588\,(RD_{it}) + 8.452197\,(Ad_{it}) + \varepsilon_{it} \qquad (22.1)$$

22.3.3.2 The Results of Hausman Tests and Regression Analysis for Model 2

Table 22.6 presents Hausman test results for model 2.

According to probability value ($p > .05$), H_0 is accepted for model 2 and random effects model is utilized for estimation. The regression results are given in the Table 22.7.

In the Table 22.7, the R&D and the advertising expenditures can explain approximately 87 % of the dependent variable. Probability value and the F statistic

Table 22.6 Correlated random effects – Hausman test

Test cross-section random effects			
Test summary	Chi-Sq. statistic	Chi-Sq. d.f.	Prob.
Cross-section random	0.466789	2	0.7918

Table 22.7 Model 2: the results of pooled regression according to random effects method

Dependent variable: sales revenue				
Method: Pooled EGLS (Cross-section random effects)				
Sample: 2009–2013				
Included observations: 5				
Cross-sections included: 13				
Total pool (balanced) observations: 65				
Variable	Coefficient	Std. error	t-statistic	Prob.
C	2.91E+08	1.11E+08	2.622536	0.0110
R&D expenditures	50.76253	25.05896	2.025724	0.0471
Advertising expenditures	8.293116	0.423083	19.60163	0.0000
R-squared	0.865940			
Adjusted R-squared	0.861615			
S.E. of regression	3.40E+08			
F-statistic	200.2393			
Prob.(F-statistic)	0.000000			

Table 22.8 Correlated random effects – Hausman test

Test cross-section random effects			
Test summary	Chi-Sq. statistic	Chi-Sq. d.f.	Prob.
Cross-section random	10.821394	1	0.0010

indicate that validity of model 2's equation is quite high. The constant values of independent variables are found positive and statistically meaningful. According to that, the large-sized businesses in Food, Beverage and Tobacco sector set the R&D and advertising budget without depending on the annual sale revenues. Thus, H_0 of the research hypotheses is rejected and the equation of model 2 is:

$$SR_{large_{it}} = 2.91E + 08 + 50.76253\,(RD_{it}) + 8.293116\,(Ad_{it}) + \varepsilon_{it} \qquad (22.2)$$

22.3.3.3 The Results of Hausman Tests and Regression Analysis for Model 3

Table 22.8 presents the Hausman test results for model 3.

According to probability value ($p < .05$), H_0 is rejected for model 3 and fixed effects model is utilized for estimation. The regression results are given in the Table 22.9.

Table 22.9 Model 3: the results of pooled regression according to fixed effects method

Dependent variable: R&D expenditures				
Method: Pooled least squares				
Sample: 2009–2013				
Included observations: 5				
Cross-sections included: 9				
Total pool (balanced) observations: 45				
Variable	Coefficient	Std. error	t-statistic	Prob.
C	65,561.33	12,865.91	5.095740	0.0000
Sales revenue	−0.000553	0.000272	−2.031740	0.0498
R-squared	0.982750			
Adjusted R-squared	0.978314			
S.E. of regression	14,503.00			
F-statistic	221.5536			
Prob(F-statistic)	0.000000			

Table 22.10 Correlated random effects – Hausman test

Test cross-section random effects			
Test summary	Chi-Sq. statistic	Chi-Sq. d.f.	Prob.
Cross-section random	1.513965	1	0.2185

According to the Table 22.9, sales revenue can explain 98 % of the dependent variable, which is determined as R&D expenditures. Probability value and the F statistic indicate that validity of model 3's equation is quite high. The constant values of independent variables are found positive and statistically meaningful. According to that, the middle-sized businesses in Food, Beverage and Tobacco sector set the R&D depending on the annual sale revenues. Thus, H_0 of the research hypotheses is rejected and the equation of model 3 is:

$$RD_{middle_{it}} = 65561.33 - 0.000553\,(SR_{middle_{it}}) + \varepsilon_{it} \tag{22.3}$$

22.3.3.4 The Results of Hausman Tests and Regression Analysis for Model 4

Table 22.10 presents the Hausman test results for model 4.

According to probability value ($p > .05$), H_0 is accepted for model 4 and random effects model is utilized for estimation. The regression results are given in the Table 22.11.

As it is seen in the Table 22.11, the sales revenue can explain 19 % of the dependent variable, which is determined as the advertising expenditures. The explanatory of independent variable is very weak. Moreover, probability value and the F statistic indicate that validity of model 4's equation is quite low. However,

Table 22.11 Model 4: the results of pooled regression according to random effects method

Dependent variable: advertising expenditures				
Method: Pooled EGLS (Cross-section random effects)				
Sample: 2009–2013				
Included observations: 5				
Cross-sections included: 9				
Total pool (balanced) observations: 45				
Variable	Coefficient	Std. error	t-statistic	Prob.
C	−127,410.0	199,599.9	−0.638327	0.5266
Sales revenue	0.010054	0.003103	3.240088	0.0023
R-squared	0.194367			
Adjusted R-squared	0.175631			
S.E. of regression	302,583.1			
F-statistic	10.37417			
Prob(F-statistic)	0.002436			

Table 22.12 Breusch-Godfrey serial correlation LM test for model 5

Variables	F-statistic	Obs*R-squared	Prob. F	Prob. chi square (2)	
Model 5	15.31351	19.24156	F(2,41)	0.0000	0.0001

Table 22.13 Correlated random effects – Hausman test

Test cross-section random effects			
Test summary	Chi-Sq. statistic	Chi-Sq. d.f.	Prob.
Cross-section random	8.754670	1	0.0031

H_0 of the research hypotheses is rejected because of the existence of the mentioned influence. According to findings, the equation of model 4 is:

$$Ad_{middle_{it}} = -127410.0 + 0.010054\,(SR_{it}) + \varepsilon_{it} \tag{22.4}$$

By the reason of the weakness of model 4, another model (model 5) is developed which tests whether middle-sized businesses act as large-sized businesses by terms of setting advertising budget. Model 5 is:

$$SR_{middle_{it}} = \theta_0 + (Ad_{it}) + \varepsilon_{it} \quad (5)$$

Breusch Pagan and Hausman tests are conducted for model 5. Their findings are presented respectively in Tables 22.12 and 22.13.

Prob. F(2, 105) and Prob. Chi-Square (2) values are found respectively .00 and .00 for model 5 $\left(X^2 > .05\right)$. According to these results, H_0 is rejected and there is serial correlation.

Table 22.14 Model 5: the results of pooled regression according to fixed effects method

Dependent variable: sales revenue				
Method: Pooled least squares				
Sample: 2009–2013				
Included observations: 5				
Cross-sections included: 9				
Total pool (balanced) observations: 45				
Variable	Coefficient	Std. error	t-statistic	Prob.
C	45,309,784	2,157,929.	20.99688	0.0000
Advertising expenditures	3.748588	4.979707	0.752773	0.4566
R-squared	0.956780			
Adjusted R-squared	0.945666			
S.E. of regression	8,932,850.			
F-statistic	86.08907			
Prob(F-statistic)	0.000000			

According to probability value ($p < .05$), H_0 is rejected for model 5 and fixed effects model is utilized for estimation. The regression results are given in the Table 22.14.

According to the Table 22.14, the advertising expenditures can explain 96 % of the dependent variable, which is determined as sales revenue. Probability value and the F statistic indicate that the validity of model 5's equation is quite high. The constant values of independent variables are found positive and statistically meaningful. According to that, the middle-sized businesses in Food, Beverage and Tobacco sector act as large-sized businesses and set the advertising budget without depending on the annual sale revenues. Thus, H_0 of the research hypotheses is rejected and the equation of model 5 is:

$$SR_{middle_{it}} = 45309784 + 3.748588\,(Ad_{it}) + \varepsilon_{it} \tag{22.5}$$

22.4 Conclusions

It is observed that the advertising and the R&D expenditures of large and middle-sized businesses that are traded on BIST in Food, Beverage and Tobacco sector-in Turkey- between the years of 2009 and 2013, have different tracking patterns. This differentiation is especially obvious in terms of R&D expenditures and leads a chaotic situation. As an explanation, the assumption that the large-sized businesses can generate satisfactory sales revenue more than relatively smaller businesses in order to set regular advertising and R&D budgets is suggested. On the other hand, because of the irregular and insufficient annual sales revenues, the relatively smaller businesses invest on the advertising and the R&D operations only in the periods that they generate sales revenue over their expectations.

Depending on these assumptions, the aim of this research is to determine whether budget strategies of R&D and advertising differ according to business size. According to that, panel regression analysis is conducted.

Before the analysis stage, 22 businesses that are in Food, Beverage and Tobacco sector between the years of 2009 and 2013 are classified depending on number of employee criteria. There are 13 large and 9 middle businesses in the sample. Then, the regression models are developed and tested.

The result of Model 1 states that the advertising (Ad_{it}) and R&D (RD_{it}) expenditures of businesses impact 87 % of their sales revenue ($SR_{sector_{it}}$). In this model, all businesses are counted in the analysis without making any classification. The result of Model 2 indicates that the advertising (Ad_{it}) and the R&D (RD_{it}) expenditures of large businesses impact similarly 87 % of the sales revenue $\left(SR_{large_{it}}\right)$. Within the context of middle-sized businesses, Model 3 presents that generated sales revenue (SR_{it}) impacts 98 % of the R&D budget ($RD_{middle_{it}}$). However the result of Model 4 is found rather weak. The sales revenue (SR_{it}) of the middle-sized businesses impacts just 19 % of the advertising budget ($Ad_{middle_{it}}$). Therefore, the hypothesis of model 4 is restructured so that the change of advertising expenditures ($Ad_{middle_{it}}$) do not effect the sales revenue ($SR_{middle_{it}}$) of *middle* businesses in Food, Beverage and Tobacco sector between the years of 2009 and 2013. The result of Model 5 presents that the advertising expenditures of middle businesses impact 96 % of the sales revenue. Surprisingly, the middle-sized businesses act similar to large-sized businesses in terms of setting advertising budget.

The results show that businesses set advertising and R&D budgets without taking into consideration the change of their sales revenues. When these businesses are classified according to their sizes, it is obviously seen that the middle and large-sized businesses act differently. Both of them accept the advertising expenditures as a regular investment instrument, though they follow different R&D strategies. These findings prove that spurious results can be obtained unless samples of businesses are classified.

Businesses in Food, Beverage and Tobacco sector in Turkey does not pay enough attention to R&D operations whereas these operations are considered as a trigger for innovation, developing markets, providing competitive advantage and naturally generating sales revenue in the developed countries and economies. It is recommended that the future researches can be conducted by focusing other attractive sectors for investors. The advertising and R&D strategies of micro and small-sized businesses can be analyzed in order to lead the venture capitals.

References

1. Aaker DA, Carman JM, Jacobson R (1982) Modeling advertising-sales relationships involving feedback: a time series analysis of six cereal brands. J Market Res 19:116–125
2. Aghion P, Howitt P (1992) A model of growth through creative destruction. Econometrica 60:323–351

3. Assmus G, Farley JU, Lehmann DR (1984) How advertising affects sales: meta-analysis of econometric results. J Market Res 21:65–74
4. Bemmaor AC (1984) Testing alternative econometric models on the existence of advertising threshold effect. J Market Res 21:298–308
5. Brooks C (2008) Introductory econometrics for finance. Cambridge University Press, New York
6. Cameron AC, Trivedi PK (2005) Microeconometrics: methods and applications. Cambridge University Press, New York
7. Cemalcılar İ (1988) Pazarlama, kavramlar kararlar. Beta Yayınları, İstanbul
8. Clarke DG (1976) Econometric measurement of the duration of advertising effect on sales. J Market Res 13:345–357
9. Day G, Fahey L (1988) Valuing market strategies. J Market 52:45–57
10. Dickey DA, Fuller WA (1979) Distribution of the estimators for autoregressive time series with a unit root. J Am Stat Assoc 74(366):427–431
11. Doi N (1977) Firm size, market power, and research and development. J Econ Kwansei Gakuin Univ 31(3):99–123
12. Johnston J, Dinardo J (1997) Econometric methods. McGraw-Hill, New York
13. Graham RC, Frankenberger KD (2000) The contribution of changes in advertising expenditures to earnings and market values. J Bus Res 50:149–155
14. Grossman GM, Elhanan H (1991) Quality ladders in the theory of growth. Rev Econ Stud 58(1):43–61
15. Hausman JA (1978) Specification tests in econometrics. Econometrica 16(6):1251–1271
16. Kamber T (2002) The brand manager's dilemma: understanding how advertising expenditures affect sales growth during a recession. Brand Manage 10(2):106–120
17. Korkmaz T, Yildiz B, Gökbulut Rİ (2010) FVFM'nin İMKB ulusal 100 endeksindeki geçerliliğinin panel veri analizi ile test edilmesi. İ.Ü. İşletme Fakültesi Dergisi 39(1):95–105
18. Leonard W (1971) R&D in industrial growth. J Polit Econ 79(2):232–256
19. Levin A, Lin CF, Chu CSJ (2002) Unit root tests in panel data: asymptotic and finite-sample properties. J Econometr 108:1–24
20. McGraw-Hill Research (1986) Laboratory of advertising performance report 5262. McGraw-Hill, New York
21. Odagiri H (1983) R&D expenditures, royalty payments, and sales growth in Japanese manufacturing corporations. J Ind Econ 32(19):61–71
22. Özbek Y, Utkutuğ ÇP (2012) Cumulative advertising effects: presentation on financial statements of long term advertising effects on sales. Int J Arts Commer 1(4):237–247
23. Palda KS (1964) The measurement of cumulative advertising effects. Prentice-Hall, Englewood Cliffs
24. Parsons LJ, Schultz RL (1976) Marketing models and econometric research. North-Holland, New York
25. Phillips PCB, Perron P (1988) Testing for a unit root in time series regression. Biometrika 75:335–346
26. Ravenscraft D, Scherer FM (1982) The lag structure of returns to research and development. Appl Econ 14:603–620
27. Republic of Turkey Ministry of Science Industry and Technology (2012) The regulation about the definition, qualifications and classifications of small and middle-sized businesses. http://www.resmigazete.gov.tr/eskiler/2012/11/20121104-11.htm. Accessed 2 July 2014
28. Romer PM (1990) Endogenous technical change. J Polit Econ 98:71–102
29. Rosenberg JB (1976) Research and market share: a reappraisal of the Schumpeter hypothesis. J Ind Econ 25(2):101–112
30. Tellis GJ, Tellis K (2009) Research on advertising in a recession: a critical review and synthesis. J Advert Res 49:9–40
31. Vaile RS (1927) The use of advertising during depression. Harv Bus Rev 3:323–330
32. Wang F, Zhang XP, Ouyang M (2009) Does advertising create sustained firm value? The capitalization of brand intangible. J Acad Market 37:130–143

33. Weiss DL, Windal PM (1980) Testing cumulative advertising effects: a comment on methodology. J Market Res 17:371–378
34. Winer RS (1979) An analysis of the time-varying effects of advertising: the case of Lydia Pinkham. J Bus 52(4):563–576
35. Yıldız G (2003) Araştırma- Geliştirme maliyetlerinin türkiye muhasebe standardı 15 (TMS-15) ve tek düzen muhasebe sistemi (TDMS) esasına göre muhasebeleştirilmesi. İktisadi ve İdari Bilimler Dergisi 17(1–2):353–370

Chapter 23
Transformation Towards Entrepreneurialism in Turkish Universities: A Dilemma and Critical Evaluation

Seçil Dayıoğlu Öcal

Abstract Due to knowledge economy and its impacts on collaborative works with industry, universities have been obliged to shift themselves into entrepreneurialism. This effect has a great impact on the research and development activities conducted and brings about cultural change in academia and institutions. The cultural change can cause conflicts and dilemma in the overall organization of the universities. University – industry collaborative activities in Turkey began in the late of 1980s and accelerated in the late of 1990s. Universities have been in the process of establishing techno-centers to promote them and take great part into the entrepreneurial activities especially in the early 1990s. Integrating the values of entrepreneurialism into universities is a chaotic and complex issue as there has been clashes with the values exited. This paper deals with this dilemma and to make a critical analysis in the case of Turkish universities.

23.1 Introduction

Entrepreneurialism is one of the most pressing issues for universities since they are the major drive to ground innovation and trigger local, regional and national economic development in the information age. High-technology products are required to have academic accumulation of knowledge and researching. This leads to emergence of the concept of university-industry collaboration. The most comprehensive definition is provided by Bruneel et al. [6], as below:

> an activity in which firms learn from experience and develop richer and more refined ways of engaging with the university sector. Indeed, overtime, the experience of collaboration should enable academics and their industrial collaborators to converge in attitudes, learning to share common norms and arrive at a mutual understanding about the nature of the collaboration and the research process. (p. 860)

S. Dayıoğlu Öcal (✉)
Hacettepe University School of Foreign Languages, Ankara, Turkey
e-mail: secild@hacettepe.edu.tr

Ş.Ş. Erçetin (ed.), *Chaos, Complexity and Leadership 2014*,
Springer Proceedings in Complexity, DOI 10.1007/978-3-319-18693-1_23

Entrepreneurialism causes universities to make some changes on their frames as structure, human resource, political and cultural frames. "Strengthening the core administration and extended administrative units" [7, p. 5–7] are the transformational changes towards entrepreneurial university, which leads to the formation of hybrid units or departments such as TTO (Technology Transfer Offices). These organizations are much more different than traditional units involving both academics and industrial representatives in terms of their structure and processes. For instance, in terms of human resource, research assistants are considered as a cheap labor by industry, which conflicts with academia [19]. This causes some faculty members to keep their students away from the university-industry research to protect them from any kind of manipulation and to train them as it should be in their own fields. However, there are some others who regard this collaborative research as "an opportunity to enhance the educational experience of students" [20]. The universities, in the political frame, can be regarded as an "organized anarchy" ([3], p. 50) by stating "the organization operates on the basis of a variety of inconsistent and ill-defined preferences" (Cohen 1972 cited in [3]). In the concern of university-industry linkage, it is a power of academia, i.e. knowledge, to be grounded or it is the power of industry, i.e. monetary values. Finally, though university and industry are two diverse worlds in terms of their values, norms and beliefs, they come together and redesign their activities in this collaboration. This collaboration reshapes university and its culture into entrepreneurial culture of the university.

23.2 University-Industry Linkage in Turkey

In Turkey, university – industry linkages are on the agenda since the 1980s. These can be easily seen in the Turkish National Development Plans, underlying the targets and aims of the Turkey in the following 5-or-7-year time. Moreover, there have been several references to the university-industry collaboration activities starting from the fifth one publicized in 1985. Some aspects of this collaboration can be listed as setting up the technology transfer offices, establishing and legalizing techno-centers, running these centers, defining the research and development studies. Since the 2000s, the university-industry activities have been highly valued so some considerable improvements have already been taken place in Turkey. For example, the Law of Technology Improvement Regions and the Implementation Regulation of Technology Improvement Regions have provided some privileges to the start-up and spin-off companies in the techno-centers and helped the establishment of techno centers in universities, triggering industry-university collaborations. Currently 32 Technology Development Region are actively established and 11 of them are in the progress of establishment. As a result, it is obvious that these plans have consolidated university-industry relations in Turkey and have accelerated establishing university-industry collaboration.

Some studies have been conducted to examine Turkish case in terms of university-industry linkages. Okay [23] found that 43 % of the academics did not participate in activities regarding university-industry relations and 57 % of them did not work in any unit involved in the service of these relations. Furthermore, it is pointed out that most of those relations were one-on-one and on issues organizing courses, seminars or training activities. Bayrak and Halis [4] carried out a study on both the academics and the industry representatives and concluded that the main reasons for this lack of collaboration include lack of communication and reliability among the parties. They explained the case with the phrase of 'waiting for something without taking any actions' (p. 66).

The ***objective*** of this study is to scrutinize the aspects of university-industry collaboration in the frame of chaos theory. The research question is considered as "how can university-industry collaboration be elaborated in the frame of the chaos theory key concepts of *nonlinearity, fractals* and *entropy* dwelling on academia?"

23.3 Chaos Theory and Key Concepts

Chaos Theory, rooted in mathematics and physical and natural sciences, deals with the alteration, disorder, or turbulence in any system from one state to the next in internal and environmental constraints [12, 15, 16]. In other words, any system that is regarded as a stable and continuous behavior fluctuates and shows unstable and discontinuous behavior when it faces with highly unexpected, unpredictable and complex changes. Since Chaos Theory presented totally different perception and understanding for the universe, it has caused a paradigmatic change in science. Nonlinearity, butterfly effects, fractals and entropy are common concepts to explain this new paradigm.

Nonlinearity is a term that has emerged from the chaos theory against Newton's physics, in which systems are assumed linear. However, linearity is very limited to comprehend the universe as well as human behavior. It is strange to try to comprehend human through a linear lenses though there are few linear explanations. "An essential contributing condition of nonlinearity" is feedback [9, p. 2]. Therefore, feedback feeds the nonlinear system and makes it sustainable as it is positively constructed completely opposed to the linear system. Moreover, it helps the system move towards the target from the initial condition. Akmansoy and Kartal [1, p. 511] pointed out "chaos theory provides refreshing insights into an education system that consists of a mechanical world in which many individuals attempt to learn information under whatever philosophy of education is popular at the time".

This theory is based on the notion that any complex system has an order in itself. In other words, there is a self-similarity at all levels of the system though they do not look like similar. These are called *fractals*. A snowflake is a common example for fractals. When it is examined under microscope, it can easily be seen that the basic pattern is continuously repeated. Also, attitudes and behavior of the people working in an organization can be considered as fractals since their attitudes and behavior become complex and diversified when the external (such as physical condition) and internal (perception of the people) factors vary for each of the people (Tuz 2004 cited in Ertürk [13]). Moreover, while system is transforming towards the state of entropy, units and particles in a system become fractals [12]. Here comes the term 'entropy'. *Entropy* refers to the balance of the system itself even when it looks there is no balance. The system adjusts and adapts itself to any current situation. The more system starts to find its entropy, the more fractals appear.

23.4 Chaos Theory and University-Industry Linkage

University-industry collaboration is *nonlinear* as it involves many dilemmas emerging tensions, diversities and unexpected changes based on two diversified world. At first glance, these issues can clearly be seen in the cultural perception in two diverse worlds. Academia has strong grounded values identified by Merton. These values are *communalism, universalism, disinterestedness,* and *organized skepticism*: *Communalism* refers to common ownership of the research without considering intellectual property. *Universalism* addresses the objectivity of scientific research – i.e. beyond the boundaries of race, gender, class, nationality, and religion. *Disinterestedness* points out that scientists act selfless. *Organized skepticism* focuses on 'rigorous and framed evaluation of community scrutiny' [20, 21].

Nowadays, academia has been transforming itself from these values towards the entrepreneurial ones, which causes universities into nonlinearity. Szlenyi and Bresonis [24] pointed out that the intersections between academic capitalism and public good may represent a new layer of normative and counter-normative values whose presence guides the research practices and behaviors of scientists and engineers. It is clearly seen in the study of Szlenyi [25] that the entrepreneurial experiences and attitudes of faculty and fellow students, organizational characteristics, and the stage of scientific training and experience heavily shape the socialization and professional identity development of doctoral students in the science and engineering fields. Tierney and Rhoads [27] highlighted the role of doctoral education as the anticipatory socialization to faculty roles. Therefore, as doctoral students engage in projects sponsored by firms and interact with industry representatives who hold a different culture from the academic culture, it is natural to question if these linkages have the potential to impact the anticipatory socialization of future faculty towards a business-like culture. If the answer is yes, then, a gradual cultural shift in the academic profession can lead to significant changes in academia including what is valued and rewarded, in particular in relation to Mertonian values.

On the other hand, Mendoza [18] revealed some hints of cultural changes among students with industrial sponsorship. For example, in the minds of these students, having industrial grants is as prestigious as traditional government grants. However, in the same study, despite the heavy presence of industry, the cultural knowledge acquired by students agreed with the core structure of the academic profession as idealized by Merton [21]. Moreover, in this case, linkages with private firms became a mechanism to achieve traditional outcomes of the academic profession by not only providing adequate funding to doctoral students and research infrastructure but as science and technology capital for ideas and connections for future research [22, 26]. In conclusion, it can obvious that academia has still strong power of transferring its culture to young scientists or younger generation through researching even though these scientists are more inclined to work with industry. All these cultural issues signify how changes in the cultural frame shape nonlinearity in academia.

Considering university-industry linkage, the entropy of these two diverse systems creates the other kind of entropy that is completely different from their own setting. For example, in academia scientists consider knowledge as common heritage to be shared and improved within the academic community. On the other hand, in industry knowledge has monetary value so it should be kept secret due to the competitive world of the business [2, 6]. Therefore, new level of entropy should be settled in coalition with two agents so as to sustain university-industry collaborative research. This process of entropy, of course, can bring about fractals which have some features evolving from both of the agents with some changes and some new forms. Another exemplary situation can be that the traditional values of academia in the process of being exposed by the values of entrepreneurialism such as commercialism and competitiveness. This creates a new need to re-define values, beliefs and norms within academic culture. Therefore, the commercial value or monetary value of the scientific research is one of the major concerns in addition to the scientific values in academia.

Though university-industry collaboration in Turkey is a new issue, this new change can cause re-generate of the dynamics of Turkish higher education system, which can lead to entropy. One of the drawbacks of this linkage can be an increased reliance of the universities on industry. In terms of pressure of the industry on the research faculty can face with the dilemma risking academic freedom, restriction in the focus of research and moves towards more short-term oriented goals and applied-science research. On the other hand, there have been other studies signifying this linkage supports that scientists can conduct basic research aligned with the applied one and there has been no such a big concern about risking academic freedom [5, 18, 20, 26]. These controversies make the comprehension of this phenomenon blurred. In other words, it can be stated that this collaboration is not considered totally black or white but it includes tones of grey. Hence, this relation causes academia find other ways of balancing itself, i.e. entropy through innovative ways of doing. This brings about fractalization. It is inevitable for universities and industries to work in collaboration with each other, regardless of all the kinds of diversification between the culture of academia and the culture of business world. Etkowitz [14] stated that universities start to take the entrepreneurial role of industry to some extent in addition to its traditional roles as education and research while start-up and spin-off companies take the role of researching. Therefore, they become fractals as a result of this collaboration. The new culture is a mixture of universities and industries, so called 'hybrid culture'. This is not stable and it is open to impacts of the industry, region and local setting [8, p. 67]. University and industry begin to get into a process of fractals as a result of hybrid culture. As a result, they both have their new and traditional roles that complement each other and cause a hybrid flora to work together.

23.5 Implications and Conclusion

It is a considerable issue that university-industry collaboration is inevitably chaotic in its nature, which leads to various implications on faulty member, higher education institutions and academic freedom of faculty members, intellectual property of the university-industry collaborative research and universities as organizations.

Initially, academic freedom is such a dilemmatic issue in this collaboration that faculty members try to keep their freedom through some strategies as extensive negotiations, keeping a diversified funding portfolio including block federal grants, and aggressively seeking funding in order to keep a healthy stream of revenue to allow room to explore their own scientific interests [20]. All of these strategies are for re-generating the balance in the nonlinear system-entropy- so that faculty can go on their researching without any interruption.

Intellectual property, pertaining to academic freedom, is another conflict of university-industry collaboration [5, 6, 10, 20]. Hayrinen-Alestalo and Peltalo [17] stated that though in the market every product has real exchange value, "universities develop on the basis of the capabilities of their researchers, teachers and students" (p. 259). In other words, in industry any product can be sold any amount of money in open market; nevertheless, it is impossible for researches, teachers and students to mention about any value as they are invaluable. This diversification lies behind the perception of knowledge. Mendoza [18] found that despite the heavy presence of industry in this particular department, the cultural knowledge acquired by students agreed with the core structure of the academic profession as idealized by Merton [21]. Moreover, in this case, linkages with private firms become a mechanism to achieve traditional outcomes of the academic profession by not only providing adequate funding to doctoral students and research infrastructure but as science and technology capital for ideas and connections for future research. This study is consistent with other findings indicating that faculty preferences are driven by academic values rather than opportunities in the market [22, 26]. Bruneel et al. [6] identified two barriers as orientation-related and transaction–related barriers, both of which have rooted in the cultural diversity and cultural conflict. For example, perception of time is divergent in academia and industry. In academia for researching, time is after quality of researching while for industry time is money. A scientist can go on the study as it should be. On the other hand, people in industry have a limited time to get the product released in a competitive world [11]. As a result, this leads to changes in both cultures and renovating new values, assumptions and beliefs, corresponding to fractalization and butterfly effects.

The value systems of university and industry is completely diverse and conflicting into each other. Secrecy of knowledge is seen as the most significant and chaotic implication to be considered in the collaboration. The first conflict here is that knowledge does not belong to anyone in academia but they all are considered universal. However, for industry, knowledge should be kept secret due to the competitive world of the business –secrecy of knowledge [2]. The other controversy is that scientists have strong intrinsic motivation on what they have been researching whereas business people have extrinsic motivation valuing more what they are getting or are going to get as a result of the research. In other words, the traditional values of academia are exposed to the values of entrepreneurialism such as commercialism and competitiveness. Bruneel et al. ([6]: 2010, p. 859) defined knowledge in academia and industry as two different systems and stated that "academics wish to create 'leaky' knowledge so that their ideas would be acknowledged by their peers while firms want the knowledge to be 'sticky' so that they can control a resource that is not available to their competitors." University researchers are also likely to choose research topics that are perceived by their peers to be interesting and valuable, while firms are likely to choose topics and problems that are perceived as being valuable for the development of new products and services for their customers. Therefore, a new value system in which both parties can be agreed should be formed by empathizing and respecting each other, and then negotiating and compromising. Bruneel ([6], p. 860) stated that "the experience of collaboration should enable academics and their industrial collaborators to converge in attitudes, learning to share common norms and arrive at a mutual understanding about the nature of the collaboration and the research process."

As a result, it is important to find a way to make academia and industry to work together. Without having any shared values, norms and beliefs it is obvious that these two diverse organizations cannot desirably interact with each other.

References

1. Akmansoy V, Kartal S (2014) Chaos theory and its application to eductaion: Mehmet Akif Ersoy University Case. Educ Sci Theory Pract 14(2):510–518. doi:10.12738/estp.2014.2.1928
2. Alves J, Marques MJ, & Saur I (2007) Co-ownership active interfaces between academia and industry. Eur Plan Stud 15(9):1233–1246
3. Bastedo MN (ed) (2012) The organization of higher education. The John Hopkins Press
4. Bayrak S, ve Halis M (2003) Öğretim elemanları ve sanayici açısından üniversite- sanayi işbirliğinin değerlendirilmesi. Sosyal Bilimler Dergisi Manas Üniversitesi Sosyal Bilimler Dergisi 5:64–85
5. Bogers M (2011) The open innovation paradox: knowledge sharing and protection in R&D collaboration. Eur J Innov Manag 14(1):93–117
6. Bruneel JD, D'Este P, Salter A (2010) Investigating the factors that diminish the barrriers to university-industry collaboration. Res Policy 39:858–868
7. Clark B (2008) On higher education. The John Hopkins University Press, Baltimore
8. Crespo M, Diridi H (2007) Intensification of university-industry relationships and its impact on academic research. High Educ 54(1):61–84

9. Cutright M (1997) Planning in higher education and chaos theory: a model, a method
10. Dayıoğlu-Öcal S (2012) A model proposal for the relations between the universities and the educational research and development companies at the technocenters. Unpublished Ph.D. Dissertation
11. Dayıoğlu-Öcal S (2013) Problems between universities and educational research development companies at techno-centers. J Educ Future (4):87–107
12. Erçetin ŞŞ (2001) Yönetimde yeni yaklaşımlar. Nobel Yayın Dağıtım, Ankara
13. Ertürk, A. (2012) Kaos kuramı: yönetim ve eğitimdeki yansımaları. Kastamonu Eğitim Dergisi 20(3):849–886
14. Etkowitz H (2003) Innovation in innovation: the triple helix of university-industry-government relations. Soc Sci Inf 42:293–337
15. Faramaznd A (2004) Chaos and transformation theories: a theoretical analysis with implications for organizational theory and public management. Public Organ Rev Glob J 3:339–372
16. Gregerson H, Sailer L (1993) Chaos theory and its implications for social science research. Hum Relat 46(7):777
17. Hayrinen-Alestalo M, Peltola U (2006) The problem of a market-oriented university. High Educ 52(2):251–281
18. Mendoza P (2007) Academic capitalism and doctoral student socialization: a case study. J High Educ 78(1):71–96
19. Mendoza P, Berger JB (2008) Academic capitalism and academic culture: a case study. Educ Policy Anal Arch 16(23). Retrieved [date] from http://epaa.asu.edu/epaa/v16n23/
20. Mendoza P (2014) Industry-Academia linkages: lessons from empirical studies and recommendations for future inquiry. Unpublished book chapter
21. Merton RK (1973) The sociology of science. Chicago University Press, Chicago/London
22. Murray F (2010) The oncomouse that roared: hybrid exchange strategies as a source of distinction at the boundary of overlapping institutions. Am J Sociol 116(2):341–388
23. Okay Ş (2009) Pamukkale Üniversitesi Öğretim Elemanlarının Üniversite-Sanayi İşbirliği Çalışmalarına Bakışları Üzerinde Bir Alan Araştırması. Teknik Bilimler Meslek Yüksekokulu Teknik-Online Dergi 8(2):94–111
24. Szenlenyi K, Bresonis K (2014) The public good and academic capitalism: science and engineering doctoral students and faculty on the boundary of knowledge regimes. J High Educ 85(1):126–153
25. Szlenyi K (2013) The meaning of money in the socialization of science and engineering doctoral students: nurturing the nest generation of academic capitalism. J High Educ 84(2):266–294
26. Thursby JG, Thursby MC (2011) Faculty participation in licensing: implication for research. Res Policy 40:20–29
27. Tierney WG, Rhoads RA (1993) Faculty socialization as a cultural process: a mirror of institutional commitment. The George Washington University, School of Education and Human Development, Washington, DC

Chapter 24
Leadership Styles and Performance of Academic Staff in Public Universities in Uganda

Rashid Kinsambwe

Abstract Leaders are perceived as persons who can single handedly create order out of chaos, navigate organizations through unthinkable environmental turbulence, bring mightiness out of mediocrity, and thrive where lesser mortals will quickly fade away. Education systems all over the world are undergoing constant changes under the effects of globalization, and the attendant rapid developments in information and communication technology. These forces demand changes within school systems across the world. These changes inevitably demand for good leadership and leadership styles for educational institutions to remain in business. This paper thus undertook to explore the relationship between various leadership styles and performance of academic staff with a bias on Public universities. The expressions made thereof are considered a light in the dark of knowledge in view of complexity engulfing both the education and leadership fields today.

24.1 Introduction

Leadership has been defined by a host of authors [10, 17, 31, 37]. In all their views, the expectations of any leader as well as key duties are fully explored. Leadership according to Yukl [38] is, "the process of influencing others to understand and agree about what needs to be done and how to do it, and the process of facilitating individual and collective efforts to accomplish shared objectives". Peter Northouse [31] on the other hand defines leadership as "a process whereby an individual influences a group of individuals to achieve a common goal". These definitions suggest several components central to the phenomenon of leadership. Some of them are as follows: (a) Leadership is a process, (b) leadership involves influencing others, (c) leadership happens within the context of a group, (d) leadership involves goal attainment, and (e) these goals are shared by leaders and their followers. The very act of defining leadership as a process suggests that leadership is not a characteristic or trait with which only a few certain people are endowed at birth. Defining leadership

R. Kinsambwe (✉)
Islamic University in Uganda, Mbale, Uganda
e-mail: shidkins@yahoo.com

Ş.Ş. Erçetin (ed.), *Chaos, Complexity and Leadership 2014*,
Springer Proceedings in Complexity, DOI 10.1007/978-3-319-18693-1_24

as a process means that leadership is a transactional event that happens between leaders and their followers.

Leaders are perceived as persons who can single handedly create order out of chaos, navigate organizations through unthinkable environmental turbulence, bring mightiness out of mediocrity, and thrive where lesser mortals will quickly fade away [17]. Education systems all over the world are undergoing constant changes under the effects of globalization, and the attendant rapid developments in information and communication technology. These forces demand changes within school systems across the world [10]. These changes inevitably demand for good leadership and leadership styles for educational institutions to remain in business.

Educational institutions are now born of complexity as reflected in the following aspects:

- Teaching and learning
- Assessment and evaluation
- Research programs
- Outreach activities
- Communication systems
- Decision making
- Delegation of authority
- Curriculum design
- Staff relations
- Organizational goals and vision

Generally therefore, a number of aspects render the education system rather complex and a summary of some is drawn from a multiplicity of researchers to the following effect [13, 24, 29, 35]:

- Integrating the commonly polarized goals of education, that is the goals that focus on transmitting knowledge with the goals that emphasize the development of an individual in the system.
- Catering for individual student needs. Running a classroom which recognizes and respects difference is first in the hands of a teacher who acknowledges that there are differences in the classroom.
- Assessment feedback and learning approach. We might consider the role of high-stakes standardized testing and assessment schemes in the present educational system as imposing an artificial fitness landscape that pulls the system toward behaviors that maximize test results rather than deep conceptual understanding.
- The dynamic nature of education system, the teaching and learning system that exists within the education system is subject to change with the changing world. Innovations and inventions must be incorporated into the system with the aim of maximizing student performance.

- The existing educational system of schooling isolates students and teachers from the wider community.
- Integrating the curriculum by developing interdisciplinary curriculum units that enable students to acquire knowledge from disciplines through unifying them while having the opportunity to contribute in different and special ways to the objectives of the integrated units.
- Dealing with misbehavior is a complex undertaking. Student misbehavior in the classroom is a tough and unavoidable task to the teachers and it takes up teachers' considerable time to deal with, which in turn affects the quality of the student's learning experiences.

Education systems are now characterized by multi-level governance where the links between multiple actors operating at different levels are to a certain extent fluid and open to negotiation [8].

Such complexities call for effective leadership and leaders since it is the said leaders who have been found to influence followers in many ways, including coordinating, communicating, training, motivating, and rewarding [39]. It is argued that effective leadership has a positive sway on the performance of organizations [4, 9] and the success of an organization depends on its people and on the leader's ability to optimize human resources [2]. A good leader understands that motivating the employees is of paramount importance in achieving organizational goals. Furthermore, it is generally accepted that the effectiveness of any set of people is largely dependent on the quality of its leadership [23].

Armstrong [1] contends that performance refers to both behaviors and results, and adjusting organizational behaviors and actions of work to achieve results or outcomes. Maicibi [27] contends that, without a proper leadership style, effective performance cannot be realized in schools. Performance appraisals are therefore an important means by which employees can discuss key work issues in a systematic manner in an effort to monitor and evaluate progress for appropriate interventions [26].

Many studies that have been conducted to show the correlation between leadership style and employee performance in wide variety of fields and in equally wide variety of settings, few studies however have focused on this relationship in the context of higher education, for example [22] in Nigeria, Hamidifar, 2010 in Iran and others in North America and Europe. Studies investigating the relationship between leadership style and organizational performance in Uganda have been limited to secondary schools [32]. In a departure from this focalization, this study will examine the Leadership styles of top university leaderships in selected universities in Uganda and explore the relationship between leadership styles and staff performance in those universities. Specifically it will concern itself with full-time academic staff in these universities and explore the effects of leadership style on the six academic staff performance indicators mentioned in Sect. 24.1.

24.2 Leadership and Leadership Styles

Although leadership has long been of interest to historians and philosophers, scientific studies only began in the early 1900s. Still, the body of knowledge has since been fast growing as attested by the more than 350 definitions of the term which scholars have come up with. Giving one specific definition of leadership is a thus very complex task [3].

Broadly speaking, leadership is an influence relationship among leaders and followers to perform in such a way to reach a defined goal or goals [7]. Whereas managers are concerned with short-term problems within an organization, leaders adopt a much broader perspective. While early leadership theories concentrated on the characteristics of successful leaders, their traits, behavior, power, influence and situational approaches (e.g.) [25] recent ones have focused on the role of followers and the correlated nature of leadership.

It is also important to note that organizations and their environments have changed rapidly over the past years and as a result a new style of leadership, one that is less bureaucratic and more democratic, is needed in order to ensure the survival of organizations [4]. Consequently, a new style of leadership has emerged in order to ensure the survival of organizations and to overcome limitations of the trait, behavioral and contingency theories of the past. This new leadership approach suggests that there are two views of leadership – transactional leadership and transformational leadership [4].

Within the context of Maslow's hierarchy of needs, transactional leadership works at the basic levels of need satisfaction, where transactional leaders focus on the lower levels of the hierarchy (Deanne N. et al.... 1997). Transactional leaders use an exchange model, with rewards being given for good work or positive outcomes. Conversely, people with this leadership style also can punish poor work or negative outcomes, until the problem is corrected. One way that transactional leadership focuses on lower level needs is by stressing specific task performance [18]. Transactional leaders are effective in getting specific tasks completed by managing each portion individually.

Transactional leaders are concerned with processes rather than forward-thinking ideas. These types of leaders focus on contingent reward (also known as contingent positive reinforcement) or contingent penalization (also known as contingent negative reinforcement). Contingent rewards (such as praise) are given when the set goals are accomplished on-time, ahead of time, or to keep subordinates working at a good pace at different times throughout completion. Contingent punishments (such as suspensions) are given when performance quality or quantity falls below production standards or goals and tasks are not met at all. Often, contingent punishments are handed down on a management-by-exception basis, in which the exception is something going wrong.

Burns [7] argues that transactional leadership entails an exchange between leader and follower. Followers receive certain valued outcomes (e.g. wages, prestige) when they act according to their leader's wishes. Taking Burns as his starting point

Bass [3] notes that leadership in research has generally been conceptualized as a transactional or cost-benefit exchange process. In many universities in Uganda, the academic staff has had a face-off with their administrations as a result of the perceived low wages. For instance, Makerere University, the largest and oldest university in Uganda, has in the past 10 (ten) years experienced over five staff strikes over wages. And as a result, the University has always suspended classes for at least 3 days, thus, affecting the staff output.

Deanne N. et al. (1997), argues that, the transformational leader on the other hand, motivates subordinates to perform as expected. The transformational leader typically inspires followers to do more than originally expected. Transformational leadership theories predict followers' emotional attachment to the leader and emotional and motivational arousal of followers as a consequence of the leader's behavior [20]. Hater and Bass [19] state that "the dynamics of transformational leadership involve strong personal identification with the leader, joining in a shared vision of the future, or going beyond the self-interest exchange of rewards for compliance".

A transformational leader is a person who stimulates and inspires (transform) followers to achieve extraordinary outcomes [37]. He pays attention to the concern and developmental needs of individual followers; changes followers' awareness of issues by helping them to look at old problems in a new way, and he is able to arouse, excite and inspire followers to put out extra effort to achieve group goals. Transformational leadership theory is all about leadership that creates positive change in the followers whereby they take care of each other's interests and act in the interests of the group as a whole (Warrilow 2012). The concept of transformational leadership was introduced by James Macgregor Burns in 1978 in his descriptive research on political leaders, but its usage has spread into organizational psychology and management with further modifications by B.M Bass and J.B Avalio [23].

24.3 Performance of Organisations

Performance is often identified or equated with effectiveness and efficiency [30]. Employee performance has been shown to have a significant positive effect on organizational performance [11]. The function that human resources can play in gaining a competitive advantage for an organization is empirically well documented [6]. One of the performance indicators of academic staff is the ability of the leader to enable staff to acquire higher degrees and to conduct research. Promoting a research culture is thus one fundamental role of academic leadership in universities. The role of research can be compared to a pivot around which the other cardinal purposes that define the essence of a university.

Research and publication form the top most criteria for academic staff promotions in universities. Therefore the onus is on the leaders to promote a culture of research by providing the necessary drive, facilities and incentives. Secondly, universities are challenged to take a lead in transforming society for the better, because they are obliged by the doctrine of corporate social responsibility. This will require that they reach out to the community to be part of the solutions to community problems. To enhance performance of staff therefore, the institutions policies on these aspects has to be very sound.

Research and publication in most Ugandan universities is at its lowest, with academic staff taking on extra teaching load, in various universities. It is very common to find a lecturer teaching in more than one university. In each university, this lecturer may have over 15 (fifteen) h per week. This leaves him with barely any time for other activities. Given the economic situation in Uganda, it would take a very good leader to convince such academic staff to focus on other activities, such as research and publication without additional monetary incentives.

Both transformational and transactional leaders are active leaders. They actively intervene and try to prevent problems. However, in this complex situation one leadership style alone may not be appropriate, as the reason why the staff are doing this is to increase on their income. This situation can become worse, if there is passive laissez-faire leadership, where a leader avoids decision making and supervisory responsibility. This type of leader is inactive, rather than reactive or proactive, which creates no difference between the presence and absence of leadership. Bass [4] reports laissez-faire leadership usually correlates negatively (−.3 to −.6) with other, more active leadership styles.

24.4 Leadership Style and Performance

It has been widely accepted that effective organizations require effective leadership and that organizational performance will suffer in direct proportion to the neglect of this [15]. Furthermore, it is generally accepted that the effectiveness of any set of people is largely dependent on the quality of its leadership. Effective leader behavior facilitates the attainment of the follower's desires, which then results in effective performance [15, 16, 33, 36]. Leadership is perhaps the most investigated organizational variable that has a potential impact on employee performance [12].

It is evident from the exiting body of literature that studies investigating the relationship between leadership styles and employee performance are limited in the Ugandan context, except for a few (see [32]). Recent leadership studies have also continued to affirm the positive relationship between transformational and transaction leadership and performance at various levels (e.g. [22, 30] and the negative relationship between Laissez-faire leadership and performance [17, 32, 34]. The relationship between leadership styles and academic staff performance is illustrated in the Fig. 24.1 below.

Figure 24.1 reflects the fact that academic staff performance, as measured in terms of: qualification, teaching/cognate experience, contribution to university and community, publications, current research/extension and administrative experience

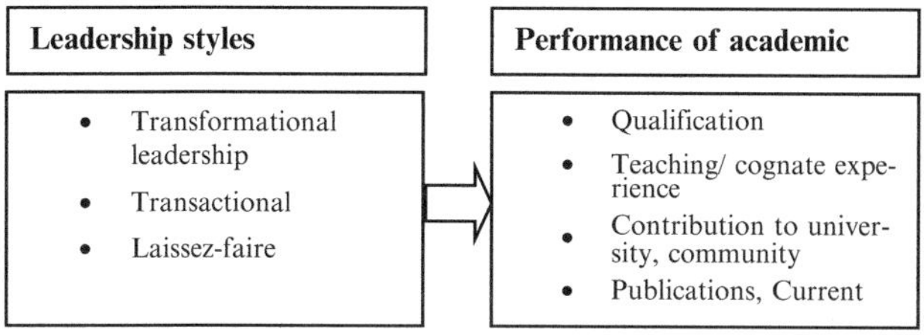

Fig. 24.1 The relationship between leadership styles and staff performance (Source: Adopted from Hamidifar [25] and modified by the author)

can be influenced by either one of the three different leadership styles considered in this article: transformational, transactional, and laissez-faire leadership. Leadership styles as measured the Multifactor Leadership and modified by [17] gives five factors (sub-variables) as characteristics of transformational leadership: idealized influence attributed to behavior, inspirational motivation, individualized consideration, and intellectual stimulation. The model also gives three factors (sub-variables) as characteristic of transactional leadership namely: contingent reward, management by exception-active, and Management by exception-passive).

Intellectuals are very difficult people to lead. Academic staff needs independent thinking in whatever they do. The leader needs to justify any decisions he chooses to make, without which, it will face a lot of resistance. For example, the Makerere University Academic Staff Association (MUASA) is a very powerful body in the management of the University, yet it is not provided for in any structures of the Makerere. On many occasions has found itself on the other side of the university administration's position, which has in most cases ended into strikes. This makes it a complex constituency to manage. Since Universities have multiple constituencies, their leadership is essentially collegial in nature where all members in the community are considered equal and make decisions in committees and consensus among colleagues [5].

Kyambogo University is one of the higher institutions of learning in Uganda, which was created in 2001, under the Universities and Other Tertiary Institutions Act 2001. This was through the merger of three institutions namely The Institute for Teacher Education Kyambogo (ITEK), Uganda Polytechnic Kyambogo (UPK) and the Uganda National Institute for Special Needs Education (UNISE). The merged institutions had different and unique missions, values, visions and cultures prompting different aspirations. Since its creation, Kyambogo University has faced a number of challenges punctuated with unrest and a series of strikes (Weekly Observer 23rd–29th November 2006 and New Vision March 6 2008). This led to indefinite closure of the institution (Monitor 1st April 2008).

The founding Vice Chancellor was retired and his position advertised twice with no applicants (New Vision 12th Jan 2008). The other key people in the top management including the University Secretary and Bursar were sent on suspension. The University Council was dissolved and a new one appointed (New Vision 13th

April 2008) and there was a cabinet committee set up to advise on the way forward (Monitor March 26th 2008). Amidst these events, there is less productivity of staff as classes are suspended in most cases, no research can be conducted under such situations, thus a manifestation of the leadership challenges in the institution, which requires the leader to adopt an appropriate leadership style.

Universities in Uganda tend to have a dual system of structure, the formal bureaucracy and the informal collegial. The collegial committee system cuts across the University with the University Council, Senate, Faculty Boards, Deans, and Departmental committees, where the administration is taken care of by the University Secretary (finances) and the Registrar (Admissions and records). However, the collegial way of doing things undermines bureaucracy which also exists in University leadership (Dear Love 1995). All this an indication of how complex it is to lead academic staff. Therefore, it is very crucial to be mindful of the leadership style under such circumstances.

24.5 Future Trends in Leadership Styles

There has been a paradigm shift in organization settings. Yet leadership has to obtain within the existing paradigm at all times. Short of this, relevance is lost and the ability to deal with prevalent issues becomes doubtable. Given the changing nature of settings within the educational institutions to date especially those of Higher learning, it is imperative to re-examine leadership styles that ought to be adopted by leaders in the said institutions.

The first step in the right direction in this case should be consideration of leadership from more specific and action laden angles than the conceptual. Such leadership styles can be viewed from the following perspective:

- People leadership
- Service leadership
- Outreach-based leadership
- Curriculum leadership
- Research leadership

Not only that but future leaders ought to understand the levels of complexity that prevail in educational leadership such as;

- Structural complexity relating to systems in the organization
- Process complexity relating to the way things are done
- Product Complexity relating to the nature of output in an organization
- Managerial Complexity relating to the way people are led

These would help in adopting and adapting to coping mechanisms of leadership in the wake of complexity within the educational institutions. That way, leadership can serve its sacred purpose of transforming school performance through proper and appropriate influence of academic staff; highly held by proponents of education.

24.6 Conclusion

The key themes of this article are performance and leadership styles, mainly focusing on the new approaches of leadership styles that were developed after the traditional ones received a lot of criticisms. This is because we are in the twenty-first century where the environment in which organizations including universities operate is so fluid, therefore the leadership styles have also evolved to take care of this kind of environment.

Although leaders are perceived to be the goal setters, and lead their followers towards achieving those goals, the way they chose to do so matters. When it comes to leading universities, the leaders have to be extremely very careful in selecting a certain leadership style. This is because, universities globally, have the highest concentration of intellectual, who cannot a no for an answer. Universities are the only places where leaders would be challenged and concede defeat. Under such circumstances, leaders could probably define situations in which highly active/inactive leadership is not necessary and maybe not even desirable. For instance, in their substitutes for leadership theory Kerr and Jermier (1978) propose several subordinate, task, and, organization characteristics that could reduce the importance of leadership. A less active role of leaders could also lead to 'empowerment' of followers which could even make for a useful component of transformational leadership.

References

1. Armstrong M (2004) Human resource management theory and practice. Bath Press Ltd., London
2. Bartlett CA, Ghoshal S (1995) Changing the role of top management: beyond systems to people. Harv Bus Rev 73(3):132–142
3. Bass BM (1985) Leadership and performance beyond expectation. Free Press, New York
4. Bass BM (1990) From transactional to transformational leadership: learning to share the vision. Organ Dyn 18(3):19–31
5. Bikas S (1995) Innovation in university management. A UNESCO Publication, Paris
6. Brewster C, Carey L, Dowling P, Grobler P, Holand P, Warnich S (2003) Contemporary issues in human resource management. Clyson Printers, Cape Town
7. Burns JM (1978) Leadership. Harper & Row, New York
8. Burns T, Wilkoszewski H (2011) Governing complex education systems: project plan 2013–2014
9. Charlton G (2000) Human habits of highly effective organizations. Van Schaik Publishers, Pretoria
10. Chinnammai S (2005) Effects of globalization on education and culture. ICDE international conference, New Delhi
11. Collis DJ, Montgomery CA (1995) Competing on resources. Harv Bus Rev 73(4):118–128
12. Cummings LL, Schwab DP (1973) Performance in organizations: determinants and appraisal. Scott, Foresman and Company, Glenview
13. Dean J (2000) Improving children's learning: effective teaching in the primary school. Routledge, London

14. Dumdum UR, Lowe KB, Avolio BJ (2002) A meta-analysis of transformational and transactional leadership correlates of effectiveness and satisfaction: an update and extension. In: Avolio BJB, Yammarino FJ (eds) Transformational and charismatic leadership: the road ahead. JAI Press, Amsterdam, pp 35–66
15. Fiedler FE, House RJ (1988) Leadership theory and research: a report of progress. Int Rev Indust Organ Psychol 19(88):73–91
16. Gabbert CC (2005) The relationship between chief executive transformational leadership and hospital performance, Capella University
17. Hamidifar F (2010) A study of the relationship between leadership styles and employee job satisfaction at Islamic Azad University Branches in Tehran, Iran. Au GSB e J 3(1):45–58
18. Hargis MB, Wyatt JD, Piotrowski C (2001) Developing leaders: examining the role of transactional and transformational leadership across contexts business. Organ Dev J 29(3):51–66
19. Hater JJ, Bass BM (1988) Superiors' evaluations and subordinates' perceptions of transformational and transactional leadership. J Appl Psychol 73(4):695–702
20. House RJ, Spangler WD, Woyke J (1991) Personality and charisma in the U.S. Presidency: a psychological theory of leader effectiveness. Adm Sci Q 36:364–397
21. Howell JM, Neufeld D, Avolio BJ (2005) Examining the relationship of leadership and physical distance with business unit performance. Leadersh Q 16(2):273
22. Irtwange SV, Orsaah S (2009) Impact of management style on performance indicators of academic staff Benue State University, Makurdi, Nigeria. Educ Res Rev 4(12):602–615
23. Jung DD, Sosik JJ (2002) Transformational leadership in work groups: the role of empowerment, cohesiveness, and collective-efficacy on perceived group performance. Small Group Res 33:313–336
24. Lemke JL, Sabelli N (2002) Complex systems and educational change: towards a new research agenda. University of Michigan. Adapted from http://www.jaylemke.com
25. Lowe KB, Kroeck K (1996) Effectiveness correlates of transformational and transactional leadership: a meta analytic review of the MLQ literature. Leadersh Q 7(3):385–426
26. Nmadu TM (1999) Human resources management: an introduction. Jofegan Associates Publishers, Jos
27. Maicibi NA (2003) Pertinent issues in employees management. M.P.K. Graphics (U) Ltd, Kampala
28. Maicibi NA (2005) Pertinent issues in management, human resource and educational management. Net Media Ltd, Kampala
29. McManus M (1993) Discipline in students exclude from school. In Varma V (ed) Management of behaviour in schools. Longman, London, pp 218–232
30. Neely A, Gregory M, Platta K (1995) Performance measurement system design. Int J Oper Prod Manag 15(4):80–116
31. Northouse PG (2010) Leadership: theory and practice, 5th edn. Sage, Thousand Oaks
32. Nsubuga YKK (2009) Analysis of leadership styles and school performance of secondary schools in Uganda, Kampala, Uganda
33. O'Reagan N, Ghobadian A, Sims M (2005) The link between leadership, strategy, and performance in manufacturing SMEs. J Small Bus Strategy 15(2):45–57
34. Pradeep DD, Prabhu NRV (2011) The relationship between effective leadership and employee performance. International conference on advancements in information technology, Singapore
35. Pintrich PR (2000) Multiple goals, multiple pathways: the role of goal orientation in learning and achievement. J Educ Psychol 92:544–555
36. Rejas LP, Ponce ER, Ponce JR (2006) Leadership styles and effectiveness: a study of small firms in Chile. J Interciencia 31(7):499–504
37. Robbins SP, Coulter M (2007) Management, 9th edn. Prentice-Hall, London
38. Yukl G (2006) Leadership in organizations, 6th edn. Pearson-Prentice Hall, Upper Saddle River
39. Yukl G (1989) Managerial leadership: a review of theory and research. J Manag 15(2):251–289

Chapter 25
Complexity of Managing Educational Resource Centres in Higher Institutions of Learning

Sanusi Sani Danmal and Abubakar Mansur DanJeka

Abstract The idea of managing resources of a group started ever since human beings started coming together and working in groups. This is essential so that the objectives of the group can be attained through proper linking of the various activities being performed by the group members. As organizations, particularly formal organizations like businesses, ministries, schools, including universities become large, the task of the manager starts getting bigger and more complex. In all these, the aim of the manager is to promote excellence. The emergent of educational technology as a discipline having much to do with how to improve the quality of educational practice come with it some new innovations of which educational resource center is one. There have been several notions and views about what should be an appropriate meaning of the "term educational resource centers" and how to effectively manage it to yield required educational target. Any problem that has a ready solution is no longer a problem. However, the solutions that are available will not become effective on the problem situation until steps are taken to implement them. Now that the problems are still very much with educational resource centers, could it be that the solution are not yet discovered or there are no initiators to set the motion. Prevention, people say, is better and even cheaper than cure. This is the reason why the maintenance culture should be imbibed by all teachers and managers of Educational Resource Centers. It has been documented in literature that instructional materials have the ability to enhance learning either by concretizing abstract concepts, by making the class lively while learners are actively involved, by making illustration of points easy for the teachers or by encouraging individualized learning. The next problem one would raise is how to get the instructional materials for use. This is the major focus of the establishment of Educational Resource Centers in the schools. Educational Resource Centers have a lot of prospects if and when established in the schools. There is some evidence

S.S. Danmal (✉)
Department of Educational Technology, Usmanu Dan Fodiyo University, Sokoto, Nigeria
e-mail: ssdanmali2011@gmail.com

A.M. DanJeka
Department of Educational Management and Administration, Federal College of Education, Zaria, Nigeria
e-mail: danjeka2@gmail.com

Ş.Ş. Erçetin (ed.), *Chaos, Complexity and Leadership 2014*,
Springer Proceedings in Complexity, DOI 10.1007/978-3-319-18693-1_25

of managerial take-up of 'complexity' as a framework for informing organizational practice of educational resource centers of Nigerian High institutions. This is still at an early stage, and take-up may or may not lead to take-off. The purpose of this paper is to contribute to a discussion on the chaos and complexity of managing educational resource centers its validity and significance for the management of organizations.

25.1 Introduction

An educational resource center (ERC) is called by many names. It is called a learning resource center, learning materials center, educational media center, educational resource center, educational services center, educational communication center, instructional technology center, a library, an audio visual center, library-media Resources center or an educational materials center (Geretti 1976). It is obvious that an educational resources center is many things.

A working definition of this term could be arrived at by matching of the meanings of the three concepts as explained earlier. The following definitions thus emerge:

(i) Place where facilities for educative purposes could be found.
(ii) Place where educational materials in the form of hardware and software are classified arranged and are retrieved for the educational purposes.
(iii) A place where books and non-book materials are provided and scientifically organized for easy retrieval by users to promote education. Abimbade [1] further educational resources center as "a setting where educational resources such as materials, tools and equipment can be designed, developed, utilized, borrowed and stored".

Educational Resources centers is thus a store house where all sorts of educational devises both small and big, simple and complex, cheap and expensive could be found, loaned, utilized by all users- students, teachers, researchers etc.

In primary, secondary and tertiary schools, it is common to find places where books are kept. These are called libraries. Science laboratories too are common features in secondary and tertiary institutions. There are places where facilities and chemicals for the teaching and learning of science subjects are kept. Both the libraries and laboratories are forms of educational resource centers. However, we need to understand what educational resource centers mean a comprehensive level.

Educational Resources Centers have been described as "a setting where educational resources such as materials, tools and equipment can be designed, developed, utilized, borrowed and stored". [1]. It should however be noted that human resources such as facilitators for consultancy services in instructional material production, technician, professionals in various fields are available at the educational resource centers.

Abimbade [1] categorized educational resource center into centralized, decentralized and coordinated types. The centralized resource center according to him is the one that function as a sole facility within a state or local government.

The decentralized resource center is the one which functions independently within the school building. Such a resource center services the needs of the immediate community where it is situated.

The third category, which is the coordinated type, operates as a system. It connects a network of schools resource centers, but it is supplemented and served by a central resource center Modern trends in the development of library and information management have introduced new nomenclatures that are found relevant to the discussion of educational resources centers. Fayose [5] wrote about "School Library Resource Centers". Elaturoti [4] used the term "School Library Media Centre" In the manner in which the two cited authors used the terms; it is obvious that these two terms are synonyms of the term educational resource centers at least in operations. According to Elaturoti [4]. The School library media center functions as a vital instrument in the education process.

It is not a separate entity from the total school programme but totally integrated into the teaching learning process of the school. It provides a variety of books and non- book materials which have been selected, acquired and organized for use in support of the entire school Programme.

25.2 Complexities Confronting Educational Resource Centres

Educational Resource Centers are expected to be established in any university that runs a first degree. Every College of Education should have it; every secondary school and Primary should at least have a well-equipped library that is to function as a library media resource center while the primary schools should have reading corners. However, this is not in reality.

Educational Resource Centers that are functional are found mostly in the universities that run degrees in education technology. Other universities may have structures lapelled as CET but which are glorified classrooms or mere stores. Where you have Centre for Educational Technology or as is commonly and wrongly called University Media Centers not many Universities have actually use them for institutional purposes as expected.

The Colleges of Education are worse hit. Educational Technology and Microteaching courses are taught theoretically and students do not have access to the electronic gadgets like projectors, computers, cameras etc. to facilitate learning and to help them in producing instructional materials for their teaching practice exercises. Even though the Educational Tax Fund (ETF) and other similar intervention are rendering supports, yet the situation is not totally resolved in many institutions.

In the secondary schools, the negative reading culture of many Nigerians still shows up in the attitude of school administrators and members of the school community towards equipment of libraries. The few schools that have libraries are the old schools. The books in these old schools too are as old as the schools. In most of them non-book materials and equipment are not available.

Some state governments are already taking some positive steps in establishing resource centers in the primary schools. But then, what can one single resource center do for a whole local government area? We may also ask whether that single resource center in the local government area is properly stuffed with the required facilities for a resource center. And so, the problems confronting Educational Resource Centers still persist across all educational levels in Nigeria.

25.3 Classification of the Complexity Challenges, Issues and Problems of Educational Resource Centers

Based on the physical evidence and many Research findings the existing problem and complexities of managing Educational Resource Centers in High institution of learning had been classified into the following categories:

1. *Managerial Complexity*
2. *Socioeconomic complexity*
3. *Educational complexity*
4. *Technical complexities*
5. *Ethics, moral and attitude toward educational resource center*

25.3.1 Managerial Complexity

The managerial complexities of educational resource center in higher institution of learning viewed in the following wings of management:

Facilities resource management: The poor maintenance culture of many Nigerian coupled with the naïve attitude of seeing government property as no-man's property shorten the life span of the few equipment procured for the Educational Resource Centers. This also brings on board the non-availability of some of the equipment/facilities in Nigeria market, and the erratic nature of power supply in Nigeria poses a great problem to the image of the electronic equipment in the Educational Resource Center.

Human resource management: Administrative bottlenecks and bureaucratic "road blocks" of school administrators who have no priority for Educational Resource Centers sometimes dampen the Morales of some resourceful teachers who like to promote the establishment of Educational Resource Centre in their schools.

Financial resource management: This might include proper budgeting, proper administration of fund effective utilization of financial resources as well as monitoring and evaluations of all incomes and expenditure this also lack dimension.

Urgency of technology response: The use and especially maintenance of technology needs constant and adequate funding because of the urgent nature of technology related problem

25.3.2 Socio-economic Complexity

Educational resource centers also couple with some complexity that are resulted from socioeconomic factor of the society these factors are:

Multi-cultural trends: The relative newness of educational technology in Nigeria still poses a problem of acceptability because the concept of educational technology is still new to many people.

Political trends: Government transition and change of educational policy from time to time by politician is also another issue militating against the progress of educational resource centers.

Religious trends: Religious has great influence in many African people especially in Nigeria, which is being referred to by IFRO 2010 [6] as most religious country in the world. So in north who are predominantly Muslims require every aspect of human to be touched and covered by Islamic principle and failure to get such will make them loose interest in their activities.

25.3.3 Educational Complexity

Education is the major background reason behind the establishment of any educational resource center yet educational activities itself faced some challenges that became complexities for educational resource center in the following areas:

Instructional complexities: These may include lack of Motivation, lack of Unique Instructional Capabilities Provision, lack of Supports for New Instructional Approaches, Decreasing Teacher Productivity, and Lack of Required Skills for an Information Age

Training and expertise: hardly do schools match the pace of the technology innovations. But there is certain level of technology a modern school cannot afford neglecting. As such schools are in constant need of training the staff and faculty on the use of new hardware and software.

Multidisciplinary resource: Interdisciplinary and multidisciplinary nature of educational activities which require educational resource center to have the resource of all disciplines for consultation

Professional development: Recruitment selection and training to develop professional that will carry out educational and consultation services is greatly lacking in many educational resource centers in Nigeria. Specialists in educational technology who could be committed to monitoring the establishment of Educational Resource Centers are very few

25.3.4 Technical Complexities

Technology as a basic tool in modern educational life deserves everything necessary to keep up with all needed by educational resource center but contextualization of these centers have not been able to be readily hand in hand with following:

Rapid technology change: Technology grows every single day while the educators are left behind dealing with old literature and techniques. Technologies such as computers coupled with appropriate applications and adequate expertise provides rich learning environment that develops necessary skills and opportunities to learners. This includes communication, collaboration, adequate research, study and practice

Purchase and upgrade: Given the constant influx and improvement of hard and software, educational resource centers technology needs funding to address compatibility challenge

Repairs and maintenance: Educational resource centers needs consistent funding in order to keep the hardware running which include repairs and maintaining and this is also lacking.

The life span of the facilities in the Educational Resource Centers depends much on the handling and usage in the personnel of the centers. It is a common say that the cost of maintenance of any facility is far less than the cost of repairing it when it breaks down. It is therefore wise to ensure a proper super vision of the handling and usage of the facilities. The maintenance of the facilities in the Educational Resource Centers shall be discussed under three sub-headings via: Storage of the facilities, Usage of the facilities and Servicing of the facilities

Storage of the Facilities: According to Afolabi [2], the pattern of storage for the facilities in the Educational Resource Centers depends on the category the facilities falls into, whether it is a printed material, photographic material, electronic equipment, wooden material and so on.

Usage of the facilities in the Educational Resource Centers: It will be abnormal for one to spend much money to procure instructional materials and then use our own hand to destroy it. The instructional materials in the Educational Resource Centers need careful handling as we use them from time to time so that they can last long.

Servicing the Facilities: Another aspect to be considered in maintaining of the longevity of the instructional materials in the Educational Resource Centers is the regular and routine servicing of the items. This is done to ensure the maintenance of the original status, colony, shape, functionality etc. of the materials for a reasonably long time [3, 7].

25.3.5 *Ethics, Moral and Attitude Toward Educational Resource Center*

In any organization Ethics, moral attitudes and perceptions of the people and stakeholder has great influence on its progress. Educational resource center in higher institution of learning also have the following issues as major complexities in Nigerian context:

Ethical issue in Technology trends: Major kinds of Moral and ethical dimensions issues like: Privacy/confidentiality Spamming/Hacking, Property/Copyright, Netiquette, Vandalism, Access, Accuracy/Trust wordiness, the new plagiarism and academic dishonesty, Illegal downloads & software

Diverse Perceptions on educational resource center: People have not yet fully what are the rationales behind establishment of educational resource centers. They therefore don't consider it to be a place for which they will consult for educational services.

Participation and Commitment: The involvement of stake holders in education is yet another issue which is to be considered. Those stakeholder include: Educators Comprising teachers students and school administrators, Technology expert, Educational Policy makers and government agencies

Community participation, Governmental and nongovernmental organization

25.4 Intervention Mechanisms to the Problem of Educational Resource Centre

There is no problem without solution. The following solutions are hereby suggested:

- Four or five schools that are in the same locality could pull their resources together to procure the so called costly equipment.
- Improvisation could be solution to the problem of non-availability of certain facilities. International bodies like USAID, UNDP, and UNICEF always assist schools to procure equipment when they are contacted.
- The use of standby generating plants could help to solve the problems of inconsistent electric power supply since power supply may not be needed all the time.
- Educational technology specialists should embark on enlightenment campaign to educate those who are ignorant of the prospects of Educational Resource Centers.

This will make school administrators to be properly educated on what the school stands to gain if Educational Resource Centers are established.

- In service training programme could be proposed for Educational
- Resource Centre workers to update their knowledge of educational technology
- These programmes are to be fully sponsored by government or instruction so that when the workers complete their courses, they are bounded to serve their sponsors for some time.
- Local factories where equipment and materials for teaching-learning are being manufactured in large scale should be empowered financially and encouraged by government, NGO and individuals.

25.5 Prospects of Educational Resource Centers

Educational Resource Center has a role to play in the teaching learning process. With the resolve of Nigeria to catch up with the rest of the world in technological advancement, the proper attention has to be paid to teacher production programmes as well as the teaching-learning process. Educational Resource Centers are very relevant to the achievement of this feat.

25.6 How the Educational Resource Centers Could Improve Teaching-Learning

A functional, well equipped Educational Resource Centre is likely to:

- Encourage teachers in the use of instructional materials since they can just walk in and avail themselves with the Instructional Materials already stored for me.
- Make the classroom interaction of training and pupils lively with the supply of instructional materials from the Educational Resource Centre.
- Many teachers have problems with how to make instructional materials but with Educational
- Resource Centre in place, they can have these instructional materials easily on loan.
- Offer to teachers the latest discoveries in teaching learning as a result of the researches conducted by experts.
- Make teachers to be better equipped for the task ahead of them since they have the opportunity of attending seminar, training and workshops organized by the Educational Resource Centers.
- Sensitize teachers at all levels to the strategies for enhancing the teaching profession. The Educational Resource Centres have a way of providing fund and consultancy services to teachers who want to carry out projects that will enhance their profession.

- Assist resourceful teachers to discover their potentials. Some teachers would not believe that they could draw, paint, make sketches, carve or mould any teaching aid. These potentials would be uncovered in them with the presence of an Educational Resource Centre in their institution.
- Improve the quality of lesson delivery and in time enhance academic performance in learners at all levels. When teachers teach well as a result of the provision of the Educational Resource Centre, the students too will learn with ease and the society will be a better literate society that it could have been without the Educational Resource Centre.

Other Factors that could pave way for the Realization of the Prospects of Educational Resource Centers include the following under description.

Government Effort The laudable objectives of Educational Resource Centers and the prospects of establishing them could become realizable if the government enacts policies that can serve as the springboard for them. Relevant government agencies could be saddled with this responsibility. For instance, the National Commission for Colleges of Education (NCCE) recently made it compulsory for all lecturers in the College of Education to become computer literate. This policy has made all lecturers, who were non-challant about computer literacy before, to pursue computer literacy with more seriousness.

Teachers' Empowerment It has been suggested that teachers should be empowered and eventually mandated to produce one teaching aid (possibly, a chart) monthly. If teachers are paid a stipend tagged Instructional Materials Allowance (IMA), teachers will have no excusable reason for failing to procure at least one improvised instructional materials in a month [2]. It is believed that by the time these instructional materials accumulate over the years, there will be enough materials to be kept in the Educational Resource Centers in the schools. This idea is expected to be implemented for teachers at all the three levels of education, that is, primary, secondary and tertiary.

Inspectorate Services It is a common routine duty to see inspectors of education insisting the primary and secondary schools. The items expected of them to supervise are already known by the head teachers and principals. Usually, these are teachers' records of work, statutory record books kept by the school, teaching-learning process etc. If these items can include the inspection of the Learning Resource Centre of the schools, it will be a step in the right direction.

Honors and Merit Award In an attempt to further create room for the realization of the prospects of Educational Resource Centers, teachers and schools that are committed to the establishment of Educational Resource Centers should be commended and honored somehow. This step will create a healthy rivalry among the schools and consequently boost the status of Educational Resource Centers in the schools.

25.7 Conclusions

Any problem that has ready solution is no longer a problem. However, the solutions that are available will not become effective on the problem situation until steps are taken to implement them. Now that the problems are still very much with us, could it be that the solution are not yet discovered or there are no initiators to set the motion. Prevention, people say, is better and even cheaper than cure. This is the reason why the maintenance culture should be imbibed by all teachers and managers of Educational Resource Centers.

It has been documented in literate that instructional materials have the ability to enhance learning either by concretizing abstract concepts, by making the class lively while learners are actively involved, by making illustration of points easy for the teachers or by encouraging individualized learning.

The next problem one would raises how to get the instructional materials for use. This is the major focus of the establishment of Educational Resource Centers in the schools. Educational Resource Centers have many prospects if and when established in the schools. We have tried to mention the problems confronting Educational Resource Centers in Nigeria. The likely solutions to the problem have also been mentioned, with these aspects in place, it is expected that the educational technology as a discipline will progress in the tertiary institutions and the teaching-learning in primary and secondary higher institution of learning be enhanced.

This article, you how to store, use and service the instructional materials in your Educational Resource Centers. In this article, the possible gains of establishing Educational Resource Centers in the higher institution of learning have been outlined. The much needed empowerment politically and financially, for the prospects to be realized, have also been discussed.

References

1. Abimbade A (2009) Principles and practice of educational technology. International Publishing Ltd, Ibadan
2. Afolabi (2005) Pattern of storage for the facilities in the Educational Resource Centers. J Bus Technol Educ 2
3. Anmon (1976) Managing and servicing resource facilities in education. Lagos, Nigeria
4. Elaturoti DF (2000) Management of media centre in Nigeria. The Abadina Media Resource Centre experience. In: Conference proceedings of the International Association of School Librarianship
5. Fayose A (2005) Teacher education in Nigeria
6. IFRO (2010) The current status of science around the world
7. Onadeko SA (2004) Servicing educational resources in Nigerian institutions. Lagos

Chapter 26
Barriers and Challenges of Distance Learning in Turkey and the Demand for Distance Education Centres (UEM)

Max Florian Hertsch, Mutlu Er, and Şefika Şule Erçetin

Abstract Today the concept of distance education is fairly well understood and the potential importance of it towards the future of education is also generally acknowledged. With advances in multimedia and communication technologies, Turkish corporations as well as Turkish universities or other Turkish education institutes are adopting distance education frequently. It is understood that the support of the traditional face-to-face approaches of education in higher education institutions continues alongside with an expansion of lifelong learning opportunities. Distance education centers have to face barriers in Turkey such as a lack of staff training in course development and technology, a lack of support for distance learning in general and inadequate faculty selection for distance learning courses. Distance education has faced several challenges in implementation during its decades in history as well. First, distance education provision constitutes a very small section of higher education provision within the federal education system. Second, many Turkish universities and its lecturers seem to be encumbered by existing rules and traditions, which they do not want to break. Third, it appears that the main motivation for offering the variants of distance education described above is financial. Another fact is the bureaucracy that distance learning has to face,

M.F. Hertsch (✉)
Institute for German Language and Literature, Hacettepe University, Ankara, Turkey

Institute of Distance Education, Hacettepe University, Ankara, Turkey
e-mail: hertsch@hacettepe.edu.tr

M. Er
Institute for German Language and Literature, Long Distance Education, Application and Research Centre, Hacettepe University, Ankara, Turkey
e-mail: mutluer@hacettepe.edu.tr

Ş.Ş. Erçetin
International Science Association (ISCASS), Ankara, Turkey

Faculty of Education, Hacettepe University, Ankara, Turkey
e-mail: sefikasule@gmail.com

Ş.Ş. Erçetin (ed.), *Chaos, Complexity and Leadership 2014*,
Springer Proceedings in Complexity, DOI 10.1007/978-3-319-18693-1_26

before it goes on track. The suggested barriers and challenges highlighted above may suit to the Turkish education sector commonly and lecturers must be aware of them if they would like to use virtual learning environments.

26.1 Introduction

Viewed as an important section of education in globalization, distance education is at the present time used by the majority of higher educational institutions around the world. The ease offered by distance education increases the number of participants worldwide and especially in Turkey, too. However, it is mainly individualized nature and hardly enables those who are not personally involved to have a clear idea of how fast and how far distance education is being developed nowadays. In the moment it is quite difficult to state accurately an exact number of those institutions in Turkey who are engaged in distance education. But it can be highlighted that the number of *Distance Educations Centres* in higher education is growing permanently.

Today the concept of distance education is fairly well-understood and the potential importance of it towards the future of education is also generally acknowledged. With advances in multimedia and communication technologies, distance education is being adopted by Turkish corporations as well as Turkish universities or other Turkish education institutes frequently.

The Hacettepe University did its first steps into the distance education society with the installation of the virtual learning environment and course management system Blackboard. Unfortunately Blackboard did not have that effect as desired. In the year 2008 Moodle (an open source software and at the same time an online platform) was installed on the Hacettepe server as a pilot project. It provides various advantages compared to Blackboard (Neumann and Logan [9]). At first, Moodle offers the University an interactive course management including all fundamental features of e-learning and distance interaction with less cost compared to Blackboard. The creation of an academic knowledge exchange or more understandable under the meaning of "open brain pool" gives academicians the opportunity to communicate and research on several topics within the system; like i.e. dissertation courses, global or national scientific exchange or projects. Conferences and symposiums can be structured and organized easily within the last segment. It provides a clear arrangement for all conferences that were and will be held at the Hacettepe University [11]. Nowadays, more than 393 Moodle websites are officially registered in Turkey, which proves the quality of the platform (Moodle.Net [7]).

But it is not only the implementation of an e-learning platform like Moodle that makes a university up to date and at the same time operational for modern lecturers. Such systems and furthermore its technology which is used for supporting higher education should be well-guided and sufficiently implemented into the academic framework at the same time. The aim of e-learning programmes and especially for the platform Moodle is to enable the University and in that case also Turkey, a further and higher education to generate an improved learning environment for all students, wherever and however they learn. The guidance of this framework must be administrated by an institution that leads and supports the profession of virtual learning, blended learning and its technology – a *Distance Education Centre.*

26.2 Barriers of Distance Education

Distance Education Centre's (tr: Uzak Eğitim Merkezi, UEM) vision is a world where students, teachers, academicians and wider institutional lobbyists use modern technology to improve the overall educational experience by improving flexibility and creativity and by encouraging comprehensive and diverse personal, high quality learning, teaching and research (cp. [5]). The vision of the *Distance Education Centre* (UEM) at the Hacettepe University can be summarized by the following statements (Gosper et al. [3]). Students and Teachers are using a combination of institutionally-provided and user-controlled technologies in a reliable and effective way. Secondly, UEM-Teams are efficiently exploiting the existing technology in all aspects like its design, development and delivery. Furthermore, the wide range of learning resources that is freely available must be discovered. The Hacettepe University has a widespread deployment of technical infrastructures that take advantage of service oriented approaches. These infrastructures must be used in an efficient way. And at last but not least the UEM must share its strategies in a national, trans-sectoral and institutional way. It is understood that the support of the traditional face to face approaches of education in higher education institutions continues alongside with an expansion of lifelong learning opportunities. UEM have to face barriers and challenges (in Turkey) in their effort such as [2]:

- a lack of staff training in course development and technology,
- a lack of support for distance learning in general and
- inadequate faculty selection for distance learning courses.

Next to these, the bureaucracy in Turkey can be barrier as well, especially the process of foundation and integration of an UEM and its platform (Moodle). Another important fact is the difficulty to persuade lecturers who use traditional methods and block themselves for innovation (or modern technology).

In general the coursework for traditional and distance students is the same. As well as for traditional courses there can be a lot of preparation in designing distance learning material as well. This can impose a burden on teachers who already have material for traditional classrooms and are not used to virtual learning environments or blended learning courses. Computers, video equipment and communication software may lead to challenges or frustrations. Faculties must know how to they use such technologies if they want to join the path of distance courses. Another problem is to train students and staff, particularly in troubleshooting problems, it is essential for success in distance learning [2]. But perhaps the biggest problem for distance programs is the lack of support for the administration. Assistance by the department or faculty is viewed as a critical instructional element in any distance education program. More than any other participant, faculty roles must change the most in administering distance learning programs. This can be difficult for teachers who want to integrate distance learning to their curriculum. Next to the change of their way of teaching they must meet the requirements of distance learners without face-to-face contact. Since the majority of distance learners are adults, teachers may need to change their curriculum accordingly.

The faculties are responsible for changing their course content to the needs and expectations of diverse students [6]. As long as a faculty feels that there is a burden associated to any distance education program, there will be little or no support for expanding distance education opportunities. There are a number of reasons for this lack of support. Lecturers may lack the basic skills or hardware to fully participate in distance education even the advent of computers, telecommunications, and the World Wide Web provides a unique opportunity for faculties and students to learn in a cooperative environment. It is interesting to note, that students respond to this changing environment more adeptly than teachers do. This happening comes along with the event of social networks [8].

Obtaining proper equipment and training is a critical demand for acceptance of distance learning. Another problem perceived by the faculties is the risk to tenure and human resource staffing. Depending on the academic department, courses taught as part of a distance program, may not always count toward tenure considerations, thus causing a lack of encouragement for participation by some non-tenured faculties [10]. Additionally, if one teacher can serve thousands of students there will obviously be fewer lecturers and fewer departments and faculties. Universities must not underestimate this resistance and should be very aware of the possibility of overburdening their faculties and staff. The final barrier is the teacher's acceptance of distance learning programs. Teachers with enthusiasm for this non-traditional coursework are best suited to teach them. One way to moderate these potentially serious problems is by selecting teachers who are relatively elder but at the same time good teachers and who like the idea of distance learning and want to participate in it. Interest and motivation are not success factors which are only reserved for students.

26.3 Challenges of Distance Education

Distance education has faced several challenges in implementation during its decades in history. First, distance education provision constitutes a very small section of higher education provision within the federal education system. Second, many Turkish universities and its lecturers seem to be encumbered by existing rules and traditions, which they do not want to break. Third, it appears that the main motivation for offering the variants of distance education described above is financial. Another fact is the bureaucracy that distance learning has to face, before it goes on track. Under such conditions, the quality and revaluation of distance education course materials and practices tend to stay static or decline over time [1]. Problems and barriers encountered by the student fall into several distinct categories:

(a) costs and motivators,
(b) feedback and teacher contact,
(c) student support and services,
(d) alienation and isolation,
(e) lack of experience, and training,

More so than conventional students, distance learners are more likely to have insecurities about learning because they are learning in a new environment. These insecurities are also founded in personal- and school related issues of students, such as financial costs of study (i.e. not everybody is able to buy a computer), disruption of family life, perceived worthlessness of their studies and lack of support from employers. These pressures often result in higher dropout rates than among traditional students [2].

A second area of distress for the distance student is the perceived lack of feedback or contact with the lecturer because there is not always a daily or weekly face to face contact with them, students may have trouble in self-evaluation. Galusha [2] believes that the separation of student and lecturer forced by distance removes a vital "link" of communication between these two parties. The link must be restored through explicit institutional efforts, so that the teaching-learning transaction may be reintegrated. Citing Tinto (1975), Nakpodia [8] hypothesized that students who did not receive adequate reintegration measures such as electronic or telephone communication, would be less likely to experience complete academic and social integration into institutional life. Consequently, such students would be more likely to drop out. These barriers can be mitigated through technological methods such as e-mail response and forums. It is important that the student receives a prompt feedback in any institutional setting, particularly in distance learning where the learner is impaired by the lack of casual contact with the teacher and other students. This is especially important for those students who live in the Anatolian countryside. They may often not have access to reliable telecommunications, or computers. The frustrations resulting from problems with communication between student and the academic institution are factors of which distance education planners should be well aware.

A third area of concern for distance students is the lack of support and services such as providing tutors, academic planners and technical assistance. The isolation that results from the distance learning process can complicate the learning process for elder students. Students need tutors and academic planners to assist them to complete courses on time and to act as a support system when stress becomes a problem. Technology costs and considerations can be a cause of budgeting problems. However, student support for distance learners should take priority.

A fourth problematic part is the feeling, as can be seen above, of alienation and isolation reported by distance students. Generally students want to be part of a larger university society, and simply a member of a correspondence course. For many traditional students, this is an important part of their social lives. Another problem encountered by students is the lack of student training, particularly in reference to technical issues. Many adult students are still not well versed in the uses of technology such as computers and the internet. Using an electronic medium in distance learning can accidentally exclude students who are not that structured with computers or writing skills, then such skills are required if computer technology is used. Students will typically be offered volumes of electronic-based information. Using this information will be a problem for some non-technical students. They must be taught how to manage, not only their study time, but the materials presented as well. If students are undertaking distance learning courses that require knowledge of computers, then the students must be taught, at a minimum, the fundamentals of operating the system of choice of the distance-taught course. If distance learning is to be successful, technical barriers must be made a non-issue.

26.4 Conclusion

The suggested barriers and challenges highlighted above may suit to the Turkish education sector commonly and lecturers must be aware of them if they would like to use virtual learning environments. Hertsch [4] highlighted the significant problems about the distance education situation at the Hacettepe University in his study about *Multimedia-Based Enrichment for Foreign Language Teaching* at the Hacettepe University in detail and figured out that the acceptance of new media in Turkish higher education sectors proceeds slowly. A main problem is the critical mindset on both sides, teachers and students, that arose against this new education technology, – because traditional education methodologies stuck in the learning behaviors of them. It is a great step forward that since 2013 Hacettepe University operates with an UEM, which was a very important and fruitful step for an educational upgrade.

For supporting distance learning, the foundation of an UEM (Distance Education Centre) at higher education institutions seems fundamental. Barriers and Challenges (as highlighted above) can be solved within its support. UEM should initiate and support tasks of media inputs and their usage, and assist higher education within virtual education at the same time. Next to the administration of e-learning platforms and their administration, UEM must provide further education for lecturers and students according new media. Furthermore, an UEM may handle innovations, scientific projects and cooperation in the field of new media. Last but not least, UEM helps the university being up to date and varies the methodology of its education.

References

1. Aghenta JA (1993) Principals and practices of educational planning: focus on the developing countries. NSEP, Benin-City
2. Galusha J (1998) Barriers to learning in distance education. University of Mississippi. http://www.nottingham.ac.uk/~ntzcl1/literature/barriers/galusha.rtf. Retrieved 2 February 2015

3. Gosper M, Malfroy J, McKenzie J (2013) Students' experiences and expectations of technologies: an Australian study designed to inform planning and development decisions. Aust J Educ Technol 29(2):268
4. Hertsch F (2012) Multi-media enrichment for foreign language teaching. Procedia - Soc Behav Sci 70(25):615–621
5. Knight S (2011) Emerging practice in a digital age. University of Bristol. http://www.jisc.ac.uk/media/documents/programmes/elearning/digiemerge/emerginpracticeword.doc. Retrieved 2 February 2015
6. Meacham D, Evans D (1989) Distance education: the design of study materials. Open Learning Institute, Charles Stuart University, Wagga Wagga, Australia
7. Moodle.Net: registered moodle sites. Internet: https://moodle.net/sites/index.php?country=TR. Retrieved 2 February 2015
8. Nakpodia ED (2010) Management of distance education in the development of Nigeria's education. Internet: http://www.krepublishers.com/02-Journals/JSS/JSS-23-0-000-10-Web/JSS-23-1-000-10-Abst-PDF/JSS-23-1-045-10-919-Nakpodia-E-D/JSS-23-1-045-10-919-Nakpodia-E-D-Tt.pdf. Retrieved 2 February 2015
9. Neumann T, Logan K (2010) Comparison of blackboard 9.1 and moodle 2.0 (2010). University of London. Internet: http://www.lkl.ac.uk/ltu/files/publications/Comparison_Bb9-Mdl2_Aug2010.pdf. Retrieved 2 February 2015
10. Oaks M (1996) Western cooperative for educational telecommunications Washington, State University. http://www.wiche.edu/telecom/techWASU.html. Retrieved 5 November 2014
11. Uygun D, Hertsch MF, Yıldız S (2010) Variations and additions for the e-learning source Moodle with a perspective towards the Turkish education system. Published in international conference on new horizons in education: proceedings book

Chapter 27
Organizational Identity Management and Chaos

Rahman Çakır

Abstract Organizational identity is focus on the organizations' definition. Because it enables the realization of the important functions in the organization, identity is an important factor for the continuation of the organization. Each organization has an identity that distinguishes itself from other organizations. However, some institutions are inadequate in managing their existing identity, to confound the minds of the target audience and therefore they are faced with the sector underperformance. This can facilitate the formation of a negative reputation. Managing chaos in the process of creating organizational identity is considered in this chapter.

27.1 Introduction

The most important concept to understand human behavior is the identity. Trying to understand who we are is not just an individual effort. Organizations as well as individuals themselves are also efforts to identify themselves. Identity affects behaviors of individuals and communities (groups and organizations) in various ways. The identity of the organization has an impact on the behavior of the organization [27].

27.2 Organizational Identity

The researchers share the view that organizational identity is sum of all the factors which defines "What is organization?", "Does organization represent what?", "Is organization deal with what?", "How does this organization do it?" and "Where does the organization go?" questions' answers and shed light on these. Organizational identity is focus on the organizations' definition.

R. Çakır (✉)
Faculty of Education, Giresun University, Giresun, Turkey
e-mail: rahmancakir28@gmail.com

Ş.Ş. Erçetin (ed.), *Chaos, Complexity and Leadership 2014*,
Springer Proceedings in Complexity, DOI 10.1007/978-3-319-18693-1_27

Table 27.1 Organizational identity definitions

Researcher	Definition
Hall and others [14]	Organizational identity is a process that has been integrated organizational goals and individuals of organization
Albert and Whetten [1]	That which members believe to be central, enduring, and distinctive about their organization
Elsbach and Kramer [6]	Organizational members' cognitive schema or perception of their organization's central and distinctive attributes, including its positional status and relevant comparison groups (p. 442)
Fiol et al. [11]	The aspect of culturally embedded sense-making that is self-focused (p. 56)
Hatch [15]	Members' experiences of and beliefs about the organization as a whole . . . Organizational identity is self-focused. It refers to how the organization's members regard themselves as an organization. (p. 257)
Hatch and Schultz [16]	The object of belonging and commitment that provides a cognitive and emotional foundation on which organizational members build attachment and with which they create meaningful relationships with their organization (p. 16)
Reger, (in Whetten and Godfrey [30])	A member's theory of who the organization is. In that definition of organizational identity, we don't make any assumptions about the content of the identity
	Members of organizations can answer the question "Who are we?" in very different ways, and it will still be identity (p. 110)
Scott and Lane [26]	A set of negotiated cognitive images that emerge out of complex, dynamic, and reciprocal interactions (Abstract section)

Source: Garmon [12]

Lee [18] is one of the first researchers that work on organizational identity. Levinson has explained the importance of identity for organizations with the following words "Before people describe themselves with their work or with the names, nowadays people define themselves by the organizations they work" (Table 27.1).

Because it enables the realization of the important functions in the organization identity is an important factor for the continuation of the organization. These organizational functions are ([3, 17]);

1. coordination, and control of organizational actors that working in different locations,
2. functioning of the working group,
3. to promote people helping behaviors except that people's installed role,
4. not lose valuable employees for the organization

According to Gerhard Regenthal [23] organizational identity concept is consist of four stages (Cited in [22]):

1. Analysis of the current status of the organization
2. Development of organizational identity goals and express a specialized founding philosophy

3. Development and application of strategically organizational identity concept or marketing precautions or organizational identity precautions
4. Completion and control of organizational identity idea or concept

There is a direct effect of social identity perceptions of the organization's members on the perception of organizational identity. There is a psychological link between organizational and social identity. If the person positively integrates himself/herself with organizational identity, wins a positive social identity. Events that can be threat to organizational identity in the same time can be threat to members' social identity [6].

Sarros et al. [25] examined their study on the potential role of leadership on the identity formation and identity protection and also searched relationship between identity formation, identity protection and work. In this study, researchers emphasized the three dimensions of work alienation; weakness (inability to get the results that individual wants), meaninglessness (inability to integrate with work that they do), away from the goal (alienation of people from the business processes).

Sarros et al. [25] stated that leadership make a strong contribution on creating meaningful work environment that will encourage personal development and leadership of individuals that form that allows labor to control their business activities. In addition, transformational leaders have been reduced work alienation by helping individuals to understand the complex relationship with the organization's goals and their work [20].

Silence and Jarzabowski [28] tried to determine how senior managers manage the organizational ties between multiple organizational identity and strategic action. They stated that organizational identity is not static process; it is changing and constantly configured process. Cheney [2] argued that organization's multiple identities within the organization is necessary to satisfy the different stakeholders in the organization. Organizational identity is an organizational structure, it can managed but it should be sought answers to the questions who and how to manage organizational identity. Senior managers cannot create identity, but they affect the identity of the restructuring process [28].

27.3 Organizational Identity Management

Each organization has an identity that distinguishes itself from other organizations. However, some institutions are inadequate in managing their existing identity, to confound the minds of the target audience and therefore they are faced with the sector underperformance. This can facilitate the formation of a negative reputation. However, some institutions are actively looked for ways for manage their organizational identity in best way [19].

According to Van Riel organizational identity is managed in two stages [10];

1. The organization's vision, mission and philosophy is accepted and disseminated within the organization and outside the organization

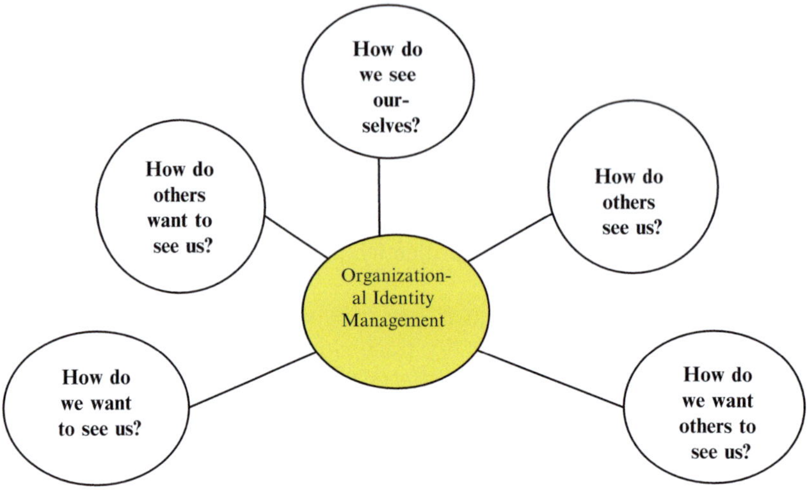

Fig. 27.1 The road map of organizational identity management process (Source: Tuna and Tuna [29])

2. Planning, implementation and continuation of organizational visual systems, marketing communications activities and behavior related to the formation to reflect organizational values.

The road map of organizational identity management process is following in Fig. 27.1.

Staffs that see the organization's successes and failures as their successes and failures would be more motivated to do good things for the success of the organization, because the result of this situation certainly will be reflected in them [4].

Organizational identification can be both positive and negative effects on individual. This process sometimes is experienced with dignity and pride, and sometimes shame and worry [5]. Researches have shown that individuals who identify better themselves with the organization will be more successful, more concerned with the business, participate in more activities for the organization [24]. The participation of these individuals is quite high. More powerful organizational identity formation is linked to behavior based on more cooperation. Therefore, managers who want to strengthen organizational identification celebrate their employees for their work and reward them. Individuals with a strong understanding of organizational identity formation are sometimes pressured other individuals for the purposes of the organization, therefore, "organizational identity" can spread faster [13].

27.4 Conclusion

Organizational identity has become very popular as a concept after realizing the strategic importance of organizational identity for an organization. Organizational identity is a concept of interdisciplinary and integrated. Organizational identity needs a stable management to provide a good reputation and competitive advantage. Managers tend to focus on the relatively easy tangible aspects of organizational identity, this means the most mentioned factors in this identity construction is that corporate communications and corporate design [21].

Leadership and chaos can be considered fundamentally interconnected, topics of chaos and leadership have been classified, unitized and graded in a variety of ways [7]. Leaders manage organizational identification process, organizational identity creates many different beliefs and identities, in this sense it includes complexity. Chaos is a source of life and also a reality for organizations and people as living and self-organizing systems [8]. Because social constructs like societies, organizations and leadership are chaotic, inter-dependent, non-linear systems that are closely tied to initial conditions [7]. Both people's and organizations' life can include chaotic situations in a normal daily routine. So, if principals cannot manage crisis situations and chaos in work environment, that organization should not function [9] and managing and creation of organizational identity does not occur. Managers should develop effective strategies, should be apply and evaluate the results. Strategies, products, service activities and communication when understood as the size of the corporate identity, a successful organizational identity as economic, social and cultural can be developed.

References

1. Albert S, Whetten D (1985) Organizational identity. In: Cummings LL, Staw BM (eds) Research in organizational behavior, vol 7. JAI, Greenwich, pp 263–295
2. Cheney G (1991) Rhetoric in and organizational society: managing multiple identities. Q J Speech 69:143–158
3. DeSantics G (1984) Attitudes toward telecommuting: implications for work-at-home programs. Inf Manage 7:133–139
4. Dutton J, Dukerich J (1991) Keeping eye on the mirror: image and identity in organizational adaptation. Acad Manage J 34:517–554
5. Dutton J, Dukerich J, Harquail CV (1994) Organizational images and membership commitment. Adm Sci Q 34:239–263
6. Elsbach K, Kramer RM (1996) Member responses to organizational identity threats: encountering and countering business week rankings. Adm Sci Q 41:442–476
7. Erçetin ŞŞ, Açıkalın ŞN, Bülbül MŞ (2013) A multi-dimensional approach to leadership in chaotic environments. In: Banerjee S (ed) Chaos and complexity theory for management: nonlinear dynamics. IGI Global, Hershey, Ch 5, pp 89–104
8. Erçetin ŞŞ, Tekin A, Açıkalın ŞN (2014) Organized and disorganized chaos a new dynamics in peace intelligence. In: Banerjee S, Erçetin ŞŞ, Tekin A (eds) Chaos theory in politics. Springer, Dordrecht, Ch 1, pp 3–16

9. Erçetin ŞŞ, Kayman EA (2014) How to be a quantum leader in an intelligent organization? In: Erçetin ŞŞ, Banerjee S (eds) Chaos, complexity and leadership 2012. Springer, Dordrecht
10. Eyüboğlu E (2008) Kurumsal kimlik tanıtımını etkileyen faktörler: Türkiye'nin ilk 500 kurumuna yönelik bir alan araştırması. Yüksek Lisans Tezi, Anadolu Üniversitesi, Eskişehir
11. Fiol CM, Hatch MJ, Golden-Biddle K (1998) Organizational culture and identity: what is the difference anyway? In: Whetten DA, Godfrey PC (eds) Identity in organizations: building theory through conversations. Sage Publications, Thousand Oaks, pp 56–59
12. Garmon M (2004) The relationship between organizational learning, culture, image, identity and identification, an empirical study. Touro University International College of Business Administration, Cypress, California
13. Gülşen MU (2010) Örgütsel adalet ve kimliklenme. Yüksek Lisans Tezi, Gebze İleri Teknoloji Enstitüsü, İstanbul
14. Hall DT, Schneider B, Nygren HT (1970) Personal factors in organizational identification. Adm Sci Q 15(2):176–190
15. Hatch MJ (1997) Organization theory: modern symbolic and postmodern perspectives. Oxford University Press, Oxford
16. Hatch MJ, Schultz MS (2000) Scaling the tower of Babel: relational differences between identity, image and culture in organizations. In: Schultz M, Hatch MJ, Holten Larsen M (eds) The expressive organization: linking identity, reputation and the corporate brand. Oxford University Press, Oxford
17. Kalemci Tüzün İ (2006) Örgütsel güven, örgütsel kimlik ve örgütsel özdeşleşme ilişkisi. Uygulamalı Bir Çalışma. Doktora Tezi, Gazi Üniversitesi, Ankara
18. Lee SM (1971) An empirical analysis of organizational identification. Acad Manage J 14(2):213–226
19. Markwick N, Fill C (1997) Towards a framework for managing corporate identity. Eur J Market 31(5/6):396–409
20. McBain R (2003) Organizational identity. Manager Updat 14(4):25–35
21. Melewar TC (2003) Determinants of the corporate identity construct: a rewiev of the literature. J Mark Commun 9:195–220
22. Okay A (2003) Kurum kimliği. Mediacat Kitapları, İstanbul
23. Regenthal G (1993) Corporate identity – Praxishilfen für das management in Wirtschaft, Bildung und Gesellschaft. Köln
24. Riketta M (2003) Organizational identification: a meta-analysis. J Vocat Behav 66:358–384
25. Sarros JC, Tanewski JA, Winter RP (2002) Work alienation and organizational leadership. Br J Manag 13:285–304
26. Scott SG, Lane VR (2000) A stakeholder approach to organizational identity [Electronic version]. Acad Manage Rev 25(1):43–62
27. Sethi R, Compeau L (2002) Social construction, organizational identity, and market orientation of the firm. American Marketing Association, 13:183–184
28. Silince J, Jarzabkowski P (2004) From organizational identity to strategic action: a process of rhetoric and rationality. Aston Business School Research Paper, Birmingham, UK
29. Tuna M ve, Tuna A (2007) Kurumsal kimlik yönetimi. Detay Yayıncılık, Ankara
30. Whetten, Godfrey PC (1998) Identity in organizations: building theory through conversations. Sage Publications, Thousand Oaks

Chapter 28
Green Supply Chain Method in Healthcare Institutions

Okan Özkan, Çağdaş Erkan Akyürek, and Şükrü Anıl Toygar

Abstract Social and environmental problems arisen with economical and technological development sometimes lead a chaos especially in healthcare industry. The current complexity of healthcare services increased in parallel with the increase of environmental awareness and scarcity in healthcare resources. Green supply chain management is defined as a process consisted of green purchasing, green manufacturing/material management, green distribution/marketing and reverse logistic process. Green supply chain management practices are mostly common in the United States when we look around the world. When we consider Turkey, on the other hand, we can say that green management approach has recently started to be recognized and practiced. In Turkey, two hospitals has focused on this topic so far: "**İstanbul Florence Nightingale Hospital**" and "**Medistate Kavacık Hospital**". In parallel with the proliferation of green hospital concept, there will be a possible increase in the practices related to green supply chain management in Turkey. With this study, it is aimed to draw attention to green supply chain, an environmentalist approach in the supply chain management of healthcare institutions and to raise awareness in managers of health institutions.

Keywords Green supply chain method • Healthcare • Healthcare institutions • Chaos

28.1 Introduction

With the acceleration of globalization, both social and environmental problems have arisen in parallel with economical and technological development. Sometimes it leads a chaos especially for healthcare industry. For this reason, in accordance

O. Özkan (✉) • Ç.E. Akyürek
Institute of Health Sciences, Health Care Management Department, Ankara Üniversity, Ankara, Turkey

Ş.A. Toygar
Institute of Health Sciences, Health Care Management, Gazi Üniversity, Ankara, Turkey
e-mail: aniltoygar@hotmail.com

Ş.Ş. Erçetin (ed.), *Chaos, Complexity and Leadership 2014*, Springer Proceedings in Complexity, DOI 10.1007/978-3-319-18693-1_28 with the increase of environmental awareness in different sectors, management approaches for environmental health have been adopted. With this approach named as green management, companies highly focus on creating environmental values and perform their work according to them. Supply chain management, integration of cash, information and material flow that enables the correct product to get through with minimum price in correct time and place, has become a field in which environmental practices are carried out.

As part of the environmental awareness approach, the process, enabling to minimize negative environmental effects of supply chain management and determine the activities as to use optimum source and to reduce complexity, is defined as "Green Supply Chain Management" [21, p. 44]. In hospitals, it is known that energy and water consumption is intense, chemical and nonchemical waste amount is too much, potentially dangerous substance output is high, and materials bought are various and plentiful also tends to be chaotic [30, p. 38]. Therefore, they should successfully perform the supply activities in order to provide an effective and fruitful health service. Adopting an environmentally friendly and environmentally conscious approach when performing supply activities is only possible with gaining an understanding of green supply chain management.

In this study, supply chain management and green supply chain management are mentioned and examples related to their implementations in health institutions are given. In line with environmental responsibility, it is aimed to create awareness in healthcare managers in subjects of green approaches of supply chain management.

28.2 Green Management

Global climate change, developing technology, a more conscious society, act of sustainable development, social responsibility projects, economical changes and legal obligations play a big role on both production and service corporations to develop environmental approaches. It also encourages managers to struggle for being a green management. Resulting from the rising awareness in environmental issues, institutional environmentalism or green management understanding has become a quite popular concept and an important way to cope with complexity and chaos in this area [6, p. 2, 2, p. 489].

Green management is a process of planning, organizing, directing and controlling organizational activities with an environmental perspective in order to protect the environment and create an environmentally-friendly organization [1, p. 105]. In other words, green management involves buying materials that can be recyclable, reusable or that have just been recycled; choosing the supplier conforming with the environmental aims; providing savings in using all types of sources; supplying products that make sourcing fruitful; by management's making environmentalist decisions; using materials, equipment and devices which do not contain hazardous material; reducing the waste materials; creating reuse or recycle opportunities; making environmental planning, attaching importance to the use of recyclable energy sources; and developing activities that improve environmentalist perspective to create ecological value [20, p. 2240].

Green Management aims to make society to which they render service with activities supporting health and healthy life sustainable [27]. Green management, a new approach for environment management system, is a process that comprise reformation aimed at environmental health [30, p. 41].

Green businesses that have adopted green management practice desire to be the creators of the ecological life quality. With this aim, the first step is generally to prevent the environmental pollution and the second step is to provide savings in all kinds of sourcing and to tend towards the renewable sources [11, pp. 191–198]. When businesses aim at increasing their productivity, profitability and market share; strengthening their competitive position and decreasing their cost; they also want to raise awareness and create a green image in society in order to protect sustainable development and ecosystem by minimizing their negative effect on the environment [15, p. 1175], Porterve van der Linde [22]).

28.3 Supply Chain Management

Supply chain is defined as a cluster of factors in which suppliers, logistic service suppliers, producers, distributors and retailers are included and among which there are constituents such as material, product and information flow [16, p. 226]. According to another definition, supply chain is an integrated process constituted by various companies such as suppliers, producers, distributors, retailers, etc with an aim of supplying raw material, making them into products and distributing these products to the retailers and ultimate users [4, p. 281, 10, p. 89].

Supply chain management is, on the other hand, the management of material/product, money and information between supplier, producer, distributor, retailer and customer, all of which are included in the value chain so that goods can get through to the customers from the raw material supply to production and distribution [9, p. 3, 19, p. 89]. Supply chain management is also defined the integration of the main work processes that provide product, service and information from supplier to the ultimate customer [18, p. 2].

Basically, supply chain comprises company, suppliers of the company and its customers. In extended supply chain, however, there is a supplier of the supplier or an ultimate supplier at the beginning, and there is a customer of the customer or an ultimate customer at the end in addition to this framework. Finally, there are some service provider companies that render service to all of the companies in supply chain (Fig. 28.1).

Main goals of Supply Chain Management can be explained as in the following [19, p. 89]:

- To increase the customer satisfaction,
- To decrease the cycle time,

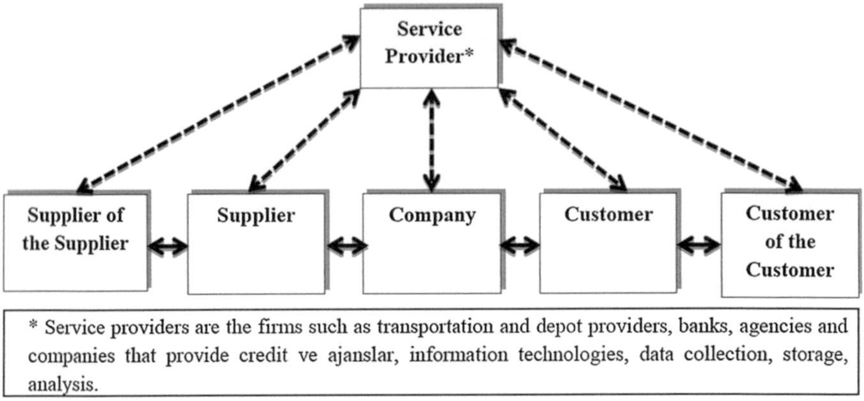

Fig. 28.1 Extended structure of supply chain (Source: Hugos [14, p. 27])

- To ensure that storage costs and costs related to storage are decreased,
- To decrease the product defects
- To decrease the activity cost

When the historical development of supply chain is analyzed, it is apparent that its emergence dates back 1960s. Physical distribution process, the first stage of supply chain management, was first mentioned by Bowersox. After the introduction of material requirements planning in 1970s, managers understood the importance of cost, quality, and delivery of the product on time; and they adopted an integrative approach. With an increase of global competition, companies switch to the second process of supply chain management, the logistics stage; and Houlian became the first person to use the term "supply chain" for this system by connecting the strategical decisions of the company and logistic facility [19, pp. 89–90]. Especially starting from the late 1990s, supply chain management became very important in both academic and business life. Units with the name of supply chain were constituted in big companies in early 2000s [25, p. 58].

Although supply chain management has a great number of advantages when it is carried out efficiently, there are many difficulties and barriers in practice. Barriers that occur in supply chain practice, difficulties encountered and the advantages that are as follows [12, p. 37]:

Forces Driving Supply Chain Collaboration

- More demanding customers
- Increasing competition
- Economic globalization
- Closer collaboration relations
- Less product life cycle
- Need for better information
- New information technologies

Implementation Barriers to an Effective Supply Chain Management

- Lack of top management support
- Not-aligned strategical and operational ideas
- Inability or unwillingness to share information
- Lack of confidence among the members of supply chain
- Inflexible organizational systems and processes
- Resisting change
- Inconsistent/insufficient performance grades
- Lack of education for new ideas and skills

Benefits of Effective Supply Chain Management

- Unique product and services
- Faster research and development activities
- Quality of Supplier
- Shorter order cycle
- Flexibility in meeting the demands of the customer
- Increasing cash-to cash velocity
- Better asset management

28.4 Green Supply Chain Management

With an effect of globalization, companies adopt reformist approaches in every stage of their activities in order to gain a competitive and sustainable edge. Nowadays, especially in the last few years, approaches for environmental awareness, which have become more and more common as the environmental pollution increases, are reflected to business activities, as well. By adopting green management with an environmentalist understanding, companies not only contribute to green management, but also discharge their social duties. Green management approaches directed companies to green supply chain management practices by being integrated to their supply management processes.

Green supply chain management is defined as a process consisted of green purchasing, green manufacturing/material management, green distribution/marketing and reverse logistic process [13, p. 334]. According to another definition, it is a method laying out supply chain that combines production process with recycling and remanufacturing [24, p. 285]. Green supply chain management is also described as integrating environmentalist approach to supply chain management in the way that it involves product design, material supply and selection, manufacturing processes, delivering the product to the customer and management of product life curve [26, pp. 54–55].

Components of green supply chain management are as follows [17, pp. 21–35]:

- *Green Purchasing*: Green purchasing decisions are the acts of buying materials that are recyclable, reusable or just recycled.

- *Green Production*: It involves a range of activities such as green production, reuse, reproduction and recycling.
- *Green Distribution*: Selection of distribution chain and storage points, administering routes, identifying which method will be used for transportation, coordination of prospective and retrospective activities and handling them in this context.
- *Green Packaging*: Having environmental awareness in packaging activities and using environment friendly and recyclable materials.
- *Reverse Logistics*: It involves a range of new activities such as exchange, return, repair, maintenance, remarketing and placement. It is the act of product transportation from the consumer to reprocessing or recycling centers.

28.5 Green Supply Chain Management in Health Institutions

Supply chain management in the sense of health sector is defined as an integrated system that manages flow of the product and information in order to provide the transaction of work processes beginning from the production to the delivery to the patients in harmony [29, p. 25]. Supply chain structure in health services is as follows in the Fig. 28.2.

Some reasons make it necessary to redesign the processes in health system. The aim of redesigning the processes in health institutions is to produce alternatives to sourcing by the help of "green" concept; to encourage the use of energy, water, and material more efficiently and effectively; to avoid all kinds of waste, to design environmentally-conscious and environmentally-friendly buildings [30, p. 38].

Green practices in supply chain take effect in all processes by giving particular importance to customer satisfaction It is a system in which environmental aspect is added to supply chain with some studies that increase the life quality of the society such as developing environmentally-friendly product, giving opportunity to recycling, using the sources efficiently, etc. [3, p. 64]. Supply system in hospitals determines reliable stock management, regular material supply, speed and quality of maintenance and service process in terms of the communication of suppliers [31].

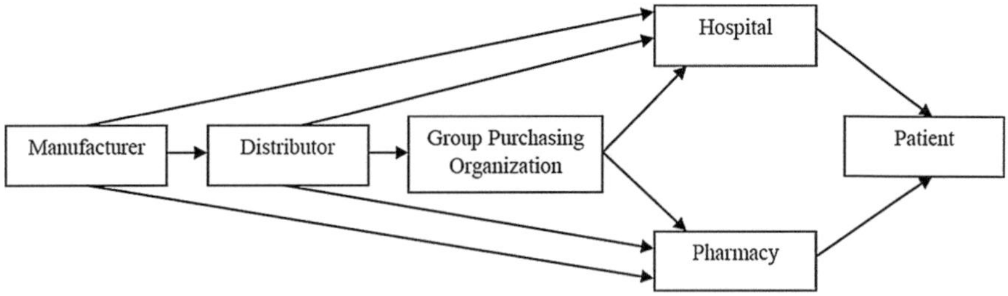

Fig. 28.2 Supply chain structure in health services (Source: Chakraborty and Zanjurne [8])

By means of adopting green management strategies; health institutions can improve supply processes and their consequences; reduce the cost thanks to the supply of reusable or recyclable products and protect the health of the environment.

Activities related to green supply management devoted to protecting the sources which are limited to reuse or recycle activities are as follows [5, 7, 23]:

- Among the important issues in Boulder Community Foothills Hospital, America are encouraging and supporting recycle of the equipment that are not used; reducing the waste materials and throwing them in a safe way; purchasing the recyclable and reusable products that include materials which give the least harm to the environment. Cleaning, disinfection, sterilization and recycle of most things such as alcohol, battery, bulb, paper, cardboard, paperboard, materials related to X-ray, old bedclothes, newspapers, etc.; and reuse of some materials such as materials used in the department of surgery, all types of packaging materials, wheel chair, crutch, print cartridge are enabled within the hospital. What is more, mercury free thermometer, blood pressure indicators, and energy saving bulbs are used.
- A cardiovascular and pediatric diseases center in America reduced the waste blood by 22 % with the help of recycling and reusing practices.
- A hospital in America confirmed that they made a profit of 30 euro per bed by enabling the recycle of every kind of materials such as bulbs and x-ray films.
- Kem Gardner, a supply chain center which operates within the body of Intermountain Health Center in America, provides medical material to numerous health institution. By using reusable packaging materials in packaging processes, this center achieved a reduction of waste materials by 20 %. Furthermore, the products and the packaging materials in the relevant center achieved saving in the storage place.
- A research was conducted to find a sustainable solution for polystyrene coolers by the complaints and demands of the patients. As a result of this research, a recyclable and cost-efficient alternative product was determined. Thanks to this green practice in supply chain management, there was a decrease in the number of waste materials and costs.

As seen in the examples, when deciding the products to use in hospitals, some points which are taken into consideration are sustainability, content, disposal requirements, package of the materials; and effect of them on the patient and personnel. Therefore, hospitals are adopting a more environmentalist purchasing policy.

Green supply chain management practices are mostly common in the United States when we look around the world. When we consider Turkey, on the other hand, we can say that green management approach has recently started to be recognized and practiced. In Turkey, two hospitals have focused on this topic so far: "***İstanbul Florence Nightingale Hospital***" and "***Medistate Kavacık Hospital***". In parallel with the proliferation of green hospital concept, there will be a possible increase in the practices related to green supply chain management in Turkey.

28.6 Conclusion and Recommendation

Supply chain involves all the processes including supply, manufacturing, shipping, storing, sale and services after sale [28, p. 243]. Green supply chain management can be defined as adopting environmentalist management principles in all the processes between enterprises which are the members of supply chain and institution activities.

Health institutions have an important role in terms of protecting environment and public health by adopting environmentalist approaches in all of the stages of supply chain. This approach, at the same time, has a process that reduces the costs and increases patient satisfaction for a continuous improvement.

With this study, it is aimed to draw attention to green supply chain, an environmentalist approach in the supply chain management of health institutions and to raise awareness in managers of health institutions. Within this framework, it can be recommended for managers of health institutions to apply environmentally-friendly green strategies to supply chain management; to develop purchasing strategies intended for environmental health; to adopt environmental policy in waste management; to raise the awareness of personnel about environmentalist approaches; and to provide them with in-service training.

References

1. Aimkij N, Mujtaba BG, Kaweevisultrakul T (2013) Green management sustainability and its economic impact. J Econ Sustain Dev 4(6):104–115
2. Banerjee SB (2001) Managerial perceptions of corporate environmentalism: interpretations from industry and strategic implications for organizations. J Manag Stud 38(4):489–513
3. Bedük M (2009) Tedarik zinciri yönetiminin işletme performansı üzerindeki etkisi: Örnek Olay çalışması. Dissertation, Selçuk University
4. Beamon BM (1998) Supply chain design and analysis: models and methods. Int J Prod Econ 55(3):281–294
5. Brannen L (2006) Preventative medicine for the environment: developing and implementing environmental programs that work, designing the 21st century hospital: environmental leadership for healthier patients and facilities. The Center for Health Design Report. http://www.healthdesign.org/sites/default/files/Designing%20the%2021s%20Century%20Hospital.pdf. Accessed 25 Dec 2014
6. Büyüközkan G, Vardaloğlu Z (2008) Yeşil tedarik zinciri yönetimi. http://www.gulcinmbuyukozkan.net/ytzy8.pdf. Accessed 25 Dec 2014
7. Carpenter D, Hoppszallern S (2010) Sustainable operations survey: green and greener: hospitals embrace environmentally sustainable practices, though laggards remain. Health Facil Manag Mag 23(7):15–21
8. Chakraborty S, Zanjurne A (2012) Supply chain management in healthcare. http://www.managementcanvas.iimindore.in/icanvas/index.php?option=com_content&view=article&id=178:supplychainmanagementinhealthcare&catid=39:operations&Itemid=58. Accessed 19 Sep 2014
9. Cuthbertson R (2011) The need for sustainable supply chain management. In: Çetinkaya B, Cuthbertson R, Ewer G, Klaas-Wissing T, Piotrowicz W, Tyssen C (eds) Sustainable supply chain management: practical ideas for moving towards best practice. Springer, Berlin, pp 3–11

10. Du TC, Lai VS, Cheung W, Cui X (2012) Willingness to share information in a supply chain: a partnership-data-process perspective. Inf Manag 49:89–98
11. Düren Z (2002) 2000li Yıllarda yönetim. Alfa Yayınları, İstanbul
12. Fawcett SE, Magnan GM, McCarter MW (2008) Benefits, barriers and bridges to effective supply chain management, supply chain management. Int J 13(1):35–48
13. Hervani AA, Helms MM, Sarkis J (2005) Performance measurement for green supply chain management, Benchmarking. Int J 12(4):330–353
14. Hugos M (2003) Essentials of supply chain management. Wiley, Hoboken
15. Kim I, Min H (2011) Measuring supply chain efficiency from a green perspective. Manag Res Rev 34(11):1169–1189
16. Kopczak LR (1997) Logistics partnership and supply chain restructuring. Survey results from the US computer industry. Prod Oper Manag 6(3):226–247
17. Korkankorkmaz N (2012) Yalın ve yeşil tedarik zinciri yönetimine ilişkin bir araştırma. Dissertation, Gebze Institute of High Technology
18. Lambert DM (2008) Supply chain management. In: Lambert DM (ed) Supply chain management: processes, partnerships, performance, 3rd edn. Supply Chain Management Institute, Sarasota, pp 1–23
19. Özdemir Aİ (2004) Tedarik zinciri yönetiminin gelişimi, süreçleri ve yararları. Erciyes Üniversitesi İktisadi ve İdari Bilimler Fakültesi Dergisi 23:87–96
20. Özkan O, Bayın G, Terekli G (2014) Hastane yönetiminde sürdürülebilir yaklaşım: Yeşil yönetim. In: Proceedings of 8th Health and Hospital Management Congress, University of Lefke Europe, Kyrenia – TRNC, 10–12 September 2014
21. Peker D (2010) Çevresel performansın geliştirilmesinde yeşil tedarik zinciri yönetimi. Dissertation, Uludağ University
22. Porter ME, Van Der Linde C (1995) Green and competitive. Harv Bus Rev 73(5):120–134
23. Practice Greenhealth, https://practicegreenhealth.org/topics/environmentally-preferable-purchasing/epp-case-studies-health-care. Accessed 25 Dec 2014
24. Pun FK (2006) Determinants of environmentally responsible operations: a review. Int J Qual Reliab Manag 23(3):279–297
25. Russell SH (2007) Supply chain management: more than integrated logistics. Air Force J Logis 31:55–63
26. Srivastava SK (2007) Green supply chain management: a state-of-the-art literature review. Int J Manag Rev 9(1):53–80
27. Stichler JF (2009) Code green: a new design imperative for healthcare facilities. J Nurs Adm 39(2):51–54
28. Tengilimoğlu D, Güzel A (2013) Hastanelerde Lojistik ve Tedarik Zinciri Yönetimi. In: Sur H, Palteki T (eds) Hastane Yönetimi. Nobel Tıp Kitabevi, İstanbul, pp 239–258
29. Tengilimoğlu D, Yiğit V (2013) Sağlık işletmelerinde tedarik zinciri ve malzeme yönetimi. Nobel Yayın Dağıtım, Ankara
30. Terekli G, Özkan O, Bayın G (2013) Çevre dostu hastaneler: Hastaneden yeşil hastaneye. Ankara Sağlık Hizmetleri Dergisi 12(2):37–54
31. Tütüncü O, Küçüküsta D (2008) The role of supply chain management integration in quality management system for hospitals. Int J Manag Perspect 1(1):31–39. http://www.bch.org/green-hospital/facts-and-figures.aspx. Accessed 25 Dec 2014

Chapter 29
Non-equilibrium Processes in the Psychological Adaptation of Developing Personality

Marina Vladimirovna Grigoryeva

Abstract The history of psychology in its entirety is connected with the search and determination of its own scientific paradigm. Modern psychology possesses methodological means allowing efficient use of both nomothemic and ideographic approaches. However given the emergence of chaos and complexity concepts within every sphere of life, things cannot be the same as before and approaches, hitherto highly effective, leave a lot to be desired. In the recent years therefore, a new line of research, synergetic approach has been forming. Psychic new formations appearing in an individual psyche under the influence of a complicated and unstable reality, acquire multidimensionality and temporality of their functioning increases, new features appear integrating history of individual and society existence, actual state of this existence and focus on the future. A determiner of psychic reality evolution and search for new grounds are necessary for adequate description of psychic reality. One of the ways could be application of the diachronic approach in psychological research. The school environment example showed that the problem of adaptational capabilities of personality must be solved in accordance with the specific character of the preceding development of the personality. Adaptational capabilities of personality play a significant role in the process of his/her interaction with the environment, they are complex multi-dimensional systemic unity where interconnection of heterogeneous psychic phenomena and personal traits provides systemic new-formation; that is adaptational readiness of personality.

29.1 Paradigms of Modern Psychological Research

Interrelations of a human to the world are diverse: as a living being he relates to nature, as a part of people's community – to the society and its history, as a subject of activity – to material objects, as a cognizing and producing subject – to the products of culture. This diversity determines the development of a complex furcate

M.V. Grigoryeva (✉)
Educational Psychology and Psycho-diagnostics Department, Saratov State University, Saratov, Russia
e-mail: grigoryevamv@mail.ru

Ş.Ş. Erçetin (ed.), *Chaos, Complexity and Leadership 2014*,
Springer Proceedings in Complexity, DOI 10.1007/978-3-319-18693-1_29

system of theoretical and practical knowledge that is becoming one of the major centers of scientific development in the modern world [1]. T.S. Kuhn, while giving the characteristics of science, describes its development as follows: "If we regard science as a combination of facts, theories, and methods… then the development of science is a gradual process in which facts, theories, and methods compose the growing bulk of achievements representing the scientific methodology and knowledge" [17, p. 5]. However, the complex furcate system of knowledge about a human can no longer exist as a simple set of facts collected on a definite stage of its development. The systematization and enlargement of basic fundamentals of research are needed. Paradigms of science arise as the achievements are approved by the whole scientific community giving to it (this community) generalized models of building up the research – universally approved instruction of organizing the research. Formation of paradigm in science according to T.S. Kuhn "is a sign of maturity of development of any academic subject" [17, p. 7].

The entire history of psychology is connected with the search and determination of its own scientific paradigm. From the XIX century the debates are going on between the adherents of scientific and humanitarian paradigms. The peculiar features of scientific paradigm are acceptance of human nature, and relying on the facts available from experiments. Humanitarians consider that psychology should study particular forms of demonstration of the psychic in humans, describe and understand them. The extreme presentations of scientific paradigms in psychological research led to reductionism, biologizing, reducing the psychic life to reflexes and physiology, and the absolute priority is given to the adherents of humanitarian paradigm – to idealism and absence of aspiration to explanation of their causes and genesis.

The search for "the code of rules and regulations creating the unified scheme for acquisition of scientific knowledge from the stage of goal setting to problem solving" [26, p. 10] that would be acknowledged by all the scientific community lasts up till now and is of a really intense nature. As an example we can give the discussion results of which are published in the journal "Problems of Philosophy", [24]. The key idea of this discussion was criticism against the "scientific mind", proof of the advantages of humanitarian paradigm (Golov, Puzyrey, Ogurtsov), proposition and grounding of transition from humanitarian to eschatological paradigm that would proceed from the idea of extreme, finite meanings of existence of a human" living within the Christian culture (B.S. Bratus), attempt to avoid confrontation and opposition of scientific and humanitarian paradigms, acceptance of scientific and logically correct paradigm (M.A. Rozov, N.I. Kuznetsova).

While characterizing the condition of fundamentals of modern psychology regulating the activity of cognizing psychologist V.A. Rybakov and A.L. Pokryshkin describe different suppositions of scientists on this issue. Some of them (T.C. Kuhn, N.I. Kuznetsova) consider that psychology is a pre-paradigm science, that in-spite of its history time span (especially regarding the part of its development within philosophy) psychology is now at the stage of collecting the factual material due to the complexity of its subject, and have not yet reached the necessity of determining the unified rules and axioms: "… psychology today is first of all a vast research zone, multiple programs and projects not correlating to one another.

This is the serious cognitive difficulty that makes us talking about crisis again and again, consider this diagnosis to be objective in spite of the activity of the researchers working here and the visible growth of particular knowledge areas" [24, p. 5]. Pre-paradigm condition of psychology can be reasoned in terms of the absence of direct links of its results and creation of material basis of society life, with provision of technical and military force of the country, which is a priority right now for almost every state. As a result psychology remains a science not as supported by the government policy in as, for instance, physics or chemistry whose achievements are directly connected to the government force and safety. Shift of priorities on the governmental level would make the psychological community define the fundamentals of the research organization and overcome the diversity of research instructions.

Other scholars consider that psychology is a multi-paradigm science, including several paradigms that correlate with major scientific theories and schools firmed up with the course of time and possessing their own scientific communities (behaviorism, cognitive psychology, psychoanalysis, gestalt-psychology, etc.). Actually, the existence of many scientific areas in psychology does no influence there acute confrontation. Criticism of the fundamentals of a scientific school from the position of another school leads to neither of the three: the destruction of the existing school being criticized, the revolutionary "overthrow", the significant loss of its adherents and clients. Moreover, some practicing psychologists efficiently use psycho-technics of different scientific schools in their work.

There is also the third group of psychologists. They are putting their effort into proving the extra-paradigm in the nature of psychology basing their reasons on the ideas of subjective nature both of its object – psychic, and the means of cognition. Psychic of a human is like an inimitable, strictly individual reality closed to the thorough objective study; its cognition can be more or less objective depending on the object and method of research. Special techniques for measuring the external demonstrations of psychic activity, the use of statistical and mathematical apparatus for data processing, standardization of the research procedure add to the objectivity of the results of the research process. Impossibility to register and standardize all the factors affecting the process and the results of the research decreases the objectivity of the results and in some cases leads to the only possibility of their description and explanation. This especially relates to the study of deeper personality structures (values, meanings, senses, etc.) that are influenced by the peculiarities of the personality's whole life path. Besides, the results of any psychological research are influenced by the factor of the cognizer's personality because the key instrument for cognition of psychic is the subjective and truly individual reality – the psychic of the researcher. In any case, as the representatives of the third group of scientists believe that more or less objective or subjective (meaning the absence of objectivity) psychological research gives the abundant material for detecting the regular patterns (in the first case) and for interpretation, explanation (in the second case). If this is correct then there is no need to define the paradigm strictly; psychologists should just follow the scientific principles laying in aspiration to the truth that is defined by the objectives of the research and the absence of objectives for any kind of manipulation.

While analyzing the scientific and humanitarian paradigms strength and weaknesses, V.A. Rybakova and A.L. Pokryshkin enumerate the following among the advantages: “scientific system of building up and verification of the scientific knowledge is the most strict, correct, and simple from the point of view of mastering and application in acquiring the new knowledge”, humanitarian instructions allow “understanding the goals and intentions of the other person, recover the meaning put into a symbol, attributed to a phenomenon” [26, p. 20]. The authors regard the following as disadvantages: danger of losing one’s own object of the research – psychic in the first case, and non-structured amorphous knowledge of meanings whose sum total equals the sum total of the adherents of humanitarian paradigm in the second case.

Studying the paradigms of modern pedagogical and developmental psychology L.F. Obukhova points out two paradigms, absolutely different and nonintersecting, from her point of view: scientific (connected with J. Piaget’s name) and cultural-historical (connected with L.S. Vygotsky’s name) [22]. The major ideas of the first are characterized above. The second one is nothing but humanitarian paradigm accepting the fact that the superior psychic functions initially develop as a form of collective behavior, as a form of cooperation with other people, i.e. as a product of culture; and as a consequence only they become individual functions of the child him-/herself.

The core problem, criterion of an absolute division of the mentioned paradigms is the question of the direction of internal changes of the child, his/her development. According to scientific paradigm of the research on psychic development it happens in the process of natural psychic maturation in definite environmental conditions. According to cultural-historical paradigm this is the track from the social to the individual during which the individual acquires social experience.

While criticizing core ideas of the scientific paradigm and considering cultural-historical paradigm to be more deep and complicated from a theoretical point of view, L.F. Obukhova analyses their differences in how they define the conditions of development: according to the scientific paradigm, conditions of development are genetic background and environment, and according to cultural-historical paradigm they are morpho-physiological peculiarities of the brain and communication [22]. However, when we compare the content of the concepts unpacking the conditions of development it can be noticed that “genetic background” and “morpho-physiological peculiarities of the brain” are the terms of the same area (natural area connecting a human to the world of organic nature): we also include social environment which major means and forms of existence are communication to the notion “environment” in modern scientific and psychological research.

N.I. Kuznetsova believes that the strategic line of the development of knowledge is the unity of the scientific and humanitarian instructions, and “separation of humanitarian sciences from the natural ones, their artificial isolation only leads to provincialism”; there is “unity of norms and ideals of scientific character” in any science, whether it is natural or humanitarian [24, p. 9].

Traditionally and historically scientific paradigm has been used for longer time in psychology to organize research and explain its results. However, the psychological knowledge based on empiric facts cannot transform into explanatory scheme, structure, conception, theory in the same empiric way. "The moment of psychotechnical change of the psychic under study by the research procedures themselves" prevent the full realization of the scientific paradigm [24, p. 15]. The requirement of experimental verification of the potential theory also turns out to be difficult for accomplishment and depends on the professional skills of the researcher.

V.A. Rybakov and A.L. Pokryshkin suggest the following ways out of this situation being accepted in the present research: "... preliminary facts gathering, their systematization, initial generalization, detection of empirical pattern, introduction of observation terms – all this preparatory work before building up a theory gives to the researcher the basis that will allow to him running "as if" experiments with theoretical reasoning, and will increase the possibility of creating a good theory" [26, p. 24].

The question of the choice of the research paradigm is considered to be crucial because the answer to it gives the guidelines to the researcher, defines directions and means of the research organization, i.e. determines its methodology and methods. If the psychologist-researcher has scientific instructions he will aspire to detect recurrent phenomena from which it is possible to point out the patterns predicting the evolution of the phenomenon. This calls for a method of research that would provide the quantitative data generalized consequentially. This nomothemic approach allows organization of facts into conceptions and theories, and pointing out general patterns of the functioning of the psychic. A psychologist having humanitarian instructions aspires to describe the psychic phenomenon observed, interpreting it, giving meaning to it, i.e. using the ideographic approach.

While nomothemic approach allows detecting general features of phenomena, ideographic approach is aimed at particular and unique features. Categories of the "universal" and "particular" are equally necessary in psychological science. Without defining the general principles of the functioning of psychic, psychology will hardly become the in-demand science within the society. Besides, any normative science (not para-science) puts precisely this goal: to define the general rules and principles of the existence and evolution of the phenomenon under study. At the same time, it is impossible to study any psychic phenomenon in accordance with the realities without consideration of the uniqueness of the individual psychic. The research becomes meaningless in its practical significance (because it is unclear how to apply the universal rules to the particular case) and in terms of the absence of meaning and explanation of the psychologist researcher directed both towards his/her own reflective thinking and the colleagues-psychologists, if there is no quantitative analysis of the qualitative data or interpretation of numeric patterns. Absolutization of ideographic approach is similarly useless as it leads to cognition of a psychic phenomenon as a single demonstration each time demanding on the new interpretation.

In this case psychology ceases its existence as a science and acts as a form of art. According to I.M. Kondakov, modern psychology possesses methodological means allowing efficient use of both nomothemic and ideographic approaches. This is the Activity theory by A.N. Leontiev and S. L. Rubinstein and cultural-historical psychology of L.S. Vygotsky. Activity theories bring to light the mechanism of representation of the internal in the external through the concept of "interiorization", and cultural-historical psychology allows understanding representation of the external in the internal and using the principle of historic development of the phenomenon in the process of study, education, and development of the child's personality. As the author of the theory A.A. Leontiev pointed out, in spite of the differences in denomination of the representation directions there are no principal distinctions in detecting the sources of development, determination of psychic phenomena in these theories [18].

Analyzing the opportunities of the paradigms for cognition of the psychic I.M. Kondakov gives the following conclusion: "... both nomothemic (with its repeatability principle) and ideographic (with its principle of feeling-in) approaches have equal rights to exist in modern psychology; here we do not talk about the choice between the two but about their integration in the context of problem solving" [16, p. 88].

29.2 Synergetic Approach to Research of Psychological Personality Adaptation

In the recent years a new line of research, synergetic approach [23, 27, 30] has been forming. Some scientists (I.V. Prangishvily, V.V. Vagurin, N.I. Sardzhveladze and others) picture it as a regular prospect for further development of a systematic approach. Researchers working in this direction study mental phenomena determination taking into account not causality, but probabilistic nature of world events happening and give a lot of significance to casual events in open system functioning. Unlike classic research paradigms, synergetic paradigm focuses on the study of "the human being in the process of becoming, rather than human being that has already become". It focuses on "dynamics of evolution processes rather than being" [30, pp. 6–7]. The categorical exception and substitution of causal determination with casualty acceptance as the only ground and scenario of a course of events, undoubtedly cannot be productive and valid for such a complex phenomenon as human psyche. The possibility (sometimes actuality) of such substitution causes fair criticism of the synergetic paradigm on the part of some philosophers and psychologists.

However B.F. Lomov pointed out that "connection between cause and consequence is not complex and straightforward", "we are forced to deal with causal connections alongside with connections determined by such concepts as "condition", "factor", "ground", "premise", "mediation" and others." [20, p. 100]. Complexity and ambiguity of the determination process lead to system instability upon certain circumstances and its development process is "blurred". Investigating the problem of mental phenomenon determination, B.F. Lomov means not only determination of evolution, but also "the evolution of determination" (Ibidem).

In other words, in order to adequately describe the psychic reality, determiner of its evolution and search for new grounds are necessary. One of the ways could be applying of self-organization of open psychic system laws interacting with probabilistic, eventful world.

Synergetic approach to the defined problem research does not exclude activity and free human will, because internal subject of interaction force, alongside external causes, are admitted and taken into account as important factors (synergetic term "attractors") and non-linear disproportional influence upon the weak reasons system is only possible when it corresponds with internal tendencies of a system development. Such approach to the study of human-environment interrelation gives an opportunity to raise a heuristic research potential. It can be expressed, according to V.A. Vagurin, "in break with old paradigms ..., emancipation of consciousness and creative imagination by a principally new self-organizational paradigm, in creative application of not only its categories, but also patterns and concepts with their following rationalization" [30, p. 9]. Contemporary psychology is at the starting point of interdisciplinary psychological and synergetic research. Its successful realization in sociological and socio-psychological study of large groups and communities gives hope for its perspective application in other fields of psychology and in psychological studies of personality adaptation in particular.

29.3 Diachronic Approach to Research of Sociological and Psychological Adaptation of Personality

Under the conditions of a contemporary and rapidly developing society, a social order is given to psychology. This order is connected to the problem of active reflection of changes in different sectors of social cooperation and relationship, and adequate effective response to them in the psyche. Psychic new formations appearing in an individual psyche under the influence of a complicated and unstable reality, acquire multidimensionality, and temporality of their functioning increases, new features appear integrating individual and society existence history, actual state of this existence and focus on the future [12]. At the same time, active and adequate influence of a subject on the environment (natural, objective, social, educational, professional etc.) is determined by the actual situation pattern, which contains all the above mentioned features integrated.

Certain contexts of a subject-environment interaction situation appear. The context of a specific subject-environment situation of interaction also depends on a subject's attitude towards different sides of this situation, and on his/her previous experience with the sides of this situation, and on the role this situation plays in his/her future life and activity. Therefore the interaction situation is paradoxical: "while including the subject, ... it stands against the subject at the same time; one ... side of the situation is living conditions, the other one is their representation by the person included in the situation" [2, p. 35].

Earlier, in our studies it was shown that the inclusion of the historical development of the situation of the subject-environment interaction requires psychological research of new methodological basis, which would allow considering the phenomenon in its development, revealing the psychological mechanisms that promote not just a statement of the past and the current state of mental phenomenon and situations of its functioning, but anticipation of their development in the future, in other words, promote scientific understanding of the origin, development, and predicting the possible dynamics of interaction between the personality and the environment in the course of time [12].

In modern psychology, there are scientific approaches to observing mental phenomenon in dynamics. They are systemic-genetic approach, functional system approach, synergetic approach, etc. All of them focus a researcher's attention to the procedural aspects of the phenomena under study. Based on the procedural aspect, socialization and adaptation studies [8, 9, 13, 28, 29] not only provide scientific data on the actual attributes of these processes, but also focus on the analysis of the historical development of a phenomenon at a certain time interval.

However, it is difficult to distinguish a scientific approach that promotes the study of psychic phenomena based on the past, the present, and the future in its development, which would allow to survey specific phenomenon in the real time period, at the appropriate level of psychological analysis and taking into account the complex system of determinants. Complexity and ambiguity of the determination process of the human psyche functioning in the "person-environment" system leads to the fact that under certain circumstances, the system becomes unstable, and the process of its development is "blurred". Investigating the problem of mental phenomenon determination, B.F. Lomov means not only determination of evolution, but also "the evolution of determination" [20, p. 100].

In other words, a determiner of psychic reality evolution and search for the new grounds are necessary for adequate describing of the psychic reality. One of the ways could be applying of the diachronic approach in psychological research. The term "diachrony" initially became popular among the historical linguistics. The literal translation of the word "diachrony" from the Greek is flowing through time, and it means the historical development of different phenomena and systems. Introduced into science by the Swiss linguist Ferdinand de Saussure, the term "diachrony" in linguistics means relations and connections of the elements (language units) in the historical sequence of their occurrence [3].

Diachrony, according to F. de Saussure, is complemented by synchrony – the existence of a current language system. Diachrony and synchrony complement each other as methods of cognition of system evolving into another system in the first case, and study of a real-time currently functioning system in the second case. Thus the basic criteria of evolution phenomena is their consistency over time (before – now, past – present, present – future, etc.) and their qualitative difference, which does not allow to combine phenomena consecutive in time in the system (ibidem). According to O.S. Razumovsky "diachrony – is the most important feature of form and architectonic of the main process which contains all the being and thought. Dynamic being without diachrony is impossible" (ibidem).

In relation to the psychological phenomenon, diachronic approach requires, firstly, defining the unit of analysis, i.e. the system, overgrowing itself in its development and turning into another system, and secondly, specifying determinants of this transition [12]. It is necessary to choose structurally complex, dynamic functioning systems complying with vital goals and combining the opportunity to account of multiple determination by a real education, for a system as a unit of psychological analysis in the diachronic approach. Such can be, for instance, the students-educational environment interaction system, the school adaptation system, the system of social interactions of personality in the process of socialization, etc. In any case, a system-forming category will be the category of interaction, as it combines features such as the activity of the personality in the process of entering the social environment, adapting to it with further integration; the possibility of realization of the personality-environment interaction idea, sequence of initiative, reactive personal and environmental influences.

Analysis of the historical development of interaction categories allows concluding its importance for the psychological research [7]. Due to personality-environment interaction every studied system that includes this interaction acquires functionality, consistency, efficiency or deteriorates because of the appearing of a whole new interaction system, functioning in the following time interval. We have shown that the personality-environment interaction category meets the basic requirement of the diachronic approach, which is organization of the research of the sequential system development process of a system development into a whole new system. The interaction category also allows integrating into a single holistic knowledge of the various and diverse factors and phenomena, which are joining in a complicated way into a system of mental functioning of a socializing and developing personality [12].

The most common determiner of a system is the inconsistency of its elements' functioning. Thus the model of adaptive interactions between personality and educational environment developed by the author [10], the development of the adaptive interactions system begins in conditions of imbalance, inconsistency, disagreement between requirements (capabilities) of the environment and capabilities (requirements) of the personality. Disagreement in the interactions with the environment system can be realized and not realized by the personality. In the first case, the dynamics of changes in the "personality – environment" system may be initiated by the personality. Mechanisms, allowing the personality to realize the disagreement with the environment can be cognitive mechanisms for distinguishing minimal changes, emotional mechanism of alarm, motivational mechanisms of striving to balance with the environment requirements, etc. If a disagreement in the "personality-environment" system is not realized by the personality, the dynamics of interactions in a given system can be initiated by the social environment with social and psychological mechanisms.

Function of the adaptive interactions system is to achieve the dynamic balance between the personality and environment requirements. In this case the achievement of dynamic balance between the personality and the environment is determined by numerous internal and external (for the personality) factors. The final result of the work of personality interactions system is readiness for changes, adaptation is expressed by the human ability to carry out top-priority activity for the given age successfully and / or personality activity in these changing conditions.

In the course of interacting with the environment the subject reaches or does not reach the objectives, which are optimal interactions. In the first case, there is an agreement with the requirements of the environment performed through internal changes or active changes of the environment, or a combination of both. From the perspective of leading activity and personal development, an agreement may be optimal or non-optimal (weak or static over time). As a result of optimal agreement, there is a successful psychological or socio-psychological adaptation in the "personality-environment" system, and new formations of the personality appear influencing the subsequent adaptation processes. Weak agreement, insufficient for successful performing the activity, as well as weak disagreement, can be considered a positive result of the adaptation process, if it gives at least a minimal or temporary opportunity to perform educational activities or personal development.

Static agreement with the environmental requirements leads to the fact that the subject does not take into account the changes of environmental conditions and as a result the system of adaptation in the given environment collapses. The interaction system in this case proceeds to the next level of development, a whole new system, functioning under different conditions and in a different time period, is formed. Excessive disagreement with the environment leads to the same result, if a person by the power of will does not "launch" the cycle of necessary interactions again.

The diachronic study of interaction systems changing in time under the influence of disagreements allows us to trace patterns of transition and predict possible ways of development of socialization and social and psychological adaptation processes in the future taking into account the past and the present [12].

Empirical evidence of the determining role of disagreements in the development of personality-environment interactions is presented to us on the example of the school interaction system [8, 9, 13]. It is shown that the whole system of these interactions is flexible and is in constant dynamics. The only things that change are the structure, the ratio of the various elements within the non-connected subsystems and intersystem relationship. Functioning of the school student – educational environment interaction system also depends on the context of the interaction situation and in accordance with the requirements of age periods.

Throughout the time of school education, interaction systems are changing qualitatively. The quality of a change is connected to the increase of the possibility of the adequate change and increase of its effectiveness, i.e. with the increase of adaptation potential of the school interaction system. It happens under the conditions of accumulation of information about a large number of variants (types) of interactions in the system. In turn, it becomes possible either through mental modeling of the set of opportunities with educational environment by the subject (a school student or a teacher), or through mastering of a set of initiatives and responses in the process of its real interaction with the educational environment. In other words, the mechanisms responsible for the optimal dynamics of student-educational environment interaction system are intellectualization of all areas of activity of a student due to the accumulation of experience in the past and at the present, and the expansion of spheres of real activity (set of activities) during its interaction with the educational environment.

Methodological requirement of the diachronic approach, which is in the necessity for interconnecting the past, the present, and the future situations or phenomena in the research, allows us to identify yet another mechanism of optimization of dynamics of the personality-environment interactions system. This is the mechanism of anticipation by the personality of the process and result of interactions with the environment. A significant role of the intellectual actions in adaptation process, beginning from the capacity for minimal differences, allowing us to notice the disagreement between its capabilities and the requirements of the environment, anticipation, and abilities for a wide transfer of effective adaptation action onto a set of situations while maintaining the flexibility of mental structures, allows us to define the intelligence of the personality as a common adaptive capacity.

As stated above, the diachronic approach allows studying the specific system selected as the unit of analysis at a certain time interval, provided that the system in this interval changes qualitatively for several times. This raises the problem of choosing the time scale. Obviously, this choice should be determined by the dynamics of an objectively existing reality, research objectives and the level of psychological analysis. In the context of the school interactions system, the time scale of existence of a certain system feature can be attached to the schooling stage – elementary, secondary and higher. However, the individual systems of school adaptation can exist in a shorter time interval and be caused by heterochrony of personality development and generalized requirements of educational environment attached to the group mode of study. In addition, at the psychophysiological level of analysis of school interactions system time scale, selected for the study, may be longer than at the psychological and socio-psychological levels. Thus, we have shown that changes in the students' individual dynamics of their mental functioning connected with the properties of the nervous system requires more time and personal contribution than a qualitative change in cognitive and socio-psychological phenomena [8, 9].

If we understand diachrony as a transition of the studied system to another level, i.e. its structural and qualitative changes and transformation into the system with other features, characteristics, goals, and synchrony as operation of the modified system, then the problem of given phenomena correlation inevitably arises. When does diachrony begin and end in its development, how does synchrony become diachrony, what are the mechanisms of this transition, what methods do we use for studying both processes and their mutual transitions?

Developers of diachronic approach in philosophy and linguistics speak about the symbiotic relationships of diachrony and synchrony in the development of a language: "... although the simultaneous analysis of language phenomena precedes the diachronic one, diachrony, nevertheless, has the basis for understanding the facts of a language in synchrony" [25], "... diachrony does not exclude the notion of a system and synchrony cannot entirely exclude the concept of evolution.

Both diachrony and synchrony form two successive phases of a diachronic (in form and structure) analysis of a language as elements of a general approach" [25].

Obviously, with regard to the system of adaptive interactions, the main objective of which is the establishment of a dynamic equilibrium in the "personality – environment" system, the answers to these questions can be sought in the analysis of the processes of agreements and disagreements of the requirements (capabilities) of the environment and capabilities (requirements) of the personality. Objective presence of disagreements between requirements and capabilities in the "personality – environment" system can be a criterion for the synchronic existence of the adaptive interaction system, as their main objective is to overcome this disagreement and pursuit of a perfect match between requirements and capabilities of the personality and the environment [12]. At the same time this situation of objective disagreement is the beginning of a diachronic existence in the "personality – environment" system as a personality in its interaction with the environment tends to its optimality and efficiency. As well as the current interactions system does not allow to achieve this objective, it is necessary to destroy it or convert into a new one.

The disagreements mentioned above and an agreement of the requirements and capabilities ratio allows us to determine a very complex and non-linear nature of the development of the adaptive interactions system. At a certain moment, the ability of a person to spontaneously generate new structures by means of internal reserves appears. Not only multiplicity of knowledge, abilities, skills, but also a complex system of interpersonal interaction with a special surrounding, readiness to openly investigate a problem, reflect an entire structure of basic elements of the problem, "generalization" by the personality of its way of life etc. can act as such new formations [12].

Maturity of any new formation in a given process can serve as a criterion for the end of the diachrony and transition of adaptation system to synchronic existence. Excessive accumulation of agreement between the requirements (capabilities) of environment and capabilities (requirements) of personality during synchrony of adaptation is also a diachrony for the "personality – environment" system, as it leads to the destruction of familiar and no longer contributing to the personal development of relations with the environment.

Alternating periods of synchrony and diachrony in the processes of interaction of an individual with the environment and adaptation can determine the qualitative difference between interaction and adaptation. Effective interactions are accomplished on the basis of successful result of the individual adaptation to the conditions of life and activity in the particular environment, i.e. on the basis of achieved equilibrium in the "personality – environment" system. Adaptation as a process begins within conditions of imbalance between the requirements (capabilities) of environment and capabilities (requirements) of the individual. However, the characteristics of the subject activity and reflexivity during the adaptation process allows to distinguish the category of "adaptive interactions", which include perceptive, intellectual, communicative, willed, etc. actions aimed at identifying the inconsistencies in the "personality – environment" system, and at modeling of its overcoming and anticipation of a result. In turn, any interaction of the individual with the environment is ultimately directed to achieving equilibrium with it via changes in the environment, self-modification, or both at the same time. All the actions of the subject related to the cognition and change of the environment and with the cognition and change of itself, either way, always match the objectives of adaptation. All this allows us to consider adaptation to be a permanent process of agreement and disagreement within the "personality – environment" system. Moreover, some of the parameters of dynamics determining the occurrence of the disagreement after the agreement can be supported in this system.

Thus, diachronic approach in psychological research allows us to trace the patterns of evolution of the studied phenomenon, the development of the system chosen as the subject of analysis, to reveal the patterns of transition of inefficient system to the desired level of development, to predict the possible trajectories of the evolutionary and the synchronic development of the system. However, beginning to use the diachronic approach for analysis of socialization and adaptation processes in psychology needs further development and empiric confirmatory research.

29.4 Adaptational Capabilities of a Personality

Adaptation has been studied for a long time in medicine and physiology. Also the present problem has specific psychological content whose meaningful core is the achievement of an optimal balance of human capabilities and requirements of the environment from the point of view of the internal and the external criteria; this is a must for psychological adaptation and development of an individual. Great theoretical significance of the psychological adaptation problem combines with its lack of conceptual development in psychology which shows in a small number of theoretical studies of the present problem, prevailing of the empirical, applied research solving the particular practical problems.

The school environment example showed that the problem of adaptational capabilities of a personality must be solved in accordance with the specific character of the preceding development of the personality [11]. Constant micro-changes in the educational environment, placing new requirements on the behalf of the educational environment not only in particular periods of studying period but in the course of study as a whole, tense and density of the new information in the educational process, orientation towards the development objectives of the students combined with the orientation towards the positive results of the Unified State Examination, – these thing actualize the problem of accordance of the requirements (capabilities) of the educational environment to the capabilities (requirements) of the students. In some of our studies we point out that school adaptation as a process goes during the whole period of studies at school; all the psychic phenomena and resources of the students are involved into this process; his/her school adaptation is determined by all the components of the educational environment and goes as an active process of self-change and transformation of the environment in the process of interaction with it [5, 6, 14, 15].

The acuteness of the problem increases when we talk about the readiness of the school leavers to the new conditions of learning, professional activities, and social interaction after school. Till the recent times scholars have been solving this question in terms of building up a certain bulk of academic knowledge with the school students that was necessary to study in a professional educational institution. However the process of the further development with the school leavers is much wider in its contents, and includes not only mastering of the new profession but also the active mastering of other areas of life and activity: social interaction, further personal self-cognition, building up family relationships, social activity, etc.

In order to successfully complete personal self-realization according to the requirement of modern dynamic society, a school leaver should possess adaptational capabilities that allow him to cognize the world adequately, understand the changing society, tune up his/her capabilities to the changed conditions, development of his/her personality, and changed environment, which is necessary to achieve socially effective results.

Adaptational capabilities develop during the whole period of studies at school and have complicated multi-level structure including cognitive, emotional, motivational, social-psychological components [11]. At each stage of schooling, in primary, secondary, and high school these component influence the result of school adaptation with different force, and have different contents. In this regard at each stage of schooling the structure of adaptational changes is unique, particular components prevail in it, and intra-structural links provide its efficient functioning. It is necessary to point out qualitative uniqueness of the structure of adaptational capabilities of the school students because the nature and peculiarities of this structure on the period of schooling influence on the result of school adaptation and provide further psychological, social-psychological, professional adaptation with the school, gymnasium, and lyceum leavers. Besides, the necessity to predict adaptational capabilities of the school leavers, there is need for development of a diagnostic criteria and coefficients, methodical apparatus, and programs for development correction. As it was mentioned before, there are almost no studies on the problems of adaptational capabilities of the general full-time school leavers (ibidem).

Development of the adaptational capabilities of a student is provided for by both his/her adaptational potential and by the educational environment at school. Adaptational potential of personality having functions similar to personal potential, as “flexible reactions to the changes of external and internal environment” [19, p. 8], possesses its own peculiar features and differs from the latter. From D.A. Leontiev point of view, personal potential is “an integral systemic characteristic of individual-psychological peculiar features of personality underlying the capability of personality to follow the stable internal criteria and guidance in his/her life and preserve the stability of meaning orientations and efficiency of the activities with the background of pressure and changing external conditions” (ibidem).

Semantic accents of the given definition point out the stability of the internal criteria and guidance of one’s own life activity on the one hand, and pressure of the external conditions on the other hand, i.e. personality’s reaction to the changes in the external environment goes towards "the change in the external world" (ibidem), and self-regulation of the subject of activity is "autonomous" (Personal potential..., p. 6). In this regard, personal potential reflects some sort of personality resistance to the external environment, ability to withstand the pressure of the outer circumstances, and characterizes the viability of personality.

It is not really important for the psychologist how and how well personality resists the circumstances (though in particular situations this trait is extremely important to the student) but how adequately the essence of the changes is understood, what kind of reaction should follow, whether the choice of means and ways of adjustment is possible, how the operative correction is happening in this process, where the borderline of the optimal balance of requirements (capabilities) of the educational environment and capabilities (requirements) of the student is. Thus it is necessary to put the accents to the non-linear dynamics of the processes of interaction of an individual and environment, and activity of the subjects of educational process.

From our point of view, the characteristics of the individual-environment interaction, educational environment included are reflected most fully in the concept "personal adaptational potential" introduced by A.G. Maklakov [21]. The author includes neuro-psychic sustainability of personality, self-esteem, social support, proneness to conflict level, experience of social interactions into the personal adaptational potential [21]. However the notion "potential" in the broad sense of the word means the alternative means the object possesses. Their effective realization is determined not by the fact of their presence but by some other phenomena determining the activity of the subject and the educational environment. This does not mean that realization of adaptational potential is only connected to need-motivation phenomena. In some studies the significant role of psycho-dynamic peculiarities of personality, its ideas, intellect, emotional processes and conditions, and also socio-psychological components of adaptational process was proved [4, 8, 9]. Specific character of interconnections between different components of school adaptation proved to be crucial. New arising systemic characteristics and traits of personality allow the students to compensate the lack of development of certain characteristics and traits, optimize other traits by the will power from the "spare channel" [6, 15], develop necessary traits from the zone of the closest development, etc.

The important things about the adaptational process described above are integrated personality traits allowing us to choose means of interaction with the environment in a fast and efficient way, providing active and developing adaptation, i.e. adaptational capabilities of personality. Customary and preferable means of personality-environment interaction in adaptational process compose into adaptational style of personality [6, 15]. Adaptational capabilities are regarded as individual-psychological peculiarities of personality expressed in the choice of the most effective ways of adaptation to the environment. The present capabilities are the result of interaction between psycho-physiological, psychic, and social-psychological phenomena functioning in the process of achievement the dynamic balance in the system "personality-environment" [11].

Development of adaptational capabilities is a very complicated, multidimensional, heterochronous, and non-linear process of movements and changes in personality accompanied by heterogeneous and heterochronous changes in the components of educational environment themselves. In this case it is necessary to point out heterochrony of the subsequent changes of personality and environment. In our research we pointed out two forms of existence of such type of heterochrony of personal changes and changes in social component of educational environment: difference in disposition of the interacting subjects and violating the temporal agreement of interactions [13]. The strongly pronounced heterochrony of personal changes and changes in educational environment itself, chronical retardation of student's capabilities from the requirements of the environment, inconsistency in dispositions of subjects participating in educational process, impossibility of conforming educational environment and requirements of the educational process participants, – these are the general reasons for decrease in school adaptation and risks of de-adaptation.

Adaptational capabilities of personality developing in the process of his/her interaction with educational environment are systemic, which means universal, and in this sense they have opportunities of broad shift to different areas of the subject's activities; also poorly developed adaptational capabilities in the schooling process have opportunities of broad limitation in other activity areas. Development of adaptational capabilities at school happens mostly spontaneously. Indeed, psychological-pedagogical support (provided not only by the school psychologist, but also by teachers and parents) can help out towards more or less purposefulness and consistency of the process of the development of adaptational capabilities, but complexity and integrity of this problem do not contribute to its solving in modern school with its regular deficit of teachers-psychologists and low psychological competence of teachers setting the priorities if their pedagogical activities in favor of the good results of the Unified State Examination.

Therefore, adaptational capabilities of personality play a significant role in the process of his/her interaction with the environment, they are complex multidimensional systemic unity where interconnection of heterogeneous psychic phenomena and personal traits provides systemic new-formation – adaptational readiness of personality.

References

1. Ananev BG (2002) Human as an object of cognition. Piter, St. Petersburg
2. Barabanschikov VA (2005) Principle of systemacity in modern psychology: grounds, problems, development trends. In: Barabanschikov VA (ed) The idea of systemacity in modern psychology. RAN, Moscow, pp 9–47
3. de Ferdinand S (2004) Course in general linguistics. Editorial URSS, Moscow
4. Grigoreva MV, Novoseltceva AA (2011) Socio-psychological characteristics middle-stage school students as internal factors of school adaptation. In: Adaptation of personality in modern world, vol 3. Science, Saratov, pp 50–63

5. Grigoreva MV (2007) Intellectual mechanisms of school students' psychological adaptation to new educational conditions. Herald of Kostroma State University named after N.A. Nekrasov Sci Methodol J Ser Psychol Sci Acmeol Educ 1:62–65
6. Grigoreva MV (2008) Developing a conceptual model of interaction between personality and environment. World Psychol 1(53):93–100
7. Grigoreva MV (2009) Interaction as the category of modern psychology. In: Proceedings of Pedagogical Institute of Saratov State University n.a. N.G. Chernyshevsky Ser Psychol Pedagog 3–4:3–12
8. Grigoreva MV (2010) Emotional processes and states in the system of dynamic interactions of a school student and educational environment. Scientific notes of Pedagogical Institute of Saratov State University n. a. N.G. Chernyshevsky Ser Psychol Pedagog 3(11):8–13
9. Grigoreva MV (2010) Psychological structure and dynamics of interactions between educational environment and a student in the process of school adaptation. Educ Psychol 6:33–42
10. Grigoreva MV (2011) Basic conceptual provisions of school adaptation research. Vector of science of Tolyatti State University. Ser Pedagog Psychol 2: 63–66
11. Grigoreva MV (2012) The role of adaptive abilities of students in the process of interaction with the educational environment. In: Proceedings of Saratov University New series. Ser Acmeol Educ Educ Psychol 3:11–15
12. Grigoreva MV (2013) Diachronic approach to research of sociological and psychological adaptation of personality. In: Proceedings of Saratov University. New Ser Ser Philos Psychol Pedagog 1:54–58
13. Grigoreva MV, Shamionov RM (2009) Temporal relationships of subjective and personal manifestations in adaptation process. In: Proceedings of Volgograd State Pedagogical University. Ser Pedagog Sci 9(43):127–130
14. Grigoreva MV (1996) Adaptation of the fifth graders to new conditions of educational activities. In: Scientific and theoretical support of students' professional training of a Pedagogical Institution, vol VIII. Science, Saratov, pp 57–60
15. Grigoreva MV (2008) School adaptation: mechanisms and factors under different educational conditions. Saratov State University, Saratov
16. Kondakov IM (2006) Paradigms "soul – body" and their methodology. Paper presented at the international conference on the history of psychology 4 Moscow meetings, RAN, Moscow, 26–29 June
17. Kuhn TS (1975) The structure of scientific. AST, Moscow
18. Leontev AA (1990) L.S. Vygotskiy. Education, Moscow
19. Leontev DA (ed) (2011) Personal potential: structure and diagnostics. Meaning, Moscow
20. Lomov BF (1984) Methodological and theoretical problems of psychology. Science, Moscow
21. Maklakov AG (2001) Personal adaptational potential: its mobilization and prediction in extreme conditions. Psychol J 1:16–24
22. Obukhova LF (1981) Jean Piaget's theory: pro and contra. MGU, Moscow
23. Prangishvily IV (2000) System approach and system-wide patterns. Sinteg, Moscow
24. Psychology and new ideals of scientific character (materials of «roundtable») (1993) Quest Philos 5:3–43
25. Razumovsky OS Diachrony URL: http://www.chronos.msu.ru/TERMS/razumovsky_diakhronia.htm. (Referral date 3 September 2012)
26. Rybakov VA, Pokryshkin AL (2005) Are psychology and natural science paradigm compatible? Hang-glider, Tomsk
27. Sardzhveladze NI (1989) Personality and its interaction with social environment. Metsniereba, Tbilisi
28. Shamionov RM (2010) Concerning some transformations of the structure of subjective well-being of a personality under different conditions of professional socialization. World Psychol 1:237–249
29. Tarasova LE (2011) Pedagogical tact in the process of students training quality. Innov Educ 1:29–46
30. Vagurin VA (2006) Synergetics of the modern society evolution. ComKniga, Moscow

Chapter 30
The Uncoordinated Proliferation of International Courts and Tribunals in the Context of Complexity Theory

Onur Uraz and Fatima Makhzoum

Abstract In today's world order, the frequency of international interactions has reached a level that was once unthinkable. As a natural consequence, international law is one of the main areas affected. Nowadays, international dispute settlement mechanisms are more vital than ever and the numbers of these mechanisms are rising sharply. However, this sudden increase has led to a phenomenon that has created concern amongst international lawyers: the so-called 'uncoordinated proliferation of international courts and tribunals.' This study has two complementary aims: First, using complexity theory to explain this phenomenon and arguing why it does not pose a threat to the international legal system; second, discussing why complexity theory should be referred to more to decide the future of international adjudication.

30.1 Introduction

Undeniably, the end of the Cold War radically changed the global order. As a natural consequence, international law was one of the main areas affected. The rule of law became an objective promoted in international relationships through means such as diplomacy, multinational and bilateral agreements, and international adjudication. The end of East-west hostility and the rise of globalisation lead to the creation of new international agreements and a sharp rise in the number of international courts

O. Uraz (✉)
University of Southampton (LL.M), Southampton, UK

The University of Glasgow (LL.B), Glasgow, UK

Gazi University, Ankara, Turkey
e-mail: onururaz215@yahoo.com.tr

F. Makhzoum
The University of Glasgow (LL.M.), Glasgow, UK

The University of Reading (LL.B.), Reading, UK
e-mail: makhzoum@hotmail.co.uk

Ş.Ş. Erçetin (ed.), *Chaos, Complexity and Leadership 2014*,
Springer Proceedings in Complexity, DOI 10.1007/978-3-319-18693-1_30

and tribunals,[1] however, this 'proliferation' of international courts and tribunals brought about some concerns. For over 20 years, much has been said and written about these concerns by academics. Most strikingly, successive presidents of the International Court of Justice (ICJ) stated their concerns about the possible negative outcomes of 'proliferation.'[2] The proliferation of international courts and tribunals has also stimulated discussion about their success and effectiveness. Studies that assess the performance of international courts and tribunals have diversified leading to different assessments.[3]

'What makes up the current structure of international adjudication?' and 'what this means for the future of international adjudication?' require answering. To answer these questions, complexity theory might be a useful tool. Complexity theory examines a circumstance when many different agents are interacting with each other in several and various ways. It aims to explain the structure, behaviour and dynamics of this system of interaction. In this context, there are certain types of systems called 'complex adaptive systems'.[4] According to the Ruhl, 'complex adaptive systems theory studies how agents interact and the aggregate product of their interactions.'[5] Complex adaptive systems 'combine qualities of coherent stability and disordered change to produce a sustaining, adaptive system.'[6] Therefore, complex adaptive systems stand between order and chaos, which make them significantly adaptable and resilient to external and internal changes.

As stated by Belcher, 'Legal systems, or a subpart thereof that deals with a specific area of law, exhibit many of the characteristics of a complex adaptive system, and arguably are a complex adaptive system. Accordingly, by applying the principles of complex adaptive systems to the interdependent laws, systems, institutions and actors that interact to form a legal system, one can better break down a system's operational format to gain a clearer insight into the how's and why's of the system's functions.'[7] The authors of this article argue that the international

---

[1]Belcher, M., and Newton, J. (2005). 'International Legal Development: A Complex Problem Deserving of a Complex Solution and Implications for the CAFTA Region', 12 Sw. J. L. & Trade Am. 189, p. 190.

[2]Statement of Judge Stephen M. Schwebel, President of the International Court of Justice, to the Plenary session of the General Assembly of the UN 26.10.1999, http://www.icj-cij.org/court/index.php?pr=87&pt=3&p1=1&p2=3&p3=1&PHPSESSID. Accessed on 07.08.2014; Statement of Judge Gilbert Guillaume, President of the International Court of Justice, to the UN General Assembly of 26.10.2000, http://www.icj-cij.org/court/index.php?pr=84&pt=3&p1=1&p2=3&p3=1. Accessed on 07.08.2014.

[3]For example. Born, G., (2012). 'A New Generation of International Adjudication', 61 Duke Law Journal, pp. 858–864, http://scholarship.law.duke.edu/cgi/viewcontent.cgi?article=1523&context=dlj. Accessed on 12.08.2014.

[4]Belcher, 'International Legal Development', *supra* note 1, p. 193.

[5]Ruhl, J. (2008). 'Law's complexity: A premier', 24 Georgia State University Law Review, p. 889.

[6]Ruhl, J. (1997). 'Thinking of Mediation as a Complex Adaptive System', Brigham Young University Law Review, p. 777.

[7]Belcher, 'International Legal Development', *supra* note 1, p. 194.

legal system exhibits many characteristics of a complex adaptive system. These characteristics are helpful to explain the current uncoordinated proliferate structure of international adjudication. Thus, considering the international legal system and international adjudication through the lens of complex adaptive systems might provide a better understanding regarding their current structure and their future.

Therefore, Sect. 30.2 of this chapter will briefly develop the principles of a complex adaptive system and explain their relationship with the international legal system. It will argue that the international legal system demonstrates certain characteristics of a complex adaptive system, meaning that the recent uncoordinated proliferation of international courts and tribunals might be better understood and assessed from this perspective. Section 30.3 will then introduce the phenomenon called 'the uncoordinated proliferation of international courts and tribunals' and its related concerns. Finally, Section 30.4 will examine the current structure of international courts and tribunals and will discuss their future in the context of complex adaptive systems.

30.2 Complex Adaptive Systems Theory as a Descriptive Tool for the Recent Structure of International Adjudication

30.2.1 What Is a Complex Adaptive System?

A complex adaptive system is 'a system in which large networks of components with no central control and simple rules of operation give rise to complex collective behaviour, sophisticated information processing, and adaptation via learning or evolution.'[8] They are a particular type of *self-organizing* system with *emergent properties* and an ability to *adapt* in reaction to a change of external situations.

Complex adaptive systems focus on the interactions between the different *agents* of a particular system. These *agents* are interconnected and affected by each other in different ways depending on the nature of their interaction (for example, through flows of energy or information). These interactions are of a *non-linear* nature.[9] Therefore, rather than a specific or singular relationship or change, the aggregation of the interactions in the system determines the system's emerging behaviours. Nevertheless, the system remains unpredictable since it is difficult to predict the outcome of the every interaction in the long term.[10]

[8]Mitchell, M. (2009). *Complexity: A guided tour*, Oxford: Oxford University Press (2009). p. 13.

[9]Meadows, D. H. (2008). Thinking in systems: A primer. White River Junction: Chelsea Green. Millennium Ecosystem Assessment. (2005). *Ecosystems and human well-being: Synthesis*. Washington, D.C.: Island Press. pp. 2–4.

[10]Belcher, 'International Legal Development', *supra* note 1, p. 194.

The core of complex adaptive system study is to focus on systems that have *complex*, *macroscopic* and *emergent properties*.[11] Accordingly, a great number of *agents* interact with each other in a system and they constantly adapt themselves to stay functional or to survive. All these activities create the cumulative behaviour of a 'system,' which is still far from being predictable in the long term due to its complex nature.[12] In the following paragraphs, the specific properties of this basic definition of complex adaptive systems will be explored, thus setting out the characteristics of a complex adaptive system.

To begin with, a fundamental distinction must be made between that the term *complex* and complicatedness. A complicated structure consists of various components, however if one of these components are taken out of the equation, the whole behaviour of the system does not necessarily, or fundamentally, change.[13] Conversely, *complexity* derives from the *interdependency, non-linear interactions* and *attributes* of the system's *components*. Removing or transforming a *component* creates a change in the system's overall behaviour beyond its actual value, function or embodiment.[14]

This *complexity* helps to make a system more *adaptive* and *resilient* against changes within or to the system, since the impact of a particular change is subtilized across the system by means of *feedback changes. Feedback changes* simply refer to the alterations of *agents'* behaviours due to feedbacks that they receive from the system.[15] When an agent changes its individual behaviour, others also react and alter theirs in respect of this change. Accordingly, while agents *adapt* themselves to these changes, their *aggregated behaviours* also make the system adaptive. The set-up of the *network of agents* and their casual relationships form the core of the system. Thus, their *adaptive, interconnected* and *heterogonous* structure creates a complex adaptive system.

Pauwelyn defines the certain features that complex adaptive systems share as a: '(i) dispersed interaction between many heterogeneous agents, (ii) leading to emergent self-organized collective behaviour, (iii) without global controller and (iv) continual adaptation with out-of-equilibrium dynamics'[16] The most common examples of complex adaptive systems are the ecosystems of biological species[17]

---

[11]Hornstein, D. (2005) 'Complexity Theory, Adaptation, and Administrative Law', 54 (4) Duke Law Journal, pp. 913–960.

[12]Ruhl, 'Law's Complexity', *supra* note 5, p. 904.

[13]Miller, J. and Page, S. (2007). *Complex Adaptive Systems: An Introduction to Computational Models of Social Life*, Princeton University Press, p. 9.

[14]Ruhl, 'Law's Complexity', *supra* note 5, p. 891.

[15]Belcher, 'International Legal Development', *supra* note 1, p. 198,199.

[16]Pauwelyn, J. (2014). 'At the Edge of Chaos? Foreign Investment Law as a Complex Adaptive System, How It Emerged and How It Can Be Reformed' 29 ICSID Review, p. 392.

[17]See. Gross, J., McAllister, R., Abel, N., Smith, D., and Maru, Y. (2006). 'Australian rangelands as complex adaptive systems: A conceptual model and preliminary results', 21 Environmental Modelling & Software, pp. 1264–1270.

and economies.[18] In this regard, Adam Smith's invisible hand theory or ant colonies can be given as standard examples, as in both of these instances highly complex collective behaviour and physical structures emerge through various interactions, without a central control mechanism.[19]

Therefore, it is evident that *agents* are the key elements of complex adaptive systems. Whilst the premise of complex adaptive systems study offers that the collective outcome or behaviour of a system is determined by the culmination of all agents' individual behaviours,[20] these agents carry out their actions based on internal and external factors such as their own operational understanding, rules, interests, experiences and targets.[21] Hence, it is this *heterogeneity* of the *agents* that is significant to the outcomes and emergent behaviours of the whole system.

Nevertheless, 'underlying all agent interactions of a complex system is often simple, deterministic rules. What makes the interactions complex is how these rules, when set in motion among the diverse components of a system, produce nonlinear relationships including reinforcing and stabilizing feedbacks.'[22] The emergence of these interactions and behaviours happen without any external and central controller, yet the system keeps its identity and functions. This is a common and defining feature of complex adaptive systems and is called '*self-organization*'.[23]

In addition, change within a complex adaptive system is perpetual and strives towards a *critical state* of dynamic behaviour, but is not homogenous in frequency or significance. In this respect, what makes a system adaptive rather than rigid or chaotic is the critical state of '*stable disequilibrium*'.[24] This hallmark feature of complex adaptive systems indicates that a system is always dynamic and in motion but also exhibits coherence under change.[25] Therefore, a complex adaptive system operates in an area entitled the '*edge of chaos*,' a significant mid-point between order and randomness. As Ruhl notes;

> We almost invariably find change occurring in *power law event distributions* in which vast numbers of small changes are punctuated by infrequent large changes. Complex adaptive systems build *adaptive capacity* based on this kind of change regime, not based on a normal distribution. An ecosystem, for example, builds *resistance* capacity to withstand

[18]See Generally, Arthur, B., Durlauf, S., and Lane, D. (1997) (eds.), 'The Economy as an Evolving Complex System II, Introduction: Process and Emergence in the Economy', Addison-Wesley, http://tuvalu.santafe.edu/~wbarthur/Papers/ADL_Intro.pdf. Accessed on 07.08.2014.

[19]Pauwelyn, 'At the Edge of Chaos?', *supra* note 16, p. 393.

[20]Newman, M. (2001). 'Complex systems', 79 American Journal of Physics, p. 800.

[21]Belcher, 'International Legal Development', *supra* note 1, p. 195.

[22]Kim, R., and Mackney, B. (2013) 'International environmental law as a complex adaptive system', International Environmental Agreements: Politics, Law and Economics, p. 7, http://link.springer.com/article/10.1007/s10784-013-9225-2/fulltext.html. Accessed on 07.08.2014.

[23]Highfield, R. and Coveney, P., (1996). *Frontiers of Complexity: The Search for Order in a Chaotic World*, Ballantine Books: New York, pp. 430–432.

[24]Ruhl, 'Law's Complexity', *supra* note 5, p. 895.

[25]Holland, J., (1995). *Hidden order: How adaptation builds complexity*, Basic Book, New York.

> environmental changes such as fire regimes, and builds *resilience* capacity to rebound from severe incidents. (. . .) Notwithstanding the remarkable capacity of complex adaptive systems to maintain their properties over time, we must return to the cold, hard truth that all systems ultimately are built on deterministic rules that cannot be violated. There is a limit to the resistance and resilience of any complex adaptive system, and if pushed hard enough or persistently enough, a system might move into a *phase transition* through which radically new network architecture is installed.[26]

Therefore, 'complex adaptive systems theory is about building models for moderate number contexts in which agent *heterogeneity* can and usually does influence outcomes, and as such it is worth exploring how it might inform our understanding of the legal system'.[27] This is also helpful to better understand the uncoordinated proliferation of international courts and tribunals, since international adjudication not only derives from the complex nature of the international legal system but is also a component of it. The international legal system exhibits many characteristics of a complex adaptive system, as it is a web of treaties and institutions that *self-organize* and display *emergent properties*.[28] Thus, the complex adaptive system characteristics of the international legal system, and consequently international adjudication, should be examined in detail before the discussion of the 'proliferation' phenomenon.

30.2.2 International Law as a Complex Adaptive System

In any legal system, different agents interact with each other in various ways. In terms of the international legal system, the interactions are more diverse due to today's highly interconnected world where endless factors are changing matters in an instant and extreme way. The international legal system consists of many diverse components and the evolution of international law is affected by a number of these *heterogeneous* components, such as: states, international organizations, international and national courts and tribunals, non-governmental organizations, individuals and so on.[29] Therefore, international law making entails a broader process than provided for in strict conventional definitions of law making because, primarily, the number of contributors or causes of law making and the processes applicable are higher than under the traditional understanding.[30]

Furthermore, in the international legal system, a law or an institution always has an interrelationship with other parts of the legal system. The interactions between

---

[26] Ruhl, 'Law's Complexity', *supra* note 5, p. 895, 896.

[27] *Ibid,* pp. 888–889.

[28] Kim, 'International environmental law as a complex adaptive system', *supra* note 22, p. 5.

[29] Pronto, A., (2008). 'Some Thoughts on the Making of International Law', 19 EJIL, pp. 601–603.

[30] See generally, Noortmann, M., and Ryngaert, C., (2010) (eds.), *Non-State Actor Dynamics in International Law: From Law-Takers to Law-Makers,* Ashgate Publication, Surrey.

all these actors (*agents*) shape international law. The interactions not only occur in the legal sphere but also in political, cultural, scientific and military spheres because the international legal system is generally a product of international relations, the continual formal and informal relationships between the aforementioned interconnected actors. Here international relations and law go hand in hand. On the one hand, international relations can shape international law. This can be seen from events such as the Cold War and its impact on both the international legal order and interactions between states and institutions.[31] On the other hand, the emergence of international legal concepts influences actors and their relationships; an example of this is the emergence of human rights and its influence on individuals, states and European politics.[32]

These bilateral and multilateral interactions, which create the international legal system, are pivotal since they prove that the creation and implementation of law is a result of multifarious interactions between different actors either in a formal or informal manner. This reveals two important complex properties of the international legal system. First, states, institutions and organizations are constantly re-defining their aims in accordance with their own dynamics as well as the 'international set-up at that moment.' Therefore, the rules of interaction between actors 'do not produce behaviours that are in continuous proportionate relationships over time; sharp tipping points and discontinuities frequently occur.'[33] This indicates the *non-linearity* of relationships between the actors of the international legal system.

Second, the various classes of the different autonomous components of international law making indicate the *heterogeneity* of the international legal system's actors. In order to achieve any significant international legal development, the support of the majority of states is needed. However, all states are legally independent and their interests and aims are varied. Moreover, with the exception of a limited number of conventions and agreements,[34] multilateral and bilateral agreements are creating independent legal islands.[35] They diverge to a significant degree in terms of their subject matters, objectives, members, regulatory mechanisms, jurisdictions, and so forth. They also frequently create their own autonomous institutional laws and give a great deal of decision-making powers, autonomous personnel and specialists to the institutions. According to complexity theory, this diversity is not

[31]Morton. J., (1999). The end of the cold war and international law: An empirical analysis, 13 Global Society, pp. 7–23; Henkin, L., (1991). 'Law and War After the Cold War', 15 Md. J. Int'l L., pp. 147–169.

[32]See generally, Brandtner, B., and Rosas, A. (1998). 'Human Rights and the External Relations of the European Community: An Analysis of Doctrine and Practice', 9(3) EJIL pp. 468–490.

[33]Ruhl, 'Law's Complexity', *supra* note 5, p. 898.

[34]The UN Charter can be regarded as an exception since it is accepted by all states. However, there are barely any other conventions or institutions that are accepted by all states, nor an international court or tribunal that's jurisdiction is accepted by all international actors.

[35]See generally, Pauwelyn, J., (2004). 'Bridging Fragmentation and Unity: International Law as a Universe of Inter-Connected Islands,' 25 Michigan Journal of International Law http://scholarship.law.duke.edu/cgi/viewcontent.cgi?article=2933&context=faculty_scholarship

random; on the contrary, it is the result of the environment of the system and the actions of the other agents.[36] This also indicates the *decentralized* characteristic of international law.

At the same time, the interconnection of states is getting more complex as economic, social and political dependency on each other increases as a natural consequence of globalism and technological advancement. In today's world, an armed conflict between two states in Europe may lead to economic crisis in America; a terrorist organization based outside of Europe can attack a European State and potentially cause a political and legal crisis; or an environmental disaster in Asia may have an effect on the Artic Region. Although, in theory, all states are independent from each other, in order to cope with all these global issues that affect them collectively, they need to work together. The number of treaties, international organizations and institutions that have been established over the past few decades exemplify this rise in co-operation between the actors of the international order. This indicates the *interdependence* of these international actors in creating international law and in creating and running legal institutions. Accordingly, international law is a system of interdependent actors in utilitarian and component interaction, as opposed to a set of uncoordinated norms and institutions.[37]

Furthermore, states and international institutions are able to learn from their past experiences and abide by lessons learnt through history as they interact. They either adapt their behaviours in accordance with past experiences or they make changes to their behaviours that were not serving their own interest. All these interactions and approaches eventually shape international law. The protection of minorities and the development of human rights are an example of this. Until the late nineteenth century, it was accepted that a state could not interfere in another state's treatment of its own citizens due to the understanding of Westphalian sovereignty.[38] However, states re-considered this approach in the late nineteenth century in pursuit of the protections of minorities in Ottoman Empire.[39] Unfortunately, this did not help prevent the disastrous experiences of the Second World War. Following this, states began aiming to protect not only minorities in other states from abuses, but all individuals, as witnessed in the Kosovo situation in late 1990s.[40]

Of course, all these experiences lead to a shift in the approaches of international actors and a consequent evolvement of international law. Our above example, has for instance, led to several different international treaties, court interpretations on law

[36]Holland, J., (1995). *Hidden order: How Adaptation Builds Complexity,* Addison-Wesley Publishing, New York, p. 27.

[37]Kim, 'International environmental law as a complex adaptive system', *supra* note 22, p. 10.

[38]Osiander, A. (2001). 'Sovereignty, International Relations, and the Westphalian Myth', 55 International Organization, pp. 251–287.

[39]Ungern-Sternberg, A. (2012). 'Religion and Religious Intervention' in B. Fassbender and A. Peters (eds.), *The Oxford Handbook of the History of International Law,* Oxford University Press, p. 310.

[40]Henkin, L. (1999). 'Kosovo and the Law of "Humanitarian Intervention"', 93 AJIL, pp. 824–828.

and the emergence of concepts like 'humanitarian intervention' or 'responsibility to protect.'[41] This demonstrates the other complex properties of the international legal system. First of all, it illustrates how past experiences alter the approaches and actions states and institutions take in relation to the creation and implementation of international law. This indicates the *feedback loops* in the international legal system to which actors then make feedback changes in order to stabilize the system. Moreover, it exhibits the *path dependency* of the international law making and applying process; the dependency of the system's new state on information that has flowed through the system in all its prior states.[42]

More strikingly, in international relations and the international legal system, parties act according to their own interests regardless of any possible impact on the system, nevertheless, these different aims, interests and acts implicitly lead to the creation of order. This evidences another complex feature of the international legal system; its *decentralized* structure. As explained in Part III, international law and adjudication suffer from decentralization because there is no centralized legislature, executive or judiciary. However, states continue creating and following international rules and there is a common agreement among states to abide by the law. The international legal system consists of independent actors who look after their own interest, but need each other to maximize their interests and provide some predictability to the system. Thus, an order is created without any higher institution or power specifically designed to provide it. This leads to a complex legal structure that is *self-organized.* In other words, the international legal order is a 'system [which] tends to organize around a set of deep structural rules that lend stability to system behaviour.'[43]

Consequently, it is essential to not individually analyse a treaty, court, tribunal or institution, but to instead analyse the links that hold the system together in order to understand the international legal system and to reform it. As Lasin argues, 'process that seems to be governed by chance when viewed at the level of the individual turns out to be strikingly predictable at the level of society as a whole.'[44] Thus, amplifying macroscopic patterns as a result of this non-reductionist approach rather than focusing on details is important to better understand the structure of international law and adjudication since the 'system behaviour *emerges* from the *aggregation* of network casual chains which cannot be explained by examining any isolated part of the system.'[45] The effectiveness of international adjudication, for example, cannot be assessed by solely focussing on a limited number of components such as the judgements of certain courts, their enforcement or the number of

[41] Background Information on the Responsibility to Protect, the United Nations, at http://www.un.org/en/preventgenocide/rwanda/about/bgresponsibility.shtml. Accessed on 01.09.2014.

[42] Ruhl, 'Law's Complexity', *supra* note 5, p. 899.

[43] *Ibid.* pp. 899–900.

[44] Lansin, S. (2003). 'Complex Adaptive Systems,' 32 Annual Review of Anthropology, p. 183.

[45] Ruhl, 'Law's Complexity', *supra* note 5, p. 899.

cases that are brought before the courts and tribunals. Instead, the effectiveness should be assessed in the context of the international legal system including all its components.[46]

This overlap and communality between the international legal system and the defining properties of complex adaptive systems are striking. However, as Ruhl points out, 'showing that complex adaptive systems theory maps well onto the legal system does not (. . .) prove that the legal system is a complex adaptive system.'[47] Despite the remarkably similar properties, there are obvious doubts regarding applying physics and biology based theory to a social system, since people are the designers of legal systems[48] and ultimately write the law. Furthermore, there are also some other problems with applying complex adaptive systems theory to international law that derive specifically from the nature of the international legal system. For example, it is not easy to identify the *deterministic rules* of the international legal system, a property that complex adaptive systems have. Ruhl provides some examples of deterministic rules applicable to law, such as court interpretations of legislation, legislature overrules, superior courts' affirmation or reverse decisions on lower court decisions, and the implementation of statutes by agencies.[49] However, the position of courts, their decisions and the structure and powers of institutions in the international legal order differ significantly to their national counterparts. International courts' decisions suffer from a lack of sufficient enforcement most of the time and institutions do not have the power to interfere in domestic issues unless a state permits it to do so. Nevertheless, as Ruhl explains, proving that a legal system is a complex adaptive system 'is not the test to which the usefulness of complex adaptive systems theory should be put. Rather, it should suffice to show that the model of complex adaptive systems provides useful design lessons for the legal system.'[50]

Thus, because of the many similarities pointed out between the international legal system and complex adaptive systems, the theory does appear to be a useful tool to understand, assess and reform international law (since international law is an outcome of the international legal system) and its sub-parts, including international adjudication. This approach has been followed by some scholars in relation to other sub-parts of international law such as international investment law,[51] international

---

[46]For related discussions see. Shany, Y. (2014) *Assessing the Effectiveness of International Courts*, Oxford.

[47]Ruhl, 'Law's Complexity', *supra* note 5, p. 901.

[48]Rudd, J. (2005). 'J.B. Ruhl's "Law-and-Society System": Burying Norms and Democracy Under Complexity Theory's Foundation', 29 Wm. & Mary Envtl. L. & Pol'y Rev., pp. 551–632.

[49]*Ibid.* pp. 898.

[50]Ruhl, 'Law's Complexity', *supra* note 5, p. 901.

[51]Pauwelyn, 'At the Edge of Chaos? Foreign Investment Law as a Complex Adaptive System,' *supra* note 16.

environmental law[52] and international legal development.[53] As a result, some have developed criticisms and suggestions using complex adaptive system theory. One suggestion for instance, is that governance of international mechanisms and institutions should behave like a complex adaptive system in order to be efficient, arguing that adaptive or decentralised governance is more suitable then the classical hierarchical structures of the international legal order.[54]

The sub-part of international law focused on in this article is international adjudication. Over the past two decades much has been said in relation to international adjudication including its significant development and, most interestingly, the uncoordinated proliferation of international courts and tribunals. Some scholars and lawyers have labelled this phenomenon a danger to the international legal order. The decentralized, fragmented, but interdependent nature of international adjudication tempts the use of the complex adaptive systems theory to explain the current situation. Thus, before the examination of the uncoordinated proliferation of international courts and tribunals in the context of complex adaptive systems theory, the phenomenon and its related discussions will be explained.

30.3 As a Phenomenon: The Uncoordinated 'Proliferation' of International Courts and Tribunals

The international order has changed considerably, particularly after the end of the Cold War. The amount international organizations has increased, existing organizations such as the European Union ('EU') have become more effective, and the rise of globalization and liberalization has created a new economic order that includes many powerful co-operational and regulative organizations and agreements such as the World Trade Organization ('WTO') and the North Atlantic Free Trade Agreement ('NAFTA'). International courts and tribunals have also been affected. Their number has significantly increased, their subject matters have diversified, and new international courts and tribunals have started to display fundamental jurisdictional differences in comparison with their older counterparts such as the International Court of Justice ('ICJ') or the International Tribunal for the Law of the Sea ('ITLOS'). Moreover, some existing courts such as the European Court of Human Rights ('ECtHR') and the European Court of Justice ('ECJ') have gained more jurisdictional powers. This recent increase in the number of international courts and tribunals has raised concerns about their 'uncoordinated proliferation.' The concerns include the creation of jurisdictional and normative conflicts that may pose threats to the international legal order, the fragmentation of international law

---

[52] Kim, 'International environmental law as a complex adaptive system,' *supra* note 22.

[53] Belcher 'International Legal Development,' *supra* note 1.

[54] Kim, 'International environmental law as a complex adaptive system', *supra* note 22, p. 6.

which may undermine its the clarity and stability, and the resulting chaos that may potentially be caused to the international legal system.

Before further explanation in relation to the 'uncoordinated proliferation' phenomenon, it is necessary to define the term 'international courts and tribunals' for the purpose of this paper since definitions provided by different international adjudicatory bodies vary.[55] The term 'adjudication' usually includes both judicial bodies and arbitral methods. Both are law-based processes that can render binding decisions. However, a judicial body usually exists before a dispute arises and its adjudicators are selected through a mechanism that does not depend on disputant parties' wills. Moreover, their powers derive from a general mandate such as a treaty, and the outcome of their process aims to protect not only the parties' interests but the public's too. Comparatively, in arbitration, the parties set up the arbitration body after a dispute arises with the purpose of solving a particular issue and the mandate of the arbitral body depends on parties' wills.[56] The parties also have control over the selection of adjudicators. For the purpose of this paper, the term 'international courts and tribunals' refers to all of the fundamentally independent international judicial bodies regardless of whether they are permanent or not. Therefore, the term refers to the bodies which are established by an international instrument, can issue binding decisions by referring to international law, follow pre-existing procedural rules on cases that at least include a state or an international organization, and are composed of independent judges.[57] Arbitral bodies are not generally included.

30.3.1 The 'Uncoordinated Proliferation'

To understand the reasons for the concerns about 'uncoordinated proliferation,' it is essential to recognize the unique nature of the international legal system, the state of its main actors as well as their changing structure, and ultimately, why international courts and tribunals are uncoordinated. As a considerably new field of law, international law shows substantial differences compared to its national counterparts. Domestically, modern states usually aim to have a system of control that is unified, clear, and coherently law driven. Hence, most states have set up a legal hierarchy between norms, sources, bodies and procedures, usually through their

---

[55]Romano, C.P.R. (1997). 'The Proliferation of International Judicial Bodies: The Pieces of the Puzzle', 31 N.Y.U. J. Int'l L. & Pol. pp. 711–723, http://www.pict-pcti.org/publications/PICT_articles/JILP/Romano.pdf. Accessed on 08.08.2014.

[56]Romano, C.P.R., Alter, K., & Shany, Y. (2014). 'Mapping International Adjudicative Bodies, the Issues, and Players' in Romano, C.P.R., Alter, K., and Avgerou, C. (eds.) *The Oxford Handbook of International Adjudication* (E-Book), p. 5, http://www.oxfordhandbooks.com.ezproxy.lib.gla.ac.uk/view/10.1093/law/9780199660681.001.0001/law-9780199660681-e-1 (Accessed on 08.08.2014).

[57]The WTO panels or Dispute Settlement Body do not fit in this definition but the Appellate Body does. However, this paper considers the WTO dispute settlement mechanism as a whole.

constitutions. The structure of the international legal system is completely different. Unlike national legal systems, international law is a 'horizontal' system that, despite the recent codification trend, is still largely uncodified. This horizontality means, aside from *jus cogens* rules or *erga omnes* obligations,[58] there are no set of rules that have supremacy over the others, unlike a constitution in a national legal system.

Moreover, there is no formal hierarchy between the primary sources of international law set out in article 38(1) of the Statute of the International Court of Justice (SoICJ). As a result, most of the bilateral and multilateral agreements or international and supranational organizations create their own autonomous and independent rules and dispute settlement means. Therefore, in the international legal system, it is possible to see the existence of more than one normative system regulating the same matter, or the existence of several courts and tribunals that have jurisdiction on same issues. This is a natural consequence of the *decentralized* nature of international law. The *decentralized* nature derives from the fact that international law has a 'contractual' characteristic rather than a 'legislative' one; international laws are usually created by bilateral and multilateral treaties or international customary law rather than by a designated primary law-making body.[59] This creates a *decentralized* legislative system that is usually unable to set a formal hierarchy between conflicting international norms or the jurisdictions of judicial bodies.

For a relatively long time, most have considered the aim of international law to be to create a united and harmonic legal order that is regulated by international legal rules and diplomacy.[60] In this regard, from its creation, some scholars have described the United Nations (UN) Charter as the 'constitution of the world community' and the UN as the body at the top of the international legal structure.[61] This was perhaps the outcome of comparing international law with its domestic counterparts and hoping for a similar structure for the international legal system. Great scholars like Kelsen or Lautherpact have pointed out the different nature of international law whilst defending the idea of a unified international legal system as a necessity. Neither lost faith in the possibility of having a unified and coherent international legal system.[62] However, both throughout the Cold war and after, their

---

[58]See Generally. Shelton, D. (2006). 'Normative Hierarchy in International Law', 100/2 AJIL, pp. 291–323, http://www.jstor.org/stable/pdfplus/3651149.pdf?acceptTC=true&jpdConfirm=true. Accessed on 07.08.2014.

[59]In this sense, the European Union is the only exception since it has a legislative body.

[60]Koskenniemi, M., & Leino, P. (2002). 'Fragmentation of International Law? Postmodern Anxieties', 15 LJIL, p. 556, http://sistemas.mre.gov.br/kitweb/datafiles/IRBr/pt-br/file/Fragmentation%20of%20International%20Law%20postmodern%20anxieties.pdf. Accessed on 07.08.2014.

[61]See. Fassbender, B. (1998). 'The United Nations Charter As Constitution of the International Community', 36 Colum. J. Transnat'l L., pp. 529–620; Sloan, B. (1989). 'The United Nations Charter as a Constitution' 1 Pace Y.B. Int'l L., pp. 61–126.

[62]Carty, A. (1998). 'The Continuing Influence of Kelsen on the General Perception of the Discipline of International Law' 9 EJIL, pp. 346–349, http://www.ejil.org/pdfs/9/2/1494.pdf. Accessed on 07.08.2014.; Lauterpacht, H. (1975). 'The Reality of Law of Nations' in Lauterpacht, E., *International Law: Being the Collected Papers of Hersch Lauterpacht, Vol.2*, Cambridge, p. 76.

wish has not materialized. From this article's point of view, it will not be possible to create a single and coherent international legal system in the near future. One of the main reasons of this is the impossibility of sustaining the traditional understanding of sovereignty. Of course, the main purpose of this paper is not to examine this topic, but in order to shed light on the other discussions in this chapter, this topic will briefly be explained.

In this regard, it no longer seems plausible to defend the traditional definition of state sovereignty, the so-called 'Westphalian Model.'[63] Before the dilution of the traditional meaning of sovereignty, states were only limited within their domestic legal system by their own national law. According to the traditional definition, state sovereignty was absolute, unitary and legitimate. Therefore, no other actor could interfere with the internal matters of a nation. However, changes in, *inter alia*, technology, international relations and international order have transformed this notion.[64] States' independence has shifted to *interdependence* and several developments have lead to this change.

Firstly, the establishment of international and supranational organizations has increased. These organizations, that mostly aim to provide stability to their members' relationships, have set up normative standards and offered dispute settlement mechanisms. Today, for instance, members of the European Union (EU) are obliged to follow certain rules set by EU bodies, whilst at the same time, the World Trade Organization (WTO) asks members to conform to the standards set by the agreement. Of course, states are involved in these organizations through consent and they can choose to stay away; however, staying out of these organizations can be a big disadvantage. Thus, states prefer to be a part of organizations like the WTO or the EU for their own interest and allow these organizations to interfere with their domestic system either directly or indirectly.

Secondly, the universalization of human rights and the rising number of independent, non-governmental, monitoring organizations have started to impact the internal matters of states. Organizations like Amnesty International and Human Rights Watch have aided in increasing the awareness of human rights and states' treatment towards human rights, and UN bodies and the European Convention of Human Rights (ECHR) have set out obligations for states to promote human rights. Despite the fact that some human rights norms have no obligatory nature and suffer from a lack of enforcement, non-obligation of human right norms puts states in tough situations in front of the international community,[65] and as a result, influences their internal acts.

---

[63]Beaulac, S. (2004). 'The Westphalian Model in Defining International Law: Challenging the Myth', 8 Austl. J. Legal Hist. p. 181 http://heinonline.org/HOL/Page?handle=hein.journals/ausleghis8&div=13&g_sent=1&collection=journals#185. Accessed on 07.08.2014).

[64]Hashm, S. H. (1997) 'Introduction' in S. H. Hashm (ed) *State Sovereignty: Change and Persistence in International Relations*, Pensilvanya State University Press, p. 1.

[65]*Ibid.* p. 3.

Thirdly, there is a growing consensus that state sovereignty does not preclude external intervention. Inter and inner state conflicts and internal humanitarian crises are often met with some form of action by members of the international community. In the recent Crimea crisis, the UN, the EU and the US got directly involved; states and international organisations such as the UN or the EU impose sanctions on states in attempting to deal with certain humanitarian crises occurring within states; and there is growing support in the international community for the concept of 'humanitarian intervention.'

Fourthly, many issues now force states to be *interdependent*. States need one another on issues such as maximizing trade profits, protecting the environment or providing a minimum standard of human rights. This, coupled with the impact of globalization and liberalization, has lead the difference between the national and the international to become blurry. Nowadays, states need to be part of international agreements and organizations to both maximise and protect their benefits.

This impact on sovereignty makes it harder for a unified international legal system to exist because international relations are now occurring in a faster and more informal manner. The actors of international relations are not just states but include, *inter alia*, international organizations and non-governmental organisations, and the number of matters that each actor can act upon is wider because of this dilution in the traditional definition of state sovereignty. Therefore, the role of states in the international legal system is changing, and the actors of international law are *diversified*.[66]

There are also other reasons why it does not seem possible for a single-handed or a hierarchical international legal system to exist. To begin with, the idea overlooks political reality. During the Cold War, it was believed that the polarized structure of the world was hindering establishment of a hierarchical structure governed by the UN. However, the rise of globalization, liberalism and the end of the East-west polarization did not change the lack of hierarchy and coherence. In the new era, the actors began competing with each other, tried to maximize their benefits and attempted to create competing normative systems.[67] Moreover, those who believed that the UN Charter would be a world constitution overlooked the source and nature of the Charter. The UN Charter was not derived from natural and legitimate sociological processes like domestic constitutions. Furthermore, with the exception of Article 103 of the UN Charter and its highly political and controversial relationship with Security Council resolutions, the UN has no formal supremacy over other international organizations or rules.[68] Also, other international rules do not take their legitimacy from the UN Charter. Therefore, the UN charter is not a constitution or constitution-like document but a pivotal multinational treaty.

Moreover, despite globalization and rising *interdependence,* there are still important economic, legal or social differences between different geographical regions.

---

[66]Koskenniemi, 'Fragmentation of International Law?', *supra* note 60, p. 557.

[67]*Ibid.* pp. 558–559.

[68]*Ibid.* p. 589.

This prevents states from accepting similar norms about even universalised issues such as human rights. In this regard, one should accept that as long as the world does not become a more harmonized place sociologically, it is not likely to have a worldwide, hierarchal and unified normative system. Thus one should embrace the current structure of international order as well as the fragmented structure of international law.

Therefore, as a result of this *decentralized* structure of international law, there is no hierarchy or coherence between international courts and tribunals.[69] This lack of hierarchy and coherence had not drawn much attention until the late 20th, when the number of international courts and tribunals began to increase. The rise of compulsory jurisdiction and the rapid growth in the number of courts and tribunals has caused concerns about the uncoordinated multiplicity of judicial dispute resolution options.[70] Consequently, the possibilities of normative conflicts or the overlapping jurisdictions of courts and tribunals have created 'the fear of chaos' in international dispute resolution.

30.3.2 The Fear of Chaos

Due to its *decentralized* and horizontal nature, the international legal system is constituted of uncoordinated norms and bodies. Treaties and organizations tend to create or choose their own norms, procedures and dispute settlement means. These so-called 'self-regulated' regimes apply their own set of rules without any formal coherence with each other. In light of this nature, the recent rise in the creation of international courts and tribunals without any formal connection to each other has caused some anxieties. One scholar has even defined the uncoordinated and incoherent aspects of the international legal system and the international courts and tribunals as 'anarchic.'[71] Thus, possibilities of normative, jurisdictional and jurisprudential conflicts between the uncoordinated courts and tribunals stand out as main concerns. The former ICJ president Guillaume emphasized these concerns;

> Overlapping jurisdiction also exacerbates the risk of conflicting judgments, as a given issue may be submitted to two courts at the same time and they may hand down inconsistent judgments. National legal systems have long had to confront these problems. They have resolved them by, for the most part, creating courts of appeal and review. (. . .) The

[69]Mégret, F. (2012) International Law as Law, in Crawford, J., & Koskenniemi, M., (eds.), *The Cambridge Companion to International Law*, Cambridge, pp. 69–73.

[70]Spain, A. (2013). International Dispute Resolution in an Era of Globalization, in Byrnes, A., Hayashi, M., and Michaelsen, C. (eds.) *International Law in the New Age of Globalization,* Martinus Nijhoff, pp. 50–54.

[71]Spelliscy, S. (2001) 'The Proliferation of International Tribunals: A Chink in the Armor', 40 Colum. J. Transnat'l L., pp. 152–153.

> proliferation of international courts gives rise to a serious risk of conflicting jurisprudence, as the same rule of law might be given different interpretations in different cases.[72]

Consequently, lawyers have questioned whether the rising number of international courts and tribunals pose a threat to the international legal system.[73] It has been argued that different interpretations on similar rules might threaten the perception of the existence of an international legal system and, if similar cases are not solved in the same means constantly, 'the very essence of a normative system of law will be lost.'[74] In this context, conflicting interpretations of courts and tribunals might lead to the fragmentation of substantive international law and threaten the credibility of international law and its unity. Moreover, different pronouncements by courts and tribunals on the same issue of law might undermine the overall legitimacy of the international legal system and be harmful to the reliability of international adjudication.[75] This could potentially make the enforcement of decisions more problematic than ever. Former ICJ presidents[76] and some authors[77] have argued that the proliferation also threatens the role of ICJ as a primary judicial organ of the UN.

In this regard, the possibility of two major problems can be identified. Firstly, different international judicial bodies might produce different jurisprudences by interpreting the same rule of law differently. Secondly, two different international courts or tribunals might have jurisdiction on the same matter leading to different judgments on the same case. These possible problems will be examined in light of the aforementioned concerns by referring to some highly cited cases.

To begin with, international judicial bodies can interpret the same legal concepts or rules differently since there is no formal system between the international courts and tribunals or no *stare decisis* principle in international judicial dispute settlement. This means neither a judicial body's own previous decisions nor other courts and tribunals pronouncements have binding force over a courts or tribunal's

---

[72]Statement of Judge Gilbert Guillaume, President of the International Court of Justice, to the UN General Assembly of 26.10.2000, http://www.icj-cij.org/court/index.php?pr=84&pt=3&p1=1&p2=3&p3=1. Accessed on 07.08.2014.

[73]Spelliscy, 'The Proliferation of International Tribunals: A Chink in the Armor', *supra* note 71, pp. 152–156.

[74]Charney, J.I. (1998) 'The Impact on the International Legal System of the Growth of International Courts and Tribunals', 31 N.Y.U. J. INT'L L., p. 699 http://www.pict-pcti.org/publications/PICT_articles/JILP/Charney.pdf. Accessed on 07.08.2014.

[75]Spelliscy, 'The Proliferation of International Tribunals: A Chink in the Armor', *supra* note 71, pp. 152–156.

[76]Statement of Judge Stephen M. Schwebel, President of the International Court of Justice, to the Plenary session of the General Assembly of the UN 26.10.1999. http://www.icj-cij.org/court/index.php?pr=87&pt=3&p1=1&p2=3&p3=1&PHPSESSID. Accessed on 07.08.2014.; Statement of Judge Gilbert Guillaume, President of the International Court of Justice, to the UN General Assembly of 26.10.2000, http://www.icj-cij.org/court/index.php?pr=84&pt=3&p1=1&p2=3&p3=1. Accessed on 07.08.2014.

[77]Abi-Saab, G. (1998). 'Fragmentation or unification: some concluding remarks' 31 NYUJ Int'l L. & Pol, p. 791.

decision. Consequently, it has been claimed that international law is facing danger of fragmentation. To illustrate the reality of this danger, scholars and the former ICJ presidents have cited some cases in which they claim this problem can be seen.

Perhaps the most cited example is the International Criminal Tribunal for the Former Yugoslavia's (ICTY) disagreement in the *Tadic* case with the ICJ's previous *Nicaragua* pronouncement on an aspect of state responsibility.[78] In fact, this was not the first time that the ICTY had taken a different position to the ICJ. In 1996, while the ICJ stated, in its *Nuclear Weapons* advisory opinion, that an armed reprisal during an armed conflict should comply with the principle of proportionality,[79] in the *Martic* case the ICTY decided that armed reprisals are fully prohibited in international law.[80] Both bodies also took different approaches on the issue of whether it is possible to examine the legality of UN Security Council resolutions. While the ICJ avoided assessing of the legality of Security Council resolutions in the *Lockerbie* case,[81] in the *Tadic* case, the Appeal Chamber of the ICTY directly examined the legality of its own creation, which was through a Security Council resolution.[82]

As mention, the decision that has been constantly referred to by scholars as well as by the former ICJ president Guillaume is the ICTY's final judgement in the *Tadic* case. In its final decision, the Appeal Chamber of the ICTY explicitly referred to the ICJ's *Nicaragua* decision and put forward a comprehensive analysis of the ICJ's jurisprudence.[83] In *Nicaragua*, the ICJ had decided that the United States had no responsibility for the acts of the 'contras,' the acts themselves had violated international humanitarian law. In its reasoning, the Court stated that the fact that the United States trained and financed the 'contras' was not ground for its responsibility. According to the ICJ, in order to confer state responsibility, there must have been 'effective control' over the specific operation in which the violation

[78] *Prosecutor v Dusko Tadic*, Judgement, Case No. IT–94–A-1, 15.07.1999. http://www.icty.org/x/cases/tadic/acjug/en/tad-aj990715e.pdf. Accessed on 04.08.2014. (Hereinafter 'Tadic'); *Case Concerning Military and Paramilitary Activities in and against Nicaragua* (Nicaragua v US) (Merits) ICJ rep. 14 1986. http://www.icj-cij.org/docket/files/70/6503.pdf. Accessed on 11.08.2014. (Hereinafter 'Nicaragua Case')

[79] *The Legality of Threat or Use of Nuclear Weapons*, (Advisory Opinion) 1996 ICJ rep. 246, at para 46. http://www.icj-cij.org/docket/files/95/7495.pdf. Accessed on 04.08.2014.

[80] *The Prosecutor v. Milan Martic*, Case No. IT-95-11-R61, T. Ch I, Decision on 08.03.1996, at para 17. http://www.icty.org/x/cases/martic/tdec/en/960308.pdf, Accessed on 04.08.2014.

[81] *Case Concerning Questions of Interpretation and Application of the 1971 Montreal Convention arising from the Aerial Incident at Lockerbie* (Libyan Arab Jamahiriya v. United States of America), (Judgement) 27.02.1998, at paras 39–41. http://www.icj-cij.org/docket/files/89/7249.pdf. Accessed on 04.08.2014.

[82] *The Prosecutor v. Dusko Tadic*, Decision on the Defence Motion for Interlocutory Appeal on Jurisdiction, Case No. IT-94-1, A.Ch., 2 October 1995, para. 20, http://www.icty.org/x/cases/tadic/acdec/en/51002.htm. Accessed on 04.08.2014.

[83] *Tadic*, *supra* note 82, para. 99–145.

was committed.[84] The ICTY disregarded the ICJ's 'effective control' test by finding its reasoning unpersuasive. The ICTY stated that, at least for organized military groups, it was sufficient to have 'overall control.'[85] This approach broadened the scope of responsibility with respect to states' extraterritorial activities compared to the ICJ's jurisprudence. In its *Celebici* decision, the ICTY explicitly stated that no hierarchical relationship existed between the ICTY and the ICJ, and therefore, the Court stressed that the ICTY is not bound by the ICJ's jurisprudence.[86]

Another example that has frequently been used to point out the 'danger of fragmentation' is the European Courts of Human Rights' (ECtHR) judgement in the *Loizidou v. Turkey* case.[87] In that case, the ECtHR discussed the validity of the reservations Turkey made in its declaration accepting the compulsory jurisdiction of the Court. Turkey argued the validity of its reservations under articles 26 and 45 of the Convention. These two articles correspond to article 36 of the ICJ Statute. An exactly similar situation has never arisen before the ICJ; however, in one of its advisory opinions, the ICJ stated that if any reservation were compatible with the object and purpose of the related convention, a reserving state would still be a party to the convention. Moreover, the Court pointed out that each reservation should be tested on its own standpoint.[88] However, in its *Loizidou* judgement, the ECtHR emphasized that it has a different legal nature to the ICJ:

> (...) unlike the Convention institutions, the role of the International Court is not exclusively limited to direct supervisory functions in respect of a law-making treaty such as the Convention. Such a fundamental difference in the role and purpose of the respective tribunals, coupled with the existence of a practice of unconditional acceptance (...) provides a compelling basis for distinguishing Convention practice from that of the International Court.[89]

The ECtHR found Turkey's reservation impermissible by stating that provisions of the Convention should 'be interpreted and applied so as to make its safeguards practical and effective.'[90] The ECtHR decision differed from the ICJ's jurisprudence and took a more restrictive interpretation regarding reservations whilst pointing out the different nature of Convention and the ECtHR to the ICJ.

---

[84] *Nicaragua Case*, *supra* note 78, para. 114–115.

[85] *Tadic Case*, *supra* note 82, para. 137.

[86] *The Prosecutor v. Zdravko Mucic et al.* ('Celebici Case'), (Decision), Case No. IT-96-21-A, A.Ch., 20.02.2001, http://www.icty.org/x/cases/mucic/acjug/en/cel-aj010220.pdf. Accessed on 04.08.2014.

[87] *Loizidou v. Turkey*, Preliminary Objections, Decision of 23.03.1995, http://hudoc.echr.coe.int/sites/eng/pages/search.aspx?i=001-57920#"itemid":["001-57920"]. Accessed on 04.08.2014.

[88] *Reservations to the Convention on the Prevention and Punishment of the Crime of Genocide* (Advisory Opinion), 28.05.1951, at para. 26, http://www.icj-cij.org/docket/files/12/4283.pdf. Accessed on 04.08.2014.

[89] *Loizidou v. Turkey*, *supra* note 87, para. 84–85.

[90] *Ibid.* para.72.

The second possible problem with the 'uncoordinated proliferation' is that different international courts and tribunals might have jurisdiction on the same matter, possibly leading to two or more judicial bodies making different decisions on the same issue. This potentially peculiar situation might endanger the legitimacy and enforceability of both verdicts. The *Swordfish* Case has been used as an example of this situation.[91] In the *Swordfish* case, Chile had closed its ports to the ships of the EU Member states and impeded EU vessels' imports to Chile. The EU claimed that the act had violated both UNCLOS rules and the GATT. Therefore, the disagreement came before both of the ITLOS and a WTO panel. [92] However, these processes were both suspended due to a peaceful agreement taken between the involved states. This case exemplifies that the same type of situation may arise in the future. Whilst the WTO panel and the ITLOS would be applying different sets of rules and assessing the cases from different perspectives, the possibility of jurisdictional conflicts and different decisions being made still caused concerns.

All these cases are used to exemplify the risks of conflicting jurisprudences and jurisdictions of the courts and tribunals. It has been claimed that situations like these can harm confidence in the courts and tribunals, the coherence of the system as a whole and the legitimacy of the specific tribunal.[93] These concerns have some valid points, however, it is questionable whether the situation is as dramatic as described by the aforementioned writers. Is the danger as big as the former ICJ presidents described? Do we need to have 'second thoughts' when establishing new courts due to 'proliferation' concerns? The authors of the article believe that the complex adaptive properties of the international legal system might dispel these concerns.

30.4 Considering International Courts and Tribunals in the Context of Complex Adaptive Systems: Dispelling the Fears Regarding 'Uncoordinated Proliferation'

As explained, the 'uncoordinated proliferation' of international courts and tribunals has led to some anxieties among international law scholars. Former ICJ presidents and some authors have stated that the 'unity' and 'coherence' of international law

[91] Oellers-Frahm, K. (2001). 'Multiplication of International Courts and Tribunals and Conflicting Jurisdiction – Problems and Possible Solutions', 5 Max Planck 68 UNYB, p. 86. http://www.mpil.de/files/pdf1/mpunyb_oellers_frahm_5.pdf. Accessed on 04.08.2014.

[92] See generally. Stoll, T. (2002). 'The Swordfish Case: Law of the Sea v. Trade', 62(1)–(2) ZAÖRV, pp. 21–36.

[93] Spelliscy, 'The Proliferation of International Tribunals: A Chink in the Armor', *supra* note 71, pp. 168–172.

and international courts and tribunals might be threatened by the 'proliferation.'[94] However, it is very doubtful whether unity or coherence has ever existed in the international legal system. If the words 'unity' and 'coherence' refer to a formal structure that has been designed and governed by rules and a set of bodies that have a definite hierarchical relationship, this type of a 'unity' or 'coherence' have never existed in the international legal system. On the other hand, if the words are intended to refer to a united mind, where all the parties have a similar approach and understanding towards the international legal system, this is also hard to find.

According to Frahm, the real hazard is to the 'consistency' and 'cohesiveness' of international law rather than to its 'unity.' He describes 'consistency' and 'cohesiveness' as the application and interpretation of international rules on the basis of legitimacy and formal standards.[95] However, it is also doubtful whether this kind of consistency and cohesiveness has ever existed.[96] In terms of substantive rules of international law, there are only a limited set of rules that have been universalized, namely *erga omnes* obligations and *jus cogens* rules.[97] There is also no uniformity in the procedural rules of the international institutions and bodies. Moreover, the relationship between the coherence of international law and its legitimacy is not linear. Therefore, the unity and coherence of international law is more likely an academic idea rather than a political reality.[98]

In this regard, approaching the international legal system as a complex adaptive system allows one to better understand and assess the current structure of international courts and tribunals. This approach also leads to a better understanding regarding any future reforms. Thus, it shall be argued that international adjudicative bodies are more efficient as they become more compatible with the complex adaptive structure of the international legal order. Moreover, considering the aforementioned anxieties through the lens of complex adaptive systems theory enables further assessment to be made regarding these concerns.

To begin with, fear of a chaotic system due to uncoordinated proliferation and fragmentation seems unwarranted since the international legal system exists in the *edge of chaos* because of its dynamical, complex and decentralized nature. 'Seeking 'the edge of chaos' is not seeking disorder or randomness but the right balance between order and flexibility. This perspective should give pause to lawyers, generally critical of fragmentation and decentralization, and intuitively in search

---

[94]Guillaume, G, (1995). 'The Future of International Judicial Institutions', 44 ICLQ, p. 862. http://journals.cambridge.org/action/displayAbstract?fromPage=online&aid=1519888&fileId=S0020589300058358. Accessed on 04.08.2014.

[95]Oellers-Frahm, 'Multiplication of International Courts and Tribunals', *supra* note 91, p. 73.

[96]Dupuy, P. (1998). 'The Danger of Fragmentation or Unification of the International Legal System and the International Court of Justice' 31 N.Y.U. J. Int'l L. & Pol., p. 792, http://www.pict-pcti.org/publications/PICT_articles/JILP/Dupuy.pdf. Accessed on 11.08.2014.

[97]*Ibid.* pp. 794,795.

[98]Koskenniemi, 'Fragmentation of International Law?', *supra* note 60, p. 558.

of order and central authority'.[99] International law is derived from a pragmatic point of view. It does not have a constitution to reflect the common purpose of states or humanity. A higher organ, acting in a systematic manner, has not created international law. Instead, international law is created by several international agreements that cover different legal issues and different actors. If one uses the terminology of complex adaptive system theory to describe this, international law, as an outcome of the international legal system, is created and affected by the *non-linear* relationships of various *heterogeneous* actors, which have many different aims and interests. Therefore, the purpose of international courts and tribunals is not to maintain the coherence of international law, but to fulfil the roles that are laid out in their specific mandates. International actors are free to accept their jurisdiction as they please.

If a state or another actor needs to refer to an international judicial body at any point, it has total freedom to do so from the beginning.[100] Of course, if actors A and B bring their case before different courts at the same time, this might create conflicting interpretations on the same issue. However, international courts or tribunals either have a different *ratione materiae, ratione loci* or set of rules to apply. Thus, whilst different branches of international law will overlap, this does not make their perspectives the same.[101] Kingsbury argues that 'the law and practice concerning provisional measures in international tribunals is somewhat chaotic (. . .)' by exemplifying the different approaches of the ICJ, the ECtHR, the Inter-American Court of Human Rights and the ITLOS.[102] On the face of it, the anxiety regarding conflicting jurisprudences seems to be valid. However, international judicial bodies have different legal perspectives, mandates, subject matters and jurisdictions. In other words, they also reflect the *heterogeneity* of international legal order. Therefore, if the same matter goes to different bodies at the same time, each one of these bodies will apply a different set of rules and interpret the law according to their own mandates, traditions and aims.

The ICTY Trial Chamber stressed this issue while it was considering whether the ICTY should follow the ICJ's jurisprudence. The Chamber stated that while the ICJ examined the state's responsibility, the ICTY was examining the individual's

---

[99] Pauwelyn, 'At the Edge of Chaos?', *supra* note 16, p. 381.

[100] Reed, L. (2002)., 'Great Expectations: Where Does the Proliferation of International Dispute Resolution Tribunals Leave International Law', 96 AJIL, p. 229 http://heinonline.org/HOL/Page?handle=hein.journals/asilp96&div=26&g_sent=1&collection=journals#231. Accessed on 04.08.2014.

[101] Pauwelyn, J. (2004). 'Bridging Fragmentation and Unity: International Law as a Universe of Inter-Connected Islands', 25 Mich. J. Int'l L. 904. http://heinonline.org/HOL/Page?handle=hein.journals/mjil25&div=33&g_sent=1. Accessed on 04.08.2014.

[102] Kingsbury, B. (1999). 'Foreword: Is the Proliferation of International Courts and Tribunals a Systemic Problem?', 31 N.Y.U. J. INT'L L. & POL., pp. 684,685. http://www.pict-pcti.org/publications/PICT_articles/JILP/Kingsbury.pdf. Accessed on 04.08.2014.

acts and responsibility. Thus, it rejected the existence of any real contradiction.[103] The Vienna Convention on the Law of Treaties[104] already provides a basic set of rules to settle normative conflicts between treaties, such as *lex specialis* and *lex posterior*. In terms of jurisdictional conflicts, as long as courts and tribunals stay within their competence and apply traditional international legal reasoning, lack of formal unity or coherence does not pose a threat to the international legal order.[105] In addition, since none of the previous international court and tribunal decisions are formally binding, even for the same body, it is peculiar to expect jurisprudence not to conflict.

Moreover, due to the complex properties of the international legal system, establishment of a systematic and formal uniformity seems unlikely. This, however, does not pose a threat to the entire judicial system of international law. A totally fragmented system is unlikely because of the *network connectivity* and *feedback loops* present in the international legal system. Even if the sets of rules that courts and tribunals apply differ from one another, they are all part of the same root, international law. As Chaney states, international courts and tribunals generally have the same understanding about the core areas of international law.[106] There is no legitimate reason to claim that the international courts and tribunals will split up from the fundamental principles of international law. Therefore, the *decentralized* structure of international courts, tribunals and laws does not necessarily mean that the system is a totally fragmented one.

In addition, the inter-judicial dialogue between international courts and tribunals and between these bodies and national courts is growing. 'The cases of institutional interactions typically involve the flow of information. Treaty and administrative bodies exchange information, both formally and informally, on shared substantive issues. They share reviews and lessons learned regarding their functioning and frequently consult each other on administrative or legal issues that arise.'[107] Despite the fact that international courts and tribunals are not bound by their own decisions or the decisions of others, they have the tendency to cite both it their judgements. This does not mean that courts and tribunals will follow one another's jurisprudence, but this evidences that they recognize one another's perspective. This is one of the instances of *feedback loops* present in international adjudication. The ICTY's approach in the *Tadic* case is a perfect example of this. Whilst the ICTY did not follow the ICJ's jurisprudence, it took it into consideration, discussed it and

[103] The Prosecutor v. Zigic et al. 'Decision on the Defence "Motion regarding Concurrent Procedures before International Criminal Tribunal for the Former Yugoslavia and International Court of Justice on the Same Questions', Case No. IT-98-30/1, T.Ch., 5 December 2000. http://www.icty.org/x/cases/kvocka/tdec/en/01205JN114797.htm. Accessed on 04.08.2014.

[104] Vienna Convention of the Law of Treaties (1969).

[105] Buergenthal, T. (2001). 'Proliferation of International Courts and Tribunals: Is It Good or Bad?' 14 Leiden Journal of International Law 267, p. 273.

[106] Charney, 'The Impact on the International. Legal System', *supra* note 74, p. 699.

[107] Kim, 'International environmental law as a complex adaptive system', *supra* note 22, p. 11.

reasoned why the ICTY would follow a different path. Nowadays, international courts and tribunals are cognisant of one another's views and decisions. Moreover, national courts have been citing international courts and tribunals' views and application of the law more than ever.[108] Perhaps the most striking example of this was the citation of an ECtHR decision by a United States Court.[109]

One should accept that international legal system is a pluralist one. There is diversity in the choices and approaches of different actors but, as stated above, the difference between the normative systems of the international law bodies and the difference in their interpretations and the interpretations of courts or tribunals only indicates a lack of formal hierarchy, but not a chaos or hazard in the system. As Burke-White puts it;

> (. . .) the pluralist conception of the international legal system recognizes – and possibly thrives on – the diversity of the system. A wide range of courts will interpret, apply, and develop the corpus of international law. States will face differing sets of obligations that may even be interpreted differently by various tribunals and may at times conflict. Possibly most significantly, national and international legal processes will interact and influence one another, resulting in new hybrid procedures, rules, and courts. Yet, these developments will occur within a common system of international law engaged in a constructive and self-referential dialogue (. . .)[110]

Burke-White's comment covers some complex properties of the international legal system and international adjudication such as the *diversity* of the actors, laws and institutions; the *decentralized* composition of the system; and the *non-linear* interactions.

His comments also put forth another important complex property of the international legal system and international courts and tribunals. Similar to other decentralized systems, the international legal system has *self-organization* qualities that 'have emerged through the interaction of its constituent components.'[111] 'Centralized institutions (. . .) run counter to the principle of requisite variety, lack sufficient flexibility, and inhibit random mutations. On the contrary, decentralized institutions (. . .) have diverse components and are constantly changing through self-organization.'[112] Approaching the international legal system as a complex adaptive system allows us to better understand how the system generally operates, as a *self-organizing* system that is built up of many interacting actors, treaties, laws and institutions. It allows order to appear in the absence of any higher institution or power specifically designed to provide it. This therefore leads to a

[108]Burke-White, W. (2004). 'International Legal Pluralism', 25 MIJL, p. 972. https://www.law.upenn.edu/cf/faculty/wburkewh/workingpapers/25MichJIntlL963(2005).pdf. Accessed on 04.08.2014).

[109]Ibid. p. 972; Lawrence v. Texas, 539 U.S. 558, 573 (2003).

[110]Burke-White, 'International Legal Pluralism', *supra* note 108, p. 978.

[111]Pauwelyn, 'At the Edge of Chaos?', *supra* note 16, p. 379.

[112]Kim, 'International environmental law as a complex adaptive system', *supra* note 22, p. 17.

complex adjudicative structure and adjudicative institutions that also exhibit *self-organizational* properties.

International courts and tribunals emerge from a series of minor or major historical events. Despite the fact that international courts and tribunals are individually designed by humans rather than having transpired organically, the underlying reasons for their establishment are the many complex social and political interactions. Thus, there are certain indicators that international law and adjudication exhibits the same *self-organized* nature of the international legal system.

The first indicator of this *self-organized* nature is that, in some areas of international law such as international investment law or international environmental law, a substantive framework has not arisen from institutional decision-making or a central international agreement, but has arisen instead from several multilateral and bilateral agreements.[113] 'The overall structure has incrementally evolved from, and is continuously shaped and reshaped by, the numerous decentralized decisions taken within individual institutions and the interaction effects arising therefrom.'[114] As a result, the system itself creates a decentralized and flexible dispute settlement understanding. Therefore, several different institutions and adjudicative and quasi-adjudicative bodies determine the substantive normative standards of these fields of international law. Of course, this set-up is not operative for some other fields of law such as human rights law or international criminal law since their normative nature does not feasibly lend itself to such a structure.

The second indicator is that multilateral regimes and their dispute settlement mechanisms have developed over time through trial and error.[115] Therefore, substantive laws, institutional powers, and jurisprudences have evolved through the agreements of several different actors in response to changes in the international legal system. To illustrate, the normative context of international criminal law has changed constantly over time. It is clear that the jurisprudence, substantive law and the structure of international criminal courts and tribunals (such as the Nuremberg Tribunal, ICTY, ICTR and ICC) are different from each other.[116] In this sense, the *network connectivity of feedback* leads to changes in the system. It does not matter that the change is deliberately imposed by parties rather than by an organic emergence. This is because various parties of the international community would have responded to feedbacks that they received from the system, and thus felt the need for a change in order to make the system more effective and stable. In other words, the parties did not create laws artificially but, instead, norms emerged from the societies and their interactions. For example, the concept of genocide

---

[113]See. Pauwelyn, 'At the Edge of Chaos?' *supra* note 16; Kim, 'International environmental law as a complex adaptive system' *supra* note 22.

[114]Kim, 'International environmental law as a complex adaptive system', *supra* note 22, p. 13.

[115]Bodansky, D. and Diringer, E. (2010). *The evolution of multilateral regimes: implications for climate change.* Arlington: Pew Center on Global Climate Change.

[116]Cryer, R. et al., (2010). *An Introduction to International Criminal Law and Procedure,* Second Edition, Cambridge.

was introduced to law only after World War II[117] and the nexus between crimes against humanity and war was removed from substantive law only after the system identified that there was no need for a war to be occurring in order for a crimes against humanity to be committed.[118]

The third indicator of this *self-organization* is that, 'more subtle and policy-driven changes in existing law may arise through the process of interpretation reflecting the notion that treaties are living instruments that should be interpreted in light of contemporary conditions. Article 31(3)(c) of the 1969 Vienna Convention on the Law of Treaties provides a powerful means in this regard. It requires that the interpreter of a treaty takes into account any relevant rules of international law applicable in relationships between the parties, and it may include other treaties, customary rules, or general principles of law. This dynamic approach to interpreting treaties provides additional adaptiveness in a way that builds a more coherent system.'[119] In fact, international courts and tribunals do not have the formal capacity to create or amend international law. However, it is certain that their interpretations on norms, in light of contemporary conditions, have an impact on both the development and creation of law (particularly customary international law).[120] There are some examples of this in international law's history such as the introduction of *erga omnes* obligations by the ICJ[121] or the emergence of the margin of appreciation doctrine in the jurisprudence of the ECtHR.[122] Thus, substantive law and the application of norms might be interpreted and re-interpreted by adjudicative bodies in order to adapt them to a current situation, and an emerging norm might have an impact on another actor, institution and the international legal system as a whole.

In light of all this, it can be argued that international adjudicative bodies should be designed to be more adaptive and flexible in order to be more efficient in a system that has complex properties. A complex adaptive system itself utilizes more adaptive

---

[117]Lemkin, R. (1944) *Axis Rule in Occupied Europe*: *Laws of Occupation – Analysis of Government – Proposals for Redress*., Washington, D.C.: Carnegie Endowment for International Peace.; Convention for the Prevention and Punishment of the Crime of Genocide, adopted Dec.9, 1948, 78. U.N.T.S. 277 (entered into force Jan. 12,1951) (hereinafter 'Genocide Convention'). https://treaties.un.org/doc/Publication/UNTS/Volume%2078/volume-78-I-1021-English.pdf. Accessed on 21.08.2014.

[118]Cryer et al., *An Introduction to International Criminal Law and Procedure, supra* note 116.

[119]Kim, 'International environmental law as a complex adaptive system', *supra* note 22, p. 15.

[120]Merrills, J. (1993) *The Development of International Law by the European Court of Human Rights* 2nd ed. Manchester University Press; Charney, J. (1998). 'Impact on the International Legal System of the Growth of International Courts and Tribunals', 31 NYUJ Int'l L. & Pol.; Lauterpacht, H. (1982).(ed.), *The development of international law by the international court*, Cambridge University Press.

[121]Tams, C., and Tzanakopoulos, A. (2010). 'Barcelona traction at 40: the ICJ as an agent of legal development', 23 Leiden Journal of International, pp. 781–800.

[122]Council of Europe, 'Margin of Appreciation', Para 1. http://www.coe.int/t/dghl/cooperation/lisbonnetwork/themis/echr/paper2_en.asp. Accessed on 07.03.2014.

institutions and shows resilience against rigidly designed ones. In this regard, it is not necessarily important whether international adjudicative institutions are organized around a single multilateral treaty, a central international organization, or by many different elements. What is important is whether adjudicative institutions and bodies are sufficiently flexible and are able to react to changes in the international order and legal system.

In this context, Kim and Mackey ask an important question while discussing the complexity of international environmental law; 'each treaty or institution may be capable of learning from the experiences of its state members in applying negotiated rules, but what about the system of treaties and institutions as a whole?'[123] This question is also valid for this article and is the underlying reason as to why the international legal system and international law and adjudication have been considered together. International law and adjudication coevolves with international relations and the international legal system in order to respond to feedbacks and the interactions of its social environment and to be able to solve problems more effectively. Of course, as exemplified in the following paragraphs, the adaptive capacity of different international institutions and adjudicative bodies varies since their legal aims and norms are different. However, in general, claiming that there is a completely distinct institution or legal framework to the all parts of the international legal system is not accurate. A change in international law or the legal order can have an impact on other actors and institutions. Disagreements on international labour law issues between the US and China might have an impact on the substance of an international investment agreement, and the emergence of a concept in international environmental law can affect international human rights or international trade law jurisprudences.

Fear of an inefficient system due to fragmented international law and the uncoordinatedly proliferated courts and tribunals appears to be an inaccurate consideration in the light of complex adaptive systems theory. The proliferation of institutions and their decentralized nature in the international legal system does not necessarily imply anarchy or chaos. On the contrary, in a complex adaptive system, more centralization can have negative impacts on the adaptation capacity of the institutions and laws, as well as on their effectiveness.

The ICJ and the ITLOS might be the best examples of adjudicative results stemming from 'deeply pathological regimes riddled with birth defects in need of drastically more coherence and structure'[124] The establishment of these international courts and tribunals was an outcome of the idea of controlling the system using a top-to-bottom approach and creating a unified international dispute understanding. The premise of the ICJ was for it to be a 'world court' under the organization of the UN, with the capacity to solve any type of disagreement between its parties. However, to date, only 71 states have recognized the jurisdiction of the Court as compulsory, and most of these states have made many reservations to their

[123]Kim, 'International environmental law as a complex adaptive system', *supra* note 22, p. 15.

[124]Pauwelyn, 'At the Edge of Chaos?', *supra* note 16, p. 381.

declarations.[125] The caseload of the court has always been an issue; the ICJ has only heard 2.4 cases per year since its establishment.[126] In addition, the implementation of the Court's judgments has proved to be problematic in cases such as *Tehran Hostages*[127] or *Nicaragua*.[128] In a similar vein, the ITLOS has suffered from very comparable problems. While it was created by the mandate of the United Nations Conference on the Law of the Sea, it has been referred to rarely; only 22 cases have ever been submitted to the ITLOS, of which most are in regards to prompt release or provisional measures.[129]

Comparatively, when adjudicative regimes are more decentralized and flexible, they become more efficient and successful in terms of their caseload and their ability to solve the disputes. For instance, the ECtHR is also an outcome of a pivotal multinational treaty like ITLOS. However, certain adaptive properties of the ECtHR make it much more effective in terms of caseload and the enforcement of its decisions. For example, in response to feedbacks from the system, parties of the ECHR constantly amend law via protocols to make the court more effective. Moreover, the ECtHR has developed an important jurisprudence, the so-called 'margin of appreciation' doctrine. This doctrine refers to 'the space for manoeuvre that the Strasbourg organs are willing to grant national authorities, in fulfilling their obligations under the Convention.'[130] It allows the ECtHR to take into account the fact that the Convention will be interpreted differently in different member states due to their varying judicial and sociological structures and public interests. Today, the ECtHR is accepted as being one of the most efficient international courts.

In the some other areas of international law, complex and adaptive properties are more prominent. For example, in international investment law, thousands of treaties, customary laws, national laws, agreements and arbitration awards co-exist without a central mechanism or multinational treaty controlling the system.[131] This decentralized structure and constant interaction between various actors enables the

[125]Declarations Recognizing the Jurisdiction of the ICJ as Compulsory, http://www.icj-cij.org/jurisdiction/?p1=5&p2=1&p3=3. Accessed on 07.08.2014.

[126]Cases Section, International Court Of Justice, (7.08.2014). http://www.icj-cij.org/docket/index.php?p1=3, Accessed on 11.08.2014.

[127]United States Diplomatic and Consular Staff in Tehran (USA v. Iran) (Judgement), the ICJ (1980), http://www.icj-cij.org/docket/files/64/6291.pdf. Accessed on 11.08.2014. (Hereinafter 'Tehran Hostages case').

[128]Case Concerning Military and Paramilitary Activities in and against Nicaragua (Nicaragua v US) (Merits) ICJ rep. 14 1986. http://www.icj-cij.org/docket/files/70/6503.pdf (Accessed on 11.08.2014) (Hereinafter 'Nicaragua Case').

[129]The List of Cases, International Tribunal for the Law of the Sea (7.08.2014). http://www.itlos.org/index.php?id=35. Accessed on 11.08.2014.; See generally, Born, G., (2012). 'A New Generation of International Adjudication', 61 Duke Law Journal.

[130]Council of Europe, 'Margin of Appreciation', Para 1. http://www.coe.int/t/dghl/cooperation/lisbonnetwork/themis/echr/paper2_en.asp. Accessed on 07.03.2014.

[131]See. Pauwelyn, 'At the Edge of Chaos?', *supra* note 16.

law to constantly adapt itself in reaction to the needs of all the components.[132] Of course, the very nature of this field of international law is different. It would not be suitable to create an international investment law like dispute settlement understanding for international human rights, international criminal law or the law of the sea since these fields require different approaches regarding their institutions and some of their significant norms are not flexible. Nonetheless, it still seems vital to make these laws and institutions as adaptive as possible to the complex nature of the international legal system in order to make them more efficient.

Nevertheless, this varied nature of the various international adjudicative bodies is a natural consequence of the fact that the international legal system operates in the *edge of chaos* and is constantly seeking the perfect balance between order and flexibility. Parties of international adjudication are looking to maximize their interests and gain more control over dispute settlement processes. At the same time, there is a need for order and predictability in the international legal system to avoid chaos. Thus, whilst some international adjudicative bodies can strike this balance such as the WTO dispute settlement mechanisms or ICSID arbitrations, international bodies, like the ITLOS, which try to create a rigid top-to-down mechanism, suffer from stagnation and deadlock. As it is stated by Pauwelyn 'resilience of the system questions the absolute need for a controlling multilateral institution or dramatically more centralization'[133] Therefore, in terms of international legal system and its sub-part international adjudication, seeking to continue operating in *the edge of chaos* is more effective and sustainable in the long run.

In light of the above analysis, the benefit international law and adjudication gains from highlighting the complex adaptive properties of the international legal system can be seen. Despite the fact that the international legal system has a fragmented structure in terms of its institutions and norms, it is not a completely chaotic, randomly organized collection of norms and institutions. On the contrary, this decentralized composition of norms and institutions creates a more flexible and effective system for the interactions between equally sovereign independent states and between them and other actors of international relations. 'Change in social systems is very often the specific intent of human intervention, in which case knowing how the system responds to change should be an important factor in the design of the instrument of change.'[134] Given this, international law and its courts and tribunals should be designed around the complex adaptive system properties of the international legal system.

As Belcher emphasises, 'in the consideration of the agents that comprise the legal system where a legal reform program can go awry – most commonly as an error of exclusion as there are numerous secondary and tertiary interested parties that

---

[132] Alvarez, J. (2009). 'A BIT on Custom', 42 New York University Journal of International Law & Policy, p. 80.

[133] Pauwelyn, 'At the Edge of Chaos?', *supra* note 16, p. 378.

[134] Ruhl, 'Law's Complexity', *supra* note 5, p. 891.

are frequently overlooked during reforms.'[135] 'Law, as a complex adaptive system, coevolves with the social systems it aims to regulate, and thereby induces changes on itself. (...) therefore, the theory of law's complexity counsels us to design law to think like a complex adaptive system.'[136] Thus, various agents and possibilities should be considered when international law is designed. To this end, organizational ties should be strengthened, such as the duty to cooperate and coordinate among treaty bodies or other institutional entities. Moreover, international courts and tribunals both in terms of their procedures and jurisprudence should search for the right balance of stability and flexibility.

For a long time, international lawyers have complained about the weak nature of the international judiciary.[137] Therefore, it is somehow peculiar to witness these 'proliferation' discussions. The 'proliferation' of international courts and tribunals is not a real danger, but instead an opportunity to strengthen the international legal system. As Buergenthal states 'the proliferation of international tribunals with specialized and regional competence has in recent decades enabled governments to experiment with and observe the effects of international adjudication involving states and their acceptance of the jurisdiction of international tribunals.'[138] In this sense, the complex adaptive properties of the international legal system indicates that different ideas and experiments are not a hazard for international law, but an opportunity to improve normative systems or to create better ones.

References

Cases

1. Case Concerning Military and Paramilitary Activities in and against Nicaragua (Nicaragua v US) (Merits) ICJ rep. 14 1986. http://www.icj-cij.org/docket/files/70/6503.pdf. Accessed 11 Aug 2014
2. Case Concerning Questions of Interpretation and Application of the 1971 Montreal Convention arising from the Aerial Incident at Lockerbie (Libyan Arab Jamahiriya v. United States of America), (Judgement) 27.02.1998. http://www.icj-cij.org/docket/files/89/7249.pdf. Accessed 4 Aug 2014
3. Lawrence v. Texas, 539 U.S. 558, 573 (2003)
4. Loizidou v. Turkey, Preliminary Objections, Decision of 23.03.1995, http://hudoc.echr.coe.int/sites/eng/pages/search.aspx?i=001-57920#"itemid":["001-57920"]. Accessed 4 Aug 2014

[135] Belcher, 'International Legal Development', *supra* note 1, p. 197.

[136] Ruhl, 'Law's Complexity', *supra* note 5, p. 908.

[137] Shany, Y. (2009). 'No Longer a Weak Department of Power? Reflections on the Emergence of a New International Judiciary' EJIL, Vol. 20 No. 1, 73–91, pp. 74–75, doi: 10.1093/ejil/chn081

[138] Buergenthal, 'Proliferation of International Courts and Tribunals', *supra* note 105, p. 272.

5. Prosecutor v Dusko Tadic, Judgement, Case No. IT–94–A-1, 15.07.1999. http://www.icty.org/x/cases/tadic/acjug/en/tad-aj990715e.pdf. Accessed 4 Aug 2014
6. Reservations to the Convention on the Prevention and Punishment of the Crime of Genocide (Advisory Opinion), 28.05.1951. http://www.icj-cij.org/docket/files/12/4283.pdf. Accessed 4 Aug 2014
7. The Legality of Threat or Use of Nuclear Weapons, (Advisory Opinion) 1996 ICJ rep. 246. http://www.icj-cij.org/docket/files/95/7495.pdf. Accessed 4 Aug 2014
8. The Prosecutor v. Dusko Tadic, Decision on the Defence Motion for Interlocutory Appeal on Jurisdiction, Case No. IT-94-1, A.Ch., 2 October 1995. http://www.icty.org/x/cases/tadic/acdec/en/51002.htm. Accessed 4 Aug 2014
9. The Prosecutor v. Milan Martic, Case No. IT-95-11-R61, T. Ch I, Decision on 08.03.1996. http://www.icty.org/x/cases/martic/tdec/en/960308.pdf. Accessed 4 Aug 2014
10. The Prosecutor v. Zdravko Mucic et al. ('Celebici Case'), (Decision), Case No. IT-96-21-A, A.Ch., 20.02.2001, http://www.icty.org/x/cases/mucic/acjug/en/cel-aj010220.pdf. Accessed 4 Aug 2014
11. The Prosecutor v. Zigic et al. 'Decision on the Defence "Motion regarding Concurrent Procedures before International Criminal Tribunal for the Former Yugoslavia and International Court of Justice on the Same Questions', Case No. IT-98-30/1, T.Ch., 5 December 2000. http://www.icty.org/x/cases/kvocka/tdec/en/01205JN114797.htm. Accessed 4 Aug 2014
12. United States Diplomatic and Consular Staff in Tehran (USA v. Iran) (Judgement), the ICJ (1980). http://www.icj-cij.org/docket/files/64/6291.pdf. Accessed 11 Aug 2014

Statements

13. Statement of Judge Stephen M. Schwebel, President of the International Court of Justice, to the Plenary session of the General Assembly of the UN 26.10.1999, http://www.icj-cij.org/court/index.php?pr=87&pt=3&p1=1&p2=3&p3=1&PHPSESSID. Accessed 7 Aug 2014
14. Statement of Judge Gilbert Guillaume, President of the International Court of Justice, to the UN General Assembly of 26.10.2000, http://www.icj-cij.org/court/index.php?pr=84&pt=3&p1=1&p2=3&p3=1. Accessed 7 Aug 2014

Books and Chapters

15. Bodansky D, Diringer E (2010) The evolution of multilateral regimes: implications for climate change. Pew Center on Global Climate Change, Arlington
16. Cryer R (2010) An introduction to international criminal law and procedure, 2nd edn. Cambridge University Press, Cambridge
17. Hashm SH (1997) Introduction. In: Hashm SH (ed) State sovereignty: change and persistence in international relations. Pensilvanya State University Press, University Park
18. Highfield R, Coveney P (1996) Frontiers of complexity: the search for order in a chaotic world. Ballantine Books, New York
19. Holland J (1995) Hidden order: how adaptation builds complexity. Addison-Wesley Publishing, New York
20. Kim R, Mackney B (2013) International environmental law as a complex adaptive system. Int Environ Agreements Polit Law Econ. http://link.springer.com/article/10.1007/s10784-013-9225-2/fulltext.html. Accessed 7 Aug 2014

21. Lauterpacht H (1975) The reality of law of nations. In: Lauterpacht E (ed) International law: being the collected papers of Hersch Lauterpacht, vol 2. Cambridge University Press, Cambridge
22. Lauterpacht H (ed) (1982) The development of international law by the international court. Cambridge University Press, Cambridge
23. Lemkin R (1944) Axis rule in occupied Europe: laws of occupation – analysis of government – proposals for redress. Carnegie Endowment, Washington, DC
24. Meadows DH (2008) Thinking in systems: a primer. Chelsea Green, White River Junction
25. Millennium Ecosystem Assessment (2005) Ecosystems and human well-being: synthesis. Island Press, Washington, DC
26. Mégret F (2012) International law as law. In: Crawford J, Koskenniemi M (eds) The Cambridge companion to international law. Cambridge University Press, Cambridge
27. Merrills J (1993) The development of international law by the European court of human rights, 2nd edn. Manchester University Press, New York
28. Miller J, Page S (2007) Complex adaptive systems: an introduction to computational models of social life. Princeton University Press, Princeton
29. Mitchell M (2009) Complexity: a guided tour. Oxford University Press, Oxford
30. Noortmann M, Ryngaert C (eds) (2010) Non-state actor dynamics in international law: from law-takers to law-makers. Ashgate Publication, Farnham
31. Romano CPR, Alter K, Shany Y (2014) Mapping international adjudicative bodies, the issues, and players. In: Romano CPR, Alter K, Avgerou C (eds) The Oxford handbook of international adjudication. Oxford University Press, Oxford
32. Spain A (2013) International dispute resolution in an era of globalization. In: Byrnes A, Hayashi M, Michaelsen C (eds) International law in the new age of globalization. Martinus Nijhoff, Leiden
33. Ungern-Sternberg A (2012) Religion and religious intervention. In: Fassbender B, Peters A (eds) The Oxford handbook of the history of international law. Oxford University Press, Oxford

Articles

34. Abi-Saab G (1998) Fragmentation or unification: some concluding remarks. N Y Univ J Int Law Polit 31(4):919–933
35. Alvarez J (2009) A BIT on custom. N Y Univ J Int Law Policy 42(1):17–18
36. Arthur B, Durlauf S, Lane D (1997) (eds) The economy as an evolving complex system II, Introduction: process and emergence in the economy. Addison-Wesley, http://tuvalu.santafe.edu/~wbarthur/Papers/ADL_Intro.pdf. Accessed 7 Aug 2014
37. Beaulac S (2004) The Westphalian model in defining international law: challenging the myth. Aust J Legal Hist 8(2):181–213
38. Belcher M, Newton J (2005) International legal development: a complex problem deserving of a complex solution and implications for the CAFTA region. Southwestern J Law Trade Am 12:189
39. Born G (2012) A new generation of international adjudication. Duke Law J 61(4):775–878
40. Brandtner B, Rosas A (1998) Human rights and the external relations of the European community: an analysis of doctrine and practice. Eur J Int Law 9(3):467–490
41. Buergenthal T (2001) Proliferation of international courts and tribunals: is it good or bad? Leiden J Int Law 14:267
42. Burke-White W (2004) International legal pluralism. MIJL 25:972
43. Carty A (1998) The continuing influence of kelsen on the general perception of the discipline of international law. Eur J Int Law 9(2):344–354

44. Charney JI (1998) The impact on the international legal system of the growth of international courts and tribunals. N Y Univ J Int Law Polit 31:697–708
45. Dupuy P (1998) The danger of fragmentation or unification of the international legal system and the international court of justice. N Y Univ J Int Law Polit 31(4):791–807
46. Fassbender B (1998) The united nations charter as constitution of the international community. Columbia J Trans Law 36(3):529–619
47. Gross J, McAllister R, Abel N, Smith D, Maru Y (2006) Australian rangelands as complex adaptive systems: a conceptual model and preliminary results. Environ Model Software 21(9):1264–1272
48. Guillaume G (1995) The future of international judicial institutions. Int Comp Law Q 44(4):848–862
49. Henkin L (1991) Law and war after the cold war. Mich J Int Law 15:147–167
50. Henkin L (1999) Kosovo and the law of "Humanitarian intervention". Am J Int Law 93:109–141
51. Hornstein D (2005) Complexity theory, adaptation, and administrative law. Duke Law J 54(4):913–960
52. Kingsbury B (1999) Foreword: is the proliferation of international courts and tribunals a systemic problem? N Y Univ J Int Law Polit 31:670–697
53. Koskenniemi M, Leino P (2002) Fragmentation of international law? Postmodern anxieties. Leiden J Int Law 15(3):553–579
54. Lansin S (2003) Complex adaptive systems. Annu Rev Anthropol 32:183–204
55. Morton J (1999) The end of the cold war and international law: an empirical analysis. Glob Soc 1:7–23
56. Newman M (2011) Complex systems. Am J Phys 7:800–810
57. Oellers-Frahm K (2001) Multiplication of international courts and tribunals and conflicting jurisdiction – problems and possible solutions. 5 Max Planck 68 UNYB
58. Osiander A (2001) Sovereignty, international relations, and the Westphalian myth. Int Organ 55(2):251–287
59. Pauwelyn J (2004) Bridging fragmentation and unity: international law as a universe of inter-connected islands. Michigan J Int Law 25:904
60. Pauwelyn J (2014) At the edge of chaos? Foreign investment law as a complex adaptive system, how it emerged and how it can be reformed. ICSID Rev 29(2):372–418
61. Pronto A (2008) Some thoughts on the making of international law. Eur J Int Law 19:601–616
62. Reed L (2002) Great expectations: where does the proliferation of international dispute resolution tribunals leave international law. Am J Int Law 96:219–231
63. Romano CPR (1997) The proliferation of international judicial bodies: the pieces of the puzzle. N Y Univ J Int Law Polit 31(4):709–751
64. Ruhl J (2005) JB Ruhl's "law-and-society system": burying norms and democracy under complexity theory's foundation. William Mary Environ Law Policy Rev 29:553–631
65. Ruhl J (1997) Thinking of mediation as a complex adaptive system. Brigham Young Univ Law Rev
66. Ruhl J (2008) Law's complexity: a premier. Georgia State Univ Law Rev 24:885–912
67. Shany Y (2009) No longer a weak department of power? Reflections on the emergence of a new international. Eur J Int Law 20(1):73–91
68. Shany Y (2014) Assessing the effectiveness of international courts. Oxford University Press, Oxford
69. Shelton D (2006) Normative hierarchy in international law. Am J Int Law 100(2):291–323
70. Sloan B (1989) The United Nations charter as a constitution. Pace Year Book Int Law 1:61–126
71. Spelliscy S (2001) The proliferation of international tribunals: a chink in the armor. Columbia J Trans Law 40(1):143–175
72. Stoll T (2002) The swordfish case: law of the sea v. trade. ZAÖRV 62(1–2):21–35
73. Tams C, Tzanakopoulos A (2010) Barcelona traction at 40: the ICJ as an agent of legal development. Leiden J Int Law 23(4):781–800

Online Resources

74. Background Information on the Responsibility to Protect, the United Nations, at http://www.un.org/en/preventgenocide/rwanda/about/bgresponsibility.shtml. Accessed 1 Sept 2014
75. Cases Section, International Court Of Justice, (7.08.2014). http://www.icj-cij.org/docket/index.php?p1=3. Accessed 11 Aug 2014
76. Council of Europe, 'Margin of Appreciation', Para 1. http://www.coe.int/t/dghl/cooperation/lisbonnetwork/themis/echr/paper2_en.asp. Accessed 7 Mar 2014
77. Declarations Recognizing the Jurisdiction of the ICJ as Compulsory. http://www.icj-cij.org/jurisdiction/?p1=5&p2=1&p3=3. Accessed 7 Aug 2014
78. The List of Cases, International Tribunal for the Law of the Sea (7.08.2014). http://www.itlos.org/index.php?id=35. Accessed 11 Aug 2014

Chapter 31
Approaches to Ecological Student's Competence Within the Transition to Green Economics

Aiman Aliyeva, Gaini Dlimbetova, Kulzhahan Bakirova, and Gulsana Shaimerdenova

Abstract All around the world there is an ecological threat to people's lives. Development of ecological competency of the future specialists must be taken into account under each specialist study. Today education defines the degree of wealth of any country, state existence, and also influences its economic development. Transition to "green" economics requires higher education systems to have ecologically competitive specialist. Today efficiency of ecological education in ecological problems solving is not enough. Under extraction of natural resources large reserves, the interest in ecological education decreases from year to year. Environmental problems are not being accepted by society as important enough though students recognize they belong to ecological problems and challenges. In this work the components of ecological competence and levels of its development are identified. So, transition to green economics requires development of ecological competence at all levels as well as formation of new competences. However this transition cannot be a bed of roses and thus poses a number of complexities here and there. Such complexity imbedded within the transition to green economics would require skills and knowledge, on one part, and a proper attitude and commitment on the other. This paper seeks to explore this process of transition and the challenges thereto.

31.1 Introduction

Troubles to people caused by ecological problems are being revealed currently with steady development. The most important factor for this steady development is ecological education. According to UNESCO (The United Nations Educational,

---

A. Aliyeva (✉) • G. Dlimbetova
Department of Social Pedogogics, L.N. Gumilyov Eurasian National University, Astana, Republic of Kazakhstan
e-mail: Gaiman-17@mail.ru; gainid@mail.ru

K. Bakirova • G. Shaimerdenova
KazNPU named after Abai, Almaty, Republic of Kazakhstan
e-mail: Bakirova59@mail.ru; gulsana1982@mail.ru

Ş.Ş. Erçetin (ed.), *Chaos, Complexity and Leadership 2014*,
Springer Proceedings in Complexity, DOI 10.1007/978-3-319-18693-1_31

Scientific and Cultural Organization) statements relating to society's steady development, knowledge and social-moral values of its professional staff are its major resources. Under that, higher education plays a key role in breeding of these qualities in future specialists. Ecological education today tries in various complex ways to solve a row of important tasks such as: nature security education, bio-sphere security, protection from ecological danger, prevention of ecological disaster, man's survival, and formation of eco-centric personality [1].

In 2011 году in Astana the 7th Conference of Ministers under the theme "Environment for Europe" was held. There were two major themes considered at the conference: "Sustainable water and water-related ecosystems management"; "Greening of economics: promotion of environment into economical development". By the end of the conference, the next declaration of ministers was affirmed: "Save water – take care of nature".

In 2012 in Rio-de-Janeiro (Brazil) the United Nations Organization conference on sustainable development "Rio +20" was conducted, where ways to steady future – future with large quantity of workplaces, more clear energy, more safe and well-deserved life level for all – were defined. In the final document of that Conference "Future we wish" it was outlined that in transitioning to a "green economics" each country has a right to choose in accordance with her national plans and development approaches and strategies considering its specific features. Thus there shouldn't be strict rules within [2].

The Republic of Kazakhstan is constructing a new economics transiting to "green" development. "Green economics" is being defined as economics with the high level of population life quality, care and rational usage of natural resources in the interests of modern and future generations. In Kazakhstan's conception of the transition to "green economics" big attention is being devoted to development of ecological culture among society, training of engineer-ecologists, as well as development of green skills of modern managers and farmers [3].

There is a new system of vision in which a new methodology is needed to form "green" economics, economics output to the way of steady, resources saving development. UNEP (The United Nations Environment Programme) formulated the law of common recommendations addressed to national governments and persons responsible for work-out of conditions and policy of transition to green economics. One of such conditions is investment in competence, study and education rise.

All economics sectors directed towards achievement of ecologically oriented economics form the basis of interest in reaching green skills. Further, the most critical problem is reflected in improvement of existing skills and introduction of steady development programs into existing education spheres. As new green sectors require new skills, these existing skills ought to be improved as major parts of sectors. Besides, problems of sustainable development will be introduced at all levels. Refusal to develop ecological competence would cardinally delay the process of transitioning to green economics. This problem can be solved through education [4].

Task of ecological competence development requires participation of many interested parties in educational processes. State, employers and educational institutions have an important role hereof. Ecological competence should be a ceaseless part of people's whole lives duration for knowledge acquirement, including different sectors such as public administration, public sector, industry, transport and agriculture.

31.2 Main Part

Ecological education is the security of the future and has to engulf each person as well as the whole society. Higher educational institutions play a unique role in this task solution, as there are required conditions for each student such as free development, formation of steady ecological position and professional competence [5]. Today efficiency of ecological education in solving ecological problems is not enough. As a result, interest in ecological education decreases from year to year. Environment problems are being understood by the population as slight problems, and not as threats to human society and environment. The fact nearly all global world ecological problems were not solved during the last years confirms insufficient efficiency of ecological education. Maybe, prevalence of theoretical education under practical education is one of the possible reasons. One of the factors defining the specific characteristic of pedagogical process is singularity of personality formed qualities. And the same analysis of pedagogical theory in the sphere of ecological education shows that there are settings in the problem of ecological knowledge and culture formation which dominate; but knowledge about pedagogical process of students, future specialist ecological competence formation at the basis of integration of all knowledge types and their inter-subjects connections is absent [6].

Ecological competence is determined by the individual characteristics of personality conformity degree to ecological education requirements: inculcation of ecological knowledge, and skills directed to form ecological consciousness, thinking, worldview, required for common orientation in ecological situation, to eliminate or restrict measures of ecological risk. Preliminary set requirement for students' ecological preparation is met under ecological competence. Ecological competence includes cognitive, motivational, social and active-professional components.

The Cognitive component is formed as a consequence of ecological knowledge in the scope of inter-subject connections (culture, legal, economical, social). Cognitive knowledge is directed to nature safety activity. System of cognitive knowledge is the basis to understand ecological world picture and encourages solutions to ecological problems.

The Motivational component has an impelling function directed to nature safety activity, presents a need for ecological knowledge, understanding of nature laws, recognition of reasons and consequences of ecological problems, recognition of own actions and motives. Social component is directed towards understanding of complex processes of social-ecological interaction. Understanding of social and individual magnitude in nature and man relations provides ecological safety and helps способствует to formation of educated specialists, who can live in harmony with nature.

The Active-professional component includes readiness to undertake nature safety activities in the professional environment. The component includes appraisal of professional ecological situation, choice and justification of own actions model, corrective actions and realization of the chosen solution.

In the bounds of education development, ecological competence of a personality can be considered as important personality characteristic, which combines ecological culture, ecological cognition, and ecological worldview. So, cognitive, motivational, social and active-professional components can be used as pedagogical criteria for ecological competence development.

A.B. Gagarin considers various criteria of personality ecological competence and ratios of its formation: Under that, levels of ecological competence development are presented as: close to ideal, sufficient, satisfactory, insufficient, unsatisfactory levels.

Ideal level: People have ecological ideas (of ecological values), anthropocentrism and eco-centrism; ability to allocate their main and most essential characteristics; their creative application for the analysis of behavior of the personality and people; emotional description of ecological knowledge; opinions about nature, aspiration to receive, looking for and processing information on objects of nature; steady position in relation to ecological values; positive behavior in interaction with the nature and personal contact with it.

Sufficient level: People have ecological ideas (about ecological values) for which anthropocentrism and eco-centrism integration is characteristic; ability to allocate their signs, to give examples about them in practice; emotional description of ecological knowledge; existence of own value judgments about nature but sometimes depending on a situation; aspiration to receive, look for and process information on objects of the nature; average intensity in relation to nature; steady position in relation to ecological values; steady tendency of positive behavior in interaction with nature; as well as personal contact with nature (aspiration is mutually advantageous interaction with nature).

Satisfactory level: People have more of anthropocentric idea than eco-centric. Existence of own value judgments about nature, and also about surrounding people as its parts, most often for self-affirmation, readiness to receive, look for and process information on objects of nature; insufficient steady position in relation to ecological values; below an average intensity of the relation to the nature of non-pragmatic modality; lack of accurately expressed certain position in behavior in interaction with nature, as well as small contact with nature (readiness is for mutually advantageous interaction with nature).

Insufficient level: People have anthropocentric idea (also of ecological values); about itself and surrounding people as its parts, and sometimes absence depending on external factors, a selfishness of the relation to the nature; practical lack of readiness to receive, look for and process information on objects of nature; low intensity in relation to the nature of non-pragmatic modality; insufficiently steady position in relation to ecology values; the tendency is observed in manifestation of negative behavior of interaction with nature and lack of personal contact with it (lack of readiness and aspiration to mutually advantageous interaction with nature).

Unsatisfactory level: People have anthropocentric ecological idea, including ecological values; absence of own value judgments about nature, and also about itself and surrounding people as its parts, non-sensibleness of the relation to nature; lack of readiness to receive, look for and process information on objects of nature; the lowest intensity in relation to the nature of non-pragmatic modality; lack of a steady position in relation to ecological values; negative behavior in interaction with nature, as well as lack of personal contact with nature (lack of readiness and aspiration to mutually advantageous interaction with the nature) [6].

We made analysis of students' creative work to determine ecological competence level. Among the suggested topics, they wish to work out intensively in their free time, while the problem of trans-border Kazakhstan rivers occupies first place. It is clear, students worry about condition of natural resources and international legal relations with neighboring countries.

Also among students there was inquiry held in which they were asked for their opinion regarding modern ecological problems. It is possible that using active educational methods at lections and also use of internet resources increases students' knowledge in the sphere of environment safety, and that lets them to realize their importance for humanity. To the question: "who should be in charge of environment safety?" The most respondents stated that belongingness of each individual person plays an important role. Governmental powers of the country, law enforcement authorities, and further international governmental organizations are in second place. To the question: "what do you know about green economics?" 60 % respondents answered that they heard about green economics from mass information means but they don't believe in the efficiency of its results. By virtue of their opinion, this is new way of imposing economic ideas. The Rest 30 % students replied that they know about the new economic development of the country and that would be one of the effective ways to solve ecological problems. Minimal quantity of students – 10 % have never heard about green economics.

Thus, we conclude that a modern student realizes his relation to the environmental problems and their complexity. But, taking into account the new economic development for sustainability, it is necessary to form new competencies such as: ecological information awareness, system thinking and creative potential.

31.3 Conclusion

As a result of the research analysis the following conclusions are being made:

1. The problem of ecological competence development of modern specialists is brought by globalization of ecological problems of all countries of the world.
2. Competencies are the grounds of personal professional maturity.
3. Now the approach based on competency is theoretical-methodological методологическим ground of common ecological education, has to predict formation of future social active people to solve ecological problems.

4. There are people with different levels of ecological values: these include anthropocentrical, eco-centric etc.
5. Transition to a "green" economics requires new competencies and methods of study and new skills development – "green" skills.

Development of such competencies as ecological information awareness, critical thinking, systematical thinking and creative potential should be appraised as ways to train specific skills. Ecological education should engulf education from early children's age and end up in education of grown-ups. There is a necessity to educate with new and additional skills in each profession (for example, professionalism in energy issues), taking into account specific features of spheres and professions (for example, organic agriculture, and technical maintenance of electronic cars).

References

1. Dlimbetova GK (2001) Ecologization of study in Universities. Almaty Scientific Research Center, Gylym
2. Bobylev SN (2012) Zakharov Green economics and modernization. Ecological-economical basis of steady development. Bulletin of the Institute of Sustainable Development of the RF Civic Chamber "Towards Sustainable Russia". № 60
3. Conceptions of transition of the Republic of Kazakhstan to green economics (2013)
4. Economical and Social Council (2011) Transitioning to a green economy: the role of education for sustainable development. E/ECE/RPM, 18th October 2011/5
5. Glyshkova LS (2001) Psychological-pedagogical conditions of young people ecological responsibility formation. Materials of Conference hold, 25–27 April 2001
6. Gagarin AV (2011) Personality ecological competency: psychological acmeological research. Monograph, Moscow

Chapter 32
Complexity in Provision of Guidance and Counseling Services in Secondary Schools in Uganda

Wahab Asad Kamonges

Abstract This scholarly article on complexity of provision of counseling services in secondary schools in Uganda focuses on attitudes and perceptions of teachers and students towards counseling services, contribution of government, implementation of counseling services and challenges met and way forward. It is expected that the submissions in this paper shall add immensely to the growing body of knowledge that seeks to explore issues of guidance and counseling in schools whilst dissecting the complexity that obtains in the modus operandi of guidance and counseling programs in schools today. The paper also opines that where complexity lies, is where effectiveness also lies. Thus proper conceptualization and appreciation of the complex nature of guidance and counseling in the schools could be the springboard for transformation of the same in the secondary schools.

32.1 Introduction

According to one of the psychoanalytic theorist Erickson, in his theory of psychosocial development, the physical social and environmental influences are intertwined as co-participants along with innate biological and psychological processes in shaping an individual's personality development. The theorist claims that an individual's life course is already decisively influenced by era, area and arrangement into which he or she is born implying that no one child can grow and develop successfully if not guided by an adult. Skinner's theory of personality stresses that; a child can develop positively or negatively due to the environmental influences. Therefore, the school setting is the most second home for the child and the role of the teachers and the school counselor is to develop the school going children to develop positively in all aspects of life.

W.A. Kamonges (✉)
Department of Educational Psychology, Islamic University in Uganda, Mbale, Uganda
e-mail: kamongesw@yahoo.com

Ş.Ş. Erçetin (ed.), *Chaos, Complexity and Leadership 2014*,
Springer Proceedings in Complexity, DOI 10.1007/978-3-319-18693-1_32

Provision of guidance and counseling services forms is part and partial of any educational system. Mutie and Ndambuki [8] defined guidance as a process, developmental in nature, by which an individual is assisted to understand accept and utilize his or her aptitudes, interests and attitudinal patterns in relation to his or her aspirations. They assert that, guidance in a school setting is a process that is aimed at leading the individuals to the achievement of desired life goals. Guidance is needed at every stage of growth that is to say from early childhood, through adolescence, adulthood and even up to old age so as to enable the individual to adjust to changing situations in life.

Counseling is part of guidance in which assistance is given to pupils with anxiety provoking problems. It involves a person to person relationship. The rapport developing through the relationship between a counselor and the counselee helps the individual to make his or her choices to resolve his or her problems in a personally realistic and meaningful way. Therefore, each secondary school is expected to design, develop and implement school guidance and counseling program intended to help students successfully overcome their physical, social, emotional, vocational and learning difficulties. Most secondary school students are in the adolescent stage. It is during this time that adolescents experience alienation which is a syndrome comprising of anxiety, distrust, pessimism, egocentrism, meaninglessness, normlessness and powerlessness. Thus, guidance and counseling is needed at this stage to assist adolescents to understand their developmental stages and adjust to school life.

Successful provision of guidance and counseling services in schools would help students to adjust to school life through understanding themselves and developing capacity to make informed decisions.

Guidance and counseling plays a significant role in the overall growth and development of a high school student and is therefore an essential part in the school curriculum. This is because high school students are at the stage of adolescence which is characterized by many physical and psychological changes which pose a number of personal, social and educational challenges [8]. According to the UNESCO Module [12], there are three major components of guidance and counseling viz; educational guidance, vocational guidance and personal social guidance.

Idowu [5] views guidance and counseling services as a process of planned intervention with in a school system by which the total development of students are stimulated in areas relating their personal, social, career, emotional and academic concerns.

The attitudes and perceptions of students towards provision of guidance and counseling services to a certain extent determine its successful implementation. The contribution made by the Government of Uganda through the Ministry of Education and Sports is equally analyzed. Whereas, provision of guidance and counseling services in secondary schools in Uganda is recognized as a major component in the educational process, its implementation has met a number of challenges which need to be addressed by all the relevant stakeholders. It's after this is critical look that the author has made recommendations for the way forward.

32.2 Guidance and Counseling in Schools: Status-Quo and the Complexity Involved

The Ministry of Education and Sports has a fully-fledged department of guidance and counseling. This department is charged with the responsibility of producing manuals and circulars which are sent to schools to help teachers in the provision of guidance and counseling services. The schools are supposed to appoint at least two teachers to be in charge of guidance and counseling. The issue of gender is supposed to be considered in the process of appointing the two teachers. Usually the senior teachers with high levels of integrity, honesty, confidentiality and who act professionally are supposed to be considered in the entire process.

The routine currently as far as provision of guidance and counseling services in secondary schools is that during assemblies' issues of guidance and counseling are supposed to be carried out. It has to be observed that in organized schools, there are even special meetings dedicated for discussion of issues which call for particular guidance and counseling services. For instance some issues may require having a session for only boys or girls. Group counseling for this matter takes the lead.

Under normal circumstances, all secondary schools are supposed to budget for guidance and counseling because it lays the foundation of a good school. The issues which are handled usually rotate around morals, discipline, personal hygiene, academic performance, benefits of education, relating with other learners and teachers among others. Some schools conduct guidance and counseling sessions after classes and this implies therefore that there is limited time devoted for these activities. Traditionally, all teachers are expected to engage themselves in the provision of guidance and counseling services in secondary schools since it is part of their training.

However, in many secondary schools much as it is budgeted for, these budgets are not usually implemented as per the budget. Though, the Ministry of Education and Sports places a lot of emphasis on provision of guidance and counseling, most secondary schools have not given it a priority. This is evidenced by the lack of office space, reference books, drawers, files, psychological test materials, career resource materials among others and this affects the delivery of counseling services in most of the secondary schools.

Emorut, April 15, [3] in the new vision reported that, shortage of professional counselors has worried the Ministry of Education as it cannot provide counseling in schools, especially in the area of teenage pregnancies and among teachers. "The Ministry of Education is struggling to have counselors in schools due to shortage of manpower," Mary Felista Apio, the senior education officer in charge of counseling and guidance in the ministry said. The Ministry has only seven professional counselors' despite the high demand for counseling services in schools. She appealed to policy makers to come up with a policy directing all institutions to have counselors' to help victims of emotional or physical abuse.

It is also observed that even the teachers whom schools have entrusted with the responsibility of providing guidance and counseling services do not do it effectively citing the large enrollment of students, lack of an office and little or no responsibility allowance. This has made them to develop a feeling that this is a neglected area in the education cycles. Whereas, all teachers are supposed to be engaged in this activity, some teachers express some level of ignorance about guidance and counseling. This assertion is in support of Ngumi [9] whose study revealed that teachers normally did not cover sufficient guidance and counseling courses in their teacher training programs to enable them effectively render guidance and counseling services. This is supported by Makinde [6] who affirms that inadequate training has hindered the development and successful implementation of guidance and counseling in high schools.

The students are the major recipients of the guidance and counseling services in secondary schools. Thus, the success of any program in school depends on the students' attitude towards it [4]. Eyo et al. further affirms that successful provision of guidance and counseling services promotes learning and boosts learners' academic achievements. Nwachukwu (2004), observes that the absence or non-utilization of guidance and counseling services in the modern school system leads to a rise in the crime rate, hostility among students, faulty career decisions and choice of subject combinations among others. In the Ugandan context, secondary schools in the recent times have witnessed a wave of strikes, rampant indiscipline cases, early pregnancies, and rise in the dropout rates among other issues. Could this be a result of our headteachers and teachers' failure to prioritize guidance and counseling services? or the problem could be the attitude that students hold about these services in our secondary schools? Validya [13] stresses that, the judgments that students make based on inadequate facts are likely to lead to wrong results and thereby develop biased attitudes. This implies that when students are not well grounded on issues of guidance and counseling, their attitude towards it would be biased making them in a way not to go for these services. The students perceptions on guidance and counseling services to a greater extent determines whether they would go for these services or not.

Ubom [11] asserts that an attitude can be said to be positive or high when the students' response to the task or program is favourable and when they show commitment to their duties and vice versa. It is therefore, vital for each and every secondary school to equip the students with the necessary information that would be of help in enabling them understand themselves and as well be in position to cope with their academic, vocational, emotional and social challenges. This can only be achieved when the headteachers and teachers show their commitment through active participation in the provision of guidance and counseling services. Studies such as those of Denga [1] and Edet [2] reported a negative attitude of school authorities to guidance services and to counselors in particular. Denga [1] reported that the principals who know little about counseling will not in any way appreciate the need for the counselor to be relieved of heavy teaching load and other co-curricular duties.

According to Musgrove [7] in his study of high school students' attitude towards guidance and counseling services found that there was no significant differences in sex and grade level but a positive overall attitude towards their guidance offices. It was interpreted that the needs and expectations of the majority of the students were being met by the guidance and counseling services. Onyejiaku [10] in his study affirmed that there were gender differences among students in many respects biologically and physically including their attitude towards guidance and counseling services and despite sex differences, the attitude of males and females to some extent are favorable towards counseling services.

32.3 Recommendations

This paper has explored the complexity imbedded in providing guidance and counseling programs in secondary schools in Uganda. As part of continuity to broaden the body on knowledge as par this field, the following recommendations are being made:

- There is need for secondary school guidance and counseling teachers to undertake refresher courses much as they did guidance and counseling as a professional course at national teachers colleges and at university.
- The Association of Secondary School Head teachers Union (ASSHU) should include guidance and counseling in their annual work programs.
- There is need for guidance and counseling teachers to be provided with responsibility allowance such that they are motivated to effectively engage in the provision of guidance and counseling services.
- There is need for government to plan and establish guidance and counseling offices and as well equip those offices with adequate resources in secondary schools.
- The Ministry of Education and Sports should provide funding to schools to cater for guidance and counseling services in particular.
- Every secondary school should timetable for guidance and counseling services and give it a priority.

References

1. Denga DI (2001) Guidance and counseling in school and non-school settings, 2nd edn. Port Harcourt Double Diamond Publications, Niger Delta, Nigeria
2. Edet IP (2008) Parents and teachers perception of vocational guidance in secondary schools in Calabar metropolis-Calabar Cross River State. An unpublished postgraduate diploma thesis, University of Calabar, Calabar
3. Emorut F (2013) New vision newspaper on shortage of professional counselors in schools, Kampala, Uganda: Vision group print press.
4. Eyo M, Joshua A, Esuong A (2010) Attitude of secondary school students towards guidance and counselling services in cross river state. Edo J Couns 3(1). Retrieved from www.ajol.info
5. Idowu AI (1990) Guidance and counselling in the national policy on education: a revisit. Ilorin J Educ 10:50–57
6. Makinde O (1987) Fundamentals of guidance and counseling. Macmillan, London

7. Musgrove WJ (1973) A survey of high school students' attitudes towards guidance services. J Stud Pers Assoc Teach Educ 12(1):25–31
8. Mutie EK, Ndambuki P (1999) Guidance and counselling for schools and colleges. Oxford University Press, Nairobi
9. Ngumi ON (2003) Assessment of the level of training and resource needs of secondary school guidance counselors in Nakuru Municipality Kenya. Unpublished MED thesis, Egerton University, Njoro
10. Onyejiaku FO (1991) Psychology of adolescent. Rapid Publishers, Calabar
11. Ubom ID (2001) Value orientations, need satisfaction and job performance of public servants in Akwa Ibom State. A PhD dissertation, Universities of Calabar, Nigeria
12. UNESCO (2000a). Guidance (Module I). Fr
13. Validya N (1989) The impact of science teaching. Oxford and IBH Publishing Co., New Delhi

Chapter 33
Leadership Training of Preschool Children in a Family Environment

Roza Bekmagambetova, Anar Alimbekova, and Lazzat Demeyova

Abstract The paper analyses the role of the family in preschool children's leadership training process. It traces changes of approaches to children's education in the family. It also examines the impact of the atmosphere the children grow in on their personality and describes how parents' behaviors determine the developing children's leadership potential. The paper describes the best ways to develop a child's leadership potential in a family environment. One major feature in the analysis is the acceptance by the author that leadership training is not a ley man's undertaking especially when meant for children. It therefore involves a lot of complexities whilst in some cases; it poses a chaotic setting that requires critical interventions. It is however this chaotic and complex setting that helps in building an effective approach to negotiating a proper program to train the children in leadership skills applicable to their home or family environments.

33.1 Introduction

The current political and socio-economic changes in Kazakhstan have had a major impact on the quality of preschool children's leadership training. Strategic public policy objectives in the field of education actualize the challenge of developing leadership qualities among the younger generation.

Modern society presents high demands to the younger generation and today we need people who are flexible, creative, enterprising, energetic, confident, ready for self-improvement with a professional qualities and are able to guide people and would occupy a worthy place in the society. Under these conditions the family role is highly increased and stretched.

R. Bekmagambetova (✉) • L. Demeyova
Department of General, Age-related and Social Pedagogy, Institute of Pedagogy and Psychology, Kazakh National Pedagogical University named after Abai, Almaty, Kazakhstan
e-mail: r.bekmagambetova@mail.ru; Laz.nr@mail.ru

A. Alimbekova
Faculty of Social Sciences, Department of Social Pedagogy and Self-Knowledge, Eurasia National University named after Gumilev L.N., Astana, Kazakhstan
e-mail: alimbekova_66@mail.ru

Ş.Ş. Erçetin (ed.), *Chaos, Complexity and Leadership 2014*,
Springer Proceedings in Complexity, DOI 10.1007/978-3-319-18693-1_33

33.2 Leadership Training of Preschool Children

Approaches to the education of children in the family have changed a lot, if earlier children had been taught to be modest and restrained, in today's society educational process places special emphasis on the developing of the child's self-respect, his dignity and self-esteem formation. This approach aims at enabling children to become leaders in future, that is why parents have increasingly began to reflect on this problem.

The desire to lead is one of the manifestations of self-centeredness children are born with. An adult's task is to help children develop their leadership potential. This has acquired particular importance in terms of market economy development when a person's success largely depends on his ability to react quickly to changes happening in society and to make appropriate and definite decisions. In this regard social development of a person and leadership development need to be started at early childhood.

At preschool age most of a child's characters are established, so this age is considered to be favorable for leadership training. To be a leader does not mean being impertinent, bossy and pushy. A child-leader is not one who runs ahead of all, pushing and neglecting his peers.

A leader is a person capable of guiding the others, able to convince and prove the righteousness of his ideas, to enthrall and inspire others with his ideas, who has his own point of view, and, most important, one who is able to complete the business he or she started.

Leadership qualities are not formed spontaneously, the family in this case plays a decisive role. It is the family that creates a child's need to be recognized, creates the ability to empathize, to be responsible for him-self and for others. No other people have more influence in a young leader's life than his parents. The best place to develop effective leadership qualities is the family the children grow in.

In fact, for successful parenting of child's leadership training, right atmosphere and environment the child grows in are extremely important. Family relationships, parents' behavior manners, the way of communicating with children largely determine whether the child is meek or has leadership potential. It should be remembered that children often copy their parents and always reflect the behavior adopted in their family. Parents seeking to raise a leader must have leadership potential themselves, and must be actively engaged in social development and ought to be successful in life.

It should be noted that there are children who have inborn leadership potential. At the age of 3 or 4, a child's leadership potential is brightly displayed. Inborn leadership qualities are visible to the naked eye, but it should be noted that this potential can be developed only with parents' painstaking and targeted educational work.

It is also important to note that the development of children's leadership qualities depends largely on the adults' targeted education and social environment and, the atmosphere surrounding the child. Adverse conditions in the family, on the contrary, will ruin the child's inborn leadership potential.

The leader-child is able to win over peers. He always takes the initiative and starts the case first, and the other kids look up to him and then imitating timidly follow him. A leader is able to set goals and to involve others in the process of their implementation.

Leader-child possesses certain properties or character and special qualities such as:

- Charm,
- Attractiveness,
- Full of energy,
- Initiative,
- Assertiveness,
- Independence and self-confidence,
- Self-discipline,
- Carefulness,
- Kindness,
- The ability to get on well with other people,
- The ability to convince others and gain their trust quickly,
- The ability to defend his opinion, which is clearly manifested in childhood
- Responsibility
- Cheerfulness,
- Well-trained voice,
- Sharp and flexible mind, a strong will-power, resourcefulness, sense of humor.

A child leader has peculiar quick thinking ability, good judgment aptitude, a thirst to learn the world around him, a high level of intelligence. A great deal of attention should also be paid to the fact that a child leader is inclined to plan and calculate his ways to act.

It is often asked what should be done by parents whose children do not have developed leadership qualities? It is important for parents to understand that any child may generate leadership qualities, it requires patience and defining the specific target position: it is necessary to make clear why you need to educate leadership qualities in a child, whether he wants to be a leader or not, and finally, what the qualities of a leader are and what the ways to do it.

On the basis of this defined target the content, methods and ways of the child's leadership training are worked out. Parents should remember clearly that the process of educating of a young leader from early childhood involves the development of a child's ability to lead people, to establish contacts with others, and on this basis to be able to organize management.

As the first stage in the leadership potential developing can serve the athletic and intellectual games organized in a family environment. The main purpose of these games/competitions is to develop the child's accuracy, agility, quick thinking ability, logic of thinking. It is desirable to conduct these games together with his friends as a team. During these games a child learns to get along with other kids and adults, he develops a sense of confidence in his abilities, insistence and respect for the game participants.

It would be doubtlessly useful if parents constantly encourage their child saying: "You can do it," "You will succeed!". Parents should always make the time for the child to participate in all of his activities, to watch his achievements and shortcomings, praise and reward each attainment. As we strengthen children's self-confidence it is advisable to complicate the task.

In the child–leader development process it is considerably important to express parent's interest in any of his activity, to support and assist him in carrying out his ideas, encouraging reasonable initiatives. Parents should be aware of the need to nurture responsibility, determination, resourcefulness, independence, self-confidence, the ability to accomplish business and allow a reasonable risk, to teach a child to solve independently the problems he faces and make decisions in difficult situations, draw conclusions, moreover, to aim a child at success. Parents should not in any case protect children from mistakes, obstacles, difficulties, it is important to teach a child not to be upset because of failures, to find a way out of difficult situations, to see and correct their own mistakes.

It is very important to teach kids to set goals, plan how to reach them, and to achieve these goals. Child-leader must be taught to analyze their results, believe in the success of the business he started and to infect others with this faith. The child should have been able to make and defend his choice since his childhood. Parents need to create and put the child into teachable situations where he could reveal his leadership potential. All of this could become fertile soil to instill both values and skills that will become a part of a young leader's life for good.

33.3 Complex Nature of Leadership Development and Training

According to Ryan [1], the life of modern-day leaders is more demanding than ever. Inside their organizations, they need to motivate a diverse group of women and men, work across organizational boundaries, improve efficiency and achieve growth. Externally, they face a complex and globalized environment. They have to manage the requirements of government, keep ahead of competitors, and exceed the expectations of other stakeholders. And within this global environment, there are many cultural considerations leaders must navigate to be effective. They must work across cultural boundaries and alongside others who, at times, are very different from them and have different ways of getting work done.

These are difficult challenges, developing leaders who can face these challenges requires a comprehensive response. Of course, some training needs and content can be derived from organizational strategy and planned change initiatives. Yet other needs and challenges arise that strategy or change initiatives do not anticipate.

The benefits of leadership development coaching and training are beyond words in many cases all is not rosy indeed. In fact it is widely believed that eighty percent of problems experienced in any organization are people-related. Twenty percent are of a technical nature. Developing the people to function at higher levels benefits the organization, and always drops to the bottom line. The people drive the bottom line. When people are valued, and are an integral part of the development process of any organization, that organization will benefit – it's that simple.

As with all programs that involve people, there will be some challenges because of the very nature of people. Here are some "people issues" [2] that may be experienced:

- Unwilling to be developed or coached
- Not open to feedback or change
- Not open to innovation
- Threatened by others (especially those below them in "rank")
- Not able or willing to try something new
- Susceptible to the three major pitfalls of "management:"
- Greed
- Need for recognition
- Need for power

The interlink leadership development framework contends that some honest "soul-searching" must be done by the leadership of teams with members who exhibit the above characteristics. If people can enter into leadership development training and coaching with a positive attitude, the results are much more dramatic. But results CAN be achieved, even if someone enters tentatively. A negative, unbending attitude, however, may force the need for an honest employee assessment.

The study thus offers a way out of the complex jig-saw by proposing a leadership training model that is not only based at home and starting at an early age but also one rich in terms of content exposed to the trainees. It is only such a leadership training that can boss the complex nature of leadership development and training for the better.

33.4 Findings

Based on the above the following can be said: we must encourage leadership qualities such as resourcefulness, initiative, responsibility, self-confidence since childhood. We should teach a child not to be upset when he fails or makes mistakes. In spite of all the errors and mistakes, a child must always believe in success, parents on their part must assist them to decide on the field of their interests. It is extremely important to praise the child for any progress he makes and to teach him to analyze and correct his own mistakes.

The main thing for a leader is to be a leader for himself, to believe in his own potential, so it is important to inspire his self-confidence, teach him how to set goals and achieve them, to demonstrate his ability to be persistent. To foster leadership qualities is very important in developing a child's ability to articulate correctly and speak beautifully. Beautiful and right speech is the way to attract people to one's side.

In this regard, parents should pay attention to:

- The enrichment of vocabulary of children through increasing of understanding of the nature phenomena and of social life,
- The development of expressive speech in the course of everyday communication,
- The enhancing the skills of the proper use of basic grammatical forms of the native language to express the ideas precisely,
- Clear pronunciation of the native language sounds,
- The use of different means of expression of speech intonation while reading poetry and retelling tales (tempo and rhythm of speech, logical accents)
- The development of coherent speech.

The main means in this direction can be songs, nursery rhymes, and fairy tales, reading and telling stories, learning poetry by heart.

33.5 Conclusion

Thus, the problem of preschool children's leadership training in a family environment is very topical at the moment. This is due to the fact that only a person with developed leadership qualities and adequate self-esteem could succeed in modern life.

References

1. Ryan JR (2013) Interlink global leadership challenges: organizations must complement their business growth strategies with sensible leadership development plans to ensure they have a talent pipeline ready when required. http://www.business-standard.com/article/management/global-leadership-challenges-113122200583_1.html
2. Interlink leadership development: what are the benefits and challenges of leadership development? http://www.interlinktc.com/public_html/benefits.html

Chapter 34
Learning, Knowledge and Impact Assessment from the Perspective of Complexity and Chaos

Tunç D. Medeni, A. Nusret Güçlü, Tolga Medeni, and Gülten Alır

Abstract The paper benefits from information on selected (public-transformation) project works, mostly carried out by Stratek R&D Ltd, a private company in Turkey, linking with existing body of academic and practical knowledge on public value and system evaluation. The construction of this link will then lead to new directions for academic and practical research, including the adaptation of a reflective (& refractive) function to satisfy the principles & dynamics of a self-organizing and self-referential complex and chaotic system. Possible practical implications for ongoing project work will also be discussed before the conclusion.

34.1 Introduction

From an academic and conceptual perspective, as McElroy [9] introduces, the New Knowledge Management (TNKM) benefits mostly from complexity theory, the complex adaptive systems theory, or CAS theory in terms of helping to explain the role and evolution of knowledge. Accordingly, it can be claimed that learning is a self-organizing process, and that the knowledge produced through such processes is then emergent, as well. These concepts of emergence, self-organization, together with self-referentiality in fact deserve more attention in terms of their applicability into public policy-making and management areas [14].

As a bridge between this academic and practical perspective, Güçlü [7] benefits from the conceptualization and application of public transformation in (electronic) public services to be based on the "value delivery" to the public. Competing for the

T.D. Medeni (✉) • T. Medeni
Department of Management Information Systems, Yıldırım Beyazıt University, Ankara, Turkey
e-mail: tuncmedeni@gmail.com; tolgamedeni@gmail.com

A.N. Güçlü
Stratek Founding Partner ve METU Informatics Institute, Stratek R&D Ltd., Ankara, Turkey
e-mail: nusretguclu@gmail.com

G. Alır
Department of Information Management, Yıldırım Beyazıt University, Ankara, Turkey
e-mail: galir@ybu.edu.tr

Ş.Ş. Erçetin (ed.), *Chaos, Complexity and Leadership 2014*,
Springer Proceedings in Complexity, DOI 10.1007/978-3-319-18693-1_34

limited public funds emphasize basic rules of Economy, Efficiency, Effectiveness, (and Environment) in complex and chaotic Public Management. A Strategic Management Model (SMM) is based on and an extensive expansion of Government Finance Statistics (GFS), which the public finance officers are familiar with, introducing simplification over complex and chaotic public financial management environment in Turkey. A specific Value Space is developed in order to provide a framework for multiple perspectives such as organizational, functional, and performance with the same set of values, by fixing related dimensions under analysis. In order to measure both direct and indirect benefits as Agency Value, User Value, Political Value, Strategic/Social Value, and Environment Value, a specific method is developed for evaluating effectiveness of information systems (IS) in the public sector. A calculation method for both different indices for these values and the combined index, called the Total Public Value (PV) is provided. Güçlü [7].

Another related important project is the Erzincan Pilot Project (EPP) for the development of the ontology model of the public services inventory. One of the answers provided by the project is the complexity of business of each service (bureaucracy level indicator) which is scored using number of documents, number of activities and the number of services given per year [7]. Furthermore, currently, for instance, a new work on impact assessment of information and communication technologies, based upon cost-benefit analysis in public administration has also been initiated [8].

Table 34.1 Overview of tools for prioritization of new investment projects and new policies/programmes

	Main application	Main indicators used	Monetization of information	Degree of quantitative skills required	Computer software requirements
Cost-benefit analysis	Investment projects	Rate of return of investment and capital net present value, social discount rate	Full	+++	Spreadsheet, some specialized software
Cost-effectiveness analysis	Broad (Investment projects + other projects)	Cost-effectiveness ratio, net benefit	Partial	++	Spreadsheet
Impact assessment	Investment projects, legislative (policy changes)	Social, environmental, quality of life impacts	Limited	++	Spreadsheet, econometric software (STATA, SPSS, EVIEWS)

(continued)

Table 34.1 (continued)

	Main application	Main indicators used	Monetization of information	Degree of quantitative skills required	Computer software requirements
Multi-criteria decision analysis	Broad scope, highly complex policy context (environment, transport)	Performance matrix, criteria weights	Full	++	Specialized software (HIVIEW, VISA, DECISION DESKTOP, MACBETH)
Mathematical programming	Broad scope, highly complex policy context (environment, transport)	Allocation variables	Full	+++	Specialized optimization software

The paper will first benefit from this information on selected project works, mostly carried out by a private company in Turkey, namely, Stratek R&D Ltd, linking it with existing body of academic and practical knowledge. The construction of this link will then lead to new directions for academic and practical research. For instance a reflective (&refractive) function will be suggested to satisfy the principles & dynamics of a self-organizing and self-referential complex and chaotic system, furthering the TNKM as a theoretical contribution. Possible practical implications for ongoing work will also be discussed before the conclusion.

34.2 Public Transformation Projects in Turkey

Melford [13] identifies the following stages in eGovernment transformation in his 2005 presentation "Beyond eGovernment, Transforming Public Services":

- Promote access and connectivity;
- Provide services online;
- Transform the enterprise (first through automating the existing processes, then transforming business processes and organization);
- Next generation government, IT enabled radically different means of providing services.

With operational fragmentation, lack of proper control & audit, it is evident that information kingdoms arise in silo/stovepipe approach. New concepts, such as value chain and Public Value emerge within the context of strategic management, covering risk management, performance management, process management and knowledge management. Based on these concepts, accountability and fiscal transparency emerge as the key approach in providing for economic, efficient and effective collection & utilization of public funds. Strategic planning is the central nervous system, transforming policy into goals, and goals into measurable objectives.

Güçlü's [7] work argues that assessment of IS effectiveness, particularly in government, still remains not well developed, mainly due to the realization of non-tangible benefits. All net benefits [3] can be merged into the concept of PV, which is a combination of Agency Value, User Value, Political Value, Strategic/Social Value, and Environmental Value, with hierarchical breakdown into key goal and performance indicators, merging the concepts from IS and public finance domains.

The developed model assumes that a strategic plan (SP) has already been developed and hence the objectives/goals, based on higher level policy papers, are identified together with Key Goal Indicators (KGI) and Key Performance Indicators (KPI). The model does not calculate the PV if the outcomes are not defined prior to the assessment, hence for instance if the environmental impact is neglected in the plan, no value will be calculated. More importantly, the model does not assess impact, but effectiveness. Although not specified in the SP, the values might be in conflict, an initiative yielding to a maximum value in one category may yield to a negative impact value on another category. Political, User and Environmental Values will often appear to be in conflict. This approach will eventually tell how much damage can be tolerated in one value (such as environment) while other values might have positive indicators (such as building a dam for energy and job creation).

Consequently, for instance, PV for Accounting Service is calculated to be as 1.0458. The model also relies on predetermined discretion and threshold by the management. Here, if the threshold set by the management is 0.80, then the TPV result is acceptable. Experience shows that the effectiveness can be calculated even at the micro level of one initiative rather that at a more global service level. Furthermore for a particular objective, how much money is spent for the related unit of performance or how much the agency had to spend in order to achieve one unit of performance can also be measured.

The developed model is also used for the project, Technical Assistance for Decision Making and Performance Management in Public Finance EuropeAid/129067/D/SER/TR. Prime Ministry, Ministry of Finance, Ministry of Development and Undersecretariat of Treasury of Turkey, as well as The Grand National Assembly of Turkey are the beneficiaries of the project. The project aims to establish a decision support system and a performance management model in order to be used firstly at the Beneficiaries and then to be disseminated to the other public administrations for the purposes of strengthening financial decision making and performance management capacity of the Beneficiaries. Stratek collaborates with Ecorys to accomplish the related activities and deliverables (Apr 2012 ongoing) (Stratek profile).

The framework that, within the given Environment, integrates the "Economy, Efficiency, Effectiveness", which can be referred to as "Value for Money" (VFM) is depicted below ("Decision Making and Performance Management in Public Finance" project deliverable report [2]):

Economy is careful use of resources to save expense, time or effort; in other words it is minimizing the cost of resources ('doing things at a low price').

Efficiency is performing tasks with reasonable effort ('doing things the right way'). Efficiency is delivering the same level of service for less cost, time or effort, i.e. attaining the most program outputs possible for each program input; and is usually expressed in monetary terms. Technical efficiency is associated with productivity, cost per unit of work done or service delivered, whereas economic efficiency is associated with benefits of a program compared to its economic costs, and is net social value of a project/program, estimated by subtracting the discounted social costs from the discounted social benefits.

Effectiveness is the extent to which objectives are met ('doing the right things'). It is focused on delivering a better service or getting a better return for the same amount of expense, time or effort. It is expected that the observed outcomes are consistent with the intended objectives.

Hence, **Value for Money** is about providing services that are of the right quality, level and cost that reflect the needs and priority of citizens, taxpayers and the wider community. VFM is a term used to assess whether or not an organization has obtained the maximum benefit from the goods and services it both acquires and provides, within the resources available to it. Some elements may be subjective, difficult to measure, intangible and misunderstood. Judgment is therefore required when considering whether VFM has been satisfactorily achieved or not. It not only measures the cost of goods and services, but also takes account of the mix of quality, cost, resource use, fitness for purpose, timeliness, and convenience to judge whether or not, together, they constitute good value. Performance management cycle evolves around these concepts; it is a normative model of organizational planning and actions that emphasizes the importance of stating clear goals and objectives, translating these into policies and programs, implementing and then assessing and reporting outcomes so that goals and objectives can be appropriately modified. ("Decision Making and Performance Management in Public Finance" project deliverable report [2]).

The project also conducts a comparative analysis or tools for assessment and prioritization of new policies and investment projects ("Decision Making and Performance Management in Public Finance" project deliverable report [2]) (Fig. 34.1).

Cost-Benefit Analysis (CBA) is especially preferable for investment projects, benefiting from main indicators such as Rate of Return of Investment and Capital, Net Present Value, and Social Discount Rate. These one-dimensional analysis and indicators, however become less useful for more complex systems evaluations such as the impact assessments of Information and Communication Technology projects or systems. Instead, for instance Multi-Criteria-Decision-Making (MCDA) that enables multi-dimensional indicators approaches can be more useful to address the complexity and chaos in the system. ("Decision Making and Performance Management in Public Finance" project deliverable report [2]).

There are certain initiatives and tools to address the arising need for such evaluations and assessments (Ex. Stratek internal report on Information System

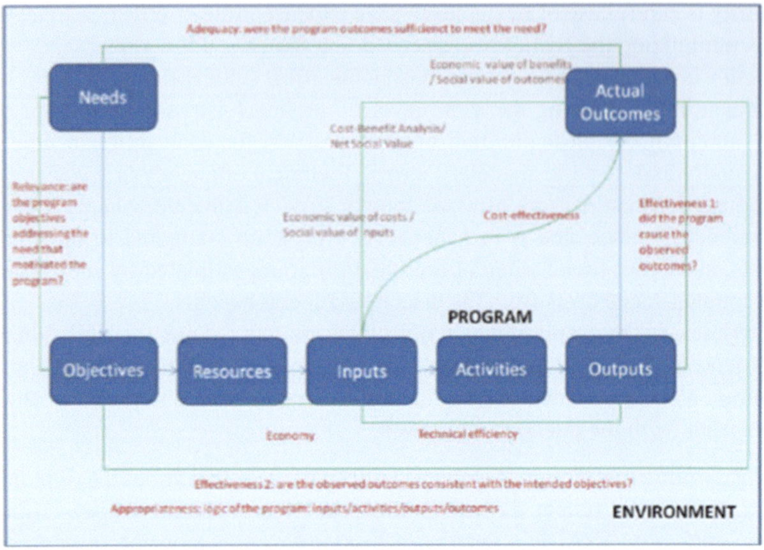

Fig. 34.1 Economy – efficiency – effectiveness (Adapted from Nagarajan, N. & Vanheukelen, M. (1997). Evaluating EU expenditure programmes: A guide, p. 25)

Evaluations [16]). While some of these methodologies and tools can be considered to be more sophisticated and useful than others, identifying and measuring more intangible benefits is still seen as a problematic issue. Furthermore, integrating the interactions among various stakeholders and other involved entities remains another important issue, among others. All these issues are significant in order to develop more useful conceptual models and practical applications to address chaotic and complex domains.

As a more real example, recently, the project of "Decision Making and Performance Management in Public Finance" provides specifically developed tools to government officers in order to help them deal with the complex and chaotic order in finance domain. The suggested programme budget modeling revitalizes the bottom-up and top-down interlinks among various planning, programming and budgeting entities. Furthermore, specific methodologies such as Data Envelope Analysis or sophisticated technologies such as Oracle Business Intelligence systems are adapted as user interfaces for agile decision support (Closure Meeting of The project of "Decision Making and Performance Management in Public Finance", [18]).

On the theoretical side, furthering conceptualization of Cost Benefit Analysis with respect to certain complex and chaotic systems features could also provide interesting implications. The next section is dedicated for such model development, benefiting from concepts of TNKM, reflection and refraction, mostly adapted from Medeni and Medeni's recent work [11].

34.3 Model Development Based upon Reflective and Refractive Interactions on Complex and Chaotic Systems

Reflection is an important concept for management of knowledge. For instance, Nonaka and Toyama [15] incorporates reflection into their conceptual framework for the conversions between tacit and explicit knowledge among different societal entities (as individual, group and organization and with environment) as part of the Socialization-Externalization-Combination-Internalization (SECI, Takeuchi & Nonaka [17]) processes of knowledge creation. However the underlying dynamics and interactions enabling such flow of knowledge among entities that can be very different deserve further analysis and articulation in order to apply these concepts into practice. Medeni [10] respectfully attempts to complement concept of reflection with a new conceptualization of refraction. Suggested as an important phenomenon in cross-border interactions among different societal entities, refraction is identified as a more cross-cultural, creative and critical types of reflection that are mostly missed in cross-cultural management and transfer of knowledge. Accordingly, reflection and refraction exists together and complement each other. Meanwhile, they function together as important dynamics for knowledge conversions between tacit and explicit knowledge.

Using the mathematical/geometric features of the ellipse, and conceptualizing the practice/management and knowing/learning both as a product and process, reflective model of experiential learning and practice can also be developed. This is somewhat related to discussions of soft and hard systems methodology [1, 12], in which, simplistically, the former stresses the process and the latter signifies the product. Initially desired, the ideal is to be able to obtain both the process and product. Second, if we place learning or knowing and management or practice as the two loci in an ellipse, the resulting figure provides a useful expression for the attainment of process and product. This elliptical diagram is also a recognition of the equal importance of both mental learning/knowing and physical managerial practice, moving beyond the perception of previous circular models like learning cycles, or singular spiral models as suggested in knowledge creation. Moving along the ellipse, one can obtain the process, then process and product together, and finally the product by itself, which is transferred to the other side with a reflective object passing through the middle of the whole figure. In this way both the spatial and temporal meanings of the possible transfer of the rich experience and knowledge gained from learning and practice can be visualized. What determines to be transferred or not is the power associated with experience and knowledge (Fig. 34.2).

Here, underlying that there are two knowledge-creating spirals instead of one spiral or two constant centers is important. These two spirals are spatial and temporal (one moves clock-wise, the other moves anti-clock wise) reflections of each other, resembling the image of a moving object in a mirror. The integration of these two reflective spirals also corresponds with lemniscates or chaordic (Chaotic & Ordered) systems, as often recalled together with Lorenz Attractor (Fig. 34.3)

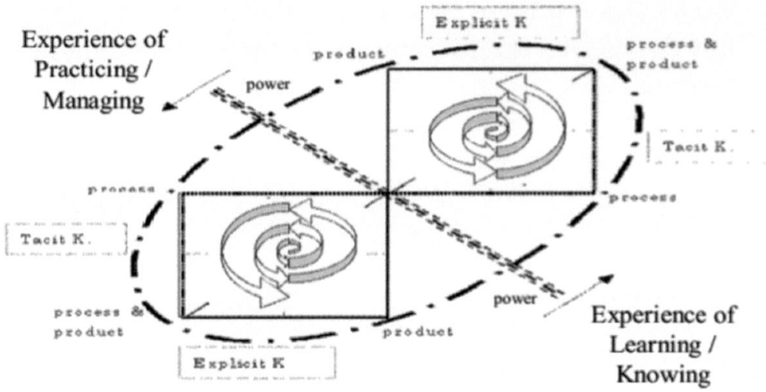

Fig. 34.2 Reflective and elliptic model of experiential learning and practice [11]

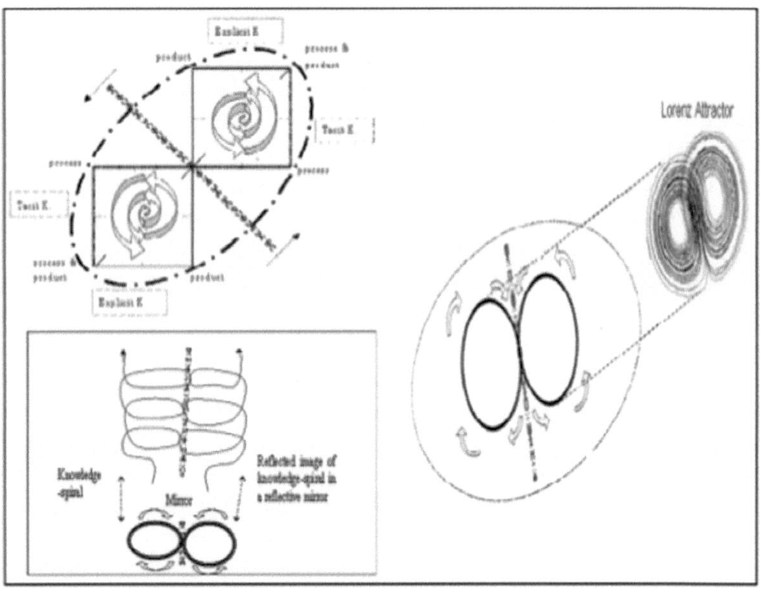

Fig. 34.3 Experienced reflections as a chaordic (Chaotic & Ordered) knowledge-creating system [11]

In this illustration, the upper part is a 3 Dimensional (3D) visualization, while the bottom part is a 2 Dimensional (2D) projection on a plain surface. The upward spiral in 3D looks like an endless cyclic movement in 2D. Finally, the cyclic movement in the left spiral/circle is clockwise, whereas in the right spiral/circle, it is anti-clockwise, which together creates a continuous horizontal-8 figure, resembling the infinity symbol, in 2D. What is also worth-noting is that the conceptualization of reflection here is not only the reflection of a physical object, but the metaphorical reflection of a knowledge-creating process that proceeds in time and produces an outcome, as well, as we have discussed above. Accordingly; if the clockwise move symbolizes the natural proceeding of time from past to present and future, then its reflected image, the anticlockwise move, symbolizes the reversed flow of time from future to past, while reflection plays the role of establishing the link between these two flows, i.e. making use of the reversed flow in order to make sense of the real flow, which could be one interpretation of reflection.

This interpretation is also in accordance with the self-referentiality aspect of complex systems. A simple reflective function can be defined, for instance, as $f(x) = -f(x)$, which also corresponds for (one type of) the self-referentiality, leading towards self- organization, as well.

What the above figure also implies is that the two knowledge-creating spirals of practice and learning are simply conceptualized as one knowledge-creating spiral and its reflected image, while their interdependence is the mere result of an exact, one-to-one reflection. However, such reflection would exhibit a simplified or ideal case, as it is more realistic to think that the reflection would be mostly refracted, and these refracted reflections would replace any exact linear correspondence with more dynamic, non-linear approximations.

In such modeling, reflection and refraction construct the practice-learning link and interaction, which includes concerns like what we learn from practice, how we apply our learning into action, or how we construct our knowledge, and accumulate our experience. These concerns highlight that how we reflect and refract is not simple, but a lot more dynamic, and chaotic; and a modeling about reflection and refraction should address this chaos. In fact, the inclusion of refraction within the conceptualization of reflection is an initial premise for such a non-linear, more dynamic modeling.

Such modeling of reflection and refraction fits also more with complex systems. For instance, as discussed above, it meets the self-referentiality principle. In fact Plato argued that all physical reality experienced by human beings in the material world are actually only imperfect and refracted reflections of a perfect world (that exists elsewhere in the universe). [19]. Benefiting from Plato and Hegel, and Heidegger' ideas, Eldred [4] discusses how (it can be seen that) the singular self is refracted on the other and thus 'broken in' on the world through the dialectic between singularity and universality. Authors like Geyer [6]) also brings the related discussions on self-reference into systems science. The model also satisfies not only self-organization but also emergence principles of complex systems, in accordance with learning and knowledge processes of TNKM, as introduced above [5]. As a result the developed model can also be used to explain the knowledge interactions in complex (and chaotic) domains.

One further step for this model development and use, which is previously suggested for knowledge creation and management learning, can also be a reinterpretation for cost and benefit analysis. Accordingly, cost and benefit could be considered as the two attractors of the system. Initially resembling for instance causal loop diagrams that illustrate the relationships between variables in system.

Such model reinterpretation could in fact be useful to formulate more measurable parameters and meaningful relations among cost and benefit variables for complex domains such as the impact assessment and public value measurement of a technological solution in a public agency.

The systems will be attracted to reach a leveling situation at the sides of both costs and benefits. These costs and benefits could be identified and measured with respect to self-referentiality, self-organization and emergence principles.

34.4 Future Work and Conclusion

The paper has presented a background on public transformation projects in Turkey, benefiting from available relevant academic and practical knowledge. Based on this, then;

1. a leading-edge public-value and cost-benefit analysis framework that is applied to real life examples is provided, benefiting mostly from Stratek projects.
2. a model development based upon reflective and refractive interactions on complex and chaotic systems is suggested.

Various implications can be drawn from these suggestions. For instance, the measurement of user efforts could be incorporated into the design of the impact analysis for specific information and/or communication technologies as real-life projects. Or assessment of individual initiatives at micro level could be conducted to complement aggregate assessments at macro level. Accordingly, inferences on the interrelation between micro and macro with respect to self-organization, emergence and co-existence aspects of complex systems [14] could be made. or, suggestions for more flexible, evolvable and maintainable organizational ontologies could be provided for Future of Internet.

It should be underlined, however, that this is mostly a conceptual work that needs to be complemented with empirical data collection in the future. Accordingly, some of the coauthors currently work on developing a cost-benefit analysis model as a part o integrated assessment for a selected information and communication technology to be applied in a government agency in Turkey, benefiting from existing literature and practical examples [8].

Impact assessment of institutional and technological systems is among the highly-appraised topics in current Turkish practice and academia, following the general trend in the world. We hope our work also contributes to these related ongoing and future works.

Acknowledgments Invaluable information and financial support has been provided by Stratek R&D Ltd, for the preparation of this work. We also would like to thank Stratek team members for their contributions.

References

1. Checkland P (1999) Systems thinking, systems practice. Wiley, Chichester
2. Decision Making and Performance Management in Public Finance (2013) EU IPA (Instrument for Pre-accession Assistance)-funded Project deliverable.
3. DeLone WH, Mclean ER (2003) The DeLone and McLean model of information systems success: a ten-year update. J Manag Inf Syst 19(4):9–30
4. Eldred M (2007) Dialectic of self and other: wrestling with Plato, Hegel, Heidegger. http://www.arte-fact.org/untpltcl/dlctcslf.html. Accessed 7 Dec.2011
5. Firestone JM (2004) The new knowledge management: a paradigm and its problems. http://www.kmci.org/media/Firestone-tnkmparadigm.pdf. Accessed 30 Jan 2014
6. Geyer F (2002) The march of self-reference. Kybernetes 31(7/8):1021–1042. Emerald
7. Güçlü N (2011) Modeling and assessment of the effectiveness of government information systems. Unpublished PhD thesis submitted to METU, Turkey
8. Güçlü N, Alır G, Öztopçu L, Medeni T (2013, 2014) Weekly group meetings. Ankara
9. McElroy MW (2003) Understanding 'The new knowledge management', 2. www.macroinnovation.com. Accessed 30 Jan 2014
10. Medeni (2008) Educating managers for the global knowledge economy: an action research into international masters program in practicing management. Unpublished PhD thesis submitted to JAIST, Japan
11. Medeni T, Medeni İT (2012) Reflection and refraction for knowledge management systems. Int J Ebus Egov Stud 4(1):55–64
12. Medeni T (2004) Reflective management learning practice model. International workshop on Business and Information (BAI), Taiwan. http://www.jaist.ac.jp/~medeni/researches/taiwanconference_reflective%20management%20learning%20model.pdf. Accessed 30 Jan 2014
13. Melford M (2005) Beyond eGovernment, transforming public services., 2005. Ministerial eGovernment conference
14. Morgül (2013) Kamu politikaları için karmaşıklık kuramı. V. Kamu Politikaları Çalıştayı, Turkey
15. Nonaka I, Toyama R (2003) The knowledge-creating theory revisited: knowledge creation as a synthesizing process. Palgrave Macmillan J 1(1):2–10
16. Stratek (2014) Stratek internal report on information system evaluation. Stratek, Turkey
17. Takeuchi H, Nonaka I (1995) Hitotsubashi on knowledge management. Wiley, Singapore
18. The project of Decision making and performance management in public finance (2014) Closure meeting, 28 January, Ankara, Turkey
19. Willner M, Hero G, Weiner J (2006) Global history volume I: the ancient world to the age of revolution Barron's. Barron's Educational Series, Inc., New York

Chapter 35
Complexity of Adopting Behaviourism Learning Theories Among Primary School Students

Jamilu Ahmad Kwari and Auwal Bala Bodinga

Abstract Learning is a relatively permanent change in behaviour as a result of reinforced practice and experience. Through learning, humans tend to become better adjusted to their environment and also become fully equipped to ensure the sustenance and development of the said environment. In Sironko (Uganda), most primary school children learn through the rote method due to a number of reasons such as language barrier. Hence, teachers strive to ensure that learners do not only memorise but also have ardent understanding of contents. Along the process, teachers adopt psychological learning theories and techniques. Behaviourism learning theory is among the most commonly used in Ugandan schools owing to it being in tandem with many aspects of the country's indigenous culture. This paper analysed the complexity of adopting behaviourism learning theories among Sironko primary school students. The researchers used a qualitative method to obtain data by interviewing teachers and students. It was found that the main impediments to adopting behaviourism learning theories include; poor funding of schools, inadequate teachers' commitment to work, language barrier, inadequate school facilities and truancy.

35.1 Introduction

Learning is a process that brings together cognitive, emotional and environmental influences and experiences for acquiring, enhancing and making changes in one's knowledge, skills, values and world views. Through learning, humans tend to become better adjusted to their environment and also become fully equipped to ensure the sustenance and development of the said environment. Changes caused

J.A. Kwari (✉)
Department of Educational Psychology and Counseling, Islamic University in Uganda, Mbale, Uganda
e-mail: jamiluahmadahmad@gmail.com

A.B. Bodinga
Department of Educational Psychology and Counseling, Usmanu Danfodiyo University, Sokoto, Nigeria
e-mail: balarabeauwal47@gmail.com

Ş.Ş. Erçetin (ed.), *Chaos, Complexity and Leadership 2014*,
Springer Proceedings in Complexity, DOI 10.1007/978-3-319-18693-1_35

by development (such as growing taller) are not instances of learning. Neither are characteristics of individuals that are present at birth (such as reflexes and responses to hunger or pain). However, humans do so much learning from the day of their birth and or even earlier [5], and that learning and development are inseparably linked. Learning to walk is mostly a developmental progression but also depends on experience with crawling and other activities. The adolescent sex drive is not learned, but learning shapes individuals' choices of desirable partners. A child's anxiety on seeing a doctor with a needle is definitely a learned behaviour. The child has learned to associate the needle with pain, and his body reacts emotionally when he sees the needle. This reaction may be unconscious or involuntary, but it is learned nonetheless [8].

However, these processes of learning are usually not spontaneous but need to follow a systematised orderly progression. Before a child learns how to read alphabet, he must have made repetitive errors until subsequently the alphabets become concrete in his mind [9]. Conversely, there often arise situations whereby children coming from the same home, attending the same class, taught by same teacher at the same time and using same teaching aids learn at a different pace. These gave rise to the complexity in learning where teachers strive to ensure optimum learning in children [2]. In doing this, numerous learning theories are used.

Like many other aspects of human development, learning has a number of theories that explains how it occurs. A theory is a system of concepts and propositions scientifically proven, that collectively presents a logical, systematic and coherent explanation of a phenomenon. A theory provides detailed systematized information of an area of knowledge. It serves as a guideline to conduct further research in the area and it produces new facts to supplement the previous facts and gives an organized explanation about a phenomenon [1]. Learning theories attempt to explain the mechanism of behavior involved in the learning process. They provide answers to the following [1];

1. Role of drill and practice in learning.
2. Utility of rewards and punishment.
3. Place of insight and understanding in the process of learning.
4. Role of transfer of learning in various situations.
5. Limits of learning with regard to the capacity of an individual in various aspects.

Behaviourism school of thought in psychology came-up with differing learning theories, some of which include; classical conditioning theory, operant conditioning theory and trial and error theory.

Learning theories are adopted and/or adapted across the world in differing patterns. In the United States, the rise of progressive education and the works of John Dewey, BF Skinner, and JB Watson helped to actuate the acceptance and utilisation of behaviourism learning theories across the country. In Germany, the dominant theory of learning was cognitivism owing to the wide spread of Gestalt psychology in the country through the works of Max Weithermer, Kurt Kofka and Wolfgang Koehler. Their approach was differentially away from the mechanistic associative view of behaviourism, that students' learning should be focused on shaping their perception rather than pairing dissimilar entities [1].

In Africa, the system was different because there were no indigenously developed learning theories; rather each country adopted what was unceremoniously transferred from the colonial masters [9]. And that because the theories were originally adopted not adapted, there were many complexities in using them across Africa.

According to Semakula [11] teachers in Uganda widely use behaviourism learning theories by pairing rewards with good performance and/or good attitude. But this was not so much effective because the rewards used were not usually complacent to the needs of the students. Similarly, cognitivism was not effectively used because the students were not offered sufficient time to identify solutions to given problems.

However, the period 2012–2014 witnessed many seminars and workshops for teachers in Uganda particularly in Sironko district. This paper therefore seek to find out if there are still complexities in adopting learning theories among primary school students in Sironko district, Uganda.

35.2 Adopting Behaviourism Among Primary School Learners in Sironko District

Behaviourism is an approach to psychology that combines elements of philosophy, methodology, and theory. It emerged in the early twentieth century as a reaction to "mentalistic" psychology, which often had difficulty making predictions that could be tested using rigorous experimental methods. The primary tenet of behaviourism, as expressed in the writings of John B. Watson, B. F. Skinner, and others, is that psychology should concern itself with the observable behaviour of people and animals, not with unobservable events that take place in their minds. The behaviourist school of thought maintains that behaviours as such can be described scientifically without recourse either to internal physiological events or to hypothetical constructs such as thoughts and beliefs [3].

Behaviourists doubted the reliability and usefulness of the method of introspection, a technique by which subjects are asked to describe their own mental processes during various tasks. They were also critical of Sigmund Freud's emphasis on unconscious motives [6]. In search of more-scientific methods, psychologists gradually turned away from research on invisible mental processes and began to study only behaviour that could be observed directly. J. B. Watson is widely credited as the *father* of behaviourism because of his experiment with *Little Albert,* it is argued that though other theorist may have preceded him, he was the first to use humans to demonstrate how conditioning typically occur [12]. Other famous behaviourists include E L Thorndike, Ivan Pavlov, and B F Skinner. These personalities came with theories of trial and error, classical conditioning and operant conditioning respectively.

35.3 Classical Conditioning Theory

Classical conditioning (also Pavlovian conditioning or respondent conditioning) is a kind of learning that occurs when a conditioned stimulus (CS) is paired with an unconditioned stimulus (US). Usually, the CS is a neutral stimulus (e.g., the sound of a tuning fork), the US is biologically potent (e.g., the taste of food) and the unconditioned response (UR) to the US is an unlearned reflex response (e.g., salivation). After pairing is repeated (some learning may occur already after only one pairing), the organism exhibits a conditioned response (CR) to the CS when the CS is presented alone. The CR is usually similar to the UR, but unlike the UR, it must be acquired through experience and is relatively impermanent [7].

A classic experiment by Ivan Pavlov (in 1906) exemplifies the standard procedure used in classical conditioning. First Pavlov observed the UR (salivation) produced when meat powder (US) was placed in the dog's mouth. He then rang a bell (CS) before giving the meat powder. After some repetitions of this pairing of bell and meat the dog salivated to the bell alone, demonstrating what Pavlov called a "conditional" response, now commonly termed "conditioned response" or CR.

Bell (NS) → No response
Meat powder (US) → Salivation (UR)
Meat powder (US) + Bell (NS) → Salivation (UR)
After repeated pairing,
Bell (CS) → Salivation (CR)

35.4 Principles in Classical Conditioning

Acquisition This is a process by which NS gains the characteristics of US to become a CS. An example of acquisition is when the bell (in Pavlov's experiment) elicits salivation. The extent of conditioning may be tracked by test trials. In these test trials, the CS is presented alone and the CR is measured. A single CS-US pairing may suffice to yield a CR on a test, but usually a number of pairings are necessary. This repeated amount of trials increases the strength and/or frequency of the CR gradually. The speed of conditioning depends on a number of factors, such as the nature and strength of both the CS and the US, previous experience and the animal's motivational state.

Stimulus Generalization This involves an organism eliciting CR for stimuli that are similar to CS. It occurs if, after a particular CS has come to elicit a CR, another test stimulus elicits the same CR [7]. Usually the more similar are the CS and the test stimulus the stronger is the CR to the test stimulus. The more the test stimulus differs from the CS the more the conditioned response will differ from that previously observed.

Stimulus Discrimination Stimulus discrimination is when an organism can distinctly identify CS. It is observed when one stimulus ("CS1") elicits one CR and another stimulus ("CS2") elicits either another CR or no CR at all. This can be brought about by, for example, pairing CS1 with an effective US and presenting CS2 in extinction, that is, with no US.

Latent Inhibition In latent inhibition, an exposure to a stimulus of little or no consequence will prevent a conditioned association with the stimulus being formed. This process will inhibit the formation of memory by preventing learning of the observed stimuli. This process is thought to prevent information overload.

Extinction This is the disappearance of a CR when a CS is repeatedly presented without the US. An example of extinction is when the bell (in Pavlov's experiment) can no more elicit salivation. In order to make a learned behaviour disappear, the experimenter must present a CS alone, without the presence of US. Once this process is repeated continuously, eventually, the CS will stop eliciting a CR. This means that the CR has been "extinguished" [4].

Spontaneous Recovery This is the reappearance of the conditioned response after a rest period. That is, if the CS is tested at a later time (for example an hour or a day) after conditioning it will again elicit a CR. This renewed CR is usually much weaker than the CR observed prior to extinction. It is a sudden reappearance of the CR after it had previously extinguished.

Other principles include; higher order conditioning, sensory preconditioning, counterconditioning, conditioned suppression, conditioned inhibition, and blocking. Classical conditioning was used to a large extent by many teachers in Sironko district. One of the teachers said during an interview with the researcher:

> Our children here have a very mechanistic life pattern such that they will like to immediately see the reward for any good work they do. Any delay in the reward may lead to such action not repeated. The children are more interested in replicating what they are taught rather than making initiative. It is for this reason and many more that we the teachers have no option than to follow that approach. Although we believe that the approach does not ultimately enhance creativity

The following are some of the complexity encountered by teachers in adopting classical conditioning theory by primary school teachers of Sironko district, Uganda;

Although classical conditioning can be used by teachers to develop students' cognitive ability through forming students into groups consisting of very intelligent and less intelligent students, the pupils in Sironko usually resist been in a group with colleagues of superior intelligence. Similarly, those with above average intelligence do not tend to appreciate been in a group with those having below average intelligence. A very bright pupil asserts to the researcher that she wouldn't like to be in other groups because "they cannot solve mathematic questions very fast".

Systematic desensitization is among the major applications of classical condition in a school setting. It is a technique used to reduce students' phobia, test anxiety etc. However, many teachers interviewed by school counsellors are not aware of how systematic desensitization is applied. The few that are aware of it claimed that any attempt to use that technique will be termed an attempt to initiate pupils into witch craft.

Classical conditioning can be used to teach abstract concepts; this is achieved through the use of instructional teaching materials. Though these materials are sparingly used in classrooms, their availability is grossly below adequate owing to the inadequate funds received by the schools [11]. The researcher inquired why the teachers do not make improvised teaching aids, one of the teachers interviewed responded by saying:

> You know the construction of teaching aids is usually time-consuming and we do not have such time because on the average a teacher in this school has 6/9 periods per day and after school hours we tend to be occupied with our secondary sources of income

Classical conditioning advocates that teachers use continuous repetition until learning become concrete. Teachers in Sironko district said they found constant repetition leads to rote learning among pupils. They explained to the researcher that the shift from constant repetition was made after it was found that there pupils having learnt multiplication in twos can tell 2×3 but cannot tell the answer for 3×2. A teacher interviewed in one of the schools said it was after these findings that they were encouraged to learning of concepts rather than memorisation.

35.5 Operant Conditioning Theory

Operant conditioning (or instrumental conditioning) is a type of learning in which an individual's behaviour is modified by its antecedents and consequences. Instrumental conditioning was first discovered and published by Jerzy Konorski and was also referred to as Type II reflexes [10]. Mechanisms of instrumental conditioning suggest that the behaviour may change in form, frequency, or strength. The expressions "operant behaviour" and "respondent behaviour" were popularized by B. F. Skinner who worked on reproduction of Konorski's experiments. The former refers to "an item of behaviour that is initially spontaneous, rather than a response to a prior stimulus, but whose consequences may reinforce or inhibit recurrence of that behaviour".

Operant conditioning proffered that behaviour can best be learnt through the use of proper reinforcement and that behaviour can also be eliminated through proper use of punishment.

35.6 Forms of Reinforcers

35.6.1 Primary and Secondary Reinforcers

- Primary reinforcer is one that satisfies a biological need e.g. food, water
- Secondary reinforcer (conditioned reinforcer) is a previously neutral stimulus that has become reinforcing through repeated association with another reinforcer e.g. praise, money, grades

35.6.2 Tangible and Intangible Reinforcers

- Material reinforcers include tangible, actual object such as food, toys
- Social reinforcers are gesture/sign that communicates positive regard e.g. smile, praise
- Intrinsic reinforcers are internal sources of satisfaction like feeling good, proud

Skinner used a rat and a specialised box *'Skinnerian box'* to conduct his experiment. The Skinner box had a grid floor, light, lever, and food cup. The animal is first deprived of food/water. It is then taken to the box where the experimenter triggers the magazine to release food pellet in a cup. This is accompanied by a click sound. With time, the animal is left on its own. A recorder is connected such each response (y) is recorded against time (x). The animal has to learn to press the lever in order to obtain food.

35.7 Principles in Operant Conditioning

1. **Free operant level/baseline**: This is the frequency of an operant before reinforcement.
2. **Terminal behaviour**: This is the form and frequency of a desired response at the end of a planned reinforcement programme.
3. **Shaping**: This is the process of reinforcing successively closer and closer approximations to the terminal behaviour until the desired behaviour is exhibited. This is when an individual's free operant level of a response is very low and may not occur, yet it has to be reinforced.

Two processes are involved, i.e. **differential reinforcement** where some responses are reinforced and others are not; and **successive approximation** where only those responses that become increasingly similar to what the experimenter wants are reinforced.

- **Chaining**: This involves teaching a sequence of responses through shaping by first reinforcing just one response, then two in a row, then three, etc. This uses the concept of secondary reinforcement. Because of their association with the primary reinforcer, the events before the primary reinforcer take on secondary reinforcing properties.
- **Extinction**: This is when a response is no longer followed by a reinforcer, it gradually decreases and returns to baseline rate.
- **Extinction burst**: after extinction, there may be some brief and sudden reappearance of behaviour. It is similar to spontaneous recovery in classical conditioning. Only that extinction burst occurs with a higher frequency.

Although both classical and operant conditioning are to some extent dependent on rewards, the forma is virtually based on contingency of the reward on the action performed while the latter is based on contiguity of the reward and the action performed.

The following are some of the complexity encountered by primary school teachers of Sironko district in adopting operant conditioning theory;

- Although operant conditioning can be used by teachers to develop students' psychomotor domain by organising competitions on crafts, arts and other skills. There after prizes are giving to winners of the competitions, this is not done within Sironko District as funds for organising such competitions are not available.
- Law of effect in operant conditioning shows that teachers should ensure that they enjoy their teaching profession because it is only through such that they can achieve job satisfaction and consequently effective work output. However, this is not what obtains in Sironko district as one of the teachers interviewed said;

  > It's only a joke to say that we are having any job satisfaction here, the meagre salary we receive is below $150/month, yet we spend the large part of our productive day time at the school. Imagine such salary to a man having a wife, five children and three dependent relatives. We only hope that things will be better in the near future . . .

  Operant conditioning advocates that school management should organise events at the end of each year to award best performing students. This will help to create healthy competitions among students. Hence, improving their performance. However, this is not what obtains in most public primary schools in Sironko although a handful of private schools awards best performing students at the end of the year.
- Contrary to classical conditioning, operant conditioning allows the use punishment for both students and staff. When effectively used, punishment can lead to the elimination of undesirable response which follows [10]. However, due to wide spread corruption, defaulting teachers are usually not punished thereby leading to laxity on the part of most teachers.

35.8 Conclusion

The aim of every teaching process is to imbibe learning on students, a process by which teachers use differing methods and techniques. Learning theories helps to explain how learning occurs and what need to be done in an effort to ensure students' learning. Behaviourism as one of the basic theories of learning explains that learning tends to occur when there is an association between stimuli and response. This paper explained the complexity encountered by teachers in adopting classical conditioning theory and operant condition theory (which are both types of behaviouristic theories).

The findings reveal that students' resistance, cultural identity, insufficient school funding, teachers' laxity and corruption are among the many complexities militating appropriate adoption of behaviouristic learning theories.

References

1. Aggarwal JC (2005) Essentials of educational psychology. Vikas Publishing House, New Delhi
2. Bandara A, Barbaranelli C, Caprara GV, Pastorelli C (1996) Multifaceted impact of self efficacy beliefs on academic functioning. J Child Dev 67:1206–1222
3. Behaviorism (2014) Retrieved August 20, 2014, from Wikipedia: en.m.wikipedia.org/wiki/behaviorism
4. Berliner DC (2009) Educational psychology. Microsoft encarta. Microsoft Corporation, Redmond
5. Bolarin TA (2011) Child growth and development. Olu-Akin Publishers, Lagos
6. Chauhan SS (2007) Advanced educational psychology, 6th edn. Vikas Publishing House, New Delhi
7. Classical Conditioning (2014) Retrieved September 12, 2014, from Wikipedia: en.wikipedia.org/wiki/classical_conditioning
8. Kassin S (2009) Learning. Microsoft encarta. Microsoft Corporation, Redmond
9. Mwamwenda T (2005) Educational psychology: an African perspective. Kratemore Publishers, Durban
10. Operant Conditioning (2014) Retrieved August 19, 2014, from Wikipedia: en.m.wikipedia.org/wiki/operant_conditioning
11. Semakula F (2008) Comparative studies of two methods of teaching on some selected concepts of magnetism at A' level in eastern Uganda. Unpublished Masters dissertation, Kampala International University, Uganda
12. Slavin RE (2006) Educational psychology: theory and practice, 8th edn. Pearson Education, New York

Chapter 36
Hampton Machine Tool Company

Almira Sinemis Şekerci

Abstract Chaos is a fact of life and accordingly engulfs almost all aspects of human society ranging from the social-economic to the political. It is thus not surprising to note that we cannot talk about any aspect of life in a chaos vacuum. Many business companies have ups and downs in their business operations. These ups and downs are typical of chaos and indeed render the activities of a company complex. They are influenced by economic conditions, political conditions (e.g. War), expansion of the export market and diminishing value of competitors due to bad economic conditions. Accordingly some companies end up closing while some resort to borrowing or taking on loans in order to survive and negotiate this complex jig-saw. Such loans have repayment schedules and conditions and given the intricate nature of the same, some companies fail to pay back as expected. There are several factors that affect the payment process mostly of which is a sharp decline in sales. This paper explored the plight of the Hampton machine tool company in the wake of a debit and loan crisis and also examines how the company was able to survive the onslaught of a chaotic and complex loan setting.

36.1 Hampton Machine Tool Company Case Analysis

Hampton machine tool Company established and started operations in 1915 in St Louis area. From their financial records it was easy to see that they were a stable and profitable company in mid 1960s, their revenues were based on military craft and automobile manufacture sector [1].

The company had ups and downs between the 1960s and 1970s due to unfulfilled demands. These ups and downs were influenced by the economic conditions, political conditions (Vietnam War), expansion of the export market and diminishing of competitors because of bad economic conditions. However HMTC survived all these obstacles at that time because they had a strong financial policy [1].

A.S. Şekerci (✉)
Hult International Business School, Sanfrancisco, USA
e-mail: almirasinemis@gmail.com

Ş.Ş. Erçetin (ed.), *Chaos, Complexity and Leadership 2014*,
Springer Proceedings in Complexity, DOI 10.1007/978-3-319-18693-1_36

In 1978 Benjamin Cowins (President of HMTC) requested a 1M USD from the bank to purchase stock from their adversary shareholders to pay it in September 1979 with the monthly interest rate of 1.5 % [1]. Vice President of the Bank; Mr. Edward allowed this loan because he had at least six solid reasons to do so. Firstly; he saw that HMTC's project sales and forecast of their financial statements are good enough; secondly; HMTC is a creditable depositor in their bank; thirdly, HMTC has an outstanding history; fourthly, they broke their profit record between 1965 and 1970 because of the strong auto market and huge number of defense spending for the Vietnam War; fifthly, they didn't have debts for 10 years and lastly the president of HMTC; Mr. Cowins is a well -known and reliable person in the business world. However things didn't go as expected and September 1979 V. President Mr. Edward, got a letter from Mr. Cowins for extension of the loan repayment on December 31, 1979 and also he asked for another loan (350,000 USD) to buy equipment.

Right now the question is; should Vice president accept the last request from Mr. Cowins or not? There are few important things that Mr. Edward needs to analyze. First of all the financial position for the upcoming months and possible calculations for forecast of the income statement, forecast balance sheet and cash flow analyses to figure it out that; if the company could be able to pay the loan and interest on time.

But, one of the most important questions is that; why did they fail to repay their loan of 1M USD in the first place? It will be really useful to understand the reasons to give more accredited decision for giving the next loan of 350.000 USD or not.

There were several factors that affected the payment process. First of all, they had a sharp decline in sales. The reason for this sharp decline was the external factors such as, when the 'Organization of Arab Petroleum Exporting Countries'(OAPEC) proclaimed an Oil Embargo, which also caused the increase of gas prices and at the end caused recession in USA!(OPEC) implements the oil diplomacy in 1973. Any nation that supported Israel in Yom Kippur War against Syria, Jordan and Egypt banned to buy oil that they sell. NY stock exchange dropped by 97 billion dollars because it was the end of the cheap oil prices. It turned out to be one of the biggest and worst recessions in USA. Normally, oil prices must be set by the oil companies however after this incident OPEC announced that they are going to determine their own prices for the gasoline that they export. After this, oil prices per barrel went up to 130 % more than last October and came to 11.65 dollars. After these events, things were settled and, HTMC's sales started to increase again. However, they needed to have a loan to purchase equipment for their backorders and the reason why they didn't have sufficient cash is because they had unfilled orders that cost 16,500,000 USD. Shipment schedules also had problems because of the supplier's that delayed to send electronic control mechanisms to the company eventually leading to reduction of cash. Another and indeed one of the main reasons is that; Stock Repurchases (75,000 shares) was a huge amount of payment (3M

USD). Between 1978 and 1979, the company's inventory purchases were 42 % of their total expenditures while the stock repurchases represented 58 % of total expenditures (Exhibit 36.2). There was an increase in their inventory holding! Hampton wanted to rebuy the shares held by dissident holders (who were preventing their rate of growth) for to reduce the numbers of shares outstanding and also wanted to increase the earnings per share. These repurchases will change the company's capital structure, which is a good move, and it could be seen as a positive thing from the eyes of investors. The company also spent $420,000 in 4 months for the raw materials before they needed them. That much of increase in inventories plays a huge part in the delay of paying the loan. In December they also needed to pay $181,000.00 tax payment and $150,000.00 dividends payment.

When we look at the ratios we can see that they were not stable however when the project time comes, they increase. Their gross profit margin which is what is going to remain from sales after a company paid CGS. Their gross profit margin is high except the September and the reason for this is the inventory reduction of 1,320,000 USD. This reduction is showing us that HMTC is going to operate with a 13 % loss in September (Exhibit 36.2). However after this reduction, forecast ratios are showing that (COGS, OP) will be stable 3 months after September, which shows us that the company will be profitable and they might be worth being given credit.

When we look at the cash flow we can see that HMTC would perform better when they get their equipment to perform their operations. They won't need further funds for their operating activities. However the problem is with their -331,500 USD cash balance (they will be 331,500 USD short), they need to have extra funds to cover their loan (Exhibit 36.1). They can also consider postponing the payment for dividends. However, even HMTC postpone paying their dividends (150,000 USD) they would still have a negative number by December. They are not in the safe financial position. They have to improve the liquidity in the company!

Meanwhile we can see that in January they had a cash balance of 933,500 USD (Exhibit 36.1). Their account receivable coming from December and they will collect their sales on January! If the company talks with the bank and asks them to extend the deadline to January rather than December they would be able to pay their debt.

Realistically it was the best decision for Mr. Edward to reject their proposal to pay at the end of December because when we see the cash flows HTMC is not able to pay their loan with the interest rate at the end of the year. Bank must re-decide their interest rate and extend time. Bank could offer 1.75 % interest rate rather than 1.5 % that means 21 % for annual and extend time to January. Increase the interest rate must have been their benefit and extending the repayment to January would be HTMC's benefit. Bank must also consider sanctioning penalties and they must ask for a security incase HTMC fails to repay again.

36.2 Exhibits of the Hampton Machine Tool Company Financial Accountability (Exhibits 36.1, 36.2, 36.3 and 36.4)

Exhibit 36.1 Cash budget Sept–Jan 1979–1980

	Sept	Oct	Nov	Dec	Jan
CASH INFLOWS					
COLLECTIONS OF RECEVİABLE	684	1,323	779	1,604	2,265
BANK LOAN	0	350	0	0	0
T.CASH INFLOWS	684	1,673	779	1,604	2,265
CASH OUTFLOWS					
PAYMENTS OF ACCOUNTS PAYABLE	948	600	600	600	600
OTHER OEPRATING OUTFLOWS	400	400	400	400	400
PURCHASE NEW EQUİPMENT	0	350	0	0	0
TAX PAYMENT	181	0	0	181	0
INTEREST PAYMENT TO BANK	15	15	20	20	0
DIVIDENS	0	0	0	150	0
T.CASH OUTFLOWS	1,554	1,365	1,020	2,701	1,000
RECONCILIATION					
BEGINNING CASH BALANCE	1,559	699	1.007	766	−331
NET CASH FLOW	−860	308	−241	−1,097	1,265
ENDING CASH BALANCE	699	1,007	766	−331	934

*IN USD ‘000

Exhibit 36.2 Pro forma income statement Sept–Dec 1979

NET SALES	Sept	Oct	Nov	Dec
PURCHASES	2,163	1,505	1,604	2,265
WIP INVENTORY REDUCTION	600(advance purchase of raw materials)	600	600	600
RAW MATERIALS REDUCTION	1,320(because of the purchase of machines)	0	0	0
COST OF SALES	105	105	105	105
GROSS PROFIT	2,025	705	705	705
DEPRECIATION	138	800	899	1,560
OTHER EXPENSES	10	10	14(350/96) = 3,64 additional depreciation expense	14

(continued)

Exhibit 36.2 (continued)

NET SALES	Sept	Oct	Nov	Dec
SALES EXPENSES	400	400	400	400
EBIT	−272	390	485	1,146
INTERESET EXPENSES	15(1M*1.5%)	15	20(350*1.5 %) 5.25 = 15 + 5 = 20 (average)	20
NET INCOME BEFORE TAXES	−287	375	465	1,126
INCOME TAXES	−138	145	223	540
NET INCOME AFTER TAXES	−149	230	242	586
DIVIDENDS	0	0	0	150
NET RETAINED EARNINGS	−149	230	242	436

*IN USD '000

Exhibit 36.3 Pro income statements for 4 months in Sept–Dec 1979

Sales	7, 5737.000 USD (TOTAL SALES FROM SEPTEMBER TO DECEMBER) (Sept;2.163USD,Oct;1.505USD,Nov;1,604 USD, Dec 2.265 USD)
Cost of sales	5,740.000 USD (−1,320 + 420 + 2,400 + 1,600)
Expenses	117,000 (depreciation + 4 month interest)
Profit before tax	1,680.000
Taxes	771,000
Profit after taxes	909,000
Dividends	150,000
Add. to Retained Earnings	759,000

*In USD

Depreciation: 350,000 USD for 8 years, 1 year 43.750 USD and 1 month is 3.650 USD

Exhibit 36.4 Pro forma balance sheet Sept–Dec 1979

	Sept	Oct	Nov	Dec
Cash	699,000	1,007,000	766,000	−331,000
Account receivable	1,323,000	779,000	1,604,000	2,265,000
Inventories				
Total current assets	5,361,000	5,020,000	5,499,000	4,958,000
Gross fixed assets	4,010,000	4,360,000	4,360,000	4,360,000
Acc. Depreciation	3,100,000	3,110,000	1,123,646	3,137,292
Net fixed assets	910,000	1,250,000	1,236,354	1,222,708
Prepaid expenses	42,000	42,000	42,000	42,000
Total assets	6,313,000	6,312,000	6,777,354	6,222,708

(continued)

Exhibit 36.4 (continued)

	Sept	Oct	Nov	Dec
NOTES PAYABLE	1,000,000	1,350,000	1,350,000	0
Accounts payable	600,000	600,000	600,000	600,000
ACCURALS	552,000	552,000	552,000	552,000
TAXES PAYABLE	160,240	305,240	528,610	888,260
CUSTOMER ADV	726,000	0	0	0
T.LIABLITIES	3,038,240	2,807,240	3,030,610	2,040,260
COMMON STOCK	428,000	428,000	428,000	428,000
TOTAL EQUITY	3,274,760	3,504,706	3,746,744	4,182,448
	2,846,760	3,076,760	3,318,744	3,754,448
T. LIABILITIES AND EQUITY	6,313,000	6,312,000	6,777,354	6,222,708
Change in cash	−860,000	−308,000	−241,000	−1.097,000

*IN USD 000's

36.3 Conclusion

The paper has fully explored the Hampton machine tool Company which was established and started operations in 1915 in St Louis area. The company encountered and indeed suffered a financial crisis which created a scenario of chaos in the said organization or company. This led to the company having to take a loan by the company and the loan itself was hard to pay given a number of complex issues especially extraneous factors. Nevertheless, the company took a loan to survive and while some of the loan was paid back, calamity had already stuck. This necessitated even another loan despite the previous one being pending. It thus took the increase in interest rates in order to rescue the company.

Reference

1. Mullins JRDW (1991) Hampton Machine Tool Company. Lecture conducted from, Boston, USA

Chapter 37
Parameter Estimation of Nonlinear Response Surface Models by Using Genetic Algorithm and Unscented Kalman Filter

Özlem Türkşen and Esin Köksal Babacan

Abstract Some of the real world problems are characterized by using nonlinear functions in the parameters. In this case, optimization of nonlinear response surface models become challenging with derivative-based optimization methods. In this study, two of the derivative free methods, Genetic Algorithm (GA) and Unscented Kalman Filter (UKF), are used for parameter estimation of complex nonlinear response surface model. A numerical example in chemical science is given to illustrate the performance of the methods.

37.1 Introduction

One of the main stage to solve a real world problem is development of an adequate functional relationship between a response of interest and a number of associated input variables. In general, such a relationship is unknown and may have uncertainty related to the structure of model parameters which characterize the model. The most common preferred basic modeling tool is regression analysis. The regression analysis investigates the response models which are linear in the parameters such as low degree polinomial models in many response surface modeling studies [2, 7]. Parameter estimation of these linear response models is achieved by using Ordinary Least Squares (OLS) method. However, many mathematical models used in scientific research contain parameters that are not expressed linearly. In this case, nonlinear regression models are preferred.

A regression model is called nonlinear in the parameters, if the derivatives of the model with respect to the model parameters depends on one or more parameters. The most commonly used assumptions for nonlinear regression is the same as assumptions for linear regression. The only exception being that the regression function is a nonlinear function of the unknown parameters instead of a linear function of the parameters. The nonlinear models have been used in many fields

Ö. Türkşen (✉) • E.K. Babacan
Faculty of Science, Statistics Department, Ankara University, Tandoğan, Ankara 06100, Turkey
e-mail: ozlemturksen@hotmail.com

Ş.Ş. Erçetin (ed.), *Chaos, Complexity and Leadership 2014*,
Springer Proceedings in Complexity, DOI 10.1007/978-3-319-18693-1_37

particularly in guidance or navigational systems, target tracking, biological and chemical sciences, economic and growth models. The models, which are nonlinear in parameters, can be made linear in parameters by suitable transformations. Even so, the transformations seem to destroy the model assumptions, e.g. assumption of variance homogeneity. A formal application of the OLS to the transformed model will not produce a model with correct statistical properties. According to this, the OLS method will be biased for parameter estimation of the transformed models. On the other hand, calculating derivatives with respect to the parameters will be resulted with normal equations which are nonlinear in the parameters. Therefore, derivative-free optimization methods will be very useful for obtaining parameter estimates of complex response problems.

The parameter estimation procedure for nonlinear response models is based on minimizing the quadratic function of difference between observed and predicted response values which is called error function. This error function is considered as objective function. There have been several derivative free methods for optimizing this complex objective function, e.g. Genetic Algorithm (GA), [4] Nelder-Mead simplex method [11], Simulated Annealing [8], Unscented Kalman Filter (UKF) [5]. In this study, two of the derivative free methods, GA and UKF, are preferred to achieve the parameter estimation procedure for nonlinear response models. In the next section, brief description about nonlinear response model is given. In Sect. 37.3, parameter estimation procedures of nonlinear response problems with GA and UKF are explained and algorithmic steps are presented. An application study is performed in Sect. 37.4. In Sect. 37.5, the conclusion is given with the obtained estimation results through performance metric.

37.2 Nonlinear Responses Model

A nonlinear response surface model is a continuous nonlinear multivariate approximation to real form of the response. Suppose an unknown response model given as the form below

$$Y_i = \eta\left(\mathbf{X}_i; \boldsymbol{\theta}\right) + \varepsilon_i, i = 1, 2, \ldots, n \tag{37.1}$$

in which Y_i is the response variable; $\mathbf{X}_i$ is a vector of input variables, $\mathbf{X}_i = [X_{1i}, X_{2i}, \ldots, X_{ki}]$; $\boldsymbol{\theta}$, is a vector of parameters, $\boldsymbol{\theta} = \left[\theta_1, \theta_2, \ldots, \theta_p\right]$; η is a nonlinear function in the parameters (at least partially nonlinear function of a η dimensional parameter vector $\boldsymbol{\theta}$) and ε_i is the error term, $i = 1, 2, \ldots, n$. In order to obtain parameter estimates of the model given in Eq. (37.1), the sum of squares for error is defined as

$$\phi\left(\boldsymbol{\theta}\right) = \sum_{i=1}^{n} \left[Y_i - \eta\left(\mathbf{X}_i; \boldsymbol{\theta}\right)\right]^2 = \sum_{i=1}^{n} \varepsilon_i^2 \tag{37.2}$$

which wanted to be minimized. The quadratic function given in Eq. (37.2) is considered as objective function for parameter estimation procedure. In order to minimize the Eq. (37.2) a well-known methodology is calculating derivatives with respect to parameter vector $\boldsymbol{\theta}$. However, the obtained normal equations can be nonlinear in the parameters. In this case, derivative-free optimization methods should be used to achieve the parameter estimates.

37.3 Parameter Estimation Procedure

In this section, two derivative-free optimization methods, GA and UKF, are explained briefly. The algorithmic steps of the methods are given in detail for parameter estimation of nonlinear response surface problems.

37.3.1 Parameter Estimation with Genetic Algorithm

Genetic Algorithm (GA) is a metaheuristic method based on natural selection and genetic mechanism. The basic principle of it is the Darwinian "survival of the fittest approach", introduced by Holland [4]. GAs search from a population points, not a single point; use objective function information, not derivatives; use probabilistic rules, not deterministic rules; can produce the solution without requiring initial solutions by searching from many search points simultaneously [6]. The algorithmic steps of the GA are given below:

Step 1: Create an initial population

A population of chromosomes is created initially. The chromosomes, represents the parameters, are candidate solutions of the problem. Each chromosome composed with genes as a string of binary digits which is called encoding. The population size, N_{pop}, is defined as the number of chromosomes in the population. Set the generation number $generation = 0$.

Step 2: Determine the fitness value of each individual

Fitness function value, which represents the objective function value, is calculated for each individual chromosome.

Step 3: Select next generation (parent population)

According to the fitness function, the strings with high fitness are selected so the best chromosomes are included in the new population by using selection functions, e.g. with replacement, roulette wheel, stochastic uniform, for reproduction.

Step 4: Perform reproduction using crossover (mating)

A locus is chosen randomly and exchanges the subsequences before and after that locus between two chromosomes with crossover probability, Pr_c, by using crossover functions e.g. single point, 2-points, uniform, for creating two offsprings.

Step 5: Perform mutation

Some of the bits in a chromosome is chosen and altered from 0 to 1 or 1 to 0 with mutation probability, Pr_m, for increasing the variability of the population.

Step 6: Replace the current population with the new population

New solutions are replaced with the current solution set. Set $generation = generation + 1$. If $generation < maxgen$ then go to *Step 2* else display results.

Each iteration of this procedure is called generation. The entire set of generations is called a run. Since randomness plays a great role in each run, different runs produce different results. So, it will be better to report statistics of results [16]. The main disadvantages of the GA stem from its computational complexity because of the N_{pop} size of the different estimated parameters in each run [3, 10]. And also, the GA does not guarantee the optimum solution, but leads to solutions acceptably close to the optimal solution [9].

37.3.2 Parameter Estimation with Unscented Kalman Filter

Kalman Filter (KF) and its derivations have been extensively used for linear and nonlinear state estimation problems [1]. The KF is an optimal estimator for linear dynamic systems. However, in real world problems, the systems are generally formulated as complex and nonlinear. For nonlinear stochastic systems and nonlinear deterministic systems KF has also been utilized but it is not optimal. The Extended Kalman Filter (EKF) is commonly used method in the field of nonlinear estimation. The EKF uses the standard KF equations to the first order approximation of the nonlinear model about the last estimate. It is very sensitive to initialization and if the arbitrary noise matrices have not been chosen appropriately filter divergence is inevitable [12]. The UKF is an another nonlinear estimation method for nonlinear state space models. The performance of the UKF estimator is equivalent to the KF for linear systems yet generalizes elegantly to nonlinear systems without the linearization steps required by the EKF. The UKF is based on the Unscented Transformation (UT). The UT is a method for calculating the statistics of a random variable which undergoes a nonlinear transformation [5, 13]. Suppose that x is a random variable which has been transformed a y by using a nonlinear function h ($y = h(x)$) and suppose that x has mean $\bar{x}$ and covariance P_{xx}. To calculate the statistics of y, the UT is used deterministically chosen $2n+1$ sample points which are called sigma points. The sigma points are chosen to guarantee that the sample mean $\overline{x}$ and sample covariance P_{xx} by the following algorithm

$$\begin{cases} \chi_0 = \overline{x} & , W_0 = \kappa / (n + \kappa) \\ \chi_i = \overline{x} + \left(\sqrt{(n+\kappa) P_{xx}}\right)_i & , W_i = 1/2 (n + \kappa), i = 1, \ldots, n \\ \chi_i = \overline{x} - \left(\sqrt{(n+\kappa) P_{xx}}\right)_{i-n} & , W_{i+n} = 1/2 (n + \kappa), i = n+1, \ldots, 2n \end{cases}$$

in which $\kappa \in R$. $\left(\sqrt{(n+\kappa) P_{xx}}\right)_i$ is the i.th row or column of the matrix square root of $(n+\kappa) P_{xx}$. W_i's are weights of i. sample point with satisfying $\sum_{i=0}^{2n} W_i = 1$. The sigma points are instantiated through the process model, $\gamma_i = h(\chi_i), i = 1, 2, \ldots, n$, and the mean and covariance of y are computed by using a weighted sample mean and covariance of the posterior sigma points

$$\overline{y} = \sum_{i=0}^{2n} W_i \gamma_i$$

$$P_{yy} = \sum_{i=0}^{2n} W_i \{\gamma_i - \overline{y}\}\{\gamma_i - \overline{y}\}^T.$$

Consider the following nonlinear discrete-time stochastic system

$$\begin{aligned} x(k) &= f(x(k-1)) + w(k) \\ y(k) &= h(x(k)) + v(k) \end{aligned} \tag{37.3}$$

where $x(k)$ is the $n \times 1$ state vector, $y(k)$ is the $m \times 1$ measurement vector at time instant k. The vector valued nonlinear functions, f and h are state transition and observation functions, respectively. $w(k)$ and $v(k)$ are uncorrelated zero-mean white noise processes with covariance

$$E\left(w(k)w^T(k)\right) = Q(k), E\left(v(k)v^T(k)\right) = R(k). \tag{37.4}$$

The UKF equations for nonlinear system given by Eq. 37.3 are summarized as follows [13, 15]

Step 0: Initialize

$$\widehat{x}(0) = E(x(0))$$

$$P(0) = E(x(0) - \widehat{x}(0))(x(0) - \widehat{x}(0))^T$$

Step 1: Calculate the sigma points

$$\begin{cases} \chi_i\,(k-1) = \widehat{x}\,(k-1) & ,\ i = 0 \\ \chi_i\,(k-1) = \widehat{x}\,(k-1) + \left(\sqrt{(n+\kappa)\,P\,(k-1)}\right)_i & ,\ i = 1, 2, \ldots, n \\ \chi_i\,(k-1) = \widehat{x}\,(k-1) + \left(\sqrt{(n+\kappa)\,P\,(k-1)}\right)_i & ,\ i = n+1, n+2, \ldots, 2n \end{cases}$$

Step 2: Prediction

$$\chi_i\,(k\,|k-1) = f\,(\chi_i\,(k-1))$$

The predicted mean and covariance are computed as

$$\widehat{x}\,(k\,|k-1) = \sum_{i=0}^{2n} W_i \chi_i\,(k\,|k-1)$$

$$P\,(k\,|k-1) = \sum_{i=0}^{2n} W_i\,(\chi_i\,(k\,|k-1) - \widehat{x}\,(k\,|k-1))\,(\chi_i\,(k\,|k-1) - \widehat{x}\,(k\,|k-1))^T + Q(k)$$

Step 3: Update

$$\gamma_i\,(k\,|k-1) = h\,(\chi_i\,(k\,|k-1))$$

$$\widehat{y}(k) = \sum_{i=0}^{2n} W_i \gamma_i\,(k\,|k-1)$$

The weighted covariance matrix of the predicted observations is given below

$$P_{yy}(k) = \sum_{i=0}^{2n} W_i\,(\gamma_i(k) - \widehat{y}(k))\,(\gamma_i(k) - \widehat{y}(k))^T + R(k)$$

and the covariance matrix between the state and the measurement is computed as follows

$$P_{xy}(k) = \sum_{i=0}^{2n} W_i\,(\chi_i\,(k\,|k-1) - \widehat{x}\,(k\,|k-1))\,(\gamma_i(k) - \widehat{y}(k))^T.$$

Then the state estimate $\widehat{x}(k)$ and the corresponding covariance matrix $P(k)$ can be updated as

$$K(k) = P_{xy}(k)P_{yy}^{-1}(k)$$

$$\widehat{x}(k) = \widehat{x}(k\,|k-1) + K(k)\,(y(k) - \widehat{y}(k))$$

$$P(k) = P(k\,|k-1) - K(k)P_{yy}(k)K^T(k)$$

where $K(k)$ is a Kalman Gain matrix.

Step 4: Repeat Steps 1–3 for the next sample.

It is clearly seen that there is no necessity to compute the Jacobian matrix in the UKF algorithm whereas it is needed in the EKF. This can be considered as an advantage of UKF against to the EKF.

37.4 Application

In this section, chemical reaction problem is considered as an application. Rational function response is given as

$$\eta(\mathbf{X};\boldsymbol{\theta}) = \frac{\theta_1\theta_3X_1}{1+\theta_1X_1+\theta_2X_2},\quad X_1, X_2 > 0;\ 1 < \theta_1 < 3,\ 10 < \theta_2 < 20,\ 0 < \theta_3 < 1$$

which models the chemical reactions of type $R \rightarrow P_1 + P$. Here, η is the speed of the reaction, X_1 is the partial pressure of the sought product P, X_2 is the partial pressure of the sought product P_1, θ_1 is the absorption equilibrium constant for P_1, θ_2 is the effective constant of the speed of reaction, and θ_3 is the absorption equilibrium constant for the reagent R [14]. The data set is given in Table 37.1.

The objective function is defined as

$$\phi(\boldsymbol{\theta}) = \sum_{i=1}^{5}\left[Y_i - \frac{\theta_1\theta_3X_1}{1+\theta_1X_1+\theta_2X_2}\right]^2.$$

Table 37.1 The data set for chemical reaction

X_1 :	0.28, 2.9, 3.2, 4.7, 5.5
X_2:	0, 0.1, 0.82, 1.1, 2
η :	0.33, 0.57, 0.33, 0.31, 0.23

Table 37.2 Initial tunable parameters of the GA and the UKF

GA	UKF
$popsize = 100$ $\Pr_{cr} = 0.90$ $\Pr_m = 0.01$ *Roulette Wheel Selection* *Single Point Crossover* *Bit flip mutation* $maxgen = 100$	$x(0) = \begin{bmatrix} 3 \\ 15.3 \\ 0.73 \end{bmatrix}$; $P(0) = \begin{bmatrix} 1 & 0 & 0 \\ 0 & 1 & 0 \\ 0 & 0 & 1 \end{bmatrix} \times 0.0001$ $R = 0.5$; $Q = \begin{bmatrix} 1 & 0 & 0 \\ 0 & 1 & 0 \\ 0 & 0 & 1 \end{bmatrix} \times 0.1$

Table 37.3 Performance metric results for the GA and the UKF

	GA	UKF
RMSE	0.02611	0.02436

The state-space model of chemical reactions for the UKF is defined as

$$x = \begin{bmatrix} 1 & 0 & 0 \\ 0 & 1 & 0 \\ 0 & 0 & 1 \end{bmatrix} \begin{bmatrix} \theta_1 \\ \theta_2 \\ \theta_3 \end{bmatrix} = \begin{bmatrix} \theta_1 \\ \theta_2 \\ \theta_3 \end{bmatrix}$$

$$y = \begin{bmatrix} \dfrac{\theta_3 X_1}{1 + \theta_1 X_1 + \theta_2 X_2} & 0 & 0 \end{bmatrix} \begin{bmatrix} \theta_1 \\ \theta_2 \\ \theta_3 \end{bmatrix}.$$

The initial parameters of the GA and the UKF are given in Table 37.2. These tunable parameters are defined according to the problem structure and expert knowledge.

The Root Mean Square Error (*RMSE*) is used as performance metric for comparison of the GA and the UKF. The *RMSE* is defined as

$$RMSE = \sqrt{\frac{1}{n-p} \sum_{i=1}^{n} \left(Y_i - \widehat{Y}_i \right)^2} \tag{37.5}$$

in which Y_i is observed response, $\hat{Y}_i$ is predicted response, n is number of observations, and p is number of parameters. In this study, n and p are 5 and 3, respectively. The performance metric results for the GA and the UKF are given in Table 37.3.

It is seen from the Table 37.3 that the UKF is slightly better than GA for parameter estimates according to the calculated RMSE values. Therefore, it can be said that the UKF can be used as an alternative method for parameter estimation of complex nonlinear response problems.

37.5 Conclusions

Nonlinear response models are most commonly used as well as linear models for modeling stage of the complex problems. In this study, response functions are considered as nonlinear in the parameters. In order to obtain parameter estimates of the nonlinear functions, two of the derivative-free optimization algorithms are used instead of derivative-based optimization algorithms. One of the method, used in the study, is GA which is an efficiently used metaheuristic method for complex optimization problems. The other one is UKF which is the most popular estimation method for nonlinear state-space models. The tunable parameters of the GA and the UKF are chosen according to the problem structure and expert knowledge. The results show that the GA and the UKF have similar performances according to the *RMSE* metric. However, the UKF is slightly better than the GA. It can be said that the UKF can be preferred as an optimization tool for complex nonlinear optimization problems.

References

1. Bar-Shalom Y, Li XR, Kirubarajan T (2001) Estimation with applications to tracking and navigation. Wiley, New York
2. Box GEP, Draper NR (2007) Response surface mixtures and ridge analysis. Wiley, Hoboken
3. Goldberg DE (1989) Genetic algorithms in search, optimization, and machine learning. Addison-Wesley, New York
4. Holland J (1975) Adaptation in natural and artificial systems. University of Michigan Press, Ann Arbor, MI
5. Julier S, Uhlmann J (1997) A new extension of the Kalman filter to nonlinear systems. In: International symposium on aerospace/defense sensing, simulation, and controls, Orlando
6. Kapanoglu M, Koca İO, Erdogmus Ş (2007) Genetic algorithms in parameter estimation for nonlinear regression models: an experimental approach. J Stat Comput Simul 77(10):851–867
7. Khuri AI, Cornell JA (1996) Response surfaces, 2nd edn. Dekker, New York
8. Kirkpatrick S, Gelatt CD, Vecchi MP (1983) Optimization by simulated annealing. Science 220(4598):671–680
9. Melanie M (1998) An introduction to genetic algorithms, a Bradford book. The MIT Press, Cambridge, MA/London
10. Michalewicz Z (1992) Genetic algorithms + data structures = evolutionary programs. AI series. Springer, New York
11. Nelder JA, Mead R (1965) A simplex method for function minimization. Comput J 7:308–313
12. Ozbek L, Babacan EK, Efe M (2010) Stochastic stability of the discrete-time constrained extended Kalman filter. Turk J Electr Eng Comput Sci 18(2), Tubitak. doi:10.3906/elk-0812-17
13. Wan EA, Merwe RV (2001) The unscented Kalman filter. In: Haykin S (ed) Kalman filtering and neural networks, vol 7. Wiley Publishing, New York
14. Wiens DP (2013) Optimla design for nonlinear models: introduction and historical overview. University of Alberta, Canada
15. Xiong K, Liu LD, Zhang HY (2009) Modified unscented Kalman filtering and its application in autonomous satellite navigation. Aerosp Sci Technol 13:238–246
16. Yao L, Sethares WA (1994) Nonlinear parameter estimation via genetic algorithm. IEEE Trans Signal Process 42(4):927–935

Chapter 38
Application of Complexity Theory as Recipe for the Chaotic Nigerian Educational System: Exploration in the Context of Sustainability

Nura S. Akilu, Mohammed Musa W. Kirfi, and Muhammad Musa

Abstract Education is seen as a means of effective change, in the social and cultural context, but as a system, it has fallen short of this requirement in Nigeria. The current educational system which enacts vocational and socializing roles or functions has led to greater emphasis on paper qualification and has made the system, highly dysfunctional. This paper, drawing its methodological strength mainly from review of relevant literature, submits that Instead of strengthening the ideal of its intrinsic value, worthy of its transformative essence, the system has cultivated instrumental values that further eroded its true purpose towards nation building. This indeed is indicative of chronic and acute problems that signify system failure and it truly calls for corrective thinking and action. The paper thus, tries to explore the situation, from the systems theory perspective, by conceptualizing the Nigerian education system as a Complex Adaptive System (CAS) by virtue of its complexity as a human and social system. It also envisages the possibility of adopting the complexity theory as an approach, in order to bring about conditions necessary for the much needed educational sustainability in Nigeria.

38.1 Introduction

Chaos is in fact an aspect of reality that proves almost present all the times and in every setting albeit with varying degree of influence in especially enabling the system discover its weaknesses hence a platform for redefinition of strategies,

N.S. Akilu (✉)
Department of Business Administration, Usumanu Danfodiyo University Sokoto, Sokoto, Nigeria
e-mail: ennesakilu@yahoo.com

M.M.W. Kirfi, PhD
Department of Accounting and Business Administration, Federal University, Kashere, Nigeria
e-mail: lawalikf@gmail.com

M. Musa
Faculty of Education, Islamic University, Uganda
e-mail: kallam934@gmail.com

Ş.Ş. Erçetin (ed.), *Chaos, Complexity and Leadership 2014*,
Springer Proceedings in Complexity, DOI 10.1007/978-3-319-18693-1_38

adoption of appropriate leadership solutions on one hand, and/or disallowing the system the utility of any possible alternative that can see the system to transforming any perceived adversity into advantage. In the same respect, it is worth noting that an environment characterized by multiple actors, either human or non-human, that allows for, mutual co-existence, interrelation, interdependence with one another at various levels of interaction and varying dimensions, has the tendency to embody complexity in all ramifications of its being. This description of the environment is consistent, with the idea of a system whose features are drastically opposed to the established idea, of linearity and feedback typical of a simple system.

Instead, such a system represents a more complex idea exhibitive of non-linear relationship of system actors or components rooted in an atmosphere of continuous interaction, between elements of the system across varying dimensions of relatedness. This becomes suggestive of the assumption that it cannot be subject to the conditions typical of linear systems.

Since social actors living within a given social environment or setting, reflect features that allow for the development of complexity as against simplicity, then the assumptions under the concepts of linearity do not apply in their case, since the laws guiding the principle of linearity, cannot be the same with those that guide non-linearity. This led to the emergence of complexity theory, whose set of assumptions and concepts are primarily meant to deal with situations of complexity, mostly associated with social systems. Most of the theory's concept borrows from assumptions already established, in their application to scientific enquiries of natural systems.

It is against this background, that this paper sought to, explore and adopt concepts that were once uniquely attributable to the science of complexity in natural sciences, with a view to providing suggestions that could be found useful in analyzing and addressing issues of concern, in the Nigerian Educational System, as a social system that is characterized by complexity, in the light of sustainability. The paper is divided into six sections. It starts with a brief abstract, and then followed by the introduction. The clarification of key term follows; we then examine the current challenges of the educational system and explore the complexity approach in the section that follows. In the next section, attempt to conceptualize complexity in the light of the education system will be made and practical implication of complexity theory will be suggested, as the paper ends with suggestive recommendations.

38.2 Conceptual Clarification of Key Terms

For the prevailing purpose, the paper discusses the following key terms with a view to situating the same in the context of intellectual argument building. At the end, a combination of their effects is expected to condition the mind of the reader to understanding the implication of the terms in explaining the connection between complexity theory and the future of education in Nigeria.

38.3 Complexity Theory

Complex adaptive systems are mostly non-linear feedback network, operating far from equilibrium, and natural and human systems have been described as such. According to Jacobson and Wilensky [6], chaos and complexity are characteristics of social systems and are triggered by the evolutionary and revolutionary changes taking place in their environment. As such, fluctuations and disturbances such as technological innovation, changing legal rules and governmental regulation force organizations or social systems to live with chaos or complexity. Complexity theory thus describes the behavior of complex human, social and natural systems in the long run.

38.4 Sustainable Education and Educational System

A change of educational culture is one which develops and embodies the theory of sustainability in a way which is critically aware. It is therefore a transformative paradigm which values, sustains and realizes human potential in relation to the need to attain and sustain social, economic and ecological well being, recognizing that they must be a part of the same dynamic [20]. There is an inherent system oriented approach in this definition that sees sustainability in the educational sense, as rather holistic and relationally consistent. The dominant paradigm of any educational system of a given society should be able to meet the transformative needs of that society, especially as it relates to sustainability. They went on, to comment the nature of the dominant paradigm of most educational system, gave a clear distinction between two opposing orientation of the system. They argue that vocationally oriented systems seem to focus more on the preparation of students to fit into particular places in the labor market. On the other hand, academically oriented systems tend towards bringing as large a proportion of students, to as high a level of general ability possible. These two orientations, though still valid, portray a picture, that is consistent with the dominant feature of most educational systems, where emphasis on qualification and socialization flourish at the expense of transformative needs of society.

An important distinction, central to the question of sustainability however, views educational values as having two distinct orientations, namely intrinsic and instrumental educational values. The latter tend to stress purpose, and product, in other words its emphasis is on outcomes and effectiveness. It is concerned more with "what education is for" rather than the nature of education. While the former, stresses process, in other words, the quality of experience of teaching and learning and is primarily concerned with "What education is" rather than what it might eventually lead to, or influence [20]. While both vocational and academic orientations of education remain valid, an educational system that promises sustainability is one that requires an integrative and relational thinking and action, while still being accommodative of synergy between the intrinsic and instrumental values.

38.5 Nigerian Education System: The Current Challenge

Based on the foregoing, it is important to state at this point, that, the analysis of the educational system in Nigeria in relation to the current challenges intended in this section of the paper, is limited in scope, to the extent that the study satisfies inquiry into the central issues of the system's characteristics and the accompanied perceived, or actual dysfunctional nature, of the system as a whole. This would be, with a view to understanding its governing dynamics, as social system and by implication, a complex system. Technical structural patterns and administrative issues as well as other components of the system across dimensions, will therefore be treated as they blend in the larger holistic framework of the system, for our purpose. This implies therefore, that the emphasis of the analysis would be in the light of systematic relationship of all its varying components, rather than in depth comprehensive account, of the multidimensional structural pattern and the relationship that characterize the system.

38.6 The System Failure

Peters [12] outlined four types of failure: Objectives not met, inappropriate objective, undesirable side effects, and designed failure. Considering the prevailing situation all of these types of failure, are contextually traceable in the Nigerian educational system. Moja [8] listed, some of the problems that are suggestive of the current state of the system.

1. Lack of adequate infrastructure
2. Lack of well-trained human resources
3. Issues of access and equity in the system
4. Deterioration of quality
5. Inadequate funding mechanism
6. Ineffective curriculum
7. Admission and examination malpractice

The above stated problems, even though obvious and reflective of the apparent system failure of the educational sector, are in reality descriptive of the symptoms that point to, even a more complex problem, inhibitive of the system, when viewed holistically from the systems perspective.

The diagrammatical illustration in Fig. 38.1, attached to this paper, is a representation of the complexity that characterizes the system, and the resulting system failure, when viewed through the lens of complexity sciences. The starting point

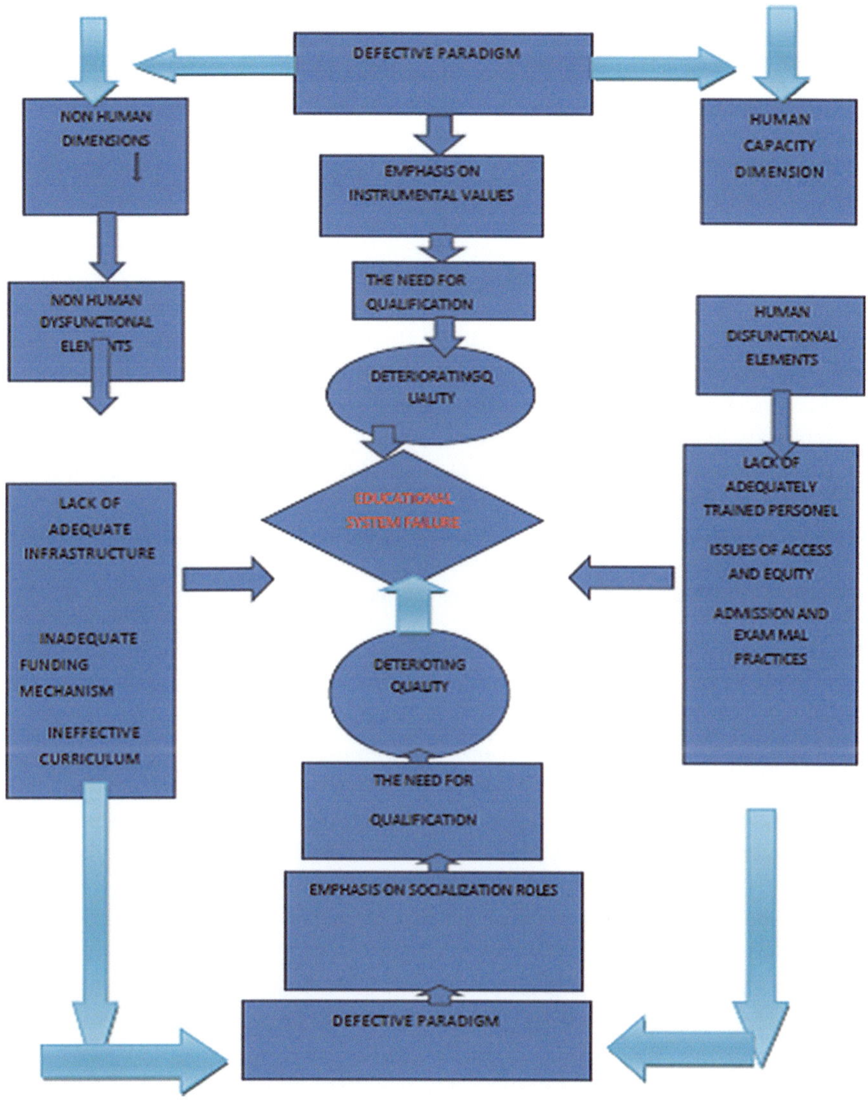

Fig. 38.1 Diagrammatical illustration of system failure (Source: The author)

of the problem as shown, lies in the defective paradigm, held by the interactive actors involved, within and outside, the complex environment where they exist and the nature of the complexity, that binds them as a system. This so called, defective paradigm, which is based on instrumental values, and which promotes a strong emphasis on paper qualification, do so, at the expense of the much needed quality learning, central to a functionally viable and operational system. The nature of interactions, due to systematic connectivity and the resulting interdependence, across dimensions and between various elements, that characterize the overall educational system, is consequential to the failure of the overall system and can best be appreciated, with increased understanding of the complexity approach that informed this very interpretation.

38.7 The Complexity Approach

The science of complexity is based upon three sets of concepts that are critical to our understanding of educational systems as phenomena in social systems. These sets of concepts, as elaborated by Ramalingam [13] represent a condensation of ten other concepts that are indispensable to the science of complexity. A close examination of the concepts, most relevant, to the current discourse, will be briefly made as presented by these authors. In their categorization of the three sets of key concepts, mostly used in the complexity sciences, Ramalingam [13] associated complexity with three other related concepts, namely; System, Change and Agency, all of which provide the basis for the applicability of the general concept, to varying situations, that are reflective of the resultant three dimensions. The three sets of complexity science concepts across these dimensions are as follows:

1. Complexity and Systems
2. Complexity and Change
3. Complexity and Agency

They listed under each of these three set of concepts, other concepts uniquely attributable to them, and all of them can provide useful insights and aid our understanding of the subject matter. The description of the concepts listed under, the first general concept will provide justifications for considering the Nigerian education system as complex system, while concepts listed, under the other two broader concepts will prove useful, when attempt would be made, to suggesting the attainability and viability of Sustainable education, as desirable end.

38.8 Complexity and System

According to Ramalingam [13], three major concepts are descriptive of a complex system; Namely: Interconnectedness and interdependence of elements or dimensions within and outside the system; feedback processes; and emergence.

Interconnection and interdependence of elements and dimensions: Nicolis and Prigogine [10] described a complex system as one made up of multiple elements (which may also be processes) which are connected to and are interdependent on each other and their environment. This implies that, Interdependence comes into being, when interconnection is firmly rooted between elements and dimensions. Commenting on the nature of such relationship in relation to diverse variables,

[22] asserts that interconnectedness may occur between individual elements of a system, between sub-systems, among subsystems, between different levels of a system, between systems and environment, between ideas, between actions, and between intentions and actions This, according to Ramalingam et al. [14] gives rise to interdependence between the elements and the dimensions of a system and gives rise to complex phenomena.

Feedback process: In a broad sense, feedback can be described as an influence or message that conveys information about the outcome of a process or activity back to its source [2]. The nature of feedback is such that, as Jarvis [7] captures it, can amplify or effect positive change, in a particular direction or of a particular kind, leading to reinforcing pressures, which can lead to escalating change in the system. By implication, feedback can also effect the other way round, such that the change, triggers forces, that counteract the initial change and return the system to the starting point, to decrease deviation from the system [14]. In complex systems, as Byrne [1] suggests, feedback is about the consequence of non linear, random change over time.

Emergence: The concept of emergence describes how overall properties of a complex system emerge from interconnections and interaction. While the nature of the entities, interactions and environment of a system are key contributors to emergence, there is no simple relationship between them. Emergence has been used to describe features such as social structure, human, human personalities, the internet, consciousness and even life itself. As one lucid account has it [9]. In complex systems, it is the interactions of the individual components, that give rise to the emergence of some kind of property which could not have been predicted from what is known of the components parts...... emergent behavior, feeds back, to influence the behaviors of the individuals that produced it (Langton, cited in [21])

38.9 Complexity and Change

Four concepts relate to different aspects of how complex systems, change overtime. This includes: Non-linearity; Sensitivity of complex systems to their starting conditions; the idea of phase space; Chaos within complex systems.

Non-linearity: Complexity science suggests, that human systems, do not work in a simpler linear fashion. Feedback processes between interconnected elements and dimensions lead to relationships that see change that is dynamic, nonlinear and unpredictable [18]. Therefore non-linearity is a direct result of the mutual interdependence between dimensions found in complex systems. In such systems, clear causal relations, cannot be traced because of multiple influences [14]

Sensitivity to initial conditions: This idea stresses the importance of history in complex systems, in that it recognizes its impact in the evolutionary process of a given systems as it moves to a different state over time. The concept of phase space (Which we shall discuss later) is much related to this idea. Phase space allows for the analysis of the evolution of systems by considering the evolution as a sequence of states in time [15]. What this implies is that, future interactions of a given complex system are being influenced by past historical processes.

Phase space: The dimension of any system can be mapped using a concept called phase space, also described as the "space of the possible" [17]. As Ramalingam et al. [14] suggests, attempting to understand the system (complex system) should start by identifying the key dimensions, then track changes in them, using it to develop a holistic picture of how the system changes and evolves. This represents the basic idea, central to the concept of phase space.

Chaos within complex systems: Chapman [3] in connection to this concept asserts that societies and all its institutions are in continuing process of transformation and that emphasis should be placed on guiding, influencing, and managing these transformations. This implies, learning to operate at the edge of chaos, which in turn encourages the acknowledgment of continues change in social systems in other words, complex systems.

38.10 Complexity and Agency

The set of concepts inclusive in this set, are: Adaptive agents; Self organized behavior; Co-evolution Stacey [19]

Adaptive agents: All living things are adaptive agents, individual people are adaptive agents, so are teams in which they work, and so are organizations. Complex systems are said to be adaptive or evolving, when individual agents, respond to forces in their environment via feedback. Regardless of size and nature, adaptive agents share certain characteristics, in that they react to the environment in different ways [5]. The ability of adaptive agents to perceive the system around them and act on these perceptions, means that, their view of the world, dynamically influences, and is influenced by, events and changes within the system [14].

Self-organization: The concept of self-organization, echoes emergent properties, and the fact that complex system, cannot be understood as the sum of its part, since it may not be discernable from the properties of the individual agents and how they may behave in large numbers. [16]. The key central idea of this concept projects the assumption that change, order, and resilience cannot be imposed from the outside or from the top down, but that these can be achieved through the adaptive tendencies of individual agents operating throughout the system. [14].

Co-evolution: This reflects the idea, that in complex systems, large-scale process through which the interaction of one or more co-evolved species with the system, results in changes so fundamental that all species in the system must and the system itself changes.

In the foregoing, a brief explanation, of the basic critical concepts, that are central to an understanding of complexity sciences has been made. The basis of which, will serve the exploratory and prescriptive objective of this paper. What follows is a logical attempt to harmonize these concepts in favor of our intent of conceptualizing Nigerian educational system as complex adaptive system, to lay the foundation of our subsequent analysis.

38.11 The Complexity of the Nigerian Education System as Adopted

In this section, attempt is made to logically harmonize all the concepts that define complexity as outlined above, in a way that fits the main assumptions of the concepts, and into the fabric of the Nigerian educational system. This is, with a view to providing a logical frame work for deductions that will eventually justify, in the final analysis, the suggestive implication of complexity theory and its applicability towards educational sustainability in Nigeria, which is the main objective of this paper.

38.12 The Education System as a Complex System

The Nigerian educational system emerged overtime, and is revealing of an inherent *interconnection and interdependence* that resonate around relationship between elements and around variables and processes from varying dimensions. From the government and administrative dimension, the system reveals high level of this *connectivity* starting from the Federal ministry of education. The ministry through various series of *activity and feedback* processes interacts with other elements along different dimensions under a unifying framework, for the purpose of coherence. This, it does, while ensuring that, through the coordination of National council of education and the advisory function of joint consultative committee on education, national policies and procedures, remain consistent with national objectives, in all states across the country [17].

Secondly, since human systems along with their varying attitudinal inclinations and preferences, represent *nonlinear* manifestations of interactive processes within and outside of their environment, transformative experiences *draw from past experiences* and are filtered through the lens of *familiar conditions*, thereby facilitating change during *chaotic* circumstances, for proper adaptation. The rapid growth of the educational system in the late seventies and early eighties was mainly in size and not quality. This problems of expansion that was highly disproportionate to the supposedly supportive intent of the then, policy implementation processes, has been a re-occurring theme, in subsequent transformative periods in the light of deliberate efforts by the government towards sanitizing the system.

Lack of planning capacity and management amongst various interplays of relational efforts at various levels and periods, seem to reflect *previous tendencies* as response to new changes [11].

The forgoing, thus illustrate, from the examples cited, the running complexity in the Nigerian educational system, by being indicative of how it resonates, with the features, embodied in the first two sets of the general concepts that define complexity, in relation to firstly; "system" and then "change" as outlined above. This provides the basis for referring to it, as: complex system.

38.13 The Educational System as Complex Adaptive System

Drawing from the foregoing deduction, as evidenced through conceptual association of the educational system of Nigeria with the tenets of the complexity sciences, a further justification of the complexity of the educational system, is the simple fact that it embodies the activity of human social interaction in social settings of varying nature, and varying dimensions that reflect a holistic sense, of *connectivity, interrelatedness* and *interdependence.* As such, it clearly manifests the possible interplay of *evolving adaptive agents* at various levels of social *interactions.* According to Ramalingam et al. [14] Complex systems made up of adaptive agents are distinguished by the term; "Complex Adaptive System" If the existence of adaptive agents within a complex system, confirms the status of such a system as Complex Adaptive System, then it follows that, such a confirmation, by virtue of the complex nature of the Nigerian educational System with its interplay of adaptive agents resonates with the description of the system as Complex Adaptive System.

38.14 Implication for Application of the (CAS) Concept to the Nigerian Education System

The need to change education paradigm, in the context of sustainability, argued Sterling [20], becomes essential, if *co evolution* towards sustainable life style is held with great importance. In his view this change in paradigm involves a change in educational culture towards transformative paradigm that values, sustains and realizes human potential. What this implies is the development of a strong will capable of affecting such a cultural shift towards desirable ends. To this end, such a will must be internally generated and must be consistent with the right kind of motivation. In other words the will must be intrinsic and ignited by a value adding motive or motives. Complex Adaptive systems that consist of adaptive agents inspired by this kind of perception are likely instrumental agents of change, when faced with situations that requires a will to change for the better. As, such, different individuals and organizations within a problem domain will have significantly different perspective, based on different histories, cultures and goals. These different perspectives have to be integrated and accommodated if effective action is to be taken by all relevant agents [3].

This concept can thus, be contextualized in a way and manner that seem consistent with the educational needs of Nigeria, and the leadership challenge that inhibits its progress towards sustainable development. This is because; the issue of leadership cannot be divorced from the scope of this current discourse by virtue of its instrumental essence towards effective transformation of any kind.

It is on the basis of these broad assumptions that the paper puts forward the following prescriptive suggestions, hoping they ignite, the right kind of curiosity, that make, their practical applicability possible, as derivatives of the foregoing analysis centered around complexity theory and sustainability and as conclusions and recommendations, derived there from. The prescriptive assumptions are also diagrammatically captured and presented in the Fig. 38.2.

38.15 Conclusion and Recommendations

Going by all that have been put across via the foregoing paragraphs, the following observations are conclusively consistent with insights derived from the analysis of all the reviewed literature thus:

The fact that educational system is a social system and thus complex, makes it receptive to the idea of Complex Adaptive System and accompanied by corresponding possible (practical) applicability.

1. The adaptive agents that characterize Complex adaptive system, through effective manipulation of their internal circle of concern, can develop and enlarge their external circle of influence, to a point of systematic self organized co-evolution, which is contagious.
2. This could lead to intrinsic processes, capable of inspiring cultural shift that defines a new set of paradigm that could be an effective tool, to be used for positive transformation that is necessarily supportive of purposeful leadership and is being supported by an effective political ideology which in turn supports prosperous economy.
3. The presence of good political ideology and the absence of poor economy will facilitate the development of effective, adaptive agency from the followers' dimension that is responsive to the contagious nature of the adaptive agents at the upper dimension. This inspires the development of paradigm shift that is intrinsically oriented towards positive transformation
4. Emphasis will now shift towards the quest, for a kind of knowledge, that is value adding, to individual adaptive agents that share common, positive, natural disposition, inspired by the common quest towards this need for positive transformation, across dimensions.

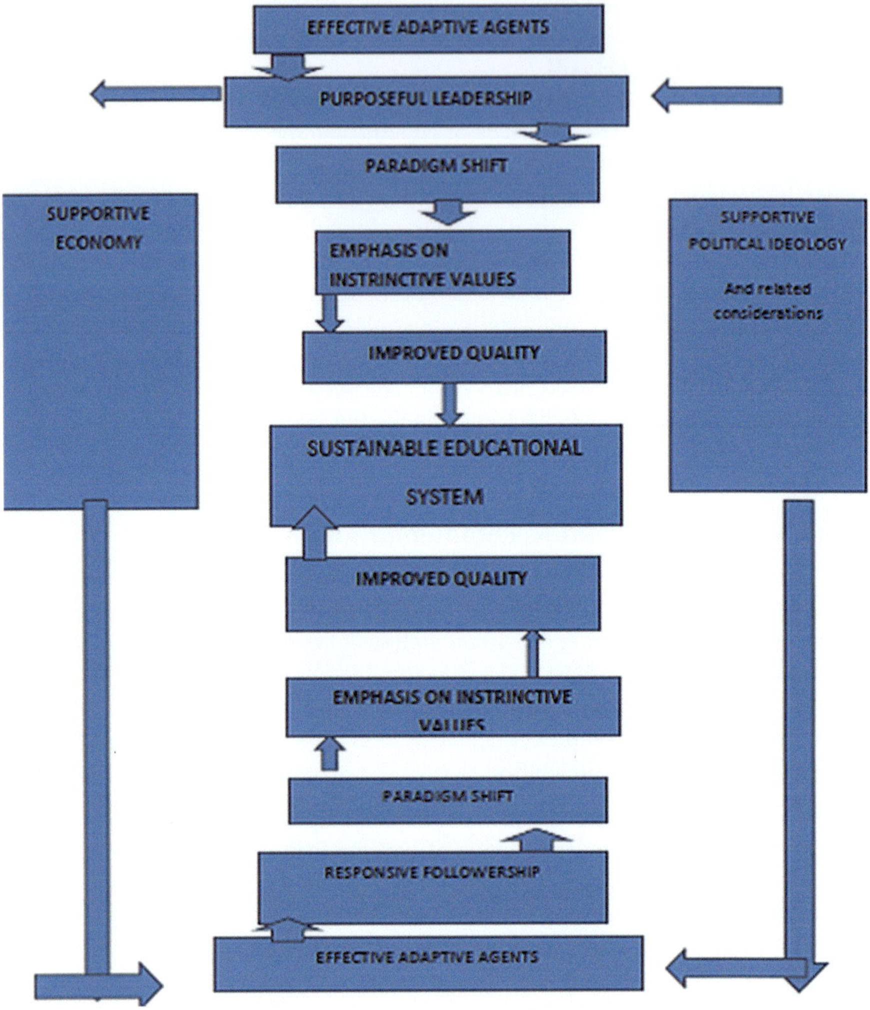

Fig. 38.2 Diagrammatical illustration of sustainable education from the point of view of complex adaptive system (Source: The author)

5. Natural positive results will become manifest in form of increased understanding of qualitative nature in all inclusive ramifications of knowledge, which through its value adding capacity broadens the intellectual horizons of adaptive agents in an emergent self organized revolutionary transformation of thought pattern, which in turn affects behavior and guides it, towards praise worthy tendency and away from behavior that discourages blame worthy dispositions
6. The continuous nature of this behavior over time, generates more number of such adaptive agents and through co evolving nature of interactive complexity, a general pattern of conformity spreads across dimensions and re enforces the need

for quality knowledge, in a rather, progressive fashion This becomes consistent with the perpetual activities of the system, while reinforcing the conditions that sustain the system in that direction towards transformation

References

1. Byrne D (1998) Complexity theory and the social sciences: an introduction. Routledge, London
2. Capra F (1996) The web of life. Flamingo/Harper Collins, London
3. Chapman J (2004) System failure: why governments must learn to think differently. Demos, London
4. Cohen J, Stewart I (1995) The collapse of chaos. Penguin, London
5. ISCID (2005) Encyclopaedia of science and philosophy. http://www.iscid.org/about.php
6. Jacobson MJ, Wilensky U (2006) Complex systems in education: scientific and educational importance and implications for the learning sciences. J Learn Sci 15(1):11–34
7. Jervis R (1997) System effects: complexity in political and social life. Princeton University Press, Princeton
8. Moja T (2006) Challenges in the internationalisation of higher education in South Africa. In: Kishun R (ed) The internationalisation of higher education in South Africa. International Education Association of South Africa, Durban
9. Newell D (2003) Concepts in the study of complexity and their possible relation to chiropractic healthcare. Clin Chiropr 6:15–33
10. Nicolis G, Prigogine I (1989) Exploring complexity: an introduction. W.H. Freeman, New York
11. Nwagwu CC (1997) The environment of crises in the Nigerian education system. Comp Educ 33(1):87–95
12. Peters JD (1999) Speaking into the air: a history of the idea of communication. University of Chicago Press, Chicago. ISBN 0-226-66276-4
13. Ramalingam B (2006) Tools for knowledge and learning: a guide for development and humanitarian organizations. ODI, London
14. Ramalingam B, Harry J, Toussaint R, John Y (2008) Exploring the science of complexity: ideas and implications for development and humanitarian efforts. Working paper 285 results of ODI research presented in preliminary form for discussion and critical comment. Overseas Development Institute, London
15. Rosen R (1991) Life itself: a comprehensive inquiry into the nature, origin and fabrication of life. Columbia University Press, New York
16. Schelling T (1978) Micromotives and macrobehavior. Norton, New York
17. SPESSA (1998) A statistical profile of education in Sub-Saharan Africa, 1993–1996. Published by the Association for the Development of Education in Africa (ADEA). In: Okpara EN (ed) Survival of the Nigerian education system: a book of readings. APQEN, Washington, DC
18. Stacey RD (1996) Complexity and creativity in organizations. Berrett-Koehler, San Fransisco
19. Stacey RD (1996) Emerging strategies for a chaotic environment. Long Range Plann 16:182–189
20. Sterling S (2001) Sustainable education – revising learning and change. Schumacher Briefings, Green Books, Dartington
21. Urry J (2003) Global complexity. Blackwell, Cambridge
22. Weick K (1976) Educational organizations as loosely coupled systems. Adm Sci Q 21:1–9

Chapter 39
A Comparative Analysis of Sukuk and Conventional Bonds

Abdurahman Jemal Yesuf

Abstract Sukuk and bonds are two kinds of financial instruments; despite their differences, they share similar responsibility of fund mobilizing from surplus (spending) units to shortage units. Sukuk can resemble conventional bonds by some of its features, but it is technically neither debt nor equity. It is complex to understand the exact nature of Sukuk and differentiating them from bonds. Conventional bonds are structured on the basis of debt whereas sukuk are basically investment certificates consisting of ownership claims in a pool of assets. Thus, claim embodied in sukuk is not simply a claim to cash flow but also an ownership claim. Although there are various fundamental differences between the sukuk and conventional bonds, both instruments try to solve the same common financial problem of raising capital for needed entities. They offer solutions in different ways to the same financial problem. Therefore, in this chapter an attempt is made to raise the awareness and knowledge base of those who have had little or no exposure to the Islamic financial instrument, such as sukuk, its complex nature and functions in today's global financial market. The chapter has compared sukuk instruments with conventional bonds based on the basic fundamental variables such as risks, return, and correlation variables. It has also shown the risk diversification potential of sukuk in a portfolio.

39.1 Introduction

In business, religion plays a significant role in determining the structure of financing and investing activities. Religions, like Islam, impose some restriction or prohibit some acts in any forms of business activities. The Islamic law (*Shariah*), for example, prohibits the charging and paying of interest. Accordingly, conventional bonds that yield interest are prohibited under the Shari'ah law. Furthermore, those who buy and sell conventional bonds are rarely interested in what is actually being financed through the bond issue, which could include activities and industries that

A.J. Yesuf (✉)
Department of Business Administration, Marmara University, İstanbul, Turkey
e-mail: abdu_5570@yahoo.com

Ş.Ş. Erçetin (ed.), *Chaos, Complexity and Leadership 2014*,
Springer Proceedings in Complexity, DOI 10.1007/978-3-319-18693-1_39

are deemed haram (forbidden) such as the production or sale of alcohol. Companies that are highly leveraged with bank debt may seek refinancing through issuing bonds, but such companies are not regarded as suitable for Muslim investors.

As a result, in countries where Muslim populations constitute an important segment of the society, considerations for Islamic law have recently hampered the advancement of competent systems for monetary control [1]. Governments of those countries are now understanding the needs for developing an equivalent means of financing and investing opportunities for these segments of their society. Among such alternative instruments introduced in Islamic finance to resemble conventional bonds are sukuk.

Sukuk, sometimes referred to as Islamic bonds, are better described as Islamic investment certificates. This distinction is crucial, as sukuk should not merely be regarded as a substitute for conventional interest-based securities. The aim is not simply to engineer financial products that mimic fixed-rate bills and bonds and floating-rate notes as understood in the west, but rather to develop innovative types of assets that comply with Shari'ah Islamic law [2].

The aim of conventional bond traders usually is to make capital gains as fixed-interest bond prices raise when variable market interest rates fall. Bond trading is therefore largely about exploiting interest rate developments and trading in paper that is usually unrelated to the value of any underlying asset. The major risk for holders of conventional bonds is of payments default, but this risk is usually assessed solely on the basis of credit ratings, with the ratings agency rather than the bond purchaser estimating the risk. Hence the bonds are regarded as mere pieces of paper with third parties estimating the risk and the purchaser, at best, only making a risk/return calculation without any reference to the business being financed [2].

As Islamic finance is by nature participatory, purchasers of sukuk securities arguably have the right to information on the purposes for which their monies are to be allocated. In other words, the funding raised through Islamic bond issues should be hypothecated or earmarked rather than used for general unspecified purposes, whether by a sovereign or corporate issuer. This implies that identifiable assets should back Islamic bonds.

Therefore, it is the aim of this paper to raise awareness and knowledge base of those who have had little or no exposure to the Islamic finance and its functions in today's global financial market. It compares sukuk instruments with conventional bonds based on basic fundamentals such as risks, return, and correlation variables.

The paper has eight sections. The first and second sections give short and brief descriptions to highlight the Islamic financial system and its features; the third part introduces the motives of doing this study and the methodologies used. The fourth section concentrates on sukuk. The fifth and sixth sections compare sukuk with conventional bonds based on their risks, yields, some other characters. The last two sections show the correlation of sukuk with other asset classes and its advantages in portfolio formation and risk diversification. The study also has a conclusion and possible recommendations that are given based on the analysis and discussions made.

39.2 Literature Review

39.2.1 Islamic vs. Conventional Finance

Islamic finance, despite its name, is not a religious product. It is however a growing series of financial products developed to meet the requirements of a specific group of people [3]. Islamic finance is a term that reflects financial business that is not contradictory to the principles of Shari'ah. On the other hand, under conventional finance, conventional products and services, such as insurance and capital markets could be based on elements that are not approved by Shari'ah principles such as uncertainty (Gharar) in insurance and interest in conventional bonds or securities. In the case of insurance, the protection provided by the insurer in exchange for a premium is always uncertain as to its amount as well as its actual time of happening. A conventional bond normally pays the holder of the bond the principal and interest.

Conventional practices could also involve selling or buying goods and services that are not lawful from a Shari'ah perspective. These might be *haram* (forbidden) foods such as pork, non-slaughtered animals or animals not slaughtered according to Islamic principles, alcohol or services related to gambling, pornography and entertainment. In short, conventional business practices could be non-compliant from a contractual structure perspective (if they are based on interest and uncertainty) and/or from a transactional perspective when they are involved in producing, selling or distributing goods and services that are not lawful according to Shari'ah ([3]: 5).

39.2.2 Components and Salient Features of Islamic Finance

As a financial system that functions on the principles of Islamic Law, called Shari'ah, Islamic financial practice has three broad components. Such as: banking, insurance and capital markets.

Islamic banking is a branch of Islamic finance that has seen the most growth to date. It is also a branch of finance that needs to be viewed from a different perspective as it cannot replicate conventional banking. This is because the most important underlying principle of conventional banking is that money creates money or that money has a premium, known as interest or usury. In the Islamic insurance, better known as Takaful, the insurer, that is the insurance company, is prohibited from providing indemnity to the insured, that is, the policyholders, as this is not acceptable to Shari'ah principles. This is because both the premium paid by policyholders and the indemnity paid by the insurer are uncertain and therefore not permissible as they contain the element of uncertainty or Gharar.

Islamic capital markets that consist of both equity investments and fixed income instruments must avoid some conventional elements and principles from both contractual and transactional perspectives. In addition to interest and uncertainty, issues such as gambling, which is a zero-sum game, investments in unlawful activities and capital guarantee elements in equity-based products are to be avoided ([3]: p. 7).

Regardless of the above classification, the main principles of Islamic finance are that[1]:

- Wealth must be generated from legitimate trade and asset-based investment. (*The use of money for the purposes of making money is expressly forbidden*).
- Investment should also have a social and an ethical benefit to wider society beyond pure return.
- Risk should be shared.
- All harmful activities (*haram*) should be avoided.

Thus, based on the above principles, Islamic finance, especially Islamic banking, enjoys certain peculiar features that are not found in conventional finance. These features are:

Avoidance any form of Interest: The payment or acceptance of interest (Riba) for a loan is absolutely forbidden. Thus, Islamic financial activities, like banking and other business activities must prima facie be free from any element of interest.

The Avoidance of Uncertainty (Gharar) or Gambling: Trading under uncertainty (Gharar) in financial transactions must be eliminated. This is because Gharar might lead to disputes caused by an unjustified term in the contract arising from misrepresentation and fraud. Accordingly, undertaking transactions with insufficient knowledge of the market or product and thereby incurring an excessive risk or interest is forbidden.

The Need for Underlying Assets: Islamic finance requires that all banking business based on sale or lease must have an underlying asset. An individual or an institution should not be able to generate income from money. This self generation of money from money is "Riba", which is absolutely forbidden in Islam. Accordingly, the trading/selling of debts or receivables (without the underlying asset) for anything other than its par is not permissible ([17], p. 7).

Profit and Loss Sharing: Instead of charging interest, profit and loss sharing is possible in some Islamic banking activities. The bank will share the profit made with its customers either on a proportionate basis or on an agreed profit sharing ratio. In the case of a loss, the loss will be borne by the bank under a Mudarabah contract or by both parties proportionately in the case of a Musharakah contract. This concept is in direct contrast to fixed-income-based products.

Unlawful Goods or Services: Any transactions (buying, selling, and distribution) that involve alcoholic beverages, pork, prohibited drugs, gambling, pornography, and weapons are forbidden. Non-involvement is not only limited to buying or selling but also includes all chains of production and distribution, such as the packaging, transportation, warehousing and marketing of these prohibited goods and services.

[1] ACCA (2011): Studying Paper F9, Section 3: Islamic Finance.

39.3 Objectives and Methodology

39.3.1 Objectives of the Paper

The main purpose of this paper is to give a short and brief description and awareness about Islamic 'bond' (sukuk) and to compare its performance and risks with conventional bonds. Thus, the purpose of this paper is twofolds: first, it presents overall features, characteristics, kinds and basic structure of sukuk. Generally, an attempt is made to give general picture of sukuk to readers of this part.

The second part of the paper attempts to compare sukuk with conventional bonds by taking into account some fundamental variables, such as yield and risk. In the comparison process, the main objective is to identify and show risks that affect the two financial instruments in general and sukuk and/or bonds in particular. The other objective of the study is to show the risk reducing and profit maximizing potential of sukuk in a portfolio.

39.3.2 Data Sources and Methods of Presentation

By taking into account the above objectives to achieve the predetermined goals of the study, the research is conducted based on secondary data and existing literatures only. In order to collect the data needed for the analysis in this paper, different sources have used. One of these sources is the Dow Jones Sukuk Index.

The Dow Jones Sukuk Index is designed to measure the performance of Global Islamic Fixed-Income Securities—also known as sukuk. The index includes U.S. dollar-denominated investment-grade sukuk issued in the global markets that have been screened for Shari'ah compliance according to the index methodology. The index was created as a benchmark for investors seeking exposure to Shari'ah-compliant fixed-income investments. The Dow Jones Sukuk Index follows the same consistent, quantitative methodology as the Dow Jones Islamic Market™ (DJIM) Indices, which are monitored to ensure their continued adherence to Shari'ah principles.

To be included in the index, a bond must pass screens for Shari'ah compliance and meet the standards issued by the [24]. It also must have a minimum maturity of 1 year, a minimum size outstanding of USD 200 million, and an explicit or implicit rating of at least BBB-/Baa3 by leading rating agencies.

In addition to the DJIM indices, some important sources of data have widely referred. Among these Bloomberg, HSBC, NASDAQ, S&P, FactSet, J.P. Morgan asset management and other institutions' reports have provided an impressive source of data about the topic on hand. The data collected from these sources are organized and presented in the form of table, graphs, charts and figures.

39.3.3 Methods of Data Analysis

The methods of data analysis used for this particular study is descriptive statistics for describing the general characteristics of the Islamic 'bonds' and conventional bonds. To describe the earning potential and risk characteristics of Sukuk as compared to conventional bond, mainly three variables are used; such as risks of the instruments (volatility ratios), risk adjusted returns, and their correlations with other asset groups and indices. Based on these variables, simple comparative analyses have made to compare the potentials and the riskiness of Sukuk as compared to their conventional market peers. In the analysis Sharpe ratio is also used to evaluate the Islamic bond rewards to investor per unit of risks investors willing to accept as compared to conventional bonds.

Sharpe's measure divides average portfolio excess return over the sample period by the standard deviation of returns over that period. It measures the reward to (total) volatility trade-off [4]. Sharpe rations are computed by using the William Sharpe's formula of:

$$S(x) = \left(r_{(x)} - R_{(f)}\right) \div \delta^2_{(x)}$$

Where x is the investment; $r_{(x)}$ is the average rate of return on x; R_f is the best available rate of return of a risk-free security (i.e. T-Bill); δ^2 is the Standard deviation of $r_{(x)}$. Therefore, in order to evaluate and compare the returns (risk adjusted returns) of Sukuk with other financial instruments, the Sharpe risk adjusted model has used. The credit quality of debts and its progress were also taken in to account.

The other figure used in the analysis to describe emerging market debt and compare them with the other asset classes is their correlation with these asset classes. Correlation is a statistical measure (between 1 and −1) that describes the relationship between two variables. The closer a correlation coefficient is to 1 indicates the likelihood that each variable will move in tandem. A negative correlation indicates that each variable will move in the opposite direction.

The correlation of return and volatility between two assets, or asset classes, refers to the degree to which the two move in step with each other and the direction of that relationship. Importantly, in this case, it does not mean that the movement of one asset causes the movement of another. Instead it is just measuring how two assets respond similarly to the same external conditions.

39.4 Sukuk

39.4.1 Concepts and Brief History

Sukuk, in general, may be understood as a Shari'ah compliant 'Bond'. Sukuk (plural of sakk) are referred to as'Islamic bonds' but the correct translation of the Arabic word of Sukuk is 'Islamic Investment Certificates'. The name sukuk is sometimes translated as certificates, or as Islamic bonds. Islamic bonds sounds a bit like a contradiction, and of course no predetermined interest rate is promised on these so-called bonds. Still, they may offer investors a steady stream of income [5]. Sukuk were extensively used by Muslims in the middle Ages as papers representing financial obligations originating from trade and other commercial activities. However, the present structure of sukuk are different from the sukuk originally used and are akin to the conventional concept of securitization, a process in which ownership of the underlying assets is transferred to a large number of investors through certificates representing proportionate value of the relevant assets.

Sukuk are essentially asset-backed instruments representing a beneficial ownership interest in the underlying asset. In its simplest form sukuk represents ownership of an asset or its usufruct. Sukuk resembles in many respects a traditional bond or asset-backed security, but is technically neither debt nor equity [6]. Under Sukuk structure, the Sukuk holders (investors) each hold an undivided beneficial ownership in the 'Sukuk assets' ([7]:154). The claim embodied in sukuk is not simply a claim to cash flow but an ownership claim. This also differentiates sukuk from conventional bonds as the latter proceed over interest bearing securities, whereas sukuk are basically investment certificates consisting of ownership claims in a pool of assets.

Unlike conventional bonds, sukuk is a recently-developed Islamic investment product that first appeared in 2002, when Malaysia issued a government-backed Sukuk, the first of its kind. According to Hans [5], first Islamic global bond issue, therefore, was floated in 2002 by the Malaysian government. The lead manager of the issue was HSBC Bank Malaysia and the sukuk paid a spread over 6-months LIBOR. Sukuk are mainly aimed at institutional investors, though there have been issues with a minimum value of each sukuk below the equivalent of €2000. Sukuk are not only issued by or on behalf of governments and quasi-sovereign agencies, but also on behalf of corporations. Sukuk issues are regularly heavily oversubscribed and the volume issued shows a steep rise. The restricting factor is supply, not demand [8].

Despite its name as Islamic, these financial instruments seem to be attractive not only to wealthy Middle East investors, but also to non-Muslim Western investors as well. They enter them in their books as their allocation of emerging market debt. Sukuk flotations are not restricted to Islamic issuers either. Borrowers in non-Islamic countries, from Germany to China, are also interested in tapping the Middle Eastern capital markets. As indicated by Hans [5], the German state of Saxony-Anhalt issued a €100 million sukuk in 2004, with Citigroup as the lead manager, and the World Bank also issued its first sukuk for 760 million Malaysian Ringgit ($202 million) in 2005. In 2006 a US private firm, East Cameron Partners, issued a sukuk for $166 million to finance off shore gas drilling in Louisiana. In his 2007 budget, the then British Chancellor of the Exchequer, Gordon Brown, announced plans to develop London as an international centre for Islamic financial products, including a secondary market for sukuk. One measure concerned offsetting the coupon payments on the securities against the company's profits for corporation tax purposes, similar to interest on conventional bonds. The first sukuk was listed on the London Stock Exchange in July 2006.

39.4.2 Salient Features of Sukuk

Sukuk can resemble conventional bonds by some of its features, but it has different underlying structure and provision. It is the trust certificate, which gives its holder an undivided proportion of ownership in the underlying project/asset and right to receive cash flows from this underlying. Returns on sukuk derive either from performance of an underlying asset or contractual agreement based on this asset. According to Wilson [9] main principles underlying sukuk issuance can be defined as follows:

all rights and obligation should be clearly defined;
the income from sukuk should be related to the project, which was financed by this issue;
Sukuk should be backed by a real asset.

The different forms of sukuk have different characteristics. Apart from the legal guise they adopt, sukuk can be fixed-rate or flexible-rate, in the latter case usually coupled to LIBOR or Euro Interbank Offered Rate (EURIBOR). They further differ as to liquidity. Murabaha and Istisna sukuk score low on liquidity, Musharakah, Mudarabah and Ijara sukuk are more attractive in this respect [5].

39.4.3 Main Sukuk Structure

The structure of Sukuk varies based on the type of contract it's based on. However, in all forms of sukuk there are three main parties involved in the Sukuk arrangement: the originator of the Sukuk who is the obligator, the issuer of the certificates who is the Special purpose vehicle (SPV) and the investors or the subscribers who invest on the Sukuk.

In general, according to Zhamal K [10], all type of sukuk structures categorized in to four types (forms). These are:

1. *Debt-Based Sukuk*: it is based on Murabahah, Ijaraa or Istisnaa contracts. This type of sukuk highly resembles conventional bonds. Unless the underlying asset is taken as collateral, rating of these bonds should be based on credit rating of an obligor.
2. *Asset-Based Sukuk*: in this case sukuk holder has some claims on assets, which were used to facilitate sukuk issuance. Rating of sukuk should be based on the credit rating of the issuer.
3. *Project-Based Sukuk*: is a new form of sukuk structure, using real projects as a base. Sukuk holder gets paid according to the profitability of the project. In this case, rating of sukuk should be based more on risks of the project, rather than rating of the obligor.

4. *Asset-Backed Sukuk*: it is the type of sukuk where all payments are solely based on performance of an asset. Rating of such sukuk should be based on the rating of the back asset.

39.4.4 Major Modes of Sukuk Issues

Based on the above structures, there will be different types of sukuk arrangements or modes. However, the major modes of sukuk arrangements are as follows:

39.4.4.1 Pure Sukuk al-Ijarah

These types of Sukuk can be used for the mobilization of funds for the development of long term infrastructure projects. This happens through the securitization of tangible assets. These tangible assets can be real state such as factory or fixed assets to be leased such as aircraft, buildings, and ships. The parties' involved in Sukuk al-Ijarah are the issuer (SPV) as issuer and as trustee, the originator as seller, lessee and obligator (under purchase undertaking and under sale undertaking) and as servicing agent as well as the subscribers whom invest in the Sukuk. The rental rates of returns on this type of Sukuk can either be fixed or floating depending on the originator.

39.4.4.2 Sukuk al-Mudarabah

These types of Sukuk are certificates that represent projects or activities managed on the Mudarabah contract principle by appointing any of the partners involved in the deal as Mudarib for the management of the business. The parties involved in the Mudarabah are the Mudarib who is the issuer, the subscribers are the investors in the Sukuk and the realized funds are the Mudarabah capital. Moreover, the Mudarabah Sukuk holders own the assets of the Mudarabah and the agreed upon share of the profits belongs to the owners of capital and they bear any loss occurred.

39.4.4.3 Sukuk al-Musharakah

These are certificates of equal value issued for the mobilization of funds to be used on the basis of partnership, were their holders become owners of the relevant project or asset as per their shares that are part of their asset portfolios.

Musharakah Sukuk is mode which can serve for the securitization of assets in big projects where huge amount of capital are required. The parties involved in the Sukuk al-Musharakah are the issuer who is the inviter to a partnership in a specific project or activity. The subscribers are the investors in the Sukuk partners in the Musharakah contract. The mobilized funds are the share contribution of the subscribers in the capital. And the certificate holders own the assets of the partnership and they are entitled to any profit realized.

39.4.4.4 Sukuk al-Salam

These types of Sukuk are based on Salam principle in which advance payment of price are made for goods to be delivered in a certain time in the future. It is certificates of equal value issued for the sale of mobilized capital that is paid in advance in the shape of price of the asset to be delivered in a certain time in the future. A Salam purchase can onward sell the Salam asset by another contract which is parallel to the first contract. Parties involved in Sukuk al-Salam are the issuer who sells the Salam asset, the subscribers are the buyers of that asset, the mobilized funds which are the purchase price of the asset, which the Salam capital and certificate holders whom are the entitled to the Salam asset, the Salam price or the price of selling on parallel Salam basis, if any.

39.4.4.5 Sukuk al-Istisna

These types of Sukuk are based on Istisna principle in which an agreement for manufacturing goods and allowing cash payment in advance and delivery at a certain time in the future. In Istisna full ownership of the constructed item is immediately transferred upon delivery of the item to the purchaser, against the differed sale price that might also include any profits which legitimately covers the cost of tying funds for the period of the repayment. It is certificates of equal value and is issued with the aim of mobilizing funds required for producing a certain item or asset.

39.5 Risk Exposure Possibilities of Sukuk and Bonds

Risks are uncertain future events that could influence the achievement of the financial institution's objectives, including strategic, operational, financial and compliance objectives (Mohamad n.d: 5). The most important thing is to identify manageable risks and try to avoid or reduce the possibilities of exposure to them. The novelty of sukuk inherently entails a higher exposure to certain market and financial risks. In conventional financial system bonds are also subject to various types of risk and some of these risks are shared by the Sukuk. On the other hand, there are also risks which are specifically related to conventional bonds or sukuk instruments only.

Thus, in this section, we will compare conventional bonds with sukuk instruments in terms of their confrontation to different types of risks. These risks have grouped into three categories as those affect both instruments, those specifically related to bond instruments, and those that are specifically affect sukuk only. According to Tahmoures [11], some of the risks that affect both conventional and Islamic bonds or either of these instruments are as follows:

39.5.1 Risks That Affect Both Sukuk and Bonds

Business/Financial Risk risk that the bond issuer will default on interest and/or face value or both. The same risk can happen in the Sukuk. However, their remedial methods are quite different. In the conventional bond, bondholders have no choice but to recourse to the issuer for unpaid amount. This often is done by a lawsuit against bond issuer and it is not clear how much of the unpaid cash flow can collected and within how long. In the Sukuk case, the Sukuk holders have recourse to the asset not to a bankrupted individual. Accordingly, more comfort is offered to the Sukuk holders than bond holders.

Liquidity Risk it is a risk that the bond is not saleable at reasonable price and reasonable time due to lack or inefficient secondary market. Liquidity risk is also vital for Islamic finance in general and sukuk in particular. This is because Islamic financial institutions have limited instruments to manage their liquidity, due to Shari'ah restriction on trade of debt and other securities. This type of risk is applicable to the non-tradable Sukuk.

Conventional bond market, while more liquid than sukuk market, is still considered as less liquid than equity market. Bond investors face the risk of not been able to trade their securities due to lack of potential buyers. Most of the trading in bond market is done "over-the-counter" rather than in organized exchanges. While traditional financial institutions have various instruments to manage their liquidity, sukuk remains one of very few options available to Islamic financial institutions. Thus, development of appropriate secondary market is crucial for sukuk more than for conventional bonds.

Inflation Risk This type of risk has opposite impact for the two instruments. Due to its fixed-income nature of conventional bonds, the investor bears the risk that inflation can be higher than the coupon payment. Thus, high inflation rate can cause the conventional bond yields to lag behind inflation rate. High inflation rate, however, has positive effect Sukuk instruments. As inflation rate goes up so does the market price of the Sukuk assets at maturity. This is indeed a return than a loss.

Foreign Exchange Risk This risk is the same for both conventional bonds and the Sukuk. It can affect bonds issued in a foreign currency, other than the issuer currency, when the unfavorable currency fluctuation decreases the initial value of investments. However, according to Tariq [12], those Sukuk which are liquid or which are relatively short term in nature will be less exposed to foreign exchange risk.

Default Risk There is a risk that the issuer would not be able to make regular payments (coupons) or to repay the principal amount. According to Tariq and Dar [13], due to the fact that Shari'ah prohibits debt trading, any rescheduling of debt for higher markup is forbidden under sukuk. This prohibition makes risk of default higher for sukuk as compared to conventional bonds, since sukuk issuers "would be more inclined to default".

Moreover, while conventional bond represents a debt obligation, sukuk is a certificate of ownership, so in case of default sukuk holders have a very limited possibility to retrieve their initial investment. The managers of sukuk can bear responsibility for any sukuk default only within the limits of their control and capabilities. Therefore, in case if default occurs due to external factors, such as "force major" or global financial crisis, all losses under sukuk will be borne by sukuk holders [14].

39.5.2 Risks Specific to Conventional Bonds

Call Risk The conventional bonds are usually subject to this risk when a big market interest rate occurs. This creates a big problem for bondholders since they will be deprived from the higher original interest rate. The Sukuk are not susceptive to the fluctuation in market interest rates as the conventional bonds.

Interest Rate Risk (Risk of Return) this type of risk is a major cause of price volatility in the bond market. As a fixed-income instrument, bond yield has an inverse relation with interest rate movement. For any change in market interest rates, two opposite changes/risks are created for existing bonds: first, there is an inverse relationship between market interest rate and price of bonds; When market interest rates grow, bond price decreases and vice versa. Second, as a market rate change so does the return bond investors will receive from reinvesting their interest/coupon. With the right duration, these risks can be eliminated. The longer is the maturity of the bond, the higher is the potential of interest rate growth and, therefore, the higher is its interest rate risk.

Downgrade Risk This is the risk that a bond price will decline due to a downgrade in its credit rating. Downgrades can come from a variety of sources; however, debt downgrades generally come from ratings agencies such as Moody's or Standard & Poor's (S&P). Thus, investor, who is willing to trade bonds on the secondary market, bears downgrade risk.

39.5.3 Risks Applied Only to Sukuk

Price Risk also known as asset redemption risk. All sukuk issues should be backed by tangible assets. At the end, the originator has to buy back the underlying assets from the certificate holder. Thus, this is the risk that the value of an asset at maturity dates will be different from market price due to fast depreciation, over usage of the asset and/or damage. The principal amount paid may not be equal to the sukuk issuance amount and, as a result, there is the risk that the assets may not be fully redeemed [12].

Shari'ah Compliance Risk This risk results from the violation of Shari'ah provisions in the Sukuk case. Shari'ah compliance is described by Tariq and Dar [13] as a risk of loss of asset value due to sukuk incompliance with Shari'ah principles. Each issue of sukuk should be approved by Shari'ah board as compliant to Islamic rulings. This type of risk became very important in light of recent criticism by some Shari'ah scholars about non-Islamic nature of most of the modern sukuk.

Consequences of issuing financial instruments non-compliant with Shari'ah can be very damaging to the reputation of issuer and may require extensive efforts to regain investors' confidence. It is also worth mentioning that some Shari'ah scholars impede development of sukuk market. One of the problems with Shari'ah boards, as mentioned by Usmani [14], is the fact that some scholars are active only at the first stage of sukuk structuring process. They issue fatwa (decision) on permissibility of issue in accordance with proposed structure and ignore remaining stages of sukuk performance. However, the new standards issued by AAOIFI [15], declares as Shari'ah boards should be active during all stages of sukuk operation, ensuring Shari'ah compliance of entire life span of sukuk.

Operational Risk it is the risk of revenue loss due to delaying in getting the benefits of the underlying assets under the lease agreement. Conventional bonds are free from the direct effect of this type of risks. Since the structures of the tradable Islamic sukuk must be based on assets and the return on these sukuk originates from these assets, then the operational risks of these assets must be carefully studied.[2]

Legal Risks Lack of standardized regulations governing the Islamic finance is a major impediment to further development of the whole market, including sukuk. Several international institutions have been established to produce the standards and regulations, including AAOIFI and IFSB. Sukuk faces regulatory risk as there is a conflict between the provisions of Shari'ah with the regulations of the country in which the Sukuk were initiated.

However, as far as Sukuk is concerned, Goldman's [16], the major risk is the violation of the six standards that were set by the AAIOFI in 2008. One of the credit rating agencies has recommended that Shari'ah scholars should take into account both a priori and posteriori compliance process carefully in order to avoid invalidation of the Sukuk over its life time.

SPV Specific Risks The Special Purpose vehicle is generally designated to be a standalone institute that is bankruptcy remote from the originator. However, there may be a notion of settlement risk involved with the SPV in that the originator will have to channel the payments through a clearinghouse. The certificate holders will then be reimbursed through the clearinghouse. It is the notion of settlement risk involved with the SPV where the originator will have to reimburse the certificate holders through a clearinghouse [17].

---

[2]The "Sukuk" Risks. Memri Economic Blog, 2008, p. 1: Available on: http://www.memrieconomicblog.org/bin/content.cgi?article=119 accessed on 21st of April, 2014.

Risks Related to the Asset The underlying assets of the sukuk certificates are subject to numerous risks such as risk of loss of an asset. This risk is minimal in case of sukuk ijarah, but it can be significant in case of construction. Furthermore, another aspect of risk is the need to maintain the structures of assets. Proper maintenance will ensure adequate returns to the certificate holder. According to Shari'ah principles, the SPV will usually be required to bear the responsibilities on ensuring asset structure maintenance.

39.6 Comparison of Yield and Risks of Sukuk and Conventional Bonds

One of the most common questions that market participants nowadays ask is whether it makes sense to buy developed countries' government bonds now that their valuations have been driven up by the market (as their yields have been driven down by the market). Developed market interest rates are at all-time low, and may remain so far longer than investors currently expect, as governments, companies and households continue to cut their borrowing, restricting potential economic growth. As a result, there is an increasing need for investors to diversify away from their traditional concentration in developed markets, and to look further afield for yield [18].

Regardless of the above mentioned unique risks, Sukuk as Shari'ah-compliant instruments provide medium to long- term fixed or variable rates of return. As compared to the global and other conventional bonds, sukuk certificates provide a better risk adjusted returns to its holders. Unlike emerging market bonds, that have a relatively high return with high volatility rate, the outperformance of sukuk is not as a result of too much volatility in its returns. Their volatility ration is also relatively low. Based on 3 year data, it is attempted to make some comparison between global sukuk and global conventional bond. The result, as depicted below (Fig. 39.1), indicates that the global sukuk outperformed with less volatility than conventional bonds. When we compare the indexes of Global Sukuk, as represented by Dow Jones Sukuk Index, with JP Morgan's GBI Broad index during the past 3 years ended December 2013, we can see how Sukuk performed better than global bonds. The volatility was also low during the periods as compared to that of GBI's.

Sukuk is still better in terms of profitability and risk when compared with the global bond market based on 2 year data from June 2011 to June 2013. Based on this analysis, during the 2-year period ended 30th of June 2013; the global Sukuk market outperformed the Global Bond market (as represented by the CITI World Broad Investment Grade (WorldBIG) Bond Index with average annualized returns of 3.71 % versus 0.57 % respectively. That means, sukuk has more than six times profitable as compared to the World Broad Investment Grade Bonds. Similar to the above comparison, this performance was also attained with less volatility. When we look at the volatility rate, as shown below (Table 39.1), the returns volatility rate of the global Sukuk market, 2.24 %, was less than half of that of the world BIG bond index which was 4.66 %.

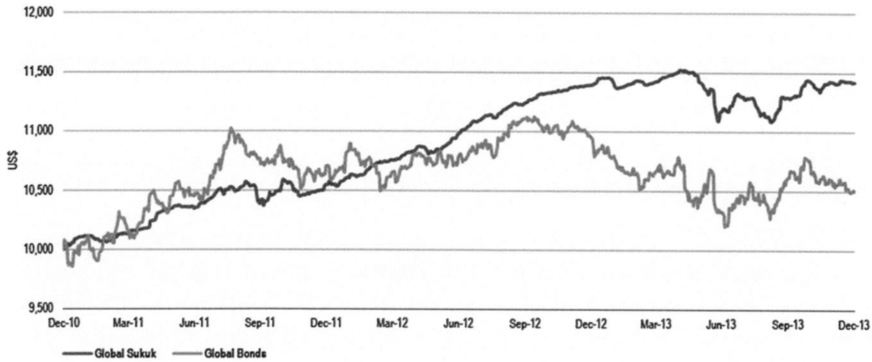

Fig. 39.1 Three-year period return ended December 31st 2013 (Source: S&P and J.P. Morgan as of 31 December 2013. Global Sukuk are represented by the Dow Jones Sukuk Index. Global Bonds are represented by the J.P. Morgan GBI Broad Index. Returns are in USD. The computation is made by FactSet Research Systems Inc)

Table 39.1 Two years annualized returns, returns volatility and Sharpe ratio of DJSI and World BIG

	Annualized returns (%)	Returns volatility (%)	Sharpe ratio
Dow Jones Sukuk Total Return Index	3.71	2.24	1.61
Citi World Broad Investment Grade (WorldBIG) Bond Index	0.57	4.66	0.10

Source: Bloomberg and CIMB-Principal Islamic Asset Management as at the 30th June 2013

Higher profitability ratio may not always give full information about an investment's performance unless it takes into account the risks attached with it. Investors always need to be properly compensated for the additional risk they take for not holding a risk-free asset. Therefore, the most important thing is that whether those investors are properly compensated for each unit of risks they are willing to accept or not. To see this, in the Table 39.1, the Sharpe ratios, as it measures the excess return (or risk premium) per unit of deviation in an investment vehicle, is also computed. Since it measures the excess return (or risk premium) per unit of risk in an investment, the ratio tells how much excess return investors are receiving for the extra volatility that they endure for holding a riskier asset. Accordingly, in examining the quality of the global Sukuk market performance, it can be compared with conventional bonds based on its risk adjusted rate of returns.

As shown above (Table 39.1), the DJ sukuk return index had 16 times higher annualized Sharpe ratio as compared to World BIG. It tells as the sukuk superior earning performance and lower volatility is further enhanced with a higher Sharpe ratio. The global sukuk have earned better average excess return per unit of risk during the periods as compared to World BIG. The world broad investment grade bond holders receive annualized return of only 0.10 per unit of extra volatility risks

Table 39.2 Returns volatility of DJSI and JP Morgan EMBIG Index for 2-year period

	Returns volatility (%)
Dow Jones Sukuk Total Return Index	2.24
JP Morgan Emerging Market Bond Index (EMBI) Global Total Return Index	7.21

Source: Bloomberg and CIMB-Principal Islamic Asset Management as at the 30th June 2013

whereas sukuk certificates offered average excess return of 1.61 per unit of extra volatility that investors endure for holding it. The improvement in the Sharpe ratio shows that the sukuk strategy can generate alpha or additional returns that better compensate for risk.

The volatility of Sukuk return is still much less as compared to the volatility of emerging market bonds. Over the 2-year period (ending the 30th June 2013), the global Sukuk market exhibited lower returns volatility compared to the emerging market bond (represented by JP Morgan Emerging Market Bond Index (EMBI) Global Total Return Index).

As indicated above (Table 39.2), the 7.21 volatility ratio of EMBI is more than twice higher than the 2.24 volatility on Sukuk returns. This signifies that global Sukuk as an asset class is relatively insulated compared to the emerging and the global bond markets. Therefore, from this data anyone can expect an increase in demand for global Sukuk portfolios since the asset class remains relatively insulated from most of the volatility in other financial markets.

39.7 Correlation of Sukuk with Other Asset Classes

Fund managers and international investors are interested to know about the relationship among investment vehicle before they invest their money and form a portfolio. Accordingly, in addition to its return and risk character, the other point that takes into account by international investors who are interested in investing on fixed income instruments to form the best combination of these vehicles is its correlation with one another. The correlation nature of an asset enables them to decide as what type of portfolio should they form and how to manage their portfolio risks. Analyzing a vehicle's correlation profiles with other asset classes is also another way of judging how effectively it can bring diversification to a portfolio.

With low correlation to other asset classes, Sukuk are a good compliment in Shari'ah compliant and conventional portfolios. As indicated in the below (Table 39.3), the correlation coefficient between the global sukuk and conventional bonds and other asset classes are relatively low which is attractive to portfolio managers. This is because given their relatively low correlations to other asset classes sukuk certificates have historically exhibited attractive diversification benefits.

As shown in the table, global sukuk has a relatively low to moderate correlation with other securities. Its relatively strong correlation was with global equity, which

Table 39.3 Three-year correlation of global sukuk with other asset classes (as of December 31st 2013)

	Global Sukuk
Global Sukuk	**1.00**
Global Shari'ah equity	0.39
Global equity	0.54
Global bonds	0.43
Commodities	0.28

Source: S&P, MSCI, J.P. Morgan, FTSE, Dow Jones as of 31 December 2013. The calculation is made by FactSet Research Systems Inc

Global Sukuk are the Dow Jones Sukuk Index. Shari'ah Global Equities are represented by the MSCI AC World Islamic Index. Global Equities are represented by the MSCI World Index. Global Bonds are the J.P. Morgan GBI Index. Commodities are represented by the Dow Jones UBS Commodity Index-Total Return

was 0.54. The weakest correlation was with the commodities, 0.28, which are represented by the Dow Jones USB Commodity index. It had also a 0.39 and 0.43 correlation with global Shari'ah equity and global bonds respectively. In general, its low correlation with these and other major asset classes provides diversification opportunities to global investors.

39.8 Portfolio Diversification Potential of Sukuk

The regular periodic income streams during the investment period with easy and efficient settlement and a possibility of capital appreciation gives unique qualities to sukuk certificates that in-turn provide diversification for those with bond portfolios that have more of a developed market assets and equities. Sukuk have an important role to play in the search for diversification, offering a significant yield premium over developed market debt, along with superior fundamentals.

In addition to providing income opportunities, an allocation of resource to sukuk offers potential diversification benefits. Among several of the most-notable features of sukuk certificates that investors motivate for considering them in asset allocation (forming portfolio) are their currency diversification benefits, attractive correlations with other assets, yields, credit quality, and wide geographical diversification.

Investing in sukuk offers strong international diversification. Since the issuer of sukuk comprises many countries from different areas of the glob more opportunities are expected from investments in these markets. Importantly, as the number of investable countries has increased, so has the regional diversification. When the DJSI formed in 2004, it represents the sukuk index of geographically diversified countries. Despite political unrest in pockets of the Middle East, the performance

Table 39.4 Principal Islamic Asset Management

Country	Percentage (%)
Malaysia	17.08
Saudi Arabia	16.21
Qatar	13.60
Indonesia	6.04
Dubai	5.53
Abu Dhabi	3.60
Total	62.06

Source: Bloomberg and CIMB-Principal Islamic Asset Management, October 2012

of the DJSI has remained resilient. This stability of the index is a reflection of the geographical diversification it offers. The sovereign representation of Sukuk issuers in the DJSI leads with Malaysia, and is followed by Saudi Arabia, Qatar, and Indonesia. According to some experts, these jurisdictions have strong macroeconomic fundamentals and offer exposure to oil- and gas related revenue streams, which offer deep reserves support to issuers (Table 39.4).

According to Islamic Finance Outlook report [19], faced with scarcity of funds in international markets, a growing number of new and emerging jurisdictions in Asia, Africa and Europe are seeking to attract Islamic finance investors, particularly from the oil-rich GCC countries, by issuing sukuk to fund their infrastructural development projects and other budgetary and corporate financial needs. In recent years, jurisdictions that have tap the global sukuk market include Azerbaijan, Turkey and United Kingdom in 2010; Hong Kong, Jordan and Yemen in 2011; France, Germany and Kazakhstan in 2012; and Luxembourg, Mauritius, Nigeria and Oman in 2013 (Table 39.5).

Since 2004 a number of debut sovereign issuance have been made by different countries. The global sukuk industry is also expected to continue its upward trajectory in 2014 as a number of high profile debut sovereign issuances are expected to take place this year. The sovereign sukuk including the United Kingdom, Ireland, South Africa, Tunisia, Mauritania, Senegal and Oman are expected to debut sovereign issuances in 2014.

The issuance of sukuk in different local currencies in addition to USD or other hard currencies also brings an advantage of reducing portfolio risks. As mentioned in the previous part of this paper, one of the risks of sukuk is related to exchange rates among currencies. Thus, the composition of assets denominated in various currencies in the pool will contribute to reduce foreign exchange risks in different ways. Hence this can be very useful tool to overcome the foreign exchange risk by diversifying the pool with sukuk of different currencies (Fig. 39.2).

Within the sukuk issuance, investors have also an alternative to form a portfolio with different sukuk structures. Investors may have a pool of Musharakah, Ijarah and some Murabaha, Salam, Istisna'a, and Ju'alah contracts. The return and risk on such securities well be depends on the chosen mix of the contracts. In general, within the sukuk issuance, the diversity by region, currency and sukuk structure should help

Table 39.5 Kuwait Finance House Research, 31st March 2014

Issuer name	Sukuk name	Structure	Domicile	Currency	Issue size (USD in Mil)
Banque Centrale de Tunisie	Tunisia Sovereign Sukuk	Unknown	Tunisia	TND	Unknown
Central Bank of Mauritania	Mauritania Sukuk	Unknown	Mauritania	–	300
Central Bank of Nigeria	Nigeria Sovereign Sukuk	Unknown	Nigeria	–	200
Government of Ireland	Ireland Sovereign Sukuk	Unknown	Ireland	–	19.493
Government of South Africa	South Africa Sovereign Sukuk	Ijarah	South Africa	USD	148.723
Republic of Tatarstan	Tatarstan Sovereign Sukuk	Unknown	Russian Federation	USD	Unknown
UK Sovereign Sukuk	UK Sovereign Sukuk		United Kingdom	GBP	328.22

Source: IFIS, Zawya, KFHR

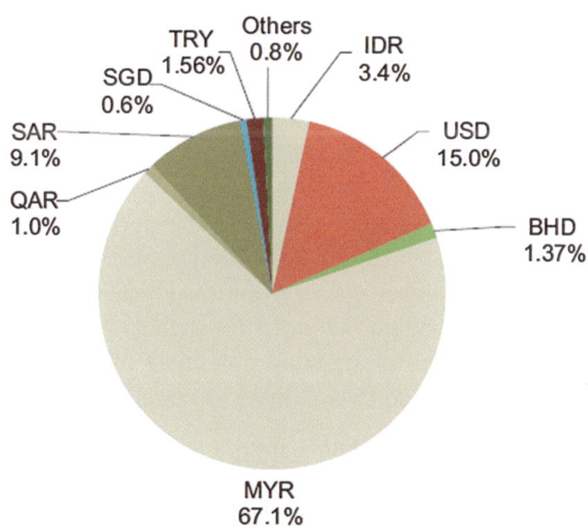

Fig. 39.2 Kuwait Finance House Research, 2013

manage volatility because the various regions and countries do not all tend to rise and fall at the same time.

We can see a case to strength the above discussion as the benefits of diversification with Sukuk securities are not just theoretical, they are demonstrable. For this, let us see two scenarios of a portfolio in CIMB Principal Islamic Asset Management. After looking in to these scenarios, one can examine the outcomes to determine whether investing in Sukuk serves as a good diversification strategy without compromising investment returns:

1. One portfolio tracks the conventional index completely.
2. A second portfolio is split, whereby 20 % tracks the DJSI and 80 % tracks the conventional index.

	Annualized returns (%)	Sharpe ratio
Portfolio 1: 100 % tracking the World Broad Investment Grade Bond Index	4.52	0.87
Portfolio 2: 20 % tracking the Dow Jones Sukuk Index and 80 % tracking the World Broad Investment Grade Broad Index	4.88	1.14

Source: Bloomberg and CIMB (**C**ommerce **I**nvestment **M**erchant **B**ank)-Principal Islamic Asset Management Sdn Bhd as of September 2012

For the 2-year period ending September 28, 2012, the annualized returns of the second portfolio resulted in a slightly enhanced return of 4.88%, which is above the 4.52 % of the first portfolio. In addition, the second portfolio's Sharpe ratio was increased significantly to 1.14 from its 0.87. Therefore, from the second portfolio, investors have not only earned better return; they have also received excess reward (1.14) per unit of every extra volatility that they endure by holding 20 % sukuk in their portfolio.

39.9 Conclusion

The nature of this study is descriptive and has intended to describe the features and the risk other characteristics of Sukuk. It is also attempted to compare and contrast Sukuk instruments with conventional bonds. Accordingly, the two instruments have compared based on the variables such as risks, returns, and volatility ration. Apart from this the diversification potentials of sukuk instruments have also discussed. Thus, from this study a number of conclusions can be drawn.

First, the two instruments are not quite identical. Although there are various fundamental differences between the sukuk and conventional bonds, both instruments tried to solve the same common financial problem of raising capital for needed entities. They offer quite different solutions to the same financial problem. Therefore, their differences are not originated from financial interests but it is from the religious backgrounds.

Second, sukuk instruments are a good mechanism to attract those investors who are away from participating in conventional system due to their religion. The Sukuk issues are available to all financial institutions and investors regardless of their religious background and believe. Which method of financing should be considered, is a matter of choice and whether the bond issuer and/or bondholder would like to benefit from the growing market share of the religiously conscientious financial market.

Third, the two financial instruments are exposed to some specific risks. Certain risks will affect only either conventional bonds or sukuk instruments. Some events which can bring a risk to conventional bonds will have a gain to sukuk instruments. One reason for this is their dissimilarities in nature. Therefore,

- If there is an active management approach supported by deep research, it can help investors to avoid undue risk.
- One of the most important risks for Sukuk is Shari'ah compliance risk. Therefore, institutions and governments should have proper Shari'ah advisory board before issuing any forms of Sukuk.
- To overcome the liquidity problem, conventional financial institutions have various instruments to manage their liquidity. However, there are only very few options available to Islamic financial institutions. Therefore, development of appropriate secondary market is crucial for Sukuk instruments.

Fourth, based on the analysis Global Sukuk has shown relatively strong performance with lesser volatility compared to conventional bonds. Sukuk were effective in achieving high returns as compared to conventional bonds. They offer better risk adjusted returns than other conventional bonds.

Fifth, Sukuk also provides diversification benefits gained from their diversity by region and currency, attractive correlation with major asset classes, and their diversity sukuk structure. Therefore, an active portfolio management can change exposures to individual countries or currency that are performing poorly, overweight countries or particular market sectors that are performing well, and may help investors expand their income opportunities by allowing them to take advantage of the diversification attributes sukuk instruments have to offer.

Sixth, the use of Sukuk instruments along with conventional bonds will create a business opportunity for bond issuers. Therefore, creating suitable regulatory environment for sukuk instruments will make easy the issuance and trading of sukuk. Creating proper Islamic financial regulatory bodies (like IIFS) in every country will be difficult, but it is possible to establish a dedicated department within existing government structures that will do at least the basic tasks of Islamic financial regulatory bodies. It will be also important to create cooperation among different regulatory bodies to form institutional frame work for sukuk within the existing system.

References

1. Sundararajan V, Marston D, Shabsigh G (1998) Monetary operations and government debt management under Islamic banking. IMF working paper, WP/98/144, Washington DC, USA
2. Wilson R (2012) Legal, regulatory and governance issues in Islamic finance. Edinburgh guides to Islamic finance, A series edition. Edinburgh University Press Ltd, Edinburgh
3. Bakar MD, Alhabshi SM (n.d.) An introduction to Islamic finance. Chartered Institute of Management Accountants (CIMA), London
4. Bodie K, Marcus (2004) Investments, 6th edn. The McGraw—Hill Companies Inc, Columbus

5. Hans V (2009) Islamic finance, principles and practice. Edward Elgar Publishing Limited, Cheltenham
6. Christopher FRICHARDSON (2006) Islamic finance opportunities in the oil and gas sector: an introduction to an emerging field. Tex Int Law J 42(119):122–154
7. Thomas A, Cox S, Kraty B (2005) Structuring Islamic finance transactions. Euromoney Books, London
8. Jobst A, Kunzel P, Mills P, Sy A (2008) Islamic bond issuance – what sovereign bond managers need to know. IMF policy discussion paper, Monetary and Capital Markets Department (MCM), Washington, DC
9. Wilson R (2004) Overview of the sukuk market Islamic bonds: Your guide to issuing, structuring and investing in sukuk. Euromoney Institutional Investor PLC, London/England, pp 3–17
10. Zhamal KN (2010) How risky Sukuk are: comparative analysis of risks associated with Sukuk and conventional bonds. Dissertation paper, the British University in Dubai
11. Tahmoures A, Afshar A (2013) Compare and contrast Sukuk (Islamic bonds) with conventional bonds, are they compatible? J Global Bus Manag 9(1):44–52
12. Tariq AA (2004) Managing financial risks of Sukuk structures. A dissertation submitted in partial fulfillment of the requirements of the degree of Master of Science at Loughborough University, UK
13. Tariq AA, Dar H (2007) Risks of *Sukuk*s: implications for resource mobilization. Thunderbird Int Bus Rev 49(2):203–223. March–April 2007
14. Usmani MT (2008) Sukuk and their contemporary applications. AAOIFI Shariah Council, Bahrain
15. Accounting and Auditing Organization for Islamic Financial Institutions (AAOIFI) (2008) Sukuk Shariah standards. Available at: www.aaoifi.com
16. Goldman (2012) Sukuk induces ownership and liquidity risks. Posted at: http://businessislamica.com/2012/03/13/goldmanssukuk-induces-ownership-andliquidrisks
17. Khalil A (2011) Sukuk: definition, structure and accounting issues. MPRA paper no. 33675, posted 25, September 2011 13:19 UTC
18. Robert S (2012) Emerging market debt: look further afield for yield. J.P. Morgan Asset Management. European Bank and Business Centre 6, route de Treves L-2633 Senningerberg, Luxembourg
19. Research Ltd KFH (2014) Islamic finance outlook 2014. Kuwait Finance House Research Ltd, Kuala Lumpur

Chapter 40
Coping Mechanisms of Effective Leaders in Chaotic and Complex Educational Settings

Ssali Muhammadi Bisaso and Fathimath Saeed

Abstract Many events that take place in the education sector can be shown as examples of chaos and uncertainty. These events indicate that the foresight we previously demonstrated linearly is insufficient and the factors that we overlook or undervalue can result in enormous and unexpected consequences. This shows that poorly managed small factors lead to large and negative consequences. Besides, regardless of the positive or negative nature of its consequences, the butterfly effect is expressed as: “small events lead to big events”. If innovation and improvement are desired in education, leaders begin from the lowest stage instead of the upper stages of the system in order to accomplish this. In this case, the leader must have further information with the aim of providing more activities and information to respond to the challenges within the system. Thus the leader both equips himself better and forces senior management to offer better opportunities. In this way improvement does not remain only in the education system, it also spreads to the other systems of the country. In this case, the leader would have achieved a better health care system, a better construction system, a better security system, etc. as well as a better education system thanks to the trigger created within the system. This paper sought to demonstrate that although chaos and complexity within systems scare and bother us, they force us to change and evolve better. For this to happen though, leaders must adopt coping mechanisms to counter chaotic and complex situations which the paper dully explored. There is no gain saying therefore, that effective leaders, can prevail over chaos and complexity laden situations and transform their systems or organizations in order to catapult them to even greater heights.

S.M. Bisaso (✉)
Islamic University in Uganda, Mbale, Uganda

Department of Educational Management and Supervision, Hacettepe University, Ankara, Turkey
e-mail: bisasossali@yahoo.com

F. Saeed
Ministry of Education, Male, Maldives

Department of Educational Management and Supervision, Gazi University, Ankara, Turkey
e-mail: fathmath_saeed@hotmail.com

Ş.Ş. Erçetin (ed.), *Chaos, Complexity and Leadership 2014*,
Springer Proceedings in Complexity, DOI 10.1007/978-3-319-18693-1_40

40.1 Introduction: The Nature of Chaos and Complexity Theory

Reigeluth [20] postulates that some of the key features of chaos theory and the sciences of complexity include co-evolution, disequilibrium, positive feedback, perturbance, transformation, fractals, strange attractors, self-organization, and dynamic complexity. As a result, chaos scares and bothers us, but thanks to those fears it forces us to change and evolve which takes us away from linearity even more and feeds chaos ([9]:863–864).

In addition to recently getting involved in social sciences, Chaos also has an interdisciplinary aspect that evaluates events from a philosophic perspective (Töremen 2000 cited by [25]). Chaos, as a theory and the disorder coming along, is a structure in which mutual relationships end in unpredictable and unexpected results. It was possible to make predictions about former changes and developments, and various evaluations could be made about the improvement.

The concepts of chaos theory can explain the way schools work [7]. For example, teachers do not exist as separate entities, but are affected by the relationships that exist within schools. It may also shed some light on how we can deal with and understand how things in our classrooms, schools, and entire communities are interrelated and all reflect in some manner upon each other. According to Gleick [12], there are several aspects of chaos theory such as nonlinearity, complexity, butterfly effect, fractals and feedback mechanisms that may have significance for educational settings.

A critical review of complexity would reveal that the term attracts a number of conceptual definitions. Accordingly, complexity refers to the condition of the universe which is integrated and yet too rich and varied for us to understand in simple common mechanistic or linear ways. Accordingly, we can understand many parts of the universe in these ways but the larger and more intricately related phenomena can only be understood by principles and patterns – not in detail. Complexity deals with the nature of emergence, innovation, learning and adaptation [26].

To many authors meanwhile, complexity theory is a theory of change, evolution and adaptation, often in the interests of survival, and often through a combination of cooperation and competition ([2, 18, 19]; Stewart 1991). Indeed some of the authors claim that complexity breaks with straightforward cause-and-effect models, linear predictability, and a reductionist, atomistic, analytically-fragmented approach to understanding phenomena, replacing them with organic, non-linear and holistic approaches, in which relations within interconnected networks are the order of the day [6, 19, 27, 28]. In one sense, many of the central tenets of complexity theory are neither new nor particularly startling; however, the bringing together of several key constructs into a more-or-less unified theory is, perhaps, what gives complexity theory its impetus and attraction especially in modern day research.

The key terms of reference for complexity theory then, and as discussed in various analyses, are: self-Organization, complex adaptive systems, non-linear change, emergence, unpredictability, diversity, differentiation and autopoiesis, networks, connectivity and relations, order without control, feedback, open systems, collectivity, distributed knowledge, autocatalysis, holism, and co-evolution [19].

40.2 Types of Complexity in Organizations

Adopting coping mechanisms to chaos and complexity situations by managers and leaders would call for a better understanding of the types of complexity abound in education systems. A number of authors have established these types depending on organizational characteristics. Some refer to them as complexity [22] while others coin the phrase contexts [23].

Shyamala [22] notes the following

- Imposed complexity includes laws, industry regulations, and interventions by nongovernmental organizations. It is not typically manageable by companies. In the times of such complexity, democratic leadership is needed for an organization.
- Inherent complexity is intrinsic to the organization, and can only be jettisoned by exiting a portion of the organization. In this situation transactional leadership is suitable for an organization.
- Designed complexity results from choices about where the organization operates, what it sells, to whom, and how. Companies can remove it, but this could mean simplifying valuable wrinkles in their organization model. During this kind of complexity, task oriented leadership is suitable for the organization.
- Unnecessary complexity arises from growing misalignment between the needs of the organization and the processes supporting it. It is easily managed once identified. In this time charismatic leadership is suitable for an organization.

Snowden and Boone [23] describe leadership in times of complexity and chaos by putting across levels of contexts. They equally provide a connection between complexity theory and leadership, suggesting that different styles of leadership are needed for varying contexts:

- Simple contexts (Domain of Best Practices) – the realm of the "known knowns" – calls for leaders to sense, categorize and respond. Established answers exist and just need to be accessed. A command and control style works best. (They warn that there are very few simple contexts within current leadership environments though).
- Complicated contexts (Domain of Experts) – the realm of the "known unknowns" – calls for leaders to sense, analyze and respond. Like the simple context, things are ordered and there is a direct connection between causes and

effects. A leadership style of listening to experts and encouraging dissenting perspectives works best.

- Complex contexts (Domain of Emergence) – the realm of the "unknown unknowns" – calls for leaders to probe first, then sense and respond. This is the realm in which answers are emergent and the whole is more than the sum of the parts. A leadership style of experimentation, patience and openness to emergent answers works best. This is the most common context in the contemporary environment.
- Chaotic contexts (Domain of Rapid Response) – realm of the "unknowables" – calls for leaders to act first to establish order, then to sense and respond. Chaotic environments are also the source of innovation if attention can be focused both on crisis management and creativity.

40.3 Chaos and Complexity Theory and Education

Educational systems, institutions and practices exhibit many features of complex adaptive systems, being dynamical and emergent, sometimes unpredictable, non-linear organizations operating in unpredictable and changing external environments [19]. These systems, institutions and practices shape and adapt to macro- and micro-societal change, and, through self-organization, respond to, and shape the environments of which they are a part [16].

Many researches have also come to the conclusion that chaos is indeed part of the education systems making educational institutions chaotic and complex ([4, 7, 15, 16]). This implies having a paradigm shift when it comes to the status-quo in our education system especially in regard to coping techniques or mechanisms of leaders in educational institutions.

As Stewart (1991 cited by [19]) dully remarks, there is co-evolution between the organism and its environment. This process occurs through learning, adaptation and development. The researches done in this area attest to this, and provide illustrations of the process at work. Indeed Morrison [19] postulates that they indicate how self-organized change and emergence of various educational 'complex adaptive systems' occur through autocatalysis, networking, relations, differentiation, co-evolution, connectivity, feedback, and distributed order, within unpredictable and unpredicted, non-linear environments.

Complexity theory also *redefines 'the basics' of education*, away from a controlled and controlling subject-based education and towards a discovered, interdisciplinary, emergent and constructivist curriculum, and a reassertion of freedom as a sine qua non of education (Doll 1993 cited by [19]). Complexity theory takes us in a direction opposite to the neatly stated, over-determined, tidy, traditional, Tylerian, externally mandated and regulated prescriptions of governments for the aims, content, pedagogy and assessment of learning and education.

Additionally, as opined by Morrison [19], complexity theory suggests the following:

- a movement towards bottom-up development and change,
- local and institutional decision-making on education,
- a re-assertion of child-centeredness and experiential,
- exploratory learning,
- a rejection of tight prescription and linear programming of teaching and learning and,
- a move towards non-linear learning and their curricular derivatives.

Premised on the above aspects therefore, complexity theory emphasizes the *process rather than the content of learning*, as the constituents of relevant and enduring curriculum content are uncertain [8]. Emergence and self-organization require room for development; tightly prescribed, programmed and controlled curricula and formats for teaching and learning, and standardized rates of progression are anathema to complexity theory. It breaks a lock-step curriculum.

Important to note is that chaotic systems take complex forms, making their precise measurement difficult if not impossible [13]. Different measurement instruments have been put in place to evaluate and compare the performance of a school. However, due to the complex nature of schools, none of these assessment methods seem to measure precisely the school performance and have very limited validity for the following reasons [7]:

Basing on the views of Glickman [13] and Cunningham [7] therefore, the prior achievement of pupils is not taken into account and this is a major factor in pupil achievement at a later stage. This also informs the nature of complexity at stake in schools as reflected in the following aspects:

- Schools are differentially effective in different subjects and with pupils of different ability, which is not reflected in a single figure.
- Schools change over time; however, the achievement data used reflects only one group and is essentially historical data.
- Student mobility between schools is not reflected in the assessment.
- Social factors, sex of students, ethnic origin and social background are not taken into account. These factors are out of the school's control.

In view of the above therefore, assessing school performance and comparing one to another have become increasingly difficult given the complex nature of today's schools. Even the established quality control tool of supervision which was highly trusted in the olden days [13] can not be trusted anymore.

It is also claimed that today's schools are more complex systems than the one-room schools of the past. However, most of the beliefs and expectations about schools today still remain the same as they were in the olden days [16]. In the one-room schools of old times, the teacher was responsible for all the instruction of all the students, the maintenance of the building, keeping the stove filled with wood and cleaning the floors [16, 17].

40.4 Coping with Issues of Chaos and Complexity in Education Systems

Chaos has been a big part of education systems in various forms [4, 7]. The three principal conditions for a chaotic system are: (1) that it operates in a non-linear way; (2) that it is iterative (the output of one cycle becomes the input of the next); and (3) that small variations in initial conditions lead to large differences in outcomes. Many systems within educational organizations appear to meet these conditions (Cunningham [7] cited by Kara [16]). Accordingly, education is an essential component in responding to emergencies after conflicts or national disasters [4]. Along the same line, complexity in education systems is on the rise due to a number of intersecting trends [15].

Research has fully explored leadership in chaos and complexity laden situations ([14]; Kelly 2005; [21, 26]). Some of the researches feature the status-quo in times of chaos and complexity [21, 26], while other researches focus on the challenges encountered by leaders during such chaotic and complex situations ([14]; Kelly 2005). Another critical group of researchers meanwhile explore the need for varying leadership elements during chaos and complexity-laden situations [10, 24].

During times of complexity writes Roussel [21],

- Analysis loses its primacy
- Long term planning becomes impossible
- Contingency (cause and effect) loses meaning
- Visions become illusions
- Consensus and strong cultures become dangerous
- Statistical relationships become dubious

It is imperative to ask key questions at this level to the effect that, what if complexity theory is applied more explicitly to the programs/learning systems being managed in schools? And if we were to take the world-view of Complexity Theory more seriously – i.e. treat it as if it were true what might be done differently in educational practice?

In this regard, Tosey [26] opines that we might:

- accept that we are not in control
- stop and inquire any time we were tempted to try to 'make' students behave a particular way
- explore how to design learning systems on 'minimal structures' of simple principles
- promote 'connectivity' and creativity among the key agents in the learning system (students and staff)
- have ways of inquiring critically into our own 'theories in use' of teaching and learning, and program management
- seek ways of valuing emergent learning and encouraging education 'at the edge of chaos'

It is also paramount to point out that leaders have always faced a certain degree of uncertainty, risk, and complexity. In recent years, however, the strength of these forces has grown dramatically, challenging even the most seasoned leaders to keep pace. Innovations continue to accelerate, impacting the pace and shape of leadership [14]. Sustainability has accordingly matured from a fringe idea into a major issue. The Millennial generation is entering the workforce and bringing a new set of values and priorities. Meanwhile, the rate of technological, geopolitical, and economic change continues to accelerate (Kelly 2005 cited by [14]).

These major challenges are not limited to the field of management in particular. Virtually all leaders today recognize an urgent need to increase the speed, adaptability, sensing ability, and competitiveness of their organizations. To achieve these goals, executives will have to move beyond traditional approaches to leadership and change that were developed in a simpler era of fewer variables and longer decision making lead times. Roussel [21] opines that truly adaptive leaders know not only how to identify the context they're working in but also how to change their behavior to match. Meanwhile, Sullivan [24] argues that leaders ought to realize that their work is to help others to deal with the changes that affect their lives and work. Focusing on the potential for change within the employees or followers is a cardinal call. Yet Garsia [10] contends that few would dispute that an essential element of leadership is acquiring the power to accomplish goals. Issues with high levels of complexity require a different leadership response to be effective than do issues of low complexity.

It is also widely believed that these days, leaders do not confront managing one or two kinds of change at a time; they must constantly adapt to an array of complex and uncertain conditions [14]. This new competitive environment requires that executives adopt new leadership practices. By using the adaptive approaches described below, we have found that leaders can improve their chances of achieving sustainable growth, increasing customer responsiveness, and re – energizing their workforce while also gaining greater clarity on their own new roles as leaders.

Studies thus have shown how leaders can adopt or adapt to different settings in the environment within their organizations by varying their approaches [1, 14, 23]. Such approaches can be effective when it comes to educational managers and leaders since educational chaos and complexity can erode prospects of effectiveness and efficiency.

Hendrickson [14] thus postulates that there are three ways leaders can be more adaptive, and therefore more successful, in their approach to implementation and change:

1. Adapt your approach to fit the challenge. One - size – fits – all doesn't work anymore. Diagnose the organizational challenge and tailor your approach to change and implementation so that it precisely fits the scope of the organizational challenge.
2. Approach every strategic initiative as an opportunity to enhance adaptive skills. Change and implementation efforts typically require a meaningful investment of organizational time and energy.

3. Shift leadership behaviors today to prepare for greater future complexity. In such a complex world, one leader cannot have all the answers. Develop a wider range of leadership approaches to engage people both inside and outside your organization. As uncertainties get larger and the stakes get higher, tap into the intelligence and energy of all your people, not just a critical few.

Appel [1] identifies two leadership essentials in times of change and uncertainty:

1. Building trust especially technical trust: Trust is crucial in a time of uncertainty and change and when dealing with complexity. When we treat people with understanding and caring, we build an emotional bond.
2. Building confidence and offering hope: It is essential for leaders to inspire confidence that together we have the capacities and abilities to surmount obstacles and successfully take on challenges; that we have the resources and talents; that we can do it and we will succeed. When leaders build confidence, it grows connections in people's brains.

Snowden and Boone [23] claim that:

1. Truly adaptive leaders know not only how to identify the context they're working in but also how to change their behavior to match the context.
2. They also claim that in the complex environment of the current world, leaders often will be called upon to act against their instincts. They will need to know when to share power and when to look to the wisdom of the group and when to take their own counsel.
3. A deep understanding of context, the ability to embrace complexity and paradox, and a willingness to flexibly change leadership style will be required for leaders who want to make things happen in a time of increasing uncertainty.

The coping techniques and mechanisms of leaders in chaos and complexity laden situations have also attracted the attention of researchers [5, 11, 14]. Most of these researchers explore the way leaders in times of chaos and complexity are supposed to behave and conduct themselves. Educational leaders equally follow suit given that educational institutions as key organizations are born of chaos and complexity.

Hendrickson [14] thus proposes the following mechanisms to adapt to complexity in leadership at all times:

- Sustain a shared vision at all levels of an organization
- Build business literacy and clear lines of sight
- Create permeable boundaries and connections
- Nurture constructive individual and organizational behavior
- Shift leadership behaviors
- The shift from control to commitment
- The shift from answers to engagement
- The shift from organizational units to organizational boundaries

The ten coping principles for leaders according to Gilbert [11] are meanwhile reflective of the following undertakings:

- Understanding that the whole is made up of parts: here a leader has to be focused on issues of fit
- All health care is local: a system operates from its point of service outward
- Adding value to a part adds value to the whole: everyone is obliged to make a contribution in one way or another
- Simple systems make up complex systems: a leader has to see his role from a perspective of the whole system despite the level, unit or department he is heading
- Diversity is a necessity of life: homogeneity is an enemy of truth. Leaders have to understand that everyone cannot think and see or say things the same way.
- Error is essential to creation and progress: leaders ought to know that error helps us to establish our position in the journey. The only unacceptable error is that which is repeated
- Systems thrive when all their functions intersect and interact: leaders should welcome interactions in the system since they are for the better
- Equilibrium and disequilibrium are always in constant tension: leaders must know that in system language, stability is another word for death. Absolute stability is the absence of life. A leader should always walk a tightrope between stability and chaos tending to favor the latter.
- Change is generated from the center outward: a leader is obliged to realize that a system can only shrive when those at the point of service own the decisions that are made there
- Evolution results from the aggregation of local changes: leaders are agents and catalysts of change. They bring the vision and context of change to the stakeholders so that the latter can develop the content of change.

Relatedly, Becker [3] identifies ten ways to live and lead forward in times of complexity and change deemed critical to leaders:

- Cultivate the power of the present moment
- Learn how to catalyze mindfulness, whole brain thinking and the power of reflection
- Refocus on what really counts. Drown out extraneous noise. Get rid of toxic thoughts & reconsider relationships with toxic people and team members. Stop focusing on complexity.
- Understand how you can use your mind to optimize your brain
- Stop searching for security in the same old places
- Use the changes you face to lead forward-yes, use them to catalyze potential. Develop coaching skills because they are a critical management, self-management and leadership competency.
- Build a community of purpose meaning that our individual and collective success will be based on our ability to collaborate, communicate with people and organizations whose values and vision are in alignment with our own.
- Use integrity, humanity, courage and incorporate them in your life and work

- Remember, going forward: sustainability will only be achieved through our ability to know that success means creating value for others
- Traditional might does not make right, it often makes wrong.

According to Becker [3] therefore, the might we need now, more than ever before is not the might of brilliance, or the might of physical force; rather it is the might of a collective consciousness, and awareness that we can become victims to what has felled every great life and empire or can be made right. Each day is a powerful opportunity to right the wrongs we face, because the ability to evolve, grow, learn, re-learn and collaborate forward is real.

In keeping with coping techniques and mechanisms to chaos and complexity situations, Chuck [5] also offers a few tips to help in leading through change:

- Watching ones attitude: in times of change and crisis, employees will often look to the leader for signals of good or bad. How a leader carries himself, as well as what and how a leader says a thing is incredibly important. Staying positive and calm will prevent organizational panic.
- Acknowledging the current situation: change and uncertainty needs to be tackled with calm and honesty. It's important to acknowledge the current situation and its impact on the people and organization.
- Establishing clear priorities: ambiguity can be a killer during change. Establishing clear and concise priorities and objectives as quickly as possible, will help force the focus and continued productivity.
- Increasing communication efforts: communication is always important. During change, it's too easy to go dark on your followers. Decisions are being made quickly, things are in a state of flux, and rumors may be rampant.
- Clarifying everyone's job: uncertainty = confusion. Taking the time to reiterate each person's job (e.g., responsibilities, goals, expectations, etc.) is the fastest way back to individual productivity. A leader should be specific, act quickly and take the time to answer questions.
- Stabilizing top players or employees: people get nervous when they are confronted with uncertainty and change in an organization. As a result, you will have people, who were not necessarily looking for a new job, start to put out feelers.
- Staying involved: staying close to the tactical action, as well as following-up on delegated items, will help keep individuals focused on the tasks at hand
- Asking people for more: when people are busy, there is less time to worry. This is an opportunity for a leader to challenge people to work smarter or in a more collaborative style. As a leader challenges his/her team, he/she will find out who will step-up and could be future stars.
- Giving customers extra love: during times of organizational change (especially during mergers and acquisitions), the customer base can become vulnerable. Accordingly a leader can remind him/her-self and the team that the reason the leader is here is because of customers.

- Motivating, motivating, and motivating: at the end of the day, uncertainty and change is hard. Leaders ought to know it and do their best to navigate through it. As a leader, people look to you.

40.5 Conclusion

In these volatile times, leaders cannot think generically about change. In fact, leaders do not face a single generic change, but rather a multitude of global, unexpected, and multivariable changes that demand a new form of leadership. In a world of constant and ubiquitous change, executives must think differently about the immediate challenges they face, diagnose how to deal with them, and decide how to lead their organizations to become more collaborative and co – creative in solving them. Such an endeavor requires personal growth in leaders as well as new capabilities in organizations.

In this article, we have described the concepts of chaos and complexity theory, explored the nature of complexity in education systems and identified the various ways in which leaders at all levels are supposed to deal with complexity if they are to turn the fortunes of their institutions into potentials for success than failure.

Future trends in the leadership arena are likely to direct our thinking and actions more seriously. The challenges leaders are likely to encounter or face are truly daunting and insurmountable. This calls for a certain level of understanding and readiness on the part of leaders if they are to survive the onslaught of unpredictability. Gilbert [11] weighs in on these futuristic tendencies in leadership by putting across the following thinking points:

- Change is endless
- Information is remarkably more available than in the past
- Knowledge is now a utility
- Technology is changing the character and content of the service relationship
- Wireless technologies now operate in a way that defies boundaries

Definition of Key Terms

Leadership: The process of influencing followers through the use of power. Using different bases of power results in different reactions from followers depending on dictates within organizations.

Chaos: Wikipedia defines the term chaos as any state of confusion or disorder while at dictionary.com it is defined as a state of utter confusion or disorder; a total lack of organization or order.

Complexity: The state of being complex or difficult with a variety of challenges and obstacles to effectiveness and efficiency. Complex situations are born of lack of a clear approach to dealing with an existing situation.

References

1. Appel W (2012) Two leadership essentials in times of change and uncertainty
2. Battram A (1999) Navigating complexity. The Industrial Society, London
3. Becker (nd) Ten ways to live and lead forward in times of complexity and change
4. Cahill K (2010) Even in chaos: education in times of emergency. Fordham University Press, New York
5. Chuck H (2011) Leading through change: leadership tactics in times of uncertainty, Management resources, internet
6. Cilliers P (1998) Complexity and postmodernism. Routledge, London
7. Cunningham R (2000) Chaos, complexity and the study of education communities. Institute of Education, Leeds
8. Doll W (1993) A postmodern perspective on curriculum. Teachers College Press, New York
9. Ertürk A (2012) Kaos kuramı: Yönetim ve eğitimdeki yansımaları. Kastamonu Educ Fac Mag 20(3):849–868
10. Garsia A (2012) Leadership during times of complexity and change. Division of Business Homecoming Presentation, Saturday, October 20, 2012, 11 a.m. Noon, Casey Building, Trading Room First Floor, Rochdale
11. Gilbert B (2002) Thriving in complexity: ten principles for leaders in the coming age. Jones and Bartlett Publishers, Boston
12. Gleick J (1987) Chaos: making a new science. Penguin, New York
13. Glickman, C. D., Gordon, S. P., & Ross-Gordon, J. M. (1985). Supervision and instructional leadership (5th Ed.). Needham Heights, MA: Allyn & Bacon.
14. Hendrickson C (2010) Leadership in complex times. Wiley, San Francisco
15. Hopper T, Ovens A, Burtler J (2013) Complexity thinking in public education: reforming curriculum pedagogy and research. Routledge Publishers, Oxford
16. Kara A (2008) Understanding today's schools with chaos theory. Fountain Magazine, Issue 62
17. Lortie DC (1975) School teacher. University of Chicago Press, Chicago
18. Morrison KRB (2002) School leadership and complexity theory. Routledge Falmer, London
19. Morrison K (2006) Complexity theory and education. APERA conference, Hong Kong, 28–30 November 2006
20. Reigeluth CM (2004) Chaos theory and the sciences of complexity. Foundations for Transforming Education. Indiana University, Indiana
21. Roussel L (2009) Leading in times of complexity and rapid cycle change. Jones and Bartlett Publishers, Burlington
22. Shyamala P (nd) Leadership during times of complexity and changes: human resource area. IOSR J Bus Manag (IOSR-JBM):52–57. e-ISSN:2278-487X, p-ISSN:2319–7668
23. Snowden DJ, Boone ME (2007) Leadership in times of complexity and chaos. The leader's framework for decision making. Harv Bus Rev, November
24. Sullivan TJ (1999) Leading people in a chaotic world. J Educ Adm 37(5):408–423
25. Tangülü Z, Ulusoy M (2015) Pivoting on chaos and disorder. Reflection on events steering the world of politics in social sciences course books. Turkey and USA example. Springer, Germany
26. Tosey P (2000) Complexity theory: a perspective on education. University of Surrey, Guildford
27. Wheatley MJ (1999) Leadership and the new science: discovering order in a chaotic world. Berrett-Koehler Publishers, San Francisco
28. Youngblood M (1997) Life at the edge of chaos. Perceval Publishing, Dallas

Chapter 41
Rethinking the Management of Secondary Schools in a Chaotic External Environment

Sa'adu Isa Bashar and Yusuf Buhari

Abstract Secondary schools as part of the totality of the open social system organizations depend largely on society for their inputs like students, teachers, finance and information as well as material resources to enable them produce outputs for the benefit of the wider society. Unfortunately, sometimes, the external environment becomes unstable, turbulent or chaotic as a result of economic, political, legal, demographic, technological, religious and or cultural factors which if not properly managed, may undermine the schools' internal operations leading to poor output. It is to the interest of this paper therefore, to justify the school as an open social system, the school external environment and the ways in which factors of the external environment compel secondary schools to complex or chaotic management imbalance. At the same time, the paper recommends that the school managers should seek to adopt both internal and external coping strategies so as to remedy the situation for the production of highly meaningful outputs.

41.1 Introduction

The primary goal of educational management is to see that the available human and material resources of educational institutions are properly mobilized, harmonized and utilized towards the production of highly meaningful and qualitative individuals for the benefit of a wider society [3]. Schools as open-social systems require effective educational managers who are capable of steering them to the attainment of that goal. The only sort of people who can undertake this job are the situation-sensitive ones who are able to relate well with the external environments of the schools and who are capable of adapting, adjusting to and protecting the institutions from being harassed by the changing situations of the environments from where

S.I. Bashar (✉)
Department of Educational Foundations, Sokoto State University, Sokoto, Nigeria
e-mail: mmbashar1984@gmail.com

Y. Buhari
Department of Business Education, Shehu Shagari College of Education, Sokoto, Nigeria
e-mail: yusufbuharig@gmail.com

Ş.Ş. Erçetin (ed.), *Chaos, Complexity and Leadership 2014*,
Springer Proceedings in Complexity, DOI 10.1007/978-3-319-18693-1_41

the resources are derived and for whom the outputs are produced. This paper thus, examines school as an open social system, the reasons why schools must relate with the external environment and how the dynamic, complex and often chaotic factors of the external environment influences the management of secondary schools and how they can be managed using relevant coping strategies for the production of highly meaningful products for the societal benefits..

41.2 Secondary School as a Social System

Secondary school is a social system that represents one part of the human habitat of students, teachers, special service personnel and administrators. It is created and controlled for the purpose of enabling young people to become more effective in the life situations in which they must participate [8]. It is an organized whole that comprises of interdependent personalities bound together in an organic relationship serving different functions to achieve the desired goals. This is indicated by Hoy and Miskel [7], where they contended that a social system is a bounded set of elements (elements) and activities that interact and constitute a single social entity.

As a social system, secondary schools are often characterized by the following features as indicated by Manga [10]:

(a) They are goal oriented and may sometimes have a multiplicity of goals. For instance, apart from preparing students for adult roles, there may be other goals at departmental and individual levels.
(b) They are comprised of interdependent parts, characteristics, and activities that may to and receive from the whole such that if one part is affected, all the system will be affected.
(c) They are made of people. Therefore people act in the varying number of roles like administrators, teachers, students, security, cooks, messengers, cleaners, attendants, etc.
(d) They are structural. They are made up of different structure of components like an organizational chart that indicates the specific action, role or function that is needed for everyone to perform for the attainment of goals. They also have physical structures which include Administrative units or offices, classrooms, laboratories, libraries, workshops, hostels, etc. where each unit is distinct for its specific function.
(e) They are normative in that every person is expected to behave in a particular manner. Similarly, formal rules and regulations are there to guide the prescribed and appropriate behavior.
(f) They are sanction-bearing. The norms for behavior are often enforced with reward and punishment. Here, formal mechanisms may include expulsion, suspension, termination, and promotion system.

Similarly, Hoy [6] added that schools as social systems cannot stay free of the following issues:

(a) Politics. This is otherwise the internal politics of the school. In secondary schools for instance, there exist a room for the formation of informal organizations like students unions, teachers' unions, non-teachers' unions, etc.
(b) Technical Core: these are the teaching-learning processes that are being executed in the school for the preparation of the young leaders of tomorrow.
(c) Environment. This is everything outside the organization or otherwise the source of inputs for the school.
(d) Output. These are simply the products of the school like the educated students, extension services, etc.
(e) Feedback. This is the information or communication that comes from the environment which monitors behavior of the school.

41.3 Secondary School as an Open System

An open system organization is that which maintains a favorable relationship with the environment thereby adapting to meet with the changing situations of that environment. An environment on the other hand, is anything outside the organization that is capable of affecting the internal operation of the organization. Environment provides resources, values, technology, demands and history all of which place constraints and opportunities on organizational actions. Similarly, environment can be external or internal. An external environment otherwise called general environment involves the general trends that may affect the internal operations of the school such as the economic, political, legal, demographic and technological changes, etc. However, specific or internal environment like parents of students, tax payers, school unions, regulatory agencies etc. may have effect on schools.

Secondary schools are open social systems in that they relate and exchange matters with their environments [10]. This relationship is relatively compulsory on the ground of both Information and resource dependency theories. From the resource dependency perspective, schools must relate with their environments so as to acquire available working resources for the smooth conduct of their operations. Those resources involve Fiscal (financial), personnel (students, teachers, administrators as well as board members), products and services (instructional materials, school facilities, electronic gadgets, etc.). Therefore, this perspective hold a strong belief that schools are unable to internally generate all the resources to maintain themselves and that resources must come from the environment. Proponents of Information model contend that schools as open social systems need to relate with their environments so as to source information and knowledge on the way they carry out their services or operations. This information is otherwise referred to as feedback. This enables them determine their strength and weaker points for making further corrections or improvements. Figure 41.1, below, demonstrates the above assertion:

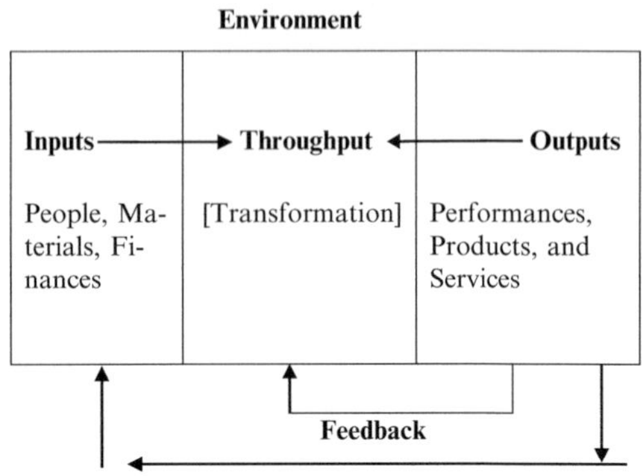

Fig. 41.1 Showing interdependent relationship between school and environment (Source: Adopted from W.K Hoy [6] and modified by the Authors)

It is clearly shown in the above table how environment affects the entire open system of school organization as a result of their interdependent relationship. Hooper and Potter [5] emphasize that since the external environment of the school is in often cases ambiguous, turbulent or unstable, there exist a need for educational institutions to be open and flexible so as to be able to adapt to the changing external conditions for them to be effective and in long term survive from entropy.

41.4 Chaotic Factors of the External Environment and Their Impact on School Management

Economic factor or trend shapes the business of education across the globe. The success of every educational institution largely depends on the magnanimity of its revenue. The little or huge the amount of financial resources of the school the few or plentiful human and material resources it acquires. The economic flexibility of an environment affects in no small measure the demands and supplies of the schools which then have to be considered by a school head so as to bring about effectiveness and efficiency in the school. For instance, the issue of inflation affects staff, students, and even the material resources of the school. As for the staff, inflation may hinder their salary from enabling them provide for themselves and their families the means of livelihood in terms of food, shelter and daily needs. This causes them to disconcentrate fully in the performance of their duties in the school thereby dodging away from the school to carry out some extra services that may fetch them some money for the conduct of their daily home operations. Students could also be affected in the way that children from the poor economic status may find it difficult to pay for registration or tuition fees as well as the acquisition of the necessary learning materials vis-à-vis payment of some money that may be required by the school for other operational issues. Material resources like the school facilities may be too difficult for the school to provide when they become highly cost such that if 20 items could be purchased with the minimum fund would then be limited to a very less number.

Politics as a trend that emanates from the environment affects the school management issues. This comes especially when there is an inconsistent government vis-à-vis the political elites support. When there is government inconsistency, it affects the consistent nature of the school management in such a way that there may be varying number of policies that may emanate directly proportional to the educational institutions from the political group. Political elites of the environment thus affect the way the schools are managed in that they help in solidifying or weakening the internal operational policies of the schools. This is because the successful implementation of every given policy especially that of education depends greatly on the mobilization of the political support of a given environment within which the policies would be implemented. Hence, if they support a school policy, it stands implementable and the reverse becomes the case when they oppose the policy. Policies of such kind therefore require the support of such kind of people for them to operate well.

On a demographical ground, schools could be affected by the population status of a given environment or society. The current population growth of African countries especially Nigeria, affects the internal operations of their educational institutions. For instance, over population leads to over enrolment of students that could result into over crowdedness of the schools and classrooms rambunctiousness which if not properly taken care of might negate effective teaching and learning. This may also pull the school managements or their supra systems to employ more classroom teachers to suit the requirement of the Student Teacher Ratio (STR) as it also necessitates building of more classrooms to accommodate all the students for a conducive learning atmosphere. This also calls for the employment of more supporting staff that would help control the atmosphere of the school from the state of disorderliness.

Legal factor which is also uncertain in an environment, changes by the passage of time. Often there may be legal rulings, laws, acts and policies on education that may be initiated and formed worldwide or by parliament, legislature, constitution or a court. These laws affect the internal operations of educational institutions in such a way that they must be fully abided. For instance, the issue of abolishing corporal punishment which was a world education law has to be wholly accepted and implemented in the schools. This and other laws make impacts on the way schools perform their disciplinary actions on their students on the basis of the gravity of their offences.

In this era of globalization, innovations of the scientific world brought about an issue of technology. Contemporarily, technology is one of the major elements of external environment that affects the operations of our educational institutions. The advent of computers, projectors and internet for instance, call for the educational institutions to adapt, provide and make use of them in the conduct of teaching and learning activities. This is necessary so as to enable the schools move with the world. This is also for the fact that they provide enormous convenience for both teachers and students in the issues of teaching and learning by making them easier and faster through facilitating information transmission, knowledge acquisition as well as dissemination and creation of a value chain [2]. These technological tools have to be used with caution since they also contribute negatively in corrupting the minds of students and even staff by exposing them to illicit and immoral issues and behaviors as they are also capable of changing people's moral perspectives and ethical values. Students come to imitate what they see whether good or bad through the net and violent behaviors are often learned through observation [9].

Religion or cultural orientations of the environment vary from one perspective to another, society to another and one religion to another. Many cultural or religious issues affect the ways schools operate. This is because the environment may be in conflict with the school practices especially when the latter operates not in consonance with the norms, values and appreciations of the former. For instance, in some environments, putting-on a veil for girl children is a cultural or religious prescription which has to be respected and adapted by the schools or else female children would remain at home. Some environments see religious acts as obligatory prescriptions which have to be adhered to by the followers of those religions. To be able to deal with this chaotic nature of the environment and to get more students, a school must be able to see that students are given adequate freedom to perform their religious rites in terms of timely prayers, fasting and religious vacations for celebrating religious festivals such as Ed and Xmas for Muslims and Christians, otherwise, the environment would be sending their children to schools where these issues are respected and duly observed.

41.5 Recommendations

To be able to deal with the chaotic, turbulent and uncertainties of the external environments of the schools, the following coping strategies identified by Bush [4] are hereby recommended by the paper for use:

Internal Coping Strategies:

- A principal should be able to protect the technical core of the school. This is by isolating or buffering the main issues of concerns of the school like the instructional activities, teaching staff and students from being negatively affected by the external forces of the environment. Economically, for instance, if there is inflation at the outside environment that may affect the financial capacities of the staff, a principal is required to employ some means of alleviating or reducing that stress of inflation from the staff. In this case, one may increase their salaries, provide them with loans, give them some financial rewards or incentives that may boost up their morale and which may also enable them to continue with their works as required. This will help prevent the staff from leaving their necessary duties to go for other businesses to earn a living. However, departments may be created for taking care of certain necessary issues of the school like personnel unit that will be in charge of recruiting staff for the school, purchasing and planning units that may be responsible for transferring services, materials, money or other necessary resources in the school from the external environments. This may reduce the chaotic pressure of the environment on the school.

- A principal should be able to adjust the internal operations of the school. This is simply a situational or a contingent adjustment of the internal activities of the school as a result of environmental dynamism. In dealing with the threat of ICT for instance, an educational manager (principal, per say) should be able to adjust the internal operations of the school so that the technology made available in the school be used in a most favorable manner that will improve teaching and learning processes and distance both staff and students from getting access to the issues of immorality identified by Adesina [1] and Bashar [3] which include among others pornography, cultism, child abuse, money laundering and international terrorism. This is clearly indicated by Bala and Umar [2] as they rightly said "Managers of education at all levels—should control how ICT (Internet and Cable) is being used by students—".
- A principal should be able to span the boundaries of the school. This is the ability of the principal to set some personnel from among the staff to serve as proxy between the school and the environment. A public relation staff, PTA chairman, Board of Governors, may be appointed for that purpose so that he/she would be feeding the school with the environmental feedback about the school and congruently protecting the school from the environmental attacks. This will also provide the school with the information about the current possible changes of the environment for adjusting and adapting to the situational trends.

Inter-organizational coping strategies:

- A principal should establish possible linkages with other sister organizations or schools. Here, one may create a favorable linkage with a good sister primary school, college or university for the supply of human resources like students, teachers, and or other supporting staff after their graduation. This minimizes the risks of employing corrupt personnel from the environment who may not be able to perform up to the demands of the tasks of the school. Similarly, a school can liaise with other manufacturing industries for the constant supply of the facilities and other material resources for the school. This may also reduce the risk of uncertainty of the dynamic nature of the environment.
- A principal should employ cooptation method. Here, one should be able to bring-in some influential members of the society to head some of the school committees. This applies to issues like advisory committee, Board of Governors or School Based Management Committee. This will also enable the people to be able to defend, protect or clear the school from some of the misunderstandings that may arise from the environment.
- A principal should employ lobbying method. Here, one should be able to lobby the political class of the society to support the policies and programs of the school. By this, it will be possible to communicate the view of the school to those members of legislation or government officials. This helps in reducing high rate of blames to be mounted on the school by the environment for some of its policies that may contradict the environmental expectations.

- A principal should employ the habit of forming Educational Associations. Here, one should be able to form some educational associations in the school which will have both professional and political missions. For instance, PTA may help the school in many different ways both academically and politically. This will also make parents feel involved in the academic issues of their children.

41.6 Conclusion

The paper so far viewed a school as an open social system which must relate with its environment for getting access to human, material, financial and information resources for the conduct of its internal operations. As a result of the environmental uncertainties, certain factors like the economic, political, legal, demographic, technological, religious and or cultural remain a great threat on the road of the schools on the ground of their flexibility. For a school function effectively, a principal has to ensure the adaptation and adjustment of the new flexibilities and the school operations. This can be achieved by following strictly the above recommendations.

References

1. Adesina OS (2012) The negative impact of globalization on Nigeria. Int J Humanit Soc Sci 2(15):193
2. Bala B, Umar M (2013) Managing education for economic security in era of globalization. A paper presented at the 32nd annual national conference of the Nigerian Association of Educational Administration and Planning (NAEAP) held at Bekandims Hotel and Resort, Tanke, Ilorin, Kwara State, Nigeria
3. Bashar SI (2014) Anomalies in the Nigerian primary schools' leadership and centenary commemoration. A paper presented at the 1st international conference and centenary commemoration (ICCC 2014) held at main lecture theatre organized by the faculty of arts and social sciences, Sokoto State University, Sokoto, Nigeria
4. Bush K (2009) Fundamental practices of educational leaderships, 4th edn. King and Quenn Publishers, Jalingo
5. Hooper A, Potter J (2001) Intelligent leadership: creating a passion for change, Revisedth edn. Random House Business Books, London
6. Hoy WK (2011) The school as a social system. Web slides, Retrieved on 13th November, 2014, from HM_PP_Slides_Ch1[compatibility mode]_Microsoft PowerPoint
7. Hoy WK, Miskel CG (2008) Education administration: theory, research and practice, McGraw Hill international edn. McGraw Hill Companies, New York
8. Jensen GE (1954) The school as a social system. Educ Res Bull 2(33):1, Taylor & Francis Ltd. Retrieved on 13th November, 2014, from www.jstor.org/discover/10.2307/1473613?uid=3738400
9. Lawal S, Ishaq IJ (2011) Domestic violence and its impact on the education of Nigerian children. Sokoto Educ Rev J 12:77
10. Manga SD (2010) Theories of administration. Usmanu Danfodiyo University Sokoto, Nigeria

Chapter 42
The Teaching of Chaotic Events in Social Studies Textbooks of Turkey and Germany: World War I

Beytullah Kaya and Oğuzhan Karadeniz

Abstract With the point where knowledge and technological power has reached, the prediction of the control and monitoring power of human events has been eliminated since 1900s. In particular, the war, immigration, and revolution and so on… many social events and phenomena have caused consequences that people can't cope with, and they have led to chaotic situations occurring in societies. Surely, the environment of this chaos did not affect every community in the same degree based on the conditions of the time. In this study, World War I, as an example to the handling of chaotic events in the textbooks taught in both Turkey and Germany, has been dealt with and has been subjected to a comparative analysis. The aim of this study is to explain how World War I which caused a chaotic environment on a large-scale holds a place in Social Studies textbooks in both Turkey and Germany. The reason as to why the researcher conduct the survey in Turkey and Germany is that they are the countries which experienced this war and the environment of chaos which this war caused most. In this study, the method of document review which is one of the qualitative research methods has been used. In this study, Social Studies textbooks in which the subject of "World War I" in Turkey and Germany is handled have been examined by comparing both countries' textbooks.

42.1 Introduction

With the increase in topics taught in the learning programs belonging to disciplines such as social studies and history, the problem of how the teaching is done has become more apparent. The question of how increasing historical information with increasing diplomatic, political, social and economic relations among the states will take place in these curriculums and how this knowledge will be taught has become the most fundamental problem of History and Social Studies curriculums.

B. Kaya (✉) • O. Karadeniz
Ereğli Faculty of Education, Department of Elementary Education, Social Studies Education Department, Bülent Ecevit University, Peak resort, 67300 Eregli, Zonguldak, Turkey
e-mail: beytul76@mynet.com; oguzhankaradeniz25@hotmail.com

Ş.Ş. Erçetin (ed.), *Chaos, Complexity and Leadership 2014*,
Springer Proceedings in Complexity, DOI 10.1007/978-3-319-18693-1_42

Since it will not be possible to consider everything in disciplines such as History and Social Studies education programs as well as teaching all of them, it is necessary to refer to the subjects which are actually taught. Students' learning situations and ages must be taken into consideration while these preferences are being made. In this context, students' age and state of readiness are among the issues to be considered in the preparation of programs. History and Social Studies curriculums have undergone many changes until 2000s as in many western countries. History and Social studies curriculums which are seen as a tool that states use to educate citizens in order to adopt their ideals, have been paid attention to in preparation of the way that reflects the state's official ideology.

The 2000s have been a period when the issue of how we teach besides what we teach, was discussed more intensively in the curriculum. In this context, as a result of the work done for curriculums in Turkey, the constructivist learning approach was gradually adopted in 2004. Constructivist learning philosophy has a flexible structure for students to configure their own information ([5]: 18, Fosnat [6]: 10, [8]: 6). However, curriculum programs which were created in Turkey have been prepared in line with traditional programs [9]. Learning situations and learning needs of the students differ from each other. In particular, these differences in subjects such as Social Studies and History have shown themselves more. In this context, the fact that curriculums are dynamic and not static can make the learning situation easier. Political, social or economic events occurring in the scope of History and Social Studies have a dynamic structure. That is, they differ in terms of their viewpoint. So, the use of dynamic methods is more suitable to the nature of training in the teaching of a dynamic subject. In this context, the use of chaotic learning programs is more convenient to the nature of learning.

The rationale for the adoption of social information in Germany resembles that of the USA because it can be said that similar social transformation process experienced in both countries in the structure of the federal government leads to the emergence of Social Studies. Social Studies emerged as a result of efforts to research solution to problems which social, cultural and economic transformation that the USA experienced caused while it was entering into the twentieth century [12]. It is observed that many ethnic and religious groups live together in Germany, and phenomena such as rapid industrialization and urbanization led up to Social Studies education in Germany in the first half of twentieth century.

When Social Studies curriculum in the United States is also examined in recent years, it is observed that the increment in people who originated from different religions especially in Muslim majority causes a discussion of new dimensions in Social Studies education and there was to be a change in the purpose and content with external and internal migration. In this context, it can be said that Social Studies curriculum takes place amongst education programs which are most affected by political, social and economic transformation ([12]: 50).

In this context, it is observed that the education programs most affected by political, social and economic transformation, call Social Studies course with a focus on "Politics" as the examples in many states such as Politics, Politics and Economics, Social Studies Politics ([12]: 54). Human activities originating from the cultural, environmental and economic activities, have been discussed in the Geography lesson. In the History lesson, the events (wars, migrations, reforms ...) which the nations and the states reach today along with the reasons for the same have been studied together today in the scope of the social-political, ideological and cultural dimensions. The values based on existing and changing social, political and economic structures, norms and rationale have been discussed in the lessons of Politics, such as in Social Studies/Politics course ([11]: 34). In the context of these topics, being discussed, many chaotic events or situations are also inevitable in social studies textbooks in Germany because, many social events include chaotic situations within them, and thus handling them in textbooks has a great importance.

It is seen that a lot of events have chaotic properties when being analyzed. That is, historical events pose a different dynamic structure on their own. When it is thought that the educational environment has a dynamic structure and it hosts chaotic principles in itself, it can be seen that History and Social Studies curriculum is suitable for the chaotic learning approach [3]. Chaotic programs are intended that students must be brought into the focus of education. It is observed that a lot of chaos experienced in the past take places particularly in the disciplines such as History and Social Studies. Moreover, these chaotic events are also taught within the scope of those courses. Historical events sometimes occur between two states, and they sometimes occur between more states or societies than one. That both events and learning situations have become more complex case affects the learning environments negatively.

The purpose of this study is to examine how World War I. is handled in Social Studies and History textbooks in Turkey and Germany. It is important to highlight how a chaotic event like World War I which is regarded by the whole world in general and the countries that participated in the war in particular can be analyzed in terms of educational programs. It has also been relevant to explore the objectives and acquisitions taking place in the subject of World War I in Social Studies and History textbooks in Turkey and how this same subject is handled for the students who are within similar age groups in Germany.

42.2 Method

This study has been conducted in order to describe how a chaotic event is handled in Social Studies and History textbooks in Turkey and Germany. For this reason, the subject of World War I which is an important turning point in the history of countries and partners as well as stakeholders has been selected for investigation. In contrast to being gotten from the common and familiar values with selected sampling method, event is a sampling method which is used in qualitative researches for uncovering phenomena and situations [19]. In this study, the textbooks utilized in education in both countries have been investigated by the method of document analysis. The findings obtained from the documents have been treated via content analysis. On the other hand, descriptive analysis has been utilized so that the findings which have been obtained from content analysis can be presented in a meaningful way. Document analysis method is a process of analysis of these resources by examining written sources related with the phenomena and events which have been investigated [13, 18]. Document analysis alone has been used as a research method in many studies done [2, 10, 18].

This study is limited to 7th grade Social Studies taught in Turkey, 8th grade Turkish Republic History and Kemalism and 11th grade History textbooks in Turkey and 9th grade History textbooks in Germany. In addition, it is limited to document analysis method which is used for collecting data in the study and content analysis methods.

42.3 Findings

It is necessary to establish proper acquisition which is appropriate to the level of the students so that events which indicate chaotic property can be learned by primary and secondary level students through interpretation. School curriculums have an expanding and spiral structure in Turkey [15]. As the level of the grade goes higher, handling of subjects in the curriculum expands. In this context, some subjects of the seventh grade are extensively processed again in the eighth grade and subsequent classes.

The subject of "World War I" is given a place as the last unit called "Global Links" in the Seventh grade curriculum. A total of four acquisitions are given a place in this unit and three acquisitions of them are related to "World War" I which is a chaotic event. The acquisitions are as follows:

1. Students associate the political and economic structure of the Ottoman Empire and European countries at the beginning of the twentieth century with the causes and consequences of World War I.
2. Students associate global issues with the purposes of the establishment of international organizations.
3. Students realize personal responsibility in the implementation of the solutions to global problems and challenges

A 6-h section in the unit of "Bridges Between Countries" in the seventh grade Social Studies Curriculum has been devoted to the World War I especially, the situation in Europe and the Ottoman Empire in the nineteenth and twentieth century has been examined. In this section, the attitudes and policies of European countries against the Ottoman Empire have been examined. Also, the causes of World War I have been examined focused on the grouping of countries [1]. The outbreak of World War I and entry into the war of the Ottoman Empire has been examined in terms of the Ottoman Empire. The analysis has been focused on the perception of events and how they evolved with a specific plan. World War I is a chaotic event. There is an equilibrium process which includes evolution of chaotic events on their own. While the seventh grade subjects in the curriculum are being handled, an evaluation hasn't been done in terms of their chaotic possibilities. World War I is being handled while the pre-war situation of Germany and some other countries is being investigated in 7th grade Social Studies book. The causes of World War I are analysed, Alliance countries are being examined, the outbreak and the section of the spread of World War I and entering into the war of the Ottoman Empire and the treaties signed after World War I have been accorded a place. That is, Germany is subjected to five different places while the subject of World War I is being handled expansively in the textbooks. Germany has been processed with only a simple description of the subject. Knowledge has been given simply and in an appropriate way to the age level of the students rather than being academic and chaotic. In the seventh grade books, World War I has been handled as though it appeared in a certain periodic process and developed in a particular order. Special attention has been paid to the approach about the level of acquisition of the students in the seventh grade books.

World War I has been handled in the unit "National Awakening: Reactions to the Occupation of Our Country". There is a total of 8 acquisitions in the unit "National Awakening: Reactions to the Occupation of Our Country" and two acquisitions of them are related to World War I [14].

1. Students evaluate the state of the Ottoman Empire in World War I in terms of the occupation and sharing of the territory as well as occupation of the territory in terms of shared value.
2. Students evaluate Ottoman administration, Mustafa Kemal and public's attitudes towards the signing of the Armistice Treaty and implementation of the same.

The seventh grade approach has been exactly maintained in 8th grade 8 Turkish Republic Revolution History and Kemalism curriculum. Subjects have been examined in the same manner but expanded a little more. While the subjects are being handled, they are handled in terms of the Ottoman administration perspective. Eighth grade acquisitions are prepared in accordance with national characteristics although seventh grade acquisitions include more universal and broad approaches. While the subjects are being handled, Germany has been mentioned in the subjects that are taking place in the seventh grade and seventh grade has been almost repeated [18].

In the classes of Turkish Republic Revolution History and Kemalism curriculum in high school, three of 7 acquisition in the unit of "Preparatory Period of the National Struggle" are related to World War I. These acquisitions are the following;

1. The students evaluate political, social, economic, and cultural status of Ottoman Empire in the early twentieth century
2. The students evaluate the state of the Ottoman Empire in World War I
3. The students describe the signing of the Armistice Treaty and implementation of it.

8th grade and 11th grade Turkish Republic Revolution History and Kemalism lesson has been written extremely similar to each other in terms of their objectives, acquisitions and content. Germany has been given a place in eight different sections in 8th grade program. The sections in which Germany is given a place are these ones: out-break of the war, the formation of groups, participating in the war of Ottoman Empire. Subjects have been evaluated only in terms of the Ottoman Empire in this section, too. Subjects have been prepared away from the chaotic approach and events have been processed as if they are parts of a whole. Likewise, 11th grade Turkish Republic Revolution History and Kemalism curriculum also have the same functionality. Being taken into consideration is students' ages in making learning easier in processing of chaotic subjects. Students have made sense of the subjects according to their age levels and their stages of historical learning [4]. Although 8th grade and 11th grade students are at the same age and at the stage of learning, handling of subjects is almost the same in both classes.

42.3.1 A Chaotic Event in the Textbooks: World War I

World War I includes the following subjects in the textbooks of Social Studies and Turkish Republic Revolution History and Kemalism: The state of the Ottoman Empire before World War I, the state of European countries before World War I, pre-war groupings, the start of the war, the battlefronts which the Ottoman Empire fought, Mustafa Kemal's achievements during the war and the results of the war. That is, textbooks have been handled chronologically and books have revealed a process scheme according to occurrence order. World War I is processed as 12 pages in the 7th grade textbooks and processed as 8 pages in the 8th grade textbooks. The subject contents of the textbooks have been almost created in the same way with each other.

> Germany: I am Germany. Once I was a poor union of states, I completed my political union and got strong. Europe began industrialization and I gave importance to industrialization immediately. I got strong in a short time. France whom we have already Alsas-Loren problem with, does not want to give me a piece from the pie (Arslan [1]: 189).
> Germany: Wilhelm II.: We established a strong union. We must spread over the lands that England and France could not reach. We can be effective by using the Balkan States and Ottoman Empire in the Middle East. While providing strength economically by investing in the Ottoman Empire, we establish friendship by sending military experts. The Baghdad railway project that makes us closer to Ottoman Empire, disturbs England (Saray [17]: 31).

Quotations related to status of the countries before World War I are given a place in the 7th and 8th grade textbooks above. The pre-war situations of the countries have been simply identified and have been discussed in terms of European countries' relations with the Ottoman Empire here. Events are subject to an assessment only in terms of political or economic relations. Moreover, in this section, the state of the Ottoman Empire has been described in detail; equally the reasons why the Ottoman Empire joined the war alongside Germany are explained. The similarities in the subjects such as the starting and reasons of the war are among the common characteristics of the books in Turkey. The starting of the war and pre-war groupings have been simply arranged and an assessment of the causes of these groupings has

Table 42.1 Pre-and post -war blocks in Turkish course books

The groups before World War I	
Triple entente (agreement) states (1907)	Triple alliance (congruence) states (1882)
England	Germany
France	Austria – Hungary
Russia	Italy
The groups after World War I	
Italy	Ottoman Empire
Greece	Bulgaria
Romania	
Portugal	
Brasil	
USA	
Japan	

Source: Arslan [1]: 191

been made. Below, the groups in the war are presented as given in the textbooks in Turkey (Table 42.1).

When the tables above are examined, it is observed that each textbook has been almost been presented similarly. It hasn't been taken into consideration that students are at different age stages and different historical stages of learning. There have been similar properties in the following chapters of the subjects in the textbooks, too. The battlefronts which the Ottoman Empire fought have been transfected with simple military information. However, only Çanakkale wars have been evaluated by being given the cause, development and results of the war. The fact that events are given within textbooks in terms of the causes and effects has been adopted as a general approach. Chaotic contents which reflect thoughts and feelings of that period have been superficially given in the textbooks.

> A Memory From Çanakkale War
>
> There was no water in which Anzac soldiers fighted. On the other hand Turkish front was rich in water. When the water-bearer (water carriers) which moved to front with rules after filled water into their boats behind Turkish lines, run off the rails, they found themselves on the face of two Anzac soldiers. These two soldiers were parching thats why they started to drink by running and dropping their rifles when they saw water-filled boats. Turkish soldier who used this opportunity, took rifles of enemy soldiers and took away them to his commander. When this soldier was come into the presence of Esat Pasha who was the commander of Western front, he said "Sir, I gave water these men and they guzzled it." Esad Pasha who was very pleased and affected by the behaviour of Turkish soldiers who weren't consent of even enemy soldiers die the lack of water, slapped on the back of soldier and gave soldier so much tip (Arslan [1]: 199).

The text such as "A Memory of the Battle of Çanakkale" in the 7th grade, "Sarıkamış Song" in the subject in Turkish Republic Revolution History and Kemalism course and "There is a letter from Gallipoli" have been demonstrated as an approach providing information about social events. Even these two examples are important for the students in that they reflect people's feelings in the face of their emotions, thoughts and chaotic events during that period. This kind of examples have made easier for students to empathize with people living at that time. The purpose of this is to provide for students' understanding the conditions of that period and sentiment which may arise in the face of chaotic events. An anecdote or text which belongs to another country has not been included in the textbooks. However, once dealt with, such an anecdote will help students to understand what another country soldiers or citizens feel. Understood how the emotions of the people who take place on the same side or the opposite side are will make it easier for the students to evaluate the events accurately in the face of chaotic events.

A negative perception about Germany in World War I hasn't been included in the textbooks in Turkey. The knowledge relating to Germany in textbooks is simple academic information. Germany's situation before the war has been investigated and the problems which Germany had with countries such as France and England have been included in the textbooks. One anecdote about pre-war Germany written by Arslan in the 7th grade Social Studies textbook is given a special place. "I, Germany, was a weak country/community before. But I have finished political unity and I have strengthened in industrialization which started in Europe before and I have given an importance to industrialization. I strengthened in a short time. France, with which I have the problem of Alsace-Lorraine, didn't want to give me a share of the cake ([1]: 189)". Similar expressions have been also included in the 8th grade textbook. These expressions have been stated in 8th grade Turkish Republic Revolution History and Kemalism textbook. "… … Germany adopted a principle of establishing good relations with the Ottoman Empire. Germany started with military help which the Ottoman Empire needed at that time. But, England was not pleased because of the problems between the two countries ([17]: 25)." In fact, expressions here have reached the findings that there is a chaotic situation between countries. Moreover, brief information such as the entering of the Ottoman Empire alongside Germany, refuging of the two warships which belonged to the Ottoman Empire, have been given a place in the following sections of the textbooks. Furthermore, there hasn't been an expression about that German officer who served in the Ottoman army although there have been such information which Germany wants that these battlefront was opened in Caucasian and the Suez Canal battlefronts, which the Ottoman Empire opened, in the textbooks in Turkey. While the Suez Canal battlefront is being explained, this expression has been used: "Germany opened this battlefront so that it could seize the Suez Canal and disconnect with the British colonial" ([1]: 194). Apart from these sections, Germany has been referred to while the treaties which were signed at the end of the war are narrated. When the results of the war are assessed in terms of the Ottoman Empire, it is observed that it has been focused on the perception that Ottoman Empire was successful but allies were unsuccessful ([17]: 31).

While the textbooks of Social Studies and Turkish Republic Revolution History and Kemalism discuss World War, they mention what Germany and the other countries mean in terms of the Ottoman Empire. In general, the subject contents have been created in accordance with an official history approach in the textbooks; it hasn't been focused on the chaotic nature of the events. One or two reading texts about information which are related to human tragedy experienced in the battlefronts or tragedy experienced by the society of that period are explored.

When the textbooks of Social Studies and History are examined, handling World War I, which is one of the most chaotic events in the history of mankind, is extremely far from reflecting chaotic perceptions of the people of that period and their thoughts and feelings in the face of events. Historical events in the textbooks are handled as if they were a series of events which happened in due course. However, World War I has a chaotic situation in its own in terms of the outbreak, the development and the results of it. Where, how and when the chaotic events will occur, how they will be able to develop, aren't fully unpredictable. So events can be shaped in the precautions which may be taken in the face of chaotic events as they evolve. There hasn't been a content which may reflect the thoughts and feelings of the people or the mangers of that period in the face of a chaotic event such as World War I in the textbooks. Investigating the solutions put forward in the face of chaotic properties of that period is important for the students to contribute to the development of problem-solving skills.

We have seen that the lesson taught under the name of Social Studies in Turkey is carried out under the name of different courses such as History, Geography, Citizenship Education and Politics, based on mono-disciplinary approach, in some states of Germany. According to mono-disciplinary approach, units consist of the subjects belonging to History, Geography, Citizenship Education, which are independently separate from one another. Thus, it is easily understood if it belongs to History, Geography or Citizenship Education ([15]: 51). In the mono-disciplinary approach, History, Geography and Citizenship Education take their places as separate lessons in the curriculum (Safran [16]: 5). German history has an important place in history lessons.

We can explain the reflection of this war in Concrete History 3 textbooks, in which it is taught in the state of Baden-Wurttemberg in German with following example.

> These statements have been given a place on page 58 of the book: From the first paragraph of the section entitled "War Threat is Increasing": In subsequent years, all government continued its preparation for war. Military authorities were developing the mobilization and combat plans for armed forces. Therefore, their suspects were growing against each other. Newspapers and politicians' speeches caused that people felt themselves under the threats of the opposing forces. The sense of nationalism in European Countries was overdeveloped because of imperialism, too: Every country wished to overcome the need of raw materials by importing from their colonies. The general consensus was that the war would break out sooner or later. It was an indication that strong countries returned repeatedly to the brink of war due to the colonies and various conflicts in the Balkans until 1914 (Hatscher et al. [7]: 58).

These expressions have been important in terms of revealing the chaotic situation that was experienced before the war. Being expressed people's sceptical feelings at that moment has revealed the ongoing impasse and people can't make predictions about the future. That is, these are an indication of a chaotic environment at that time.

These statements have been given a place on page 60 of the book:

> On August 6th, the Austro-Hungarian Empire declared war on Russia. People who joined the army were sent off with public demonstrations of joy from their hometowns. German declared that they didn't recognize Kaiser and only saw him as a German. Opponents of the war were thought to be "traitor". Most people hoped the war to be short term (Hatscher et al. [7]: 58).

These expressions are important in terms of showing that people are in a chaotic state regarding when the war will end. War subjects have been handled in detail and more fluency expression has occurred by adding people's thoughts at that time in the given texts.

The harsh realities of the war have been displayed in the following text which is taken from the novel called "Stop the Guns" on page 61 of the book. By explaining what happened with all the clarity, it has been shown how people live in a complex situation. It has been explained that a chaotic event such as World War I creates a chaotic state in people, too.

> The person who keeps alive with limbs blown away on the battlefield after the combat defeated without being there and dying by decaying in hunger, thirst, and in indescribable pain for four or five days and nights-and while dying, being aware of that this country didn't win anything and abundance despair remains what people love with this death. What I want to know, did he die gladly for the whole time with that certain saying (for the sake of the country) (Hatscher et al. [7]).

A poem describing the love of the country and the necessity of the war has been given a place after the above text. It is provided that students must find out these arguments between these two approaches. The questions have been asked about how these two texts may have influenced the students and they have been tempted to think about this chaotic situation with the question 'what would you think if you were'. Such texts and questions have great importance in terms of learning that we get opinions about the event from the students and how they should act in case of a possible case of turmoil which they may experience. In these two texts, students have been made to live as if at that moment.

We have seen what happened in 1916 and when the war continued what took place has been handled in detail.

> 46 villages have been destroyed; nobody never lived in 9 villages. Forests, fields, streets and rail lines have ruined, soil and water poisoned because of gas, explosive, and rotting corpses (Hatscher et al. [7]).

The above sentences are important in terms of showing large and the devastating impact of the war on people. The destruction of their village and being unable to meet the needs of food and water have been expressed here and how people were in a chaotic situation has been described.

War is described in a fluent style while the subject of World War I is expressed. Describing the events experienced is given importance instead of the classic knowledge. Figuring out soldiers who are blind due to toxic gases used during the war, poems inspired patriotism, being painted of torture experienced during the war are important documents for the students to visualize that moment.

This information which is given on page 64 has been revealed as a chaos and confusion laden environment experienced during the war one more time

> They have attempted the five-day to a punitive action. Cities have been razed, arson has been conducted, and the people who are seen as a partisan person have been shot and killed in masse. In addition, there has been mass rape. Similar situations have occurred in many cities occupied. German troops have killed total 6000 Belgian and French civilians in this action.

In general, it is observed that World War I has been processed in detail in Concrete History 3 textbook which is taught in the state of Baden-Wurttemberg in German. Scientific data have been supported by visual, it has been seen that the dramatic events experienced have been explained with photos, poems and by describing the event itself. Such an expression style has been thought to have great importance for the student in terms of recognizing the chaotic events and learning how to act in such situations.

42.4 Conclusion

It is not easy to predict when and where a chaotic case will appear and how it will change in the process. As it is complex and complicated by the efforts of world nations, First World War is a chaotic case itself and it is not easy to be understood via all its dimensions by the students. In this sense, as it is one of the main contexts of Turkish national history curriculum at 7th, 8th and 11th grades, it is aimed to make some details more clear in order to get the students aware and beware of the facts by touching some other dimensions and participant countries' points of views.

The social studies, Turkish Republic Revolution History and Kemalism programs being taught in Turkey are not prepared in accordance with the chaotic properties. Programs can only be evaluated in terms of the Ottoman Empire and the subjects are thought of from the perspective of the Ottoman Empire. Other states fought in the war handled through their relations with the Ottoman. The events in World War I were selected, as if they were a certain section of the whole book as a teaching method. However, the start of World War I and the process is exactly an unpredictable chaotic event. In this regard it is important to touch these details in the lessons. In addition, the details about outbreak and development of the war were not given in the sense of perspectives of the participating states in programs in Turkey. It becomes hard to understand World War I as a chaotic event by students by all its aspects. On the other hand, programs in Turkey have an expanding structure according to the spiral structure. Some programs repeat previous ones and expand in upper grades. But teaching methods do not change. Chaotic perceptions are ignored [15].

While World War I is being processed in Concrete History 3 textbook, visual materials and the dramatic events experienced have been given a place and the dramatic events experienced have been fluently expressed. The chaotic situation which people feel has been reflected in the content of the textbooks. By asking the questions such as "how would you act in this chaotic environment?" it is to aim that students' problem solving skills can develop in the face of the chaotic situation. It is observed that complexes experienced at that time are reflected in detail but chaotic events are not being reflected in a reason and effect relationship within the textbook.

In this context, it can be suggested that chaotic events may be widely expressed in accordance with chaotic properties. Being given a place as a historical event in view of not only your country but also the other countries may make the students find it easier to read and understand correctly. Therefore, it can be suggested that the viewpoints of different countries are also reflected in the chaotic events.

References

1. Arslan M (2013) İlköğretim sosyal bilgiler 7 öğretmen kılavuz kitabı. Anıttepe yay, Ankara
2. Balcı A (2001) Sosyal bilimlerde araştırma-yöntem, teknik ve ilkeler. Pegem A, Ankara
3. Bülbül MŞ (2007) Kaos ve eğitim. Beyaz Kalem, Ankara
4. Dilek D (2002) Tarih derslerinde öğrenme ve düşünce gelişimi. (2. baskı). Pegem A, Ankara
5. Erdem E (2001) Program geliştirmede yapılandırmacılık yaklaşımı. Yayınlanmamış yüksek lisans tezi, Hacettepe Üniversitesi Sosyal Bilimler Enstitüsü
6. Fosnat CT (1996) Constructivism: a psychological theory of learning. In: Fosnat CT (ed) Constructivism: theory perspectives, and practice. Teachers College Press, New York, pp 8–33
7. Hatscher RC, Huneke MJ, Kniep K (2004) Geschichte Konkret 3. Schroedel. Schroedel Verlag GmbH, Berlin
8. Jonassen DH (1991) Objectivism versus constructivism: do we need a philosophical paradigm. Educ Technol Res Dev 39(3):5–14
9. Kaya B (2008) Oluşturmacı yaklaşıma gore hazırlanan TC.inkılap tarihi ve atatürkçülük dersinin tarihsel öğrenmeye etkisi: Bireylem araştırması. Yayınlanmamış doktora tezi, Marmara Üniversitesi, Sosyal Bilimler Enstitüsü, İstanbul
10. Karasar N (2002) Bilimsel Araştırma Yöntemi: Kavramlar, İlkeler ve Teknikler. Nobel Yayıncılık, Ankara
11. Kaya E (2006) Türkiye'nin AB ile bütünleşmesinde sosyal bilgiler dersinin işlevselliği. Anadolu Üniversitesi Yayınları, Eskişehir
12. Kaya E (2011) Farklı ülkelerin sosyal bilgiler öğretim programları. In: Cemil Öztürk (ed) Almanya. Pegem Akademi, Ankara
13. Mason J (1996) Qualitative researching. Sage, London
14. MEB (2008) Sosyal Bilgiler 7. Sınıf Çalışma Kitabı. MEB Yayınları, Ankara
15. Öztürk C ve Dilek D (2003) Hayat bilgisi ve sosyal bilgiler programları. İçinde: C Öztürk ve D Dilek (Der.). Hayat bilgisi ve sosyal bilgiler öğretimi (2. baskı). Pegem A, Ankara
16. Safran M (2008) Sosyal bilgiler öğretimine bakış In: Tay B, Öcal A (ed) Sosyal bilgiler öğretimi, PegemA, Ankara, pp 2–17
17. Saray M (2010) İlköğretim T.C. İnkılap Tarihi ve Atatürkçülük 8 Öğretmen Kılavuz Kitabı Devlet Kitapları. İstanbul
18. Ürküt M (2013) İlköğretim T.C. İnkılap Tarihi ve Atatürkçülük 8 öğretmen kılavuz kitabı. In: Sevtap G (ed) Netbil, İstanbul
19. Yıldırım, Şimşek (2005) Sosyal bilimlerde nitel araştırma yöntemleri. Seçkin Yayınları, Ankara

Chapter 43
A Research on the Reviews of the Social Science Teachers Who Serve in Turkey and Germany About Chaotic Situations They Encounter in Class

Oğuzhan Karadeniz, Harun Er, and Selçuk Turan

Abstract Today's world has a consistently dynamic structure in constant change and development. We live in a world where we can encounter positive and negative unexpected situations and everything can happen every moment. That such environments in which people and communities live change constantly and quickly shows that people live in a chaos laden environment. The social science lesson which has a very important function in terms of adapting to social life and socializing has an important part in the education of individuals. Thus, the teachers who give lessons regarding the lesson which holds such importance state chaotic situations they encounter in the teaching-instruction process and the evaluation of their responses consists the core of our study. In this study, the aim is the comparative evaluation of the chaotic situations which social science (history, geography, politics, citizenship education) teachers, who serve in Turkey and Germany, encounter in the course. The study using qualitative research methods was carried out by a total of 20 social sciences teachers including 10 in Turkey and 10 in Germany working in 5 different locations (history-geography, politics, citizenship education). As a result of the study, the chaotic situations that teachers who serve in Turkey and Germany, encounter are different in specific ways but basically similar. While teachers in Germany pointed out student-centered problems, the teachers in Turkey tended to non- students factors mostly.

O. Karadeniz (✉)
Ereğli Faculty of Education, Department of Elementary Education, Social Studies Education Department, Bülent Ecevit University, Peak resort, 67300 Eregli, Zonguldak, Turkey
e-mail: oguzhankaradeniz25@hotmail.com

H. Er
Faculty of Education, Department of Elementary Education, Social Studies Education Department, Bartın University, Bartın, Turkey
e-mail: erharun06@gmail.com

S. Turan
Faculty of Education, Institute of Educational Science, Hacettepe University, Ankara, Turkey
e-mail: selcukturan25@yahoo.com

Ş.Ş. Erçetin (ed.), *Chaos, Complexity and Leadership 2014*,
Springer Proceedings in Complexity, DOI 10.1007/978-3-319-18693-1_43

43.1 Introduction

Teaching, one of today's most important public professions, is brimming with the problems in which all the problems of complex social life and the globalizing world are directly felt. Together with the changing world and the person, perception of teaching and the problems are also changing. Hence in this profession with the public nature, current detection of problems and solution seeking continue to be very important phenomena not to be neglected. Because human and problem are essential phenomena existing always together. Wherever there are people, problems are never lacking. If so, what should be done should not be denying the problem, trying to stay away from the problem or targeting the smoothness. Problems must be resolved with the least possible damage. The proper solution of the problem is only to get the opinion of the people experiencing those problems at school where the problem occurs [12].

Teachers are obliged to see the chaotic aspects of the education system. In the opinion of Çobanoğlu ([2]: 118), in organizations where everything is in constant change, having the necessary information is impossible to make plans for the future. Together with the environmental factors, organizational factors are also changing. In this respect, instead of being managed with a mechanical approach, educational organizations should also need to benefit from the creativity of individuals with flexible, always reorganizing and constant new teams.

By their very nature, educational institutions have a complex structure and exhibit a non-linear situation ([1]: 153). Therefore, for an educational institution to be successful, it must define the education phenomenon as a chaotic phenomenon ([6]: 14). In Radford's opinion ([8]: 184), schools are organizations under the influence of different effects of multiple implicit or explicit elements and also intending to create different effects. In these organizations there is no single and unchanging formula for success. So there are many factors affecting the success of those organizations. According to Garmston and Wellman [5], schools are also under the influence of their own feedbacks. As a child, these feedbacks increase resonantly and exponentially. Thus they create radical changes. So in order to get the desired result, it is necessary to note that the effect on the system shall be the right effect for the long term. Reilly refers this as follows: ([9]: 424) "the linear proportional relationship can not be established in the education system" and he gives the following example: " in order to assess whether educational investments are successful, looking the ratios of inputs and outputs is the result of a linear thought and this is a faulty assessment." Sullivan [10] also emphasizes this chaotic order in educational institutions. He also implies that the administrators can get the opportunity to bring a new dynamic to the school using the environment of educational institution and the network of relationships in chaotic environment. In other words, it is possible for the administrator to turn the chaos into an opportunity by taking more advantage of the environment.

Trygestad [11] indicates that classes are systems that are non-linear, chaotic and having unpredictable processes. Teachers respond to chaos with the classification method in order to reduce instability and undesired behaviours and to increase stability and rate of behaviour. Educators usually indicate that order exists and they deny or underestimate disorder and the errors. They also acknowledge those errors and disorders as a coincidence. But errors are factors requiring investigation for analysis of a system. Errors that we accept as a coincidence in education include information needed for restructuring of the system. When we think in this context, we can consider that multiple intelligence theory has been developed as a model some more suitable for the chaotic environment in classes. In classical education, teacher regards the students as monotype and teaches in a single format in the class. In accordance with and at the rate of their capacity and tendencies, students can receive the information given by the teacher or not. However in multiple intelligence theory, the teacher admits that students have different abilities, aptitudes and capacities and accordingly classifies each student differently and guides him or her to benefit more from the course. The teacher does not accept the improper behaviour and information deemed as a coincidence also during the evaluation process. To correct those behaviours and information, he/she treats them as a feedback.

Many events that take place in the education sector can be shown as examples of chaos and uncertainty. Today, the violent movies and serials published mostly in TV channels are expressed to be the cause of violence acts at schools. However, when those violent movies and serials first started to be published, they were much in demand in society and not considered to lead to violence among school-age children. These events indicate that the foresight we previously demonstrated linearly is insufficient and the factors that we overlook or undervalue can result in enormous and unexpected consequences. This example shows that poorly managed small factors lead to large and negative consequences. Besides, regardless of the positive or negative nature of its consequences, the butterfly effect is expressed as: "small events lead to big events". In this context, it's possible to express an assumption that well-managed small factors can lead to large and positive consequences. For example, if an innovation and improvement s desired in education, we can begin from the lowest stage instead of the upper stages of the system in order to accomplish this. For example, let's suppose that we make students believe that they can have more knowledge and skills thanks to educational system. This belief of students means that the student will expect more from the system. This situation results in a way that the student will be much curious and ask more questions. In this case, the teacher must have further information with the aim of providing more activities and information to respond to the requests of students. Thus the teacher both equips himself better and forces senior management to offer better opportunities. Let's assume that this effect continues successively. In this way improvement does not remain only in the education system, it also spreads to the other systems of the country. In this case, we will have achieved a better health care system, a better construction system, a better security system, etc. as well as a better education system thanks to the trigger we have created among students ([4]: 861–862). As a result, chaos scares and bothers us, but thanks to those fears it forces us to change and evolve which takes us away from linearity even more and feeds chaos ([4]: 863–864).

In today's rapidly developing and changing world, the way of people's life and thinking, their values, expectations and problems have also changed and with the impact of globalization it has entered into the process in which people influence each other more. When people are considered to be in the center of education and training, one of the occupational groups affected most by those developments is the teaching profession, and that is an undeniable fact. Therefore in this profession with the public nature, the current detection of the problems and seeking of solutions continue to be very important phenomena not to be neglected. Dwelling on the problems especially starting from the training process of teachers will have a positive impact on the success of education which is a long-term investment ([3]: 110).

Like in other fields of teaching, there are also certain problems experienced by social studies teachers in the classroom. These problems can vary from cities to cities as well as varying from country to country. The comparative researches done in Germany, an important and powerful country of the EU, of which Turkey is aiming to be a member and which has a close relationship with Turkey both politically and economically are very important for education. In this research done for this purpose, the evaluations of the social sciences teachers (history-geography, politics, citizenship education) working in Germany and Turkey regarding chaotic situations they encounter in the lesson have been included.

43.2 Method

43.2.1 The Model of the Research

This research has been carried out with qualitative approach and its data has been gathered with the non-directive semi-structured interview technique [7]. Yıldırım and Şimşek [13] describe the qualitative research as "a research in which qualitative data gathering methods like observation, interview and document analysis are used and a qualitative process aimed at introducing the perceptions and events in a natural environment in a realistic and integrative way is followed." A semi structured interview form that includes open ended questions has been used to define the views of the social science teachers (history-geography, politics, citizenship education) participating in the research about the determination of the chaotic situations that they encounter.

43.2.2 The Work Group of the Research

The research has been carried out in a total of 5 residential areas, 3 from Turkey and 2 from Germany. 10 from Turkey and 10 from Germany in total 20 social science teachers (history-geography, politics, citizenship education) have participated in the research. Easily accessible situation sampling has been chosen as sampling method. This sampling method attracts the attention since it provides the researcher with speed and practicality. In this method, the researcher chooses the individuals who are close and easily accessible [13]. The enthusiasm of those who have attended the research has been taken into consideration.

43.2.3 Data Gathering Tools and the Data Analysis

According to the screening of the relevant article, a lot of data were obtained about the chaotic situation encountered in training as well as the chaos term. As a result of this data, 5 questions for the interview were created by researchers. These questions, being created, are examined by 2 in domain and assessment experts 3 in social sciences education. Also being applied to two teachers in Turkey and Germany, the functionality of the questions were examined. As a result of all these evaluations, after the number of questions is decreased to 3, the final interview form was done.

Descriptive analyses which are frequently used in the qualitative studies have been utilized on the data gathered from the teachers. The data gathered from the teachers has been entered into the process of descriptive analysis separately. In the analysis of the open ended questions, the answers of the participants have been read line-by-line a few times and some codes have been formed about them. In the process of coding the views of the participants, the words and sentences that are parts of a meaningful whole have been taken into consideration. The obtained codes have been presented in a table with their prevalences below the interview questions in which they have been stated.

43.3 Findings and Comments

This chapter includes respectively the definitions of social science teachers working in Turkey and Germany regarding the chaotic situation, chaotic situations they experienced in the course of lessons and the data analysis concerning positive and negative remarks of chaotic situation assessments.

43.3.1 What Is "Chaotic Situation" According to You?

The answers of social science teachers in Turkey to the question "what is chaotic situation?" are as follows:

T1: "*The lack of certainty about what will happen, the situation of being suspicious, being under stress and being sometimes pessimistic and sometimes happy*".

T2: "*Complex situation, a can of worms*".

T3: "*It means the complex situations of which symptoms we can never predict resulting in many unexpected situations.*"

T4: "*A situation of complexity, uncertainty*".

T5: "*A tough situation happening to the person, the difficulties he is facing meanwhile and his struggling to get out of this situation*".

T6: "*A state of uncertainty. They are the situations encountered suddenly and arising unexpectedly.*"

T7: "*The state of being in a mess*".

T8: "*The situation of being unable to foresee what should be done with encountered condition*".

T9: "*It can be explained as difficulty, crisis and chaos. Besides, there is always a system in each chaos*".

T10: *"A chaotic situation is disorder. The non-applicability of the process sequence planned*".

As can be seen from the above statements, the social science/history, geography, politics, citizenship education teachers working in Turkey define chaotic situation mostly with such expressions as "difficulty, uncertainty, complexity, disorder, chaos, crisis and the situation encountered suddenly and arising unexpectedly". T5 in remarks of the teacher evaluating the chaotic situation from various aspects has perceived the chaotic situation as an effort, struggle and transition from a negative situation to a positive situation with the statement *"A tough situation happening to the person, the difficulties he is facing meanwhile and his struggling to get out of this situation"*. T10 has specified the chaotic situation as disorders emerging, non-fulfilment of the foreseen factors with his statement *"A chaotic situation is disorder. The non-applicability of the process sequence planned"*. And T1 who has evaluated the chaotic situation as a both positive and negative situation has described this situation explicitly with his definition *"The lack of certainty about what will happen, the situation of being suspicious, being under stress and being sometimes pessimistic and sometimes happy."*

The answers of social science teachers (history-geography, politics, citizenship education) working in Germany to the question "what is chaotic situation?" as follows:

T1: "Chaotic situation occurs in the turmoil where there is an actual or ostensible structural defect. In such cases there is no order, system, structure, logic or reasonable legal order".

T2: "It means the interruption of lessons, violation of the rules and teachers' losing their authority".

T3: "I think "a chaotic situation" occurs when all students speak at the same time, bustle around and nobody fulfils the teacher's instructions".

T4: "I describe a situation as chaotic which doesn't contain an order namely legality because of the fact that it involves a number of units and communication complexity".

T5: "Chaotic situation is a state of turmoil which means unhappiness and where people feel uneasy, their expectations are not met".
T6: "I think chaotic situation means a state of disorder where the people in the community don't know what to do and unforeseen situations come about".
T7: "Chaotic situations arise when communication is disrupted; people have difficulty in understanding each other".
T8: "Even though we express the chaotic situation with such words as complexity, disorder and the deterioration of the system, we can also accept it as termination of negativity and initiation of new things".
T9: "Teachers not knowing what subjects he should teach, students sleeping during the class and being uninterested and indifference to the lessons."
T10: "It is a situation, where law and logic don't work; authority is weakened."

As can be seen from the definitions, social science teachers (history-geography, politics, citizenship education) in Germany have defined the chaotic situation in general terms with such concepts as "corrosion, rule violation, lack of communication, plan and system and the weakening of authority". The teachers assess the chaotic situation from different points of view. T3 expresses the chaotic situation with negative examples occurring in the education process by stating *"I think a chaotic situation occurs when all students speak at the same time, bustle around and nobody fulfils the teacher's instructions"*. T8, with his statement: *"Even though we express the chaotic situation with such words as complexity, disorder and the deterioration of the system, we can also accept it as termination of negativity and initiation of new things"*, remarks that chaotic situation could also contribute to the development in a positive way as well as in a negative way. And T1 considers the event in the systemic sense as the deterioration of law by explaining with these words: *"Chaotic situation occurs in the turmoil where there is an actual or ostensible structural defect. In such cases there is no order, system, structure, logic or reasonable legal order"*.

43.3.2 Which Chaotic Situations Have You Experienced in Social Science Lessons?

The distribution about the themes regarding the chaotic situations social science teachers in Turkey experience during lessons is shown in Table 43.1.

When Table 43.1 is analysed, it can be said that the chaotic situations, which social science teachers experience in lessons, have been collected under the titles "curriculum, student, school's physical facilities, lesson materials, parent factor, lesson hours and school administration and management". Teachers state that the problems resulting especially from curriculum being too comprehensive and far from up-to-dateness (5), lesson hours being inadequate (5), inadequacy of materials (4), inadequacy of school books (4), parents' inability and indifference in guiding (4) and the lack of classroom a system cause chaotic situations.

Table 43.1 Distribution of the themes regarding the chaotic situations social science teachers working in Turkey experience

	Answers	f	Total
Student	Discipline disorder	3	9
	Attending the class without preparation	3	
	Absenteeism	1	
	Lack of attendance to the lesson	2	
Lesson materials	Inadequacy of lesson materials	4	8
	Inconsistency of lesson materials with the curriculum	2	
	The problems experienced in providing material	2	
Curriculum	Attainments not being clear enough and appropriate for the class level	2	12
	The curriculum being too comprehensive and far from being up-to-dateness	5	
	Inadequacy of school books	4	
	Inconsistency of the attainments in school books with the activities in work books	1	
Parent factor	Parents' inability and indifference in guiding	4	7
	Parents trying to dominate over the school and teacher	3	
Lesson hours	Inadequacy of lesson hours	5	5
School administration and management	Pressure of administration	3	5
	Fear of control	2	
School's physical facilities	Lack of a classroom system	4	9
	Lack of places to do activities	3	
	Overcrowding of classes	2	
Total			55

Some of the answers that the teachers have given about the chaotic situations they experience during the social science lessons are as follows:

T1: "The fact that the physical facilities of class don't make it possible to teach the social science lesson in a funny way is one of the reasons of chaos."

T3: "That the curriculum is too comprehensive and doesn't contain current events and developments sufficiently fails to attract students' attentions and meet their needs of learning."

T4: "The chaotic situation I mostly experience in classes is the mess caused by being unable to teach the lesson completely due to the inadequacy of lesson hours."

T9: "Discipline disorder of students, their attending the class without preparation, the lack of attendance could lead to confusion over time".

As can be seen from the responses of teachers, the chaotic situations experienced in social science lessons vary; but mostly result from the reasons about curriculum, students and school's physical facilities. The distribution about the themes regarding

Table 43.2 Distribution of the themes regarding the chaotic situations social science teachers (history, geography, politics, citizenship education) working in Germany experience

	Answers	f	Total
Student	Discipline disorder	4	16
	Attending the class without preparation	4	
	Lack of attendance and interest	4	
	Different achievement levels between students	2	
	Comprehension problem	2	
Lesson materials	Inadequacy of lesson materials	3	3
Curriculum	School books being old and impractical	3	7
	The curriculum being too comprehensive and far from being up-to-dateness	2	
	Inadequacy of school books	2	
Parent factor	Parents' inability and indifference in guiding	2	2
Lesson hours	Inadequacy of lesson hours	3	3
School administration	Pressure of administration	2	2
School's physical facilities	Insufficient physical facilities in schools	2	2
Total			35

the chaotic situations social science teachers (history, geography, politics, citizenship education) in Germany experience during lessons is shown in Table 43.2.

When Table 43.2 is analysed, it is seen that the chaotic situations that social science teachers (history, geography, politics, citizenship education) in Germany experience in lessons have been collected under the titles "student, lesson material, curriculum, parent factor, lesson hours, school administration and school's physical facilities". Teachers state that the problems resulting mostly from discipline disorder (4), attending the class without preparation (4), lack of attendance and interest (4), inadequacy of lesson materials (3), school books being old and impractical (3), inadequacy of lesson hours (3) cause chaotic situations.

Some of the answers that the teachers have given about the chaotic situations they experience during the social science lessons are as follows:

T8: "Students often think that social science is only a chat lesson and thus they become undisciplined or they fail".

T3: "Students attend the classes substantially without preparing and behave frivolously".

T6: "Most of the students lack of interest and attention during lessons and because they can't deal with this situation, they don't fulfil their responsibilities."

T5: "School books on social science/politics are usually old and therefore they are impractical for me. I only use the materials of online-publishers".

As can be understood from the answers, chaotic situations that teachers experience in social science lessons are associated with the problems resulting especially from student and curriculum-oriented factors.

Table 43.3 The frequency distribution of positive remarks by subcategories on the assessment of chaotic situations that the teachers working in Turkey experience during lessons

Subcategories of positive remarks on the assessment of chaotic situations that social science teachers experience during lessons	f
1. Because chaos creates uncertainty, it may generate a positive impact	2
2. Chaos raises the level of teacher's and student's alertness	1
3. Students' asking challenging questions makes the teacher more prepared for the lessons	2
4. Pressure of administration and fear of control force the teachers to work	3
5. Fear of getting high or low grades impels the students to study	2
6. Problems concerning the inadequacy of materials make the teacher become a good improviser in difficult circumstances	3
Total	13

43.3.3 Positive Remarks on the Assessment of Chaotic Situations Encountered in Social Science Lessons

Positive remarks on the assessment of chaotic situations that the teachers working in Turkey experience during lessons are shown in the Table 43.3 by subcategories.

As can be seen in the Table 43.3, within the subcategories of positive remarks on the assessment of chaotic situations that social science teachers working in Turkey experience during lessons the statement "pressure of administration and fear of control force the teachers to work" (3) has been mostly mentioned; this has been followed respectively by the statements: "Problems concerning the inadequacy of materials make the teacher be a good teacher in difficult circumstances" (3), "Because chaos creates uncertainty, it may generate a positive impact" (2), "Students' asking challenging questions make the teacher more prepared for the lessons" (2), "Fear of getting high or low grades impels the students to study" (2) and "Chaos raises the level of teacher's and student's alertness". Within the frame of these opinions, it can be inferred that the social science teachers don't regard all chaotic situations they experience in the classes as negative.

Some of the positive remarks on the assessment of chaotic situations teachers experience during lessons are given as an example below:

T3: "Uncertainty in the class often impels the student to study".

T5: "Even though the pressure of administration and controls of inspectors disturb the teachers psychologically, they in a sense force the teachers to work".

T7: "The problems concerning the inadequacy of materials in schools make me think that I would be a good teacher in difficult circumstances as well".

Positive remarks on the assessment of chaotic situations that social science teachers (history, geography, politics, citizenship education) working in Germany experience during lessons are shown in the Table 43.4 by subcategories.

Table 43.4 The frequency distribution of positive remarks by subcategories on the assessment of chaotic situations that social science teachers (History, Geography, Politics, Citizenship Education) working in Germany experience during lessons

Subcategories of positive remarks on the assessment of chaotic situations that social science teachers (History, Geography, Politics, Citizenship Education) experience during lessons	f
1. Students' reluctance forces the teachers to teach the lesson in different ways	2
2. Inadequacy of school books forces the teachers to use different and various sources	2
3. The problems concerning the inadequacy of materials make the teacher become a better improviser in difficult circumstances	1
4. The inadequacy of lesson hours may increase the teacher's performance	1
Total	6

As can be seen in the Table 43.4, within the subcategories of positive remarks on the assessment of chaotic situations that social science teachers (history, geography, politics, citizenship education) working in Germany experience during lessons the statements: "Students' reluctance force the teachers teach the lesson in different ways" (2), "Inadequacy of school books forces the teachers to use different and various sources" (2), "The problems concerning the inadequacy of materials make the teacher be a good teacher in difficult circumstances" (1), "The inadequacy of lesson hours may increase the teacher's performance" have been expressed. Within the frame of these opinions, it can be inferred that the social science teachers (history, geography, politics, citizenship education) do not regard all chaotic situations they experience in the classes as negative.

Some of the positive remarks on the assessment of chaotic situations teachers experience during lessons are given as an example below:

T2: "The fact that students aren't interested in the lessons sufficiently leads the teacher to seek different ways".

T5: "Being unable to find the materials we are looking for from time to time allows us to give the lesson in different ways".

T9: "The fact that the school books are too old led me to use electronic sources. Therefore I give all my lessons by using electronic sources".

43.3.4 Negative Remarks on the Assessment of Chaotic Situations Encountered in Social Science Lessons

Negative remarks on the assessment of chaotic situations that the teachers working in Turkey experience during lessons are shown in the Table 43.5 by subcategories.

As can be seen in Table 43.5, within the subcategories of negative remarks on the assessment of chaotic situations that social science teachers working in Turkey experience during lessons the expression "Inadequacy of lesson hours poses

Table 43.5 The frequency distribution of negative remarks by subcategories on the assessment of chaotic situations that the teachers working in Turkey experience during lessons

Subcategories of negative remarks on the assessment of chaotic situations that social science teachers experience during lessons	f
1. Inadequacy of lesson hours poses problems for attaining the objectives of the lesson	5
2. Physical structure of the school and the intensity of curriculum cause adversities	4
3. Teacher's personal problems and lack of motivation affect the education adversely	2
4. The fact that the attainments aren't appropriate and explicit causes adversities for the teacher and students	2
5. Discipline problems and lack of a reward system affects the education adversely	2
Total	15

problems for attaining the objectives of the lesson" (5) has been mostly mentioned; and this has been followed respectively by the statements: "Physical structure of the school and the intensity of curriculum cause adversities" (4), "The fact that the attainments aren't appropriate and explicit causes adversities for the teacher and students" (2), "Teacher's personal problems and lack of motivation affect the education adversely" (2) and "Discipline problems and lack of a reward system affects the education adversely" (2). Within the frame of these opinions, it can be inferred that the social science teachers don't regard all chaotic situations they experience in the classes as positive.

Some of the negative remarks on the assessment of chaotic situations teachers experience during lessons are given as an example below:

T9: "The fact that the social science lessons are insufficient affects adversely the teacher and students for attaining the objective of the lesson".

T2: "The fact that the physical structure of schools is not sufficient enough and the curriculum is too intense brings about a situation like being unable to teach all the subjects".

T4: "The current discipline problems in schools and lack of a reward system are major shortcomings in terms of motivation".

Negative remarks on the assessment of chaotic situations that social science teachers (history, geography, politics, citizenship education) working in Germany experience during lessons are shown in the Table 43.6 by subcategories.

As can be seen in Table 43.6, within the subcategories of negative remarks on the assessment of chaotic situations that social science teachers (history, geography, politics, citizenship education) working in Germany experience during lessons the expression "Discipline problem and reluctance may reduce the interest in the lesson" (4) has been mostly mentioned; this has been followed respectively by the statements: "The inadequacy of lesson hours poses problems for attaining the objectives of the lesson" (2), "Chaotic situations may adversely affect the course of the lesson" (2), "The physical structure of schools adversely affects the lessons" (2) and "The fact that the school books are too old and impractical causes adversities in transferring the content of the lesson" (2). Within the frame of these opinions, it

Table 43.6 The frequency distribution of negative remarks by subcategories on the assessment of chaotic situations that social science teachers (History, Geography, Politics, Citizenship Education) working in Germany experience during lessons

Subcategories of negative remarks on the assessment of chaotic situations that social science teachers (History, Geography, Politics, Citizenship Education) experience during lessons	f
1. Discipline problem and reluctance may reduce the interest in the lesson	4
2. The inadequacy of lesson hours poses problems for attaining the objectives of the lesson	2
3. Chaotic situations may adversely affect the course of the lesson	2
4. The physical structure of schools adversely affects the lessons	2
5. The fact that the school books are too old and impractical causes adversities in transferring the content of the lesson	2
Total	12

can be inferred that social science teachers (history, geography, politics, citizenship education) do not regard all chaotic situations they experience in the classes as positive.

Some of the negative remarks on the assessment of chaotic situations teachers experience during lessons are given as an example below:

T9: "The fact that some of the students don't obey the rules within class sufficiently negatively impacts other students that are interested in the lesson".

T6: "Lesson hours sometimes hinder us from completing the subjects, being unable to attain the objectives of the lesson is a bad situation".

T2: "The students speaking at the same time and the disobedient students affect negatively the course of the lesson".

43.4 Conclusion and Recommendations

In this study where the assessment of social science teachers (history, geography, politics, citizenship education) working in Turkey and Germany concerning the chaotic situations they have experienced has been analysed comparatively, the following conclusions have been reached:

While the social science teachers in Turkey have defined the chaotic situation with generally known aspects, the teachers in Germany have tried to explain this situation by giving the examples mostly in terms of educational aspect. This is especially important to display the difference in the perspectives of the chaotic situation of teachers in Turkey and Germany.

Whereas the social science teachers in Turkey put an emphasis on curriculum, school books, content, insufficient lesson hours and lack of materials regarding the chaotic situations they've experienced during lessons, the teachers in Germany expresses the chaotic situations they have experienced on student basis with discipline problems, lack of interest, attending the lessons without preparation and difference in the students' success levels. When compared, it is seen that the chaotic situations in Turkey result mostly from the need for material factors, and in Germany, contrary to popular belief; it rather results from immaterial factors and is shaped around the student's axis who is the subject of education process.

In their assessment on the chaotic situations they have experienced, the social science teachers working in Turkey state that especially the chaotic situation concerning the lack of materials and the pressure of chiefs in a manner of administration may create a positive effect. The teachers in Germany have expressed the opinion that the chaotic situation of students' reluctance and the lack of school books may create a positive effect. Besides that; the social science teachers working in Turkey have mostly agreed upon the opinion that the inadequacy of lesson hours, the physical structure of schools and the intensity of curriculum cause chaotic situations; and the teachers in Germany have reached a consensus on the opinion that the chaotic situations resulting from discipline problems and reluctance cause adversities. When these are compared, it can be said that the teachers in Turkey and Germany accept the adversities occurring in different points as a transition to favourableness and again that they prefer accepting the chaotic situations occurring in different points as a tendency towards adversity.

References

1. Baker SB (1995) Chaos theory in educational systems: principals' perceptions of sensitive dependence on initial conditions. Unpublished doctoral thesis. East Tennessee State University. www.libraries.etsu.edu/search. Accessed 4 Mar 2010
2. Çobanoğlu F (2008) Değisim mantığını anlamak: Akış ve dönüşüm olarak örgüt. Pamukkale Univ Educ Fac Mag 23(1):110–119
3. Demir MK, Arı E (2013) Öğretmen sorunları – çanakkale ili örneği. Ondokuz Mayıs Univ Educ Fac Mag 32(1):10
4. Ertürk A (2012) Kaos kuramı: yönetim ve eğitimdeki yansımaları. Kastamonu Educ Fac Mag 20(3):849–868
5. Garmston R, Wellman B (1995) Adaptive schools in a quantum universe. Educ Leadersh 52(7):6–12
6. Gunter H (1995) Jurassic management: chaos and management development in educational institutions. J Educ Adm 33(4):5–30
7. Karasar N (2004) Bilimsel araştırma yöntemleri. Nobel, Ankara
8. Radford M (2006) Researching classrooms: complexity and chaos. Br Educ Res J 32(2):177–190
9. Reilly DH (1999) Non-linear systems and educational development in Europe. J Educ Adm 37(5):424–440
10. Sullivan TJ (1999) Leading people in a chaotic world. J Educ Adm 37(5):408–423
11. Trygestad J (1997) Chaos in the classroom: an application of chaos theory. Roundtable session of chaos and complexity theory. American Educational Research Association, University of Minnesota, March 24–28. www.eric.ed.gov (Eric Documant: 413289) (05.02.2008)
12. Yapıcı M, Yapıcı S (2003) İlköğretim öğretmenlerinin karşılaştığı sorunlar. Bilim, Eğitim ve Düşünce Dergisi 3(3):9
13. Yıldırım A, Şimşek H (2005) Sosyal bilimlerde nitel araştırma yöntemleri. 5. baskı. Seçkin Yayınevi, Ankara

Chapter 44
Management of Students' Discipline in a Drug and Substance Abuse-Laden School Environment

Abubakar Sadiq Lawal and Buhari Marafa

Abstract This study intended to examine the influence of drug and substance abuse on the management of discipline in secondary school. Today drug abuse or drug addiction is one of the most vexing and pervasive problems that almost all countries including Nigeria; are grappling with in as far as administration and management of schools is concerned. The consequences on the youth are disastrous and devastating especially the negative effects in schools and the communities. This is has been aggravated by the rapid social and technology changes which have had a corrupting influence on the youth in schools. It has also translated to an impediment in as far as the management of schools is concerned. Drug abuse is a global health and social problem. It is one of the major problems affecting the youth both in school and out of school. It seems to have impacted negatively on the academic, social, psychological, economical and physiological development among the abusers. Therefore it has become a matter of grave concern to educational stakeholders, medical practitioners, sociologists, religious leaders, counselors and parents thus needing urgent attention. Educational institutions the world over seem to be threatened by the global phenomenon of drug use; abuse and the abuse of other substances. Whereas their aim is to transmit knowledge beliefs, values, norms, they seem to be over loaded by the monster of drug and substance abuse. This paper therefore, explored the management of discipline in a drug and substance abuse-laden school environment.

A.S. Lawal (✉)
Department of Business Studies, Shehu Shagari College of Education, Sokoto, Nigeria
e-mail: sadiqmaigoro69@gmail.co

B. Marafa
Department of Social Studies, Shehu Shagari College of Education, Sokoto, Nigeria
e-mail: marafabuhari09@gmail.com

Ş.Ş. Erçetin (ed.), *Chaos, Complexity and Leadership 2014*,
Springer Proceedings in Complexity, DOI 10.1007/978-3-319-18693-1_44

44.1 Introduction

Discipline is concerned with the establishment and maintenance of order and the harmonious functioning of a society. A school is also a society on a small scale, and discipline within school serves the purpose of ensuring that learning can take place. Within this, the rights of the individual and of all members of the school society are protected. In most schools, a set of rules which act as a code of conduct is drawn up for students to conform to. Such rules should be as few as possible, and should be reasonable.

Discipline is a necessity for the proper functioning of a school and is essential for effective learning and the quality of school life. Effective school discipline strategies need to be sought to encourage responsible behaviour and to provide all students with a satisfying and fruitful school experience by discouraging misconduct. School discipline has three main goals.

44.2 Factors that Influence Drug and Substance Abuse in Schools

Studies on the factors that influence drug and substance abuse in schools have been an attraction to several researchers [3, 6, 9, 10, 16]. The increase in the abuse of cannabis, hard drugs and volatile solvents is attributed to unemployment, social upheavals, family disruptions as well as high rates of drop-outs from school. Increased production and trafficking of cannabis has led to increased availability of the drug. Some sedatives are also ingested by students in schools. Multiple drug abuse like volatile solvents mixed with alcohol, and combinations of cannabis and volatile solvents with alcohol have been reported by NGOs as common in schools across Africa as habits the youth get in schools. Cannabis is smoked and dissolved in water for consumption. Khat is sold openly and chewed by youth in urban centres and the habits have cropped in schools. The major reason is pass time and eventually addiction.

In an attempt to control sleep or energise themselves, most adolescents and young ones start experimenting with tobacco, alcohol, ephedrine and other caffeinated substances such as Nescafe and red bull. Some of the reasons for the drug abuse, as identified by Ajayi and Ekundayo [3], are to reduce pain, anxiety and tension, ignorance and misinformation, parental background, urge to commit crimes, peer group influence, isolation and loneliness.

The Nigerian schools and Educational institutions are dominated by drugs used by students such as morphine, heroin, tobacco, ephedrine, valium five and Chinese capsules that the youth commonly use. Oshodi et al. [16] reported that, despite worldwide concern and education about psychoactive substances, many adolescents have limited awareness of their adverse consequences. They further explained that curiosity, social pressure and peer group influence are noted to be primary reasons for substance and drug misuse. Makanjuola et al. [10] lamented that a substantial percentage of the national budgetary allocation is utilized for treatment and rehabilitation of people and youth with substance use problems in Nigeria [2] and [5].

Linhadt [9] also noted that students see the use of stimulants in positive terms for relief from pain and problems, elevation of mood, wakefulness, increased confidence, feeling and psychomotor activities and athletics, and feeling of euphoria. McCrystal et al. [13] confirmed that for many adolescents, drug abuse has now become a part of their lives and perhaps may have now contributed to their academic failure. At the center of the habit of drug use inconveniences school administration because of instances of breaking the rules by students and lawlessness and anti-social behavior that is not commensurate to the standards and needs of schools. According to Kasule [6] Prevalence research revealed that cannabis/khat use was 15.6 %/34.6. About 75.7 % of the respondents indicated that what influenced drug and substance use in schools in Uganda, low education level, family influence, use by friend, religion and the environment the schools were located in.

The motive behind drug abuse of drugs and substances are sociological; that is status-seeking, peer pressure, the news media or a substance-oriented society. Students also use drugs for, psychological reasons; that is to banish pain or discomfort, to attain euphoria, fantasy or to escape from unpleasant reality, out of curiosity, boredom, to alleviate fear, derive sexual and physical pleasures, or family background [4]. Despite that drug abuse has adverse effects on the youth involved by changing their brain perception of difficulties and problems, the number of students that use or abuse stimulants has steadily increased in recent years. Murray [15] reported that students who witnessed dissatisfaction with life due to anger, frustration and boredom, resorted to drugs to find comfort. In most schools across Nigeria abuse of drugs appears to be increasing frequency. The reasons for their use is include the need to belong, expectancy, mental set, sex drives and the need to fit in peer groups [17]. Oshodi (2010) also confirmed that students use drugs for relief of stress and for self-medication at night in order to study. Adamson and Ogunwale [18] specified that in Kenya and South Africa, drugs of abuse have been known to be alcohol, cannabis, tranquilizers, and tobacco for the same reasons.

Murray [15] however argues that drug and substance abuse is a result of; lack of family and school role models, peer pressure, poor school performance, conflicts between the young people and their parents, easy availability of drugs and substances, poor parenting seen as the most and major problem as the youth reported, conflict between school system and family values among others.

Zullig et al. [23] also reported that in Nigeria as a nation the reasons individuals, including students, often give as a reason for stimulant usage include the need to belong, expectancy, mental set, sex, certain drives, integrative use, ceremonial use, hedonistic use, utilitarian use and disintegrative use [17]. Oshodi et al. [16] also confirmed that students use drugs for relief of stress and for self-medication at night in order to study.

44.3 The Nature of Indiscipline Attributed to Drug and Substance Abuse

Researchers such as Banda [5], Maleka [13], Matsoga [12], Keorang [8], Moswele [14], have shown interest in the study of the nature of indiscipline attributed to drug and substance abuse. This is because the promotion or maintenance of effective discipline is essential if organised group action is to be effective or productive whether the group is a club, society, a union, a company, a business; school or industrial concern or a nation. The word discipline connotes that the members or a group should reasonably conform to the rules and regulations, which is the code of behaviour which have been formed for it or by it, so that everyone may benefit by them. People's morale or school functions succeed maintenance of discipline. Failure of any members of a group do not abide by the rules of the organisation, exacerbates indiscipline common in schools such as chaos, confusion, disobedience, disloyalty and antisocial activities associated with strikes; drug use and general disobedience.

In the words of Spriegel [20] "discipline is the force that prompts an individual or a group to observe the rules, regulations which are seemed to be necessary to the attainment of an objective". It is a factor, which restrains an individual from doing certain things, which are deemed to be disruptive for the group objectives. It is also the exercise of restraint or the enforcement of penalties for the violation of group regulations. Thus, discipline can be said as an attitude of the mind, a product of culture and a particular environment which promotes an individual to willingly co-operate in the observance of the rules of the organizational to which he belongs. He argued that the occurrence of indiscipline such theft; strikes, disobedience; drug use and abuse; violence; destruction of property or vandalism in schools is as a result of specific individuals failing to comply with school regulations.

According to Matsoga [12] indiscipline related to drug use in Nigerian schools is associated with violence and misbehaviour. This lack of discipline which interferes with the teaching and learning process, manifests itself in various ways including bullying, vandalism, alcohol and substance abuse, truancy, inability or unwillingness to do homework etc. Moswele [14] and Matsoga [11]. Vandalizing school property is at rampart and this has influenced the government to introduce school fees in order to mend, that which was broken such as window, panes, furniture and walls.

Theft is also common in schools and students break into laboratory to steal ethanol chemicals with the drive of wanting to sell for money [5]. In another senior secondary school, a 19-year-old boy committed suicide after fighting with another student over a borrowed plate Maleka [13]. Such acts and incidents sparked the debate on the use of corporal punishment in schools which concluded that Nigerian schools cannot do away with it, but it has to be used guardedly [8, 13]. Experience had it that, teachers may ask for transfers; while parents may withdraw their children from schools with numerous cases of student misconduct such as the one mentioned above. Notably most the students who misbehaved used drugs as a way of gaining courage and boldness.

The report on the indiscipline of Students Unrest and Indiscipline in Nigerian Secondary Schools (2009) listed types of indiscipline related to drug use. The report states that unrest and indiscipline can take place in the following forms: truancy, further chronic absenteeism, rudeness and disrespect, unacceptable verbal expression of dissatisfaction and abuse of drugs and alcohol taking, non-compliance to rules and regulations, destruction of property, bullying fellow students, boycotts, riots, assault and indecent behavior for example rape and arson. All these types of indiscipline impact negatively on the achievement of the schools aims and objectives for they are detrimental to good academic performance.

44.4 The Challenges School Managers Face in Mitigating Drug and Substance Abuse Related Cases of Indiscipline

Studies on the challenges faced by managers in mitigating drug and substance abuse related cases of indiscipline was conducted by several researchers Adeyemo [1], Riley and Docking [24], Mukherjee [18], Sieber and Wilder [25]. Indeed discipline is an aspect of school function which if not well maintained can render the school system ineffective. School discipline as a matter of fact is seen as a vital element in the process by which students are enabled to function in the society. The general idea underlying this fact is that if the school is situated in the society and it is hoped that the products of these schools will be absorbed into the society, the students must therefore be made to develop rule-following and law, adding habits so that they conform to the general social expectations of the main culture absorbing its basic attitudes and beliefs, Sieber and wilder (2010) pointed out that a society without rules is inconceivable, and rules without attitude of disapproval towards them are inconceivable. On the other hand, thinking of discipline in terms of training for society, Riley and Docking [24] says it may encourage schooling for subordination where teachers use their disciplinary authority to satisfy some unfulfilled need within themselves so that they are to view children as a means rather than as ends. However he emphasizes that school managers ought to use all available means to them to help reduce indiscipline for the benefit of schools as well as society.

School discipline is often seen as an important ingredient in the process by which children are enabled to function in the society because living in the society entails living in association with certain agreed rules, which govern one's behaviour. Thus, keeping order in the school is a multi-faced problem associated with range of interacting factors, such as the child himself, home and neighbourhood influence, changing societal values and expectations, the school and its natural environment, and the individual teacher.

The problem of drug related indiscipline is more apparent among secondary school students all over the world. Indiscipline among them has attracted serious attention of scholars and administrators. These scholars and administrators opine that when students notice certain biological changes signalling maturity in the course of the growth and development, they tend to misbehave by faulting school rules and regulations. Therefore school administrators ought to put strict mechanism and realistic measures to wrestle the influence of indiscipline.

The consequences on the youth are disastrous and devastating especially the negative effects in schools and the communities. This is has been aggravated by the rapid social and technology changes which have had a corrupting influence on the youth in schools. The monster is a real threat to the youth; parents; schools and the economies of the world countries that could rely on the valuable youth potential that is being robbed by drugs. It has also translated to an impediment in as far as the management of schools is concerned. Drug abuse is a global health and social problem. It is one of the major problems affecting the youth both in school and out of school. It seems to have impacted negatively on the academic, social, psychological, economical and physiological development among the abusers [22].

Therefore it has become a matter of grave concern to educational stakeholders, medical practitioners, sociologists, religious leaders, counsellors and parents thus needing urgent attention. Educational institutions the world over seem to be threatened by the global phenomenon of drug use; abuse and the abuse of other substances. Whereas their aim is to transmit knowledge beliefs, values, norms, they seem to be over loaded by the monster of drug and substance abuse. The world over the practice is affecting school systems through the commonly abused drugs being cocaine, Indian hemp (marijuana) Kuber; cigarettes and petrol sniffing. Drug abuse has been identified as a serious constraint to effective teaching and administration of schools and learning process.

Involving parents in managing indiscipline is not just informing them about their children, progress in school or the education system; it is about increasing understanding of the potential that might otherwise remain hidden. The parents need to understand why the school has detected the indiscipline and its nature. Teachers can bridge the gap between the school and the community by guiding students on correct ways of behavior. Ukeje et al. [28] point out that in spite of the fact that some parents do not like to be bothered by the authorities with their children's behavior challenges, nevertheless it has been found to be very useful in many cases where pupils have become sources of unusual challenges for someone who is skilled in such matters to confer with one or both parents and to see if together they can discover what is the root cause of deviance and what best remedy can be. During conferencing the best principles and strategies and techniques for conferences should be utilized. No sign of prejudice, antagonism, suspicion, unfriendliness or blame should be apparent. What should be transparently obvious is everybody's concern for the student's welfare. Both parties must be prepared to give and take complaints and see each other's point of view in good faith [21]. However this may sometimes not be the case beach sides.

Indiscipline is a mode of life not in conformity with rules and non-subjected to control. By extension, the term connotes the violation of school rules and regulations capable of obstructing the smooth and orderly, functioning of the school system Adeyemo [1]. School rules and regulations in most cases affect students more than any other thing because they are made by the school authorities in order to guide and protect the students while in school. Therefore school managers ought to emphasize of school regulation more than anything else in managing indiscipline. However the challenge is that more often than not school regulations are ignored by students; parents as well as some teachers especially from schools in urban areas. Parents may not have enough time to detect that their children are involved in drug use while other parents from the community are the very ones selling the drugs to the students.

The head teacher is responsible for the overall management, control and maintenance of standards in the school. The head teacher is therefore accountable for all that happens in the school. The head teacher has a charge over a community of teachers and students and it is to him that they look for guidance and direction. He plays leadership role of steering the school community towards realization of educational goals. Kochhar [7] emphasizes the importance of head teacher whom he notes as the keystone in the arch of school administration and has the steering wheel in his hands. However; head teachers perform better if indiscipline is controlled in the schools. Some head teachers have been victims of drug related indiscipline staged by students in their schools. Some teachers and parents have often taken sides with the students.

44.5 Conclusion

Discipline is a necessity for the proper functioning of a school and is essential for effective learning and the quality of school life. On the factors that influence drug and substance abuse in schools studies have shown that the increase in the abuse of cannabis, hard drugs and volatile solvents is attributed to unemployment, social upheavals, family disruptions as well as high rates of drop-outs from school. Increased production and trafficking of cannabis has led to increased availability of the drug. On the nature of indiscipline attributed to drug and substance abuse, in Nigerian schools studies show that indiscipline related to drug use in Nigerian schools is associated with violence and misbehaviour. This lack of discipline which interferes with the teaching and learning process, manifests itself in various ways including bullying, vandalism, alcohol and substance abuse, truancy, inability or unwillingness to do homework etc. Vandalizing school property is rampart and this has influenced the government to introduce school fees in order to mend, that which was broken such as windows, panes, furniture and walls.

School rules and regulations in most cases affect students more than any other thing because they are made by the school authorities in order to guide and protect the students while in school, on the challenges of managing indiscipline therefore the school managers ought to emphasize of school regulation more than anything else in managing indiscipline. The head teacher is responsible for the overall management, control and maintenance of standards in the school. The head teacher is therefore accountable for all that happens in the school. The head teacher has a charge over a community of teachers and students and it is to him that they look for guidance and direction. He plays leadership role of steering the school community towards realization of educational goals.

Operational Definition of Key Terms

Cannabis; a tall annual dioecious plant

Khat; a shrub Catha edulis whose leaves are used as mild stimulants when chewed or brewed as tea.

Ephedrine; an alkaloid found in some species of ephedra shrubs or prepared synthetically, used as a sympathomimetic drug.

Morphine; A crystalline alkaloid, extracted from opium, the salts of which are soluble in water and are used as analgesics, anaesthetics and sedatives, it is one of the group of morphine alkaloids.

Heroin; a powerful and addictive drug derived from opium producing intense euphoria classed as an illegal narcotic in most of the world.

References

1. Adeyemo M (2005) Prevent the discipline for effective teaching and learning. A source book for teachers and administrators. Allyn and Beacon, Reston.
2. Aina OF, Olorunsola DA (2008) Alcohol and substance use portrayals in Nigerian video tapes: an analysis of 479 films and implications of public drug education. Int Q Community Health Educ 28(1):63–71
3. Ajayi IA, Ekundayo HT (2010) Contemporary issues in educational management. Bolabay Publications, Lagos
4. Alan E (2003) Analysis of the use, dependence and source of knowledge of stimulants among Nigerian university undergraduates. Kamla Raj Anthropol 11(3):213–218
5. Banda A (2004) Parents order whipping of Masunga SSS students. Mmegi Newsletter. Gaborone 19 May
6. Buddy W (2009) Pushing the drug war in the South West. Daily Trust, 30 November, 2008. On-line edition
7. Kasule A (2011) Factors influencing performance among primary and secondary pupils in Central Uganda: a policy study. Unpublished research paper, Kampala, Uganda
8. Keoreng E (2004) Research blames corporal punishment for school drop-outs. Mmegi Newsletter. Gaborone 12 November
9. Kochhar J (1988) Prevent the discipline for effective teaching and learning. A source book for teachers and administrators. National Association of Secondary School Principals, Reston
10. Lonngvist L (2010) Monitoring the Future national survey results on drug use, 1975–2003: volume 1, secondary school students (NIH Publication, No. 04-5507)
11. Linhardt RJ (2001) Drug in world book encyclopedia D, vol 5. Chicago World Book Inc. a Scott Fetzer Company, Cleveland, USA
12. Makanjuola AB, Daramola TO, Obembe AO (2007) Antisocial behaviour in adolescent guide. Acto Psychiat Scand 8(9):167–173
13. Maleke L (2003) Head teachers prescribe the stick. Mmegi Newsletter. Gaborone 17 October
14. Matsoga A (2003) Challenges principals face in enhancing student discipline in secondary school in Tigania District, Kenya. Unpublished M.ED thesis, Chuka University College
15. Matsoga L (2008) School discipline policies and procedures. A practical guide, Revised edn. Alexandria

16. McCrystal P, Higgins K, Percy A (2007) Exclusion and marginalization in adolescence: the experience of school exclusion on drug use and antisocial behavior. J Youth Stud 10(1):35–54. Retrieved 15 Dec 2011. www.drugsandalcohol
17. Moswele K (2009) Educational research: an introduction, 7th edn. Allya and Bacon, Boston
18. Mukherjee B (2008) Open access scholarly publishing in library and information science. Ann Libr Inf Stud 55(3):212–223
19. Murray H (2009) A handbook for elementary school teachers. Ally and Bacon, Boston
20. Oshodi OY, Aina OF, Onajole AT (2010). Substance use among secondary school students in an urban setting in Nigeria: prevalence and associated factors. *African Journal of Psychiatry* 1(3):52–57
21. Oizimende (2009). A study of challenges faced by teachers in teaching primary science in Kahawa Zone, Mathare Division, Nairobi. Unpublished research paper, Kenyatta University
22. Ogunwale B (2010) Challenges in secondary school administration, Nigeria. http://www.secondary.school.admn.edu/html
23. Rathod, as cited in Egbochuku, Aluede, Oizimende (2009) Family type and juvenile delinquency, issues among secondary school students in Akwa Ibom State, Nigeria: counseling implications. Kamla-Raj J Soc Sci 23(1):21–28
24. Riley K, Docking J (2002) Perceptions of schooling among disadvantaged pupils. Paper presented at the BERA annual conference. September 2002, in Exeter, UK
25. Sieber SD, Wielder DE (eds) (1973) The school in society. Studies in the sociology of education. Collier MacMillan Publishers, London
26. Spriegel H (1997) A case study of strategies for improvement of- discipline in five secondary schools within Meru District, Nairobi. Unpublished research paper, Kenyatta University
27. Ukeje BO, Akabogu GC, Ndu A (1992) Educational administration. Fourth Dimension Publishing Co. Ltd., Enugu
28. Ukeje C, Odebiyi A, Sesay A, Aina O (2009) Oil and violent conflicts in the Niger Delta. Cedcoms, Ile-Ife
29. West L, Graham C (2005) A survey of substance prevention efforts at Virginia's colleges and universities. J Am Coll Health 54(3):185
30. Zullig K, Haebue E, Parton J, Murray K (2009) The brief multidimensional students' life satisfaction (Scale-College Version). Am J Health Behav 339(5):483–493

Chapter 45
Efficiency of Private Pension Companies in Turkey Using Data Envelopment Analysis (DEA)

Abdu Seid Ali

Abstract This paper tries to examine the efficiency of private pension companies in Turkey using Data Envelopment Analysis (DEA) covering the year 2008–2012. Total stockholders' equity and operating expenses of the pension companies have been taken as input variables whereas investment income and fund administration charges were considered as output variables. The findings imply that there are technical efficiency gains in some private pension companies in Turkey. Though there is no continuous trend towards an improvement in technical efficiency, important changes over time have been witnessed. This may be due to the immaturity of the private pension industry given that it has launched operations 10 years ago. Note ought to be taken that management of the private pension scheme is a complex undertaking and in some instances, really chaotic. This is premised on the fact that some end up missing out on the true value of the scheme and that they even lose trust of the same. This paper thus explores the efficiency imbedded within private pension companies using Data Envelopment Analysis.

45.1 Introduction

Pension funds are the prominent sources of retirement income for millions of people both in the developed and developing world. They are also a significant contributor to the GDPs of countries and a considerable source of capital in financial markets. Pension funds bring together retirement savings from workers and their employers and invest this money in a broad assortment of assets. Since pension funds deal with the money of up to millions of individuals they are major group of actors in the financial markets. Pension fund is a kind of fund established by an employer to bolster and put in order the investment of employees' retirement funds contributed

A.S. Ali (✉)
Lecturer at Wolaita Sodo University, Wolaita Sodo, Ethiopia

PhD candidate, Department of Business Administration, Marmara University, Bahçelievler/Istanbul, Turkey
e-mail: abduseid_ali@yahoo.com

Ş.Ş. Erçetin (ed.), *Chaos, Complexity and Leadership 2014*,
Springer Proceedings in Complexity, DOI 10.1007/978-3-319-18693-1_45

by the employer and employees. It is a common asset pool intended to produce unwavering growth over the long term, and offer pensions for employees when they arrive at the end of their working years and start retirement.

Pension funds are commonly orchestrated by some sort of financial intermediary for the company and its employees, although some larger corporations operate their pension funds in-house. They manage relatively huge sum of capital and signify the largest institutional investors in many nations. A private pension is an investment scheme into which individuals contribute from their earnings, which then will pay them a private pension after retirement. It is a substitute to the state pension. Usually individuals invest funds into saving schemes or mutual funds, run by insurance companies. This can coined as a voluntary pension system intended to offer additional income for individuals and to guarantee they keep on earning the income level that they used to earn during their working lives, due to the fact that it will be supplementary to the social security system.

45.2 Private Pension Funds in Turkey

The crucial measures have been commenced concerning the Social Security Reform in 1991 for the purpose of unraveling the existing financial, institutional and administrative structural predicament in the social security sector of Turkey. In line with this, in October 2001, Private Pension Savings and Investment System Act became effective ([17, 22]). The legal and institutional framework of the Turkish Private Pension System was consummated in 2002. Subsequent to enactment of the law together with additional legislations that adds to robustness of system, Turkish Individual Pension System inaugurated on October 27, 2003 with the involvement of six pension companies [13, 17]. The main objectives are reducing burden of state social security and increasing scope of social security, improving welfare level, contributing to economic development by creating long term resources and increase employment level and deepening capital markets.

According to the Central Bank of Turkey financial stability report [6], overwhelming improvement is observed in private pension system in Turkey despite the fact that it was launched in recent times. Basically, individual pension system aspires to expand the coverage of social security, to elevate the wellbeing of individuals during the retirement period and to advance the living standards in the long term. In addition, it has irrefutable contribution towards the macro economy and financial stability on account of the capacity and maturity structure of funds pull together in the system. The implementation of private pension system is estimated to boost domestic saving rates thereby plays an imperative role in the dissemination of savings into the financial system and the augmentation of maturities [19].

The establishment of Turkish private pension system is the most important step in the advancement of the public social security system and Turkish insurance market and therefore a phase towards resolving the space in public finance and generating fresh funds for capital and money markets which are the critical challenge for the

Turkish government in the progress of its monetary and fiscal policy. Besides, it engenders extra returns for pensioners, wealth for new investments which can boost prolific capacity and maturity broadening in the financial segment [13].

According to OECD [18] report, over the period 2008–2012 the nominal annual rate of return of 16 OECD countries had been taken. Turkey came through the global instability with the best results in nominal terms, with a return equal to 11.6 %. Conversely, after taking into account inflation, Turkey turned into third after Denmark and the Netherlands which performed the best over the period. Pension funds in Turkey recently experienced positive cash flows, amounting more than 15 % of total investment in the country.

Pension funds efficiency relied robustly on competition and the regulation environment which may have an effect on inducement for consolidation and portfolio restrictions. Dynamic investment management plays an ever increasing role to make the market well organized and to ascertain the flow of funds to the most triumphant ventures, over and above contributing a vital role in the allotment of resources within the economy.

The analysis of efficiency in the private pension system and the companies that have engaged in the sector in Turkey is a subject matter that has allured some researches in the past. Most of the studies widely employed Data Envelopment Analysis (DEA) to measure the efficiency of these private pension funds and the companies which are managing them in different period of time.

This paper tries to examine the efficiency of private pension companies in Turkey using Data Envelopment Analysis (DEA) covering the year 2008–2012.

45.3 Theoretical Framework of the Study

45.3.1 Models of Industry Efficiency

In economic literature, two contending models of industry efficiency exist which try to measure the performance of companies. The strategic group theory [5] elucidates differences in efficiency scores as being attributable to differences in the structural characteristics of units within an industry, which sequentially direct to variations in performance. In the case of the pension funds management company, units with comparable asset patterns follow analogous strategies with similar outcome in terms of performance [20]. Due to the existence of different strategic options in the different sectors of an industry, because of mobility obstruction, it is difficult to find all alternatives to each pension funds management company, causing a stretch in the efficiency scores of the industry. On the other hand, the resource-based theory [1, 21, 23] gives an explanation for different efficiency scores in the form of heterogeneity of resources and potential on which retailers bases their strategies. These may not be flawlessly mobile across the industry, resulting in a competitive advantage for the best-performing retailers.

Purchasable assets cannot be regarded as to signify sources of sustainable efficiency. In fact, critical resources are not available in the market. Rather, they are built up and accumulated on the pension funds management company's premises, their non-imitability and non-substitutability being dependent on the specific traits of their accumulation process. The difference in resources thus results in barriers to imitation [21] and in the pension fund managers' inability to alter their accumulated stock of resources over time. In this context,

45.4 Literature Review

Several empirical studies have been made focusing on the efficiency of pension funds in general and private pension funds and companies in particular. The outcomes of those studies are multifarious and inconclusive. This study opts for the inquiries conducted recently on emerging and developing countries in order to compare and contrast their results with the study at hand. Barrientos and Boussofiane [2] appraised the technical efficiency of pension fund managers in Chile over time using Data Envelopment Analysis Techniques. Their findings imply that there are considerable probable technical efficiency gains in the pension fund management market in Chile and there are significant changes over time even though there is no uninterrupted trend towards the development in technical efficiency. On the other hand, [3] have projected a stochastic frontier cost model to probe the existence of economies of scale and scope in a sample of 12 Portuguese pension funds companies over the period 1994–2003. Costs rise as output mounts but with minimum rate related to the economies of scale of pension companies. On the other hand, joint estimation of outputs reduces cost due to complementary technologies.

In their technical efficiency and heterogeneity analysis of Argentina pension funds, Baross et al. [4] suggested that policy proposition coming up from their results is that benchmarking analyses are desirable to promote the Argentinean pension funds management companies to enhance relative efficiency. The improvement should be based on the balance of inputs and outputs and enlarged market share. Moreover, Njuguna and Arnolds [16] examined the efficiency of pension funds in Kenya using DEA and revealed that pension fund governance, leadership and regulations do not have positive impact on the financial efficiency of these funds. Nevertheless, smaller pension funds are more financially efficient than larger ones.

The other significant and recent study was made by Musalem and Pasquini [15] on performance of private pensions systems across 27 countries over 17 years illustrating the pension fund performance and the structure of a country's private pension industry and the design of its pension schemes. The study's core findings consist of higher returns are linked with size, type, and number of pension funds. Moreover, lower volatility in pension system returns is coupled with older systems, voluntary systems, systems with restrictions on foreign investing, and systems with minimum return guarantees.

Some studies have been carried out to analyze the efficiency of private pension companies in Turkey. Kurtaran et al. [14] employed DEA to gauge the efficiency of private pension system and uncovered that foreign companies are more efficient than domestic ones due to their advantage of applying modern technology, getting easier and cheaper resources. Besides, small-scaled companies have higher efficiency as compared to large-scaled companies. The other empirical result is contributed by Dalkılıç and Ada [10] focusing on scale efficiencies of 20 life insurance companies and life insurance and/or private pension companies in Turkey. The efficiency of those companies has decreased in 2011 as compared to 2010. Diverse results have been witnessed with regard to the input and output efficiency of the companies.

Ruzgar and Akkaya [22] scrutinized efficiencies of private pension companies in Turkey through data envelopment analysis from the year 2004–2008 and found that both the technical efficiency average and the scale efficiency average of the Private Pension Companies operating in Turkey are in the same levels approximately 80 %. They deduced that the companies couldn't attain sufficient level of efficiency even though their average isn't too low to be ignored. Moreover, they came across companies with lowest market shares registered high efficiency.

45.5 Data Adopted by the Study

To illustrate the technical efficiency, the study employed panel data on Turkey's private pension funds management companies for the years from 2008 to 2012 (10 companies $\times$ 5 years $=$ 50 observations). Data Envelopment analysis (DEA) entails the identification of inputs (resources) and outputs (transformation of resources). In order to attain the objective of the study which is measuring the efficiency of private companies, annual data have been collected from the data base of the Under secretariat of Treasury under the Prime Ministry of Turkey.

45.6 Methodology

Pension funds, like many other organizations, can be viewed as open systems which obtain inputs, convert these inputs into outputs and distribute these outputs to stakeholders. Pension funds get inputs (scarce financial resources in the form of contributions and investment funds) and change these inputs to outputs (pension fund value and retirement benefits) [11]. A pension fund would be considered as efficient if it succeed in maximizing financial outputs by the efficient use of the financial resources (inputs) [7].

Thus, the study used 2 inputs and 2 outputs to measure the efficiency of pension companies through data envelopment analysis (DEA). Total stockholders' equity and operating expenses of the pension companies have been taken as input variables whereas investment income and fund administration charges considered as output variables. In order to analyze efficiencies in production, Data Envelopment Analysis Program (DEAP) Version 2.1 software has been used.

45.7 Data Envelopment Analysis (DEA)

Data Envelopment Analysis (DEA) is a relatively new "data oriented" approach for assessing the performance of a set of peer entities called Decision Making Units (DMUs) which convert multiple inputs into multiple outputs. It is a nonparametric technique which has been used to compare the technical efficiency of relatively homogeneous sets of production units. It was initially built up to compare the technical efficiency of public sector and not-for-profit production units [8], however, more recently; it has been implemented fruitfully in the financial sector.

The method received wide attention after the paper by Charnes et al. [8] which coined the term Data Envelopment Analysis (DEA). There have since been a large number of papers which have extended and applied the DEA methodology. The envelopment surface will differ depending on the scale assumptions that underpin the model. Two scale assumptions are generally employed: Constant Returns to Scale (CRS), and Variable Returns to Scale (VRS). CRS reflects the fact that output will change by the same proportion as inputs are changed. The benefits of the approach are that it recognized efficient projected points which have input and output mixes which are as similar as possible to those of the inefficient points, and that it is also invariant to units of measurement [9].

A key merit of DEA over other approaches is that it more easily accommodates both multiple inputs and multiple outputs. Further, a specific functional form for the production process does not need to be imposed on the model (as is required in the use of the SPF approach). It is therefore appropriate to the task of analyzing the technical efficiency of private pension companies in Turkey through CRS DEA input oriented model.

45.8 Results of the Analysis

Farrell [12] considered an input orientated production process with two inputs x_1 and x_2 and one output y under the assumption of constant returns to scale. Knowledge of the unit isoquant of the fully efficient firm, represented by SS' in Fig. 45.1, permits the measurement of technical efficiency. If a given firm uses quantities of inputs, defined by the point P, to produce a unit of output, the technical inefficiency of that firm could be represented by the distance QP, which is the amount by which all inputs could be proportionally reduced without a reduction in output, this is usually expressed in percentage terms by the ratio QP/OP, which represents the percentage by which all inputs could be reduced. The technical efficiency of the firm is most commonly measured by the ratio $TE_I = OQ/OP$, which is equal to one minus QP/OP.

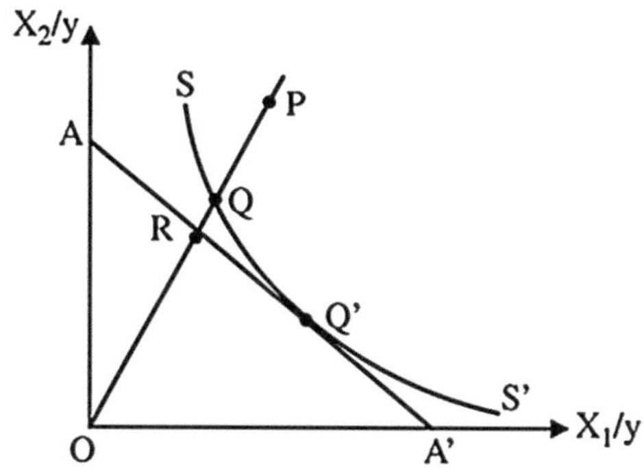

Fig. 45.1 Technical and allocative efficiencies (Source: Coelli et al. [9])

Table 45.1 Summary of efficiency scores from DEAP version 2.1 based on input oriented CRS DEA model from 2008 to 2012

Private pension companies	Technical efficiency (TE)
Aegon Emeklilik ve Hayat A.Ş.	0.324
Allianz Hayat ve Emeklilik A.Ş.	1.000
Anadolu Hayat Emeklilik A.Ş.	1.000
Avivasa emeklilik ve Hayat A.Ş.	1.000
BNP Paribas Cardif Emeklilik A.Ş.	0.459
Ergo Emeklilik ve Hayat A.Ş.	0.670
Finans Emeklilik ve Hayat A.Ş.	1.000
Garanti Emeklilik ve Hayat A.Ş.	0.751
ING Emeklilik	0.715
Vakıf Emeklilik A.Ş.	0.975
Mean	0.789

It will take a value between zero and one, and hence provides an indicator of the degree of technical inefficiency of the firm. A value of one indicates the firm is fully technically efficient. For example, the point Q is technically efficient because it lies on the efficient isoquant.

Data that have been gathered as inputs i.e. operating expenses and stockholders' equity and outputs investment income and fund management charges are analyzed by the input oriented CRS DEA model.

Table 45.1 depicts the abridged version of the efficiency report, where efficiency scores of all ten private pension companies are reported. This two-input and two-output model shows that four of the ten private pension companies are efficient using these four dimensions. Furthermore, we observe that the efficiency of one additional pension company is closer to an efficiency frontier whereas two more companies registered below the former but above 50 %. The last two have shown the least efficiency becoming closer to zero.

According to the above table, four pension companies namely Allianz Hayat ve Emeklilik A.Ş., Anadolu Hayat Emeklilik A.Ş. Avivasa emeklilik ve Hayat A.Ş. and Finans Emeklilik ve Hayat A.Ş. scored the highest and become technically efficient for the period under study. Vakıf Emeklilik A.Ş. has scored 0.975 which is closer but inefficient. The other two companies i.e. Garanti and ING Emeklilik attained 75.1 % and 71.5 % efficiency scores respectively while the least scores are recorded by Ergo, BNP Paribas and Aegon Emeklilik with scores 67 %, 45.9 % and 32.4 % respectively.

Mergers, acquisition and changes of the titles of the pension companies have taken place during the study period. Among the changes, in the year 2008 the capital structure of Ankara Emeklilik changed and it became AEGON Emeklilik, Koç Allianz Emeklilik, only Allianz Hayat ve Emeklilik and finally Oyak Emeklilik changed into ING Emeklilik. Fortis Emeklilik acquired by BNP Paribas Cardiff in October 2010 (Fig. 45.2).

As we can observe in the above figure, nearly 50 % of the private pension companies achieved utmost efficiency based on inputs and outputs employed in this study. Moreover, three companies achieved scores above half in spite of their inefficient performances. The remaining two became below 50 % score.

The above table illustrates the efficiency scores of private pension companies for 5 years. Avivasa Emeklilik scored 100 % across all years and become the most efficient pension company during the study period. This result is confirmed by the analysis of Ruzgar and Akkaya [22] which Avivasa became efficient in 2004, 2007 and 2008. Moreover, Vakıf and ING Emeklilik got 100 % score for 3 years while Finans, Garanti and BNP Paribas Cardif Emeklilik registered for 2 years. Allianz, and Anadolu Emeklilik each remained efficient only once in the year 2008 whereas Ergo Emekilik in 2010. The average efficiency score of all companies under investigation is approximately 79 % that is close to 80 % score consistent with study of Ruzgar and Akkaya [22] for the period 2004–2008 (Fig. 45.3).

Table 45.2 was taken from the “Slack” sheet of the DEA run results. Here, we observe that none of the efficient pension companies have any slacks. Slacks exist only for those pension companies identified as inefficient. Nonetheless, slacks correspond to only the residual portions of inefficiencies; after proportional reductions in inputs or outputs, if a DMU cannot reach the efficiency frontier (to its efficient target), slacks are required to push the DMU to the frontier (target) (Table 45.3).

Aegon Emeklilik ve Hayat A.Ş. should augment its one input and one output variable i.e. Investment income and Stockholders’ equity by about 1064.691

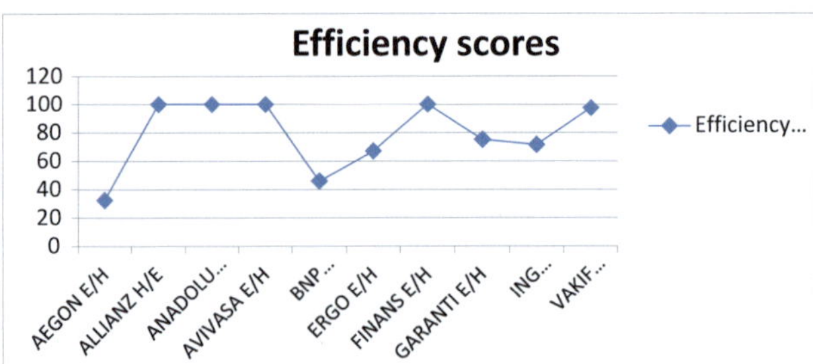

Fig. 45.2 Total efficiency scores of private pension companies for the period 20082012 (Source: Author’s calculation based on the efficiency scores from DEAP 2.1 programs)

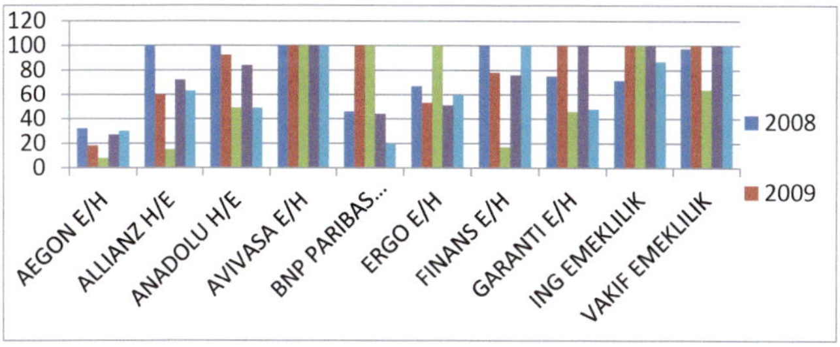

Fig. 45.3 Annual efficiency scores of the pension companies from 2008 to 2012

Table 45.2 Annual efficiency scores of the pension companies from 2008 to 2012

Pension companies	2008	2009	2010	2011	2012
AEGON E/H	0.324	0.179	0.076	0.270	0.302
ALLIANZ H/E	1.000	0.602	0.149	0.724	0.633
ANADOLU H/E	1.000	0.923	0.493	0.842	0.493
AVIVASA E/H	1.000	1.000	1.000	1.000	1.000
BNP PARIBAS CARDIF EMEKLILIK	0.459	1.000	1.000	0.442	0.198
ERGO E/H	0.670	0.532	1.000	0.508	0.600
FINANS E/H	1.000	0.783	0.166	0.760	1.000
GARANTI E/H	0.751	1.000	0.459	1.000	0.476
ING EMEKLILIK	0.715	1.000	1.000	1.000	0.868
VAKIF EMEKLILIK	0.975	1.000	0.639	1.000	1.000
Mean	0.789	0.802	0.598	0.754	0.657

TL and 4160.372 TL respectively in order to achieve efficiency. Besides, BNP Paribas Cardif Emeklilik A.Ş. required enhancing its output Investment income by approximately 3842.419 TL. However, despite this increase it can't attain efficiency unless it reduces its operating expenses by 2423.364 TL which is an input variable. This finding is similar with Dalkılıç and Ada [10] results which were focusing on scale efficiencies of 20 life insurance companies and/or private pension companies in Turkey. They inferred that eight companies are supposed to reduce their input variables so as to attain efficiency. Ergo Emeklilik ve Hayat A.Ş. need to raise its fund management charges by 18,578.711 TL while ING Emeklilik expected to uplift the other output variable investment income nearly by 13,052.070 TL.

45.9 Conclusion

The paper has investigated the technical efficiency of the private pension companies in Turkey from 2008 to 2012 using two inputs and two outputs employing Data Envelopment Analysis (DEA) techniques. The concept of efficiency concerns pension companies' ability to produce outputs (such as investment income and fund management charges) via the use of inputs such as operating expenses and

Table 45.3 Summary of input and output slacks from DEAP version 2.1 based on input oriented CRS DEA model

Private pension companies	Output slacks		Input slacks	
	Investment income	Fund management charges	Operating expenses and	Stockholders' equity
Aegon Emeklilik ve Hayat A.Ş.	1064.691	0.000	0.000	4160.372
Allianz Hayat ve Emeklilik A.Ş.	0.000	0.000	0.000	0.000
Anadolu Hayat Emeklilik A.Ş.	0.000	0.000	0.000	0.000
Avivasa emeklilik ve Hayat A.Ş.	0.000	0.000	0.000	0.000
BNP Paribas Cardif Emeklilik A.Ş.	3842.419	0.000	2423.364	0.000
Ergo Emeklilik ve Hayat A.Ş.	0.000	18,578.711	0.000	0.000
Finans Emeklilik ve Hayat A.Ş.	0.000	0.000	0.000	0.000
Garanti Emeklilik ve Hayat A.Ş.	0.000	0.000	0.000	0.000
ING Emeklilik	13,052.070	0.000	0.000	0.000
Vakıf Emeklilik A.Ş.	0.000	0.000	0.000	0.000
Mean	1795.918	1857.871	242.336	416.037

capital. The findings imply that technical efficiency gains have been observed in some private pension companies in Turkey. There are important changes over time, but no continuous trend towards an improvement in technical efficiency. This may be due to the immaturity of the industry since it has started operations 10 years ago. According to the study, most pension companies' inefficiency scores were due to their failure in producing/converting the output variables especially investment income. Therefore, those companies should invest in areas which they collect more return in order to achieve efficiency. In addition, high operating costs are detrimental towards achieving efficiency. This happened particularly as a result of failing to utilize their inputs and reducing those costs that enable the companies accomplish their target.

References

1. Barney J (1991) Firm resources and sustained competitive advantage. J Manag 17:99–120
2. Barrientos A, Boussofiane A (2005) How efficient are pension funds managers in Chile? Revista de Economia Contemporânea 9(2):289–311
3. Barros CP, Garcia MTM (2007) Analyzing the performance of pension funds industry with a stochastic cost frontier: a Portuguese case study. Geneva Pap Insur Theory Pract 32:190–210, Switzerland
4. Barros C, Pestan FG, Romero C (2008) Technical efficiency and heterogeneity of Argentina pension funds. Working papers, Technical University of Lisbon, Portugal
5. Caves R, Porter ME (1977) From entry barriers to mobility barriers: conjectural decisions and contrived deterrence to new competition. Q J Econ 91:241–261
6. Central Bank of Turkey (2011) Financial stability report. Central bank of Turkey, Turkey. http://www.tcmb.gov.tr/wps/wcm/connect/tcmb+en/tcmb+en/main+menu/publications/reports/financial+stability+report
7. Chansarn S (2005) Efficiency in Thai financial sector after the financial crisis. University Library of Munich, Munich, Germany
8. Charnes A, Cooper W, Rhodes GE (1978) Measuring the efficiency of decision making units. Eur J Oper Res 2(6):429–444
9. Coelli TJ, Rao DSP, O'Donnell CJ, Battese GE (2005) An introduction to efficiency and productivity analysis, 2nd edn. Springer, New York
10. Dalkılıç N, Ada AA (2014) Efficiencies of life/pension insurance industry in Turkey: an application of data envelopment analysis. J App Finance Bank 4(1):181–191
11. Davis PE (2005) The role of pension funds as institutional investors in emerging markets. Brunel University and NIESR, London
12. Farrell MJ (1957) The measurement of productive efficiency. J R Stat Soc Ser A 120:253–290. G Turkey
13. KPMG Turkey (2013) Turkish pension system amendments. Yetkin Yeminli Mali Müşavirlik A.Ş, Istanbul-Turkey. http://www.kpmgvergi.com/PDF/Yayinlar/KPMG-Global-Yayinlar/Turkish-Pension-System-Amendments.pdf
14. Kurtaran A, Karakaya A, Dağlı H (2013) Improvement of private pension system in Turkey and measurement of its efficiency with DEA. Int J Econ Finance 5(11):163–173
15. Musalem AR, Pasquini R (2012) Private pension systems: cross-country investment performance. Working paper, Social protection and labor world Bank, No. 1214
16. Njuguna AG, Arnold C (2010) Improving the financial efficiency of pension funds in Kenya. Nelson Mandela Metropolitan University, Port Elizabeth
17. Olgaç K (2006) Private Pension System of Turkey, discussion on the development of Private Pension System of Turkey. OECD/IOPS Global Forum on Private Pensions, Istanbul
18. Organization for Economic Co-operation and Development (OECD) (2013) Pension markets in focus. OECD, Paris, France. http://www.oecd.org/pensions/PensionMarketsInFocus2013.pdf
19. Özgür Ö, Yalçın C (2013) The effect of private pension reforms on domestic saving rates in developing countries: cross-country empirical evidence. CBT research notes in economics. Central Bank of Turkey, Ankara
20. Porter ME (1979) The structure within industries and companies' performance. Rev Econ Stat 61:214–227
21. Rumelt R (1991) How much does industry matter? Strateg Manag J 12(2):167–185
22. Ruzgar B, Akkaya A (2009) Examining efficiencies of private pension companies in Turkey with data envelopment analysis. In: IWW2009-proceeding of the IVth international workshop on applications of wavelets to real world problems, At Kocaeli, Turkey
23. Wernerfelt B (1984) A resource-based view of the firm. Strateg Manag J 5(2):171–180

Chapter 46
Pivoting on Chaos and Disorder, Reflection on Events Steering the World of Politics in Social Sciences Course Books (Turkey and USA Example)

Zafer Tangülü and Melek Ulusoy

Abstract Knowledge gets more competitive and expands this competition day by day in the current century. But there is nonesuch dynamic structure constantly in the constant change and development before. Paradigms are continuously effected by unpredictability, sustainability, change, uncertainty, chaos and confusion. The social, political, economic and cultural issues about chaos and confusion take part in the textbooks. This study examines the visual design and expression of the politically dominating events that occured in the past which took part in Turkish and American student textbooks. Qualitative research methods are used in order to define the situation. The case study methodology is used at the entire research. Documents are analyzed carefully and parameters are compared by their visual design and manner of telling. After achieving the results these parameters gone through interpretation process. Interpretations showed us that Turkish textbooks contains more visual materials than American one. However these visual materials turn out as limitation of the lecturing. It is found out that American textbooks have plainer way of teaching. These results show that Turkish textbooks should have homogenous subject structure.

46.1 Introduction

In Greek mythology, 'khaos' means emptiness, openness, indefiniteness, outer space [4], expresses, complexity, uncertainty, disorder and anarchy. There is a raw pureness which tells the meaning of the World [14], an order in the disorder [16]

Z. Tangülü (✉)
Faculty of Education Department of Elementary Education, Social Studies Education Department, Muğla Sıtkı Kocman University, T Block 48000, Kötekli, Muğla, Turkey
e-mail: zafertangulu@gmail.com

M. Ulusoy
Social Studies Education Department, Educational Sciences Institute, Niğde University, Niğde, Turkey
e-mail: melekulusoy01@gmail.com

Ş.Ş. Erçetin (ed.), *Chaos, Complexity and Leadership 2014*,
Springer Proceedings in Complexity, DOI 10.1007/978-3-319-18693-1_46

at chaos. In the present world where technology and science are improving rapidly, there are uncertainties and unrests that could cause chaos and disorder. Since such situations had been experienced before, necessary precautions are being taken. A report named "Global Trends 2025; Changing World" was published in 2008 by the American National Intelligence Council. In this report, the possibility of occurrence of events by 2025 that effect the whole world; like globalization, immigrations, population increase, studies of international institutes, geopolitical evaluation of energy consumption, global warming and climate changes, political relationships of countries; and their potential results. In other words, attention is being drawn on issues that may cause chaos and disorder in the world by 2025. The Santa Fe institute, pioneering disorder researches, defines disorder as: Disorder is the universe's state of being integrated as well as being rich and abundant which we cannot understand through accustomed mechanical or linear aspects. These aspects define many pieces of the universe, and express big internal relationships and wider phenomena with patterns and principles. Disorder, between order and chaos, is regarded as the pre-chaos state which is the precursor of chaos. Disorder is related to the nature of self-discovery, and learning [12] and questions how complex structures of pieces come together [21]. The system that works dynamically, lasts its existence with the possibilty of chaos [4].

While there are three classes in dynamic systems – statical state, order, chaos – is added as 4th class complexity [30]. Science fields like education, management and sociology have contributed to alter paradigms in society by making use of chaos and complexity. In education, transformation of schools, program developments and regulations are carried out with the help of the chaos theory. Trends that explain events and phenomena in a cause/effect relationship and approach topics from a linear aspect are insufficient in explaining complex events and phenomena caused by chaos. Positivist paradigm is the product of the nineteenth century scientific approach and this approach lacks the qualities of understanding our present world. The chaos theory provides a new aspect of solving ambiguity and its reasons. In light of the developments shed by researches on education, there have been some changes in educational paradigms, and new approaches have been put forward in the analysis and synthesis of events in the past.

French mathematician Jules Henri Poincare started the first studies about the notion of chaos towards the end of the nineteenth century. Poincare, the best scientist expressing chaos among physicians and mathematicians, states that multi-variable structured systems may not have permanent solutions, or they may have infinite solution options. Small unseen details may cause big results, and people think this is a result of coincidence. Seemingly trivial details that occur at the initial circumstances and in the forthcoming time frame effect the flow of events and create big changes in the result [21].

The Chaos Theory takes its place like other science fields [16]. This theory focuses issues philosophically with an interdisciplinary approach [27]. Edward Lorenz states that the small changes at the structure of variables causes unpredictable and suprising results ([21], p. 127, [10], pp. 15–19). Disorder and chaos emerge as a reaction to occurring actions in a universe where everything is

Table 46.1 Traditional science model and new science model within changing paradigms

Traditional paradigm	New science model
Mostly statical, linear, Newtonist, mechanical world view	Fluid, non-linear, complex, living systems
Balance	Complexity
Statistics used to discriminate the incidental and uncontrolled from the predictable by making use of processes like probabilities	Chaos and complexity mathematics used to determine certainties and predictables by making use of results like probabilities
Focusing on qualities and mechanisms	Focusing on processes, patterns, potentials, varieties
Reductionism	Interdisciplinary interaction
With rational selection parameters, individual as a unit of solutions	With systematic selection parameters, synergic, individuals or institutes that develop cooperation

Source: Erçetin [8]

dependent on each other and small effects cause big results. In positively expected situations; while the realization of the probability to become positive in result is regarded as "coincidence", the negative result is referred to as "chaos" [2]. Chaos and disorder are a pausing point for asking the question "Why do we always have to watch like this?" and seeking its answer; where a result is tried to be attained by various inquisitions [12].

Disciplines like teaching, management and sociology changed the paradigms of the society by touching chaos and confusion subjects. Chaos theory is used at education when developing programmes and transforming schools and arrangements. Linear approach remains incapable when defining the facts that caused by chaos. Dynamic education system requires non linear approach and thinking [27]. The school age kids learn war, conflict, migration, destruction from internet or live in person. A new perspective should be provided to children in order to maintain the tendency to search and learn like Einstein said: "the most important thing is to keep asking questions" (Table 46.1).

Positivist paradigm has been deficient in responding to questions asked to explain social events, universe and nature. These are important features of scientific studies carried out in the postmodern era [25]:

1. The universe has a holographical structure. Every piece in the universe carries a piece related to the whole and everything that constitutes the whole is interrelated.
2. Reality presents a complex structure. All systems have their own distinctive structures. Changeability has the features of mutual interaction and variety.
3. Since it is impossible to determine the future and direction, there is no exact result. Only probabilities can be known and predictions can be made. Relationships emerge and persist with mutual reasoning interaction on a non-linear platform. Events or phenomenons may be the cause of another event, in addition, two events or phenomenons may interact and change, and form a new appearance.

4. Change possesses a morphogenic structure. A new order environment may emerge from disorder. Since moving with single discipline will be insufficient, there is a need for an interdisciplinary study and multi-dimensional point of view.

Realized changes demolish some paradigms and bring new transformations along. Children at school age see and learn political events like war, rage, battle, destruction from books, internet and media channels; or they experience these in person. Such situations raise questions in minds of children who especially lack analytical thinking, analyzing and synthesizing skills, and sometimes cause them to experience an emotional destruction. Education, science, such as management and sociology, leaning on the chaos and confusion issues [28], have contributed to the change in paradigms that exist in society. In the field of education; transformation of the school, the program benefited from the development of chaos theory and regulations. Chaos created by the complex events and upcoming terms linear in explaining the phenomenon, describing the relationship cause-and-effect current events is insufficient (Table 46.2).

Learning is a complex process. Interaction should be increased with chaotic learning environments like brain storming. Different ideas sparks each others minds and creates different structures. Chaos and confusion subjects at textbooks provides a quizzical thinking which improves cognitive processes [7]. Its crucial that teachers extract these chaotic subject to students. The missing transfer of the knowledge makes it harder for students to follow new subjects. The small changes may end up big changes. The success of education relies on complex feedback [27].

Table 46.2 Basic intellectual features related to descending and ascending paradigms

Positivist/mentalist paradigm	
Positivist, mentalist, modern, empirical, structuralist, newtonist	Metapositivist/interpreter metapositivism, metastructuralist, postmodern, metafunctionalist, interpretative
Mechanical world view	Holographic world view
Predictability	Unpredictability
Generalizability	Situationalist
Universality	Subject oriented
Objective reality	Subjective reality
Big speeches, big theories, single truth	Pluralist
Perfect information	Imperfect information
Objectification	Perspective
Reduction	Wholism
Measuring	Participation
Quantification	Qualitative analysis
Universal law	Situational findings
Pure results	Value-added results
Experimental processes	Participation based processes
Information is discovered and unveiled	Information is interpreted and generated

Source: Şimşek [25]

As well as in all fields, chaotic understanding has become dominant in the field of education. Societies are getting involved into a big and swift shift in the twenty-first century world, an event happening kilometers far away, can have an effect on other societies within seconds. The uncertainty principle defined by Heisenberg in the new century has not allowed scientists to make measurements by making estimations. Heisenberg argues that it is impossible to know all the features of a piece, and a measurement can change only according to the general state of the piece. As a result of this, it is understood that time, direction and speed are relative (cited: Say [23], p. 6).

The chaos theory on education imitates the physical and mathematical chaos theory. This theory emphasizes that random and ignored facts may cause critical results in the long term [13]. For example; Second World war was destructive and caused sad events. After the war all values and beliefs were re-evaluated and revised. Crises and conflicts influenced everybody and caused chaos all over the World. Confusion in a country influenced neighbours and spread even overseas. Students have opinions about other countries, cultures and societies when they search about them. Social Sciences contains law, economics, geography, history, cultural anthropology to provide students to understand these facts as a whole [26]. Social Sciences aims to raise responsible and effective citizens [6]. That is the need of a democratic society and a good heritage for future generations [9]. Social movements and cultural diversity lead to conflicts and chaos at USA in twentieth century [1]. Social Sciences added as a course to educational institution to avoid this confusion [11]. Social sciences defined as the knowledge of people and society and the improvement of them [19]. Two main purposes of the course were to improve the integration-coordination and to prepare lecture contents regarding to the needs of children [22] (Table 46.3).

Multiple learning areas in the unit constituting the Social Studies learning areas can be located. Although nine of the Social Studies curriculum learning areas are in 6th and 7th grade unit containing 7. "Time, Continuity and Change" and "Groups, Institutions and Social Organizations" learning area in the 6th and 7th grade and is considered a separate unit because it has not been established in other areas of learning. In our country, the constructivist approach of students

Table 46.3 Social sciences learning fields and primary disciplines

Learning field	Primary discipline
Individual and society	Psychology
Culture and inheritance	Anthropology
People, places and environment	Geography
Production, distribution and consumption	Economy
Time, persistence and change	History
Science, technology and society	Interdisciplinary
Groups, institutes and social associations	Sociology
Power, management and social politics	Citizenship
Global links	Interdisciplinary

Source: MEB [20]

prepared for primary and secondary school program approach “to count”, “say the names”, “define” “list to” like memorization rather than work that requires learning to “discuss”, “analyze”, “classes to” “create”, as higher-level cognitive activities are aimed at the creation of instructional media can be used as active.

The main goal of Social Science is to implement the existence of an individual. Moreover it also aims to evaluate the physical and social environment of an individual on a past-present-future basis [20]. Social Sciences help children to understand and learn the World [3], and guides them to make reasonable decisions and to solve problems [5]. Its important to solve the problems effectively and to know the historical culture to understand present while making future plans. Textbooks are frequently used materials to teach children in Turkey. They are the main tools to maintain education process and used as a bridge for transferring basic values and skills. Textbooks let students to harden their knowledge, study with desired speed and prepare for the exams ([20]: pp. 93–94).

46.1.1 The Purpose of the Research

This research has a qualitative pattern. At qualitative analysis cases and perceptions are occured with realistic and holistic methods such as observations, interviews and document analysis. Case study is a method that examines cases with their own environment [29]. It also examines social events with their historical and current features and peripheral relations particularly. So that both countries books' are compared with their own environment. So that, this research tries to find the answers of;

1. What are the differences and similarities of visual design at both Turkish and American text-books ?
2. What are the differences and similarities
   (a) unit layout
   (b) subject headings
   (c) concepts of lecturing and expression of chaotic topics' ?

46.1.2 Restrictions and Assumptions

1. Social Sciences textbooks at Turkey and USA are published for common general purposes.
2. The book that represents American samples is universal.

46.2 Methodology

This research has a qualitative pattern. At qualitative analysis cases and perceptions are occured with realistic and holistic methods suc as observations, interviews and document analysis. Case study is a method that examines cases with their own environment [29]. It also examines social events with their historical and current features and peripheral relations particularly. So that both countries books' are compared with their own environment.

46.2.1 Nature and Sample of the Research

The nature of the research is elementary school Social Science textbooks. Samples for this research are picked from MEB Social Science textbooks which is accepted by the Turkish Minister of Education to teach 7th grade students for 5 years. Sample from American case is CK-12 U.S. History Sourcebook.

46.2.2 Analysis of the Data

The chaotic events that leads World political issues involved both Turkish and American textbooks considered as data for this research. To achieve proper outcomes, there are also expert opinions for gaining mutual data.

46.3 Findings and Comments

46.3.1 Visual Design

It became a necessity to use visual tools on Daily life and also pedagogical purposes. Virtual design means to illustrate verbal information visually for education purposes [24]. A qualified textbook should fit students understanding on language and content basis. The books should be able to express students' feelings, ideas, happiness and worries with their language ([15]: 10) The books are favoured materials for lectures in Turkey. So these books should be designed wisely and eye-pleasing.

Social Science textbooks for 7th grade students in Turkey have small sized letters around pictures. However these small letters cause difficulties while reading it. 68th page of the exact book is about Ottoman Empire: Conquer of the Istanbul and Defeat of Byzantine. There are four pictures at the same page and these pictures represent the preperation of the Ottoman Empire for the war, after war and pictures from our current time. On the 69th page you can see 5 pictures which is way to much to tell the story but informs students about past and present. Moreover these pictures prevent students from anachronical mistakes and make it concrete. Students are asked questions in order to improve their knowledge about the topic.

Renaissance and reforms are examples that dominates World politics. At page 109 and 100 these examples are illustrated with pictures. There is a Martin Luther picture and speech balloon with his own words inside. In this section students are asked about scientist and artists who lived on Renaissance Period and their effects to current age. The pictures represent the mentioned period but not enough to make it concrete. U.S. History Book has a simple visual design comparing with Turkish Social Science textbook. The pictures are related with the topic but they fit the page and appropriate for page size. Less pictures are used in order to express chaotic events. At 58–59th page you can see Texas Declaration of Independence with one picture seperated but two pages. There is the original picture of the declaration.

There are significant differences at both books if compared visually and qualitatively. There are more speech balloons, diagrams, pictures, graphics at Turkish textbooks in quantity. Both books have convenient pictures related with topic. But the size of visual materials on Turkish book is not fitting the page. The pictures are too big at some pages and sometimes you can see too small pictures. However the pictures used in History Sourcebook are placed more carefully in order to fit the page and the size of the pictures are consistent.

46.3.2 Lecturing

46.3.2.1 Unit Layout

Topics are mentioned chronologically at both of the books. Both books have a viable layout for the chaotic events that dominates the World politics. First titles of the topic are written with big fonts and sub-titles are shown with a diagram that makes it easy to notice. In Turkish textbook you can see the terminology with explanations for each topic at the beginning. For example before beginning First World War, there is a terminology diagram that explains by asking students why is it called World War (p. 166) etc. There is a picture that represents real life look after the war and lights up a question 'Why does this city look like that?' They are also asked 'What are the pros and cons of the war for colonialists? That question aims to help students to gain historical empathy [20]. Moreover, the book targets to improve students' verbal thinking by poems and memories of the witnesses.

U.S. History Sourcebook tell of basic history of USA. The Declaration of Independence is known all over the World. The declaration that accepted by the delegates of 13 American colonies gathered at 4th of July 1776 at Philadelphia, dominated the World politics. You can see the Declaration of Independence at 6th grade Turkish Social Science textbook mentioned in the history and improvement of democracy section. The freedom and independence take part at both of these books with pictures and information. Woman has equal rights with man today. There has been chaotic events before this equality. Woman rights at Turkey is explained with success stories of famous womans. Stories continue with the importance and place of woman in society (page 177–179). Likewise at U.S. History Sourcebook Basic, page 161) you can find information about The Movement of Women Rights 1848.

The Civil War in USA, the Foundation of USA and Lincoln takes significant place at American History textbooks. Mustafa Kemal Ataturk and the foundation of Turkish Republic is told in a similar way at Turkish Social Science textbooks. Both of these significant people created a whole new Republic in a chaotic environment of First World War. These peoples lives, ideas and achievements are taught to students (Social Sciences 7, 194–197). It is stated in the U.S. History Sourcebook that, during the Russian Revolution Tzar went down and communist government replaced him. This caused unemployment, inflation and strikes which lead radical ideas like communism, socialism, anarchism that effect other countries. The ones who tried to spread these ideas got arrested.

Woodrow Wilson was one of the attempters that tried to create a new world order. He is known with Wilson Principles which guide strategies of countries after the war. This principles can be found both of Turkish and American textbooks. Crucial decisions made to create a new World order with League of Nations and Paris Peace Conference. League of Nations is underlied at American books however League of Nations stated as unsuccessful to create peace at Turkish textbook. It is also mentioned that League of Nations couldnt stop the Second World War at Turkish.

Both books are available in a consistent internal arrangement between the units. This is positive for the teachers and students. Skills that will be given directly to students with seven units in the book;

- Communication and human relations unit, communication skills.
- unit of population in our country, charting skills.
- ride on the Turkish unit, Ability to distinguish historical facts and interpretations.
- Economic and social life of the unit, the historical empathy skills.
- Time in science unit, time and chronology of detection capability.
- Living Democracy unit, decision making skills.
- Countries bridges between units, the ability to realize stereotypes.

Skills between the unit has internal consistency. These skills facilitates the students' perception of chaotic and complex events.

46.3.2.2 Subject Headings

Social Studies lessons in book 7 chaos and confusion axis of world politics that shape the narrative of events began with the main topic. Search this main topics relevant to the subject headings are followed. Search irrelevant does not overlap with the main topic in the book title is used. Topics are listed in chronological development results at the entrance axis. Visual elements used in the chapter headings give tips on the subject. Subject processing of the main and intermediate students before heading out to çıkarat ideas and views on the subject in question is situated immediately before and after.

US textbook unit names and titles are listed below. Not entered in the search for a topic. Questions were asked after the first treatment of the topics and issues, students were asked to answer to questions. Topics in the US textbook few, simple, and consistent with each other. While many sub-title textbooks in Turkey, the number of sub-headings in the US course kştab less. Subjects were told by a main title and subtitles attached. "Colonical America" topics in the unit; Early Maps, Pocahontas, Passenger Lists to The New World, The Pritans, Kinf Philip's War, The Salem Witch Trials were identified as The First Great Awakening. Topics are described in chronological integrity and internal consistency. Chaotic and complex situations in events close to the present day, starting from when the remote is described in plain language. Under each topic related to the subject of the story, memoir, biography, and so on are described. This provides students with empathy by understanding the chaotic and complex events.

Social Studies 7 on page 166 in the First World War the issue of a "World" War began with the main title. For a better understanding of the war, the world is emphasized words. The unit will eventually get a Peace Judge world title is used. A better understanding of this topic is provided in the chaotic and complex events that shaped world politics. Students will pay attention to what is emphasized. American textbook page 110–115't to populism is described. To migration in America has led to chaotic and complex situations. Women, men, black, white people do to promote equality in society, in short, is described where. Better understanding of the events giving place to the story of the people among the topics is provided. Both books under each logical chaotic and complex topics were followed a holistic and chronology. Topics include memoirs, biographies, letters, enriched with texts such as newspaper articles. Chaotic events are handled in the cause-effect relationships.

46.3.2.3 Concepts

Social Studies 7 textbooks at the beginning of each unit, the main concepts are illustrated with a template of the main topics. Templates are blue. students readiness ready for one of the students explained this concept is provided. Students begin the course, knowing what they would earn in the course concepts. This is positive for the teachers and students. Thus, the concept is prevented from experiencing confusion. "A World War" concepts to be learned in the topic; are expressed in peace and solidarity. The definition of the concept of peace is made. "How did the scientific know-how" in the main title; energy, exploration, secularism, market reform, renaissance, industry, capital concepts are given. This concept of secularism is described. Be related to the concept of many events that cause chaos and confusion is positive. Students understand the events and it is easier to reconcile with our day. US main concepts in the textbook on the subject, is given by the instructions before starting the topic. For example, on pages 14 and 15; posterity: future generations, prosperity: wealth, heathen: a term used at this time to describe anyone uncivilized and who did not believe in God, appoint: assign decide on it, inhabitants: people who live in a certaine the place concepts are explained. In both manual concepts are suitable for target acquisition.

Social Studies lessons in book 7 that the direction of the axis of world politics chaos and confusion about the concept of evolving events; public, climate, population, location, area, geographic location, export, import, urbanization, investment, migration, rights, patriotism, citizenship, employment, contract, state, conquest, settlement, political power, war, language, religion, tolerance, culture, reform, science, invention, energy, secularism, exploration, industry, capital, Renaissance, reform, freedom, taxes, economy, civil society organizations, constitution, democracy, sovereignty, elections, monarchy, oligarchy, independence, freedom, national sovereignty, the concept of environmental pollution. Prosperity in the book in the United States, religion, migration, compels, allegiance, secure, instituted, reconciliation, rational, anarchy, remedy, fanatic, tyranny, enfranchisement, heathenism, degraded, franchise, defiantly, aliens, banish, aspiration, menace, territorial integrity, monuments, redemption, war, peace, freedom, citizenship, climate, government, industry kavramlarıdır.h concepts are similar in the two books.

Chaotic events in other countries, should occur has also affected in a similar way to people in other regions. A chaotic events that shape the world of politics and need not take place in a very old history to be complicated. Emerging with globalization, all countries to environmental issues affecting the entire world receive common measures. 7 Social Studies textbook pages 176–179 in the recently emerging environmental pollution problems are described as struggling with the whole country. Chaos and confusion caused by environmental pollution has created a new order. This scheme also international treaties and cooperation, international organizations, projects and created civil society organizations. This period is described by the documents that go to students why the Chinese Exclusion Act. Thus, students of the events in that period were evaluated with the conditions of that period.

46.4 Conclusion and Recommendations

In Turkey and USA the chaotic situations that effect World politics are taught due to the attendance of the students. Students' gainings are supported with pictures, maps and documents. Chaotic cases are being teached by memories, biographies and videos to help students gain predictive ability. So that students wont be anachronically mistaken. In both of these case studies, countries seems to teach more than just wars and political issues. In contrast they aim to give information about heroic people and their ideas, inventions, national and cultural values, differences and respect to these differences, tolerance and peace for humanity. Students must be aware that sometimes same actions may result different in different countries. So that students can have a sophisticated view.

Both books contain visual elements that are needed by each audience. Turkish Social Science text books contain quite a few visuals that are relevant to the issue. These images shouldnt distract students attention but rather help to focus on the subject. The subjects are places homogeneously at Social Science 7 textbook. The subjects should increase in quantity and widen. Writers should put more effort on process and result evaluation for both of the books. The evaluation that mentioned here must be placed at the end of each section in order to provide stable learning. The results and outcomes of these chaotic events should be transfered as a knowledge to the students. That helps students to gain predictive ability. Holistic method will be useful to help students gain a better digestion. In the end that will provide a interactive book-student connection. Expression of the events that shaped world politics in chaos and confusion axis is important for students to gain multiple perspectives. Students will evaluate the conditions of history that day. When they fall into error. Topics and concepts of the book in chaotic and complex events that shaped world politics shows an integrated structure in terms of internal consistency. All concepts are explained with the description given in the US book one of the books in Turkey. The study of concepts themselves, empathy. to gain significance in the subject, prediction, narrative and chronology of their ability to develop procedures, events and images are available in both book. Simplifying a bit more visually Social Studies books in Turkey course inspired the students will be more effective.

References

1. Bilgili AS (2006) Geçmişten günümüze sosyal bilimler ve sosyal bilgiler. In: İsmail Hakkı Demircioğlu (ed) Sosyal bilgilerin temelleri. Hegem, Ankara, pp 3–56
2. Bülbül MS (2007) Kaos ve egitim. Beyaz kalem, Ankara. Retrieved from http://books.google.com/books/about/Kaos_ve_Eğitim.html?hl=tr&id=0mos0i1T0p8C
3. Çengelci T (2005) Sosyal Bilgiler Dersinde Beyin Temelli Öğrenmenin Akademik Başarıya ve Kalıcılığa Etkisi. Yayımlanmamış Yüksek Lisans Tezi. Anadolu Üniversitesi Eğitim Bilimleri Enstitüsü, Eskişehir
4. Cramer F (1998) Kaos ve düzen. Alan, İstanbul
5. Doğanay H (2002) Coğrafya Öğretim Yöntemleri (Ortaöğretim Coğrafya Eğitiminin Esasları), Aktif Yayınevi, Erzurum s. pp 147–173
6. Dönmez C (2003) Sosyal bilimler ve sosyal bilgiler. In: Şahin C (ed) Konu alanı ders kitabı inceleme kılavuzu sosyal bilgiler. Gündüz, Ankara
7. Erçetin ŞŞ, Açıkalın ŞN, Bülbül MS (2012) A multi-dimensional approach to leadership in chaotic environments. In Banerjee S (ed) Chaos and complexity theory for management – nonlinear dynamics. Springer, Netherlands, pp 89–104
8. Erçetin ŞŞ (2001) Yönetimde yeni yaklaşımlar. Nobel, Ankara
9. Garcia J, Mıchaelis J (2001) Social studies for children. Allyn and Bacon, Boston
10. Gleick J (2005) Kaos. Tübitak Popüler Bilim, Ankara
11. Gündem S (1995) Sosyal bilgiler öğretimine genel bir bakış. İlköğretim kurumlarında sosyal bilgiler öğretimi ve sorunları türk egitim dernegi XIII. toplantısı (No: 13). Türk Eğitim Derneği, Ankara
12. Gürsakal N (2007) Sosyal bilimler karmaşıklık ve kaos. Nobel, Ankara
13. Harshbarger B (2007) Chaos, complexity and language learning. ICU Lang Res Bull 22:17–31

14. İnam A (2006) Kaos kavramı üstüne. Bilim ve Ütopya 149:4–6
15. İpşirlioğlu Z (1993) Çocuk icin iletişim aracı. Yapı Kredi Yayınları, Istanbul
16. Kendirli S (2006) Portföy yönetiminde kaos teoremi. Journal of İstanbul Kültür University 2:171–180
17. Köstüklü N (1999) Sosyal bilimler ve tarih öğretimi. Günay, Konya
18. Küçükahmet L (2001) Konu alanı ders kitabı inceleme kılavuzu. Nobel, Ankara
19. Latif H (2002) Stratejik boyutuyla modern yönetim yaklaşımları. Beta, Istanbul
20. Milli Eğitim Bakanlığı (MEB) (2005) İlköğretim sosyal bilgiler dersi öğretim program ve kılavuzu. Milli Eğitim, Ankara
21. Öge S (2005) Düzen mi düzensizlik (kaos) mi? Örgütsel varlığın sürdürülebilirliği açısından bir değerlendirme. Selçuk Üniversitesi Sosyal Bilimler Enstitüsü Dergisi 13:271–284
22. Safran M ve Ata B (2003) Öğrencilerin tarih metinlerinden anlam çıkarmalarına yönelik araştırrmalara bir bakış. In: Şahin C (ed) Konu alanı ders kitabı inceleme kılavuzu sosyal bilgiler. Gündüz, Ankara
23. Say Ö (2008) Kimlik ve sosyal bilimlerde görelilik ve belirsizlik ilkesinin açmazı. In: M Safran ve D Dilek (ed) 21. Yüzyılda kimlik vatandaslık ve tarih eğitimi. Yeni İnsan, İstanbul, pp 154–160
24. Semerci C (2004) İlköğretim türkçe ve matematik ders kitaplarını genel değerlendirme ölçeği. Cumhuriyet Üniversitesi Sosyal Bilimler Dergisi 28(1):49–54
25. Şimşek H (1997) Pozitivizm ötesi paradigmatic dönüşüm ve eğitim yönetiminde kuram ve uygulamada yeni yaklasımlar. Kuram ve Uygulamada Eğitim Yönetimi Dergisi 3(1):97–109
26. Sönmez V (2010) Sosyal bilgiler öğretimi ve öğretmen kılavuzu. Anı, Ankara
27. Töremen F (2000) Kaos teorisi ve eğitim yöneticisinin rolü. Kuram ve Uygulamada Egitim Yönetimi Dergisi 22:203–219
28. Yeşilorman M (2006) Kelebek kanadını kimden yana çırpıyor? Birleştirilmiş bilimin kıyısında kaos ve sosyal bilimler, (2006/3). Kültür Üniversitesi, İstanbul, pp 77–86
29. Yıldırım A, Şimşek H (2003) Sosyal bilimlerde araştırma yöntemleri. Seçkin, Ankara
30. Waldrop MM (1997) Karmaşıklık düzen ve kaosun eşiğinde beliren bilim (Zülfü Dicleli tr.). Türk Henkel Dergisi Yayınları, İstanbul

MIX
Papier aus verantwortungsvollen Quellen
Paper from responsible sources
FSC® C105338

If you have any concerns about our products,
you can contact us on
ProductSafety@springernature.com

In case Publisher is established outside the EU,
the EU authorized representative is:
Springer Nature Customer Service Center GmbH
Europaplatz 3, 69115 Heidelberg, Germany

Printed by Libri Plureos GmbH
in Hamburg, Germany